N

D L

A guide hemes

MICRO-HYDRO DESIGN MANUAL
A guide to small-scale water power schemes

by
Adam Harvey

with
Andy Brown, Priyantha Hettiarachi
and Allen Inversin

Intermediate Technology Publications Ltd
trading as Practical Action Publishing
Schumacher Centre for Technology and Development
Bourton on Dunsmore, Rugby,
Warwickshire CV23 9QZ, UK
www.practicalactionpublishing.org

Reprinted with corrections 1993
Reprinted in 1998, 2000, 2002, 2005, 2008

ISBN 978 1 85339 103 3

A catalogue record for this book is available from the British Library.

Since 1974, Practical Action Publishing has published and disseminated books and information in support of international development work throughout the world. Practical Action Publishing (formerly ITDG Publishing) is a trading name of Intermediate Technology Publications Ltd (Company Reg. No. 1159018), the wholly owned publishing company of Intermediate Technology Development Group Ltd (working name Practical Action). Practical Action Publishing trades only in support of its parent charity objectives and any profits are covenanted back to Practical Action (Charity Reg. No. 247257, Group VAT Registration No. 880 9924 76).

Printed by Replika Press Pvt. Ltd

Preface

Many people around the world live in areas where streams and rivers are potential sources of energy for lighting, communications and processing industries. This is a very valuable natural resource, which can be exploited by the building of small hydro-power schemes.

Unfortunately no two hydro plants are ever precisely the same, and thus designing and building them requires a broad range of skills and experience. A further obstacle to widespread implementation has been the scarcity both of designers and of local manufacturers of equipment. This in turn has largely been due to unfamiliarity with the technology, and the absence of infrastructural support.

Intermediate Technology has sought to relieve this situation over recent years by developing comprehensive design guidelines which are presented here in this book. It is hoped that this guidance will encourage familiarity with the technology and will assist manufacturers to develop their local markets. It is hoped that it will increase the activities of local design consultants in developing countries. Finally it is hoped that the financial guidance and proposed report structures given here will help introduce micro-hydro and other renewable energy sources more firmly into routine rural development planning around the world.

Our experience of micro-hydro has been the outcome of field work in Peru, Nepal and Sri Lanka. This work has taught us the potential for micro-hydro as a relatively low-cost, locally-managed source of productive energy. To date our experience has been shared largely through a series of international training courses in Asia and Latin America, which have been attended by engineers from many developing countries around the world. This manual provides one of the bases for our design courses on micro-hydro.

This book is intended to give quick and reliable methods of planning a scheme, assessing viability and sizing and selecting components. The main requirement has been for easy-to-use design procedures, rather than for refined and complex ones. So, while it is not possible to cover every conceivable design problem, these procedures will be adequate for most situations and will provide a valuable reference to support practical training and experience. Users will quickly build up personal experience and develop improved procedures to suit local circumstances.

Please send any corrections, refinements or case study information to the IT address. Your comments and your recommendations for alternative or improved designs and planning methods will be of great value.

Adam Harvey

Acknowledgements

This book originated in the work of Intermediate Technology in Sri Lanka in the 1980s. A number of disused hydro schemes on tea estates were rehabilitated and converted from line shaft sites to electrical power generation sites, usually in the 30 to 100kW range. A series of training notes were prepared by Andy Brown and Greg Wishart for use in courses in Nepal and Sri Lanka. Since 1988, the notes have been revised and extended by Adam Harvey with help from a number of contributors.

Thanks are specially due to:

Andy Brown
Intermediate Technology, Rugby, UK

John Burton
Reading University, UK

Priyantha Hettiarachi
Systems Engineering, Sri Lanka

Ray Holland
Intermediate Technology, Rugby, UK

Allen Inversin
National Rural Electric Co-operative Association, USA

Keith Pratt
Nottingham Polytechnic, UK

Nigel Smith
Nottingham Polytechnic, UK

Richard Vearncombe
East Warwickshire College, UK

Arthur Williams
Nottingham Polytechnic

Greg Wishart
Intermediate Technology, Rugby, UK

Thanks are also due to Lahiru Perera (IT Sri Lanka), Mark Waltham (IT UK), Steve Fisher (IT UK), Kelvin Mason (IT UK), Nicky Eaton (IT UK), Bikash Pandey (IT Nepal), Aliza Kahn (Exeter University), Edinburgh University Department of Mechanical Engineering (for Chapter 4), Gerry Pope (GP Electronics), Ranil Senaratne (Fentons, Colombo), Juvenal Medina (IT Peru), Ian Smout (Water, Engineering and Development Centre, UK), Sunith Fernando (Consultant), Shavi Fernando (Ceylon Electricity Board), Anne Wesselink (Institute of Hydrology, UK), to Andy Smith, Les Price and Matt Whitton for their work on diagrams, and to Stephen Young for typesetting.

Summary of Contents

Contents

Introduction

1

1.1

Micro-hydro power

This book gives information on the design of small-scale water power schemes. These are usually classified into three levels of size: full-scale, mini and micro.

Full-scale hydro schemes produce enough electricity for large towns and extensive grid supplies. For instance the Kariba dam supplies both Zambia and Zimbabwe with grid electricity. A full scale hydro scheme produces more than 10 MW of power. A megawatt (MW) is a million watts, enough power for 20,000 light bulbs. A kilowatt (kW) is one thousand watts. One kilowatt is enough electricity for 5 households each using 4 light bulbs of 50 watts each.

Mini-hydro schemes make a smaller contribution to national grid supplies, typically in the range of 300 kW to 10 MW. Sometimes the higher end of this range, 3–10 MW, is referred to as 'small hydro' power.

Micro-hydro schemes are smaller still, and usually do not supply electricity to the national grid at all. They are used in remote areas where the grid does not extend. Typically they provide power to just one rural industry or one rural community. They range in size from 200 watts, just enough to provide domestic lighting to a group of houses through a battery charging arrangement, to 300kW, which can be used for small factories and to supply an independent local 'mini-grid' which is not part of the national grid.

Photo 1.1.1 A micro-hydro electrical scheme in Nepal. The forebay tank in the foreground can be used for limited storage of water. (IT/Jeremy Hartley)

Photo 1.1.2 Water is drawn from the waterfall crest and returned via the tailrace of the powerhouse. (IT/Jeremy Hartley – Sri Lanka)

In many cases micro-hydro power schemes do not generate electricity. Grain mills, for instance, are often driven directly from the turbine shaft. The design guidance given in this book can be used both for this kind of scheme (direct mechanical drive), and for electricity generation schemes. It is quite common for an installation to drive both an electrical generator and mechanical processing machinery.

There is an increasing need in many countries for power supplies to rural areas, partly to support industries, and partly to provide illumination at night. Government authorities are faced with the very high costs of extending electricity grids. Often micro-hydro provides an economic alternative to the grid. This is because independent micro-hydro schemes save on the cost of grid transmission lines, and because grid extension schemes often have very expensive equipment and staff costs. In contrast, micro-hydro schemes can be designed and built by local staff and smaller organizations following less strict regulations and using 'off-the-shelf' components or locally made machinery. This kind of approach is known as the localized approach. Fig 1.1.3 shows how significant a difference this can make to the cost of the electricity produced. It is hoped that this book will help to promote the localized approach. Some governments have already adopted formal policies which encourage localization, and it may be that many more will do so in the near future.

It is useful to distinguish between 'run-of-the-river' schemes (Fig 1.1.1) and 'storage' schemes (Fig 1.1.2). A storage scheme makes use of a dam to stop river flow, building up a reservoir of water behind the dam. The water is then released through turbines when power is needed. The advantage of this approach is that rainfall can accumulate during the wet season of the year and then release power during some or all of the drier periods of the year.

A run-of-the-river scheme does not stop the river flow, but instead diverts part of the flow into a channel and pipe and then through a turbine. Micro-hydro schemes are almost always run-of-the-river. The disadvantage of this approach is that water is not carried over from rainy to dry seasons of the year. The advantage is that the scheme can be built locally at low cost, and its simplicity gives rise to better long term reliability. Run-of-the-river schemes are preferable from the point of view of environmental damage since seasonal river flow patterns downstream of the installation are not affected and there is no need for flooding of the valleys upstream of the installation.

Storage schemes with dams also have the disadvantage of being more complex and expensive. They can encounter severe problems, for instance, the reservoirs will often fill with silt after some years. When this happens it is often found that it is too expensive to dredge the reservoir clean again. The scheme then ends up delivering less than the expected energy output.

Although a micro-hydro scheme never has a full-scale dam it may sometimes be designed with a small reservoir to accumulate water on a daily basis. This reservoir is usually an enlarged version of the 'forebay tank' (Fig 1.1.1) in schemes using a channel. In micro-hydro schemes which do not need a channel, the reservoir can be accommodated by the weir which then acts both as a weir and as a very small dam.

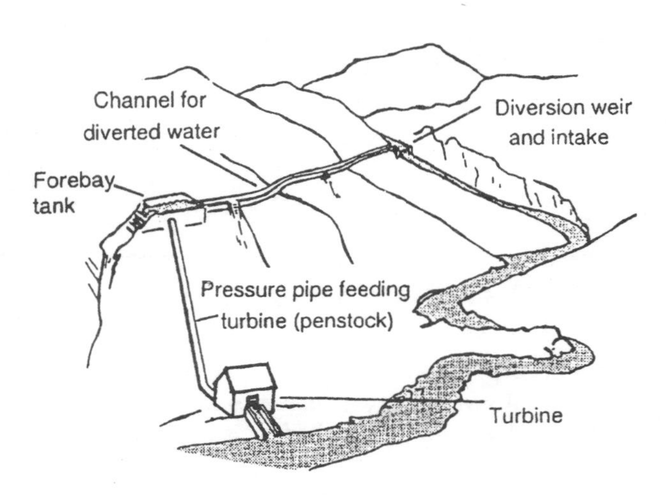

Fig 1.1.1 Run-of-the-river micro-hydro scheme
A diversion weir causes minimal environmental impact to the river and does not change the seasonal flow patterns. Some micro-hydro schemes accumulate water in the forebay tank on a daily basis. This can be useful if there is a high level of power demanded for only a few hours each day.

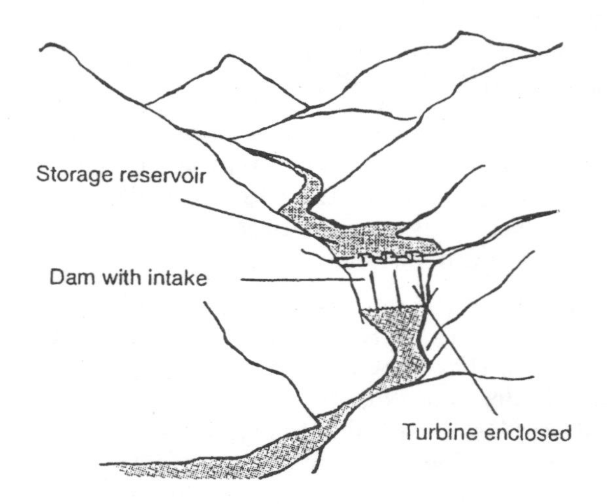

Fig 1.1.2 Hydro scheme with storage A dam causes accumulation of water by flooding the valley upstream. The water is released later in the year, so changing the down-stream river flow patterns. Silt accumulation in the reservoir can cause severe problems.

Photo 1.1.3 and 1.1.4 Rice milling is the main use of micro-hydro power in Nepal. (IT/Jeremy Hartley)

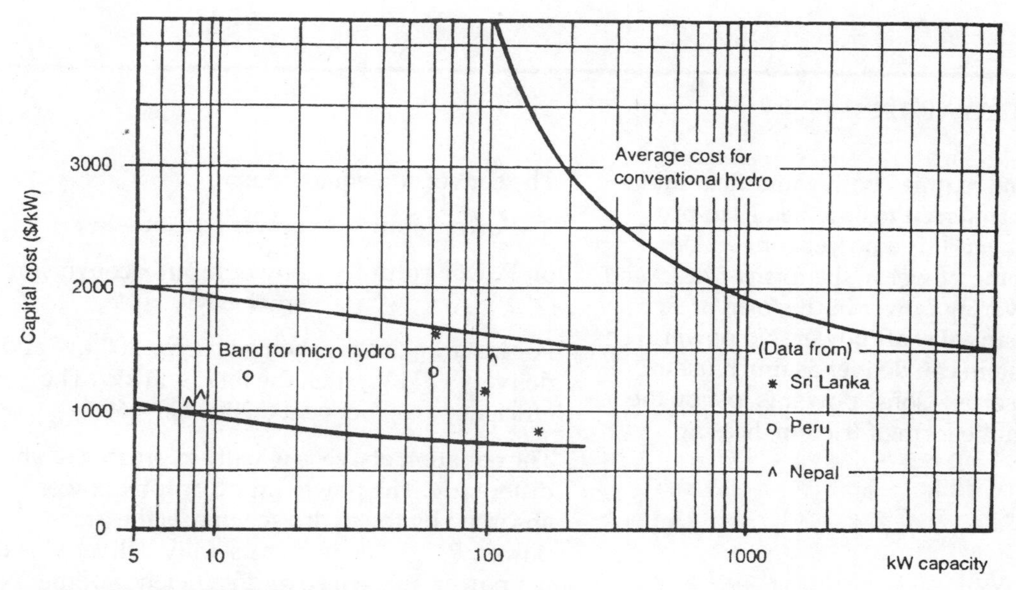

Fig 1.1.3 Micro-hydro's economy of scale The graph above (based on data in 1985) shows the overall costs of a number of micro-hydro schemes compared to mini and larger schemes. It clearly shows that the localized approach to micro-hydro can lead to low cost.

1.2 The components of a scheme

Fig 1.2.1 shows the major components of a typical scheme.

The weir acts to divert water through an opening in the river side (the 'intake' opening) into an open channel. A settling basin is used to remove sand particles from the water. The channel follows the contour of the hillside so as to preserve the elevation of the diverted water. The water then enters a tank known as the 'forebay' and passes into a closed pipe known as the 'penstock'. This is connected at a lower elevation to a waterwheel, known as a turbine. The turning shaft of the wheel can be used to rotate a mechanical device (such as a grain mill, oil expeller, wood lathe and so on), or to operate an electrical generator. The machinery or appliances which are energized by the hydro scheme are called the 'load'. In Fig 1.2.1 the load is a saw mill.

There are of course many variations on this design arrangement, for example, the saw mill could be driven directly by the rotating shaft of the turbine, without any need for electricity. Another possibility is that the channel could be eliminated, and a penstock run directly to the turbine from the first settling basin. Variations like this will depend on the characteristics of the particular site and the requirements of the user of the scheme.

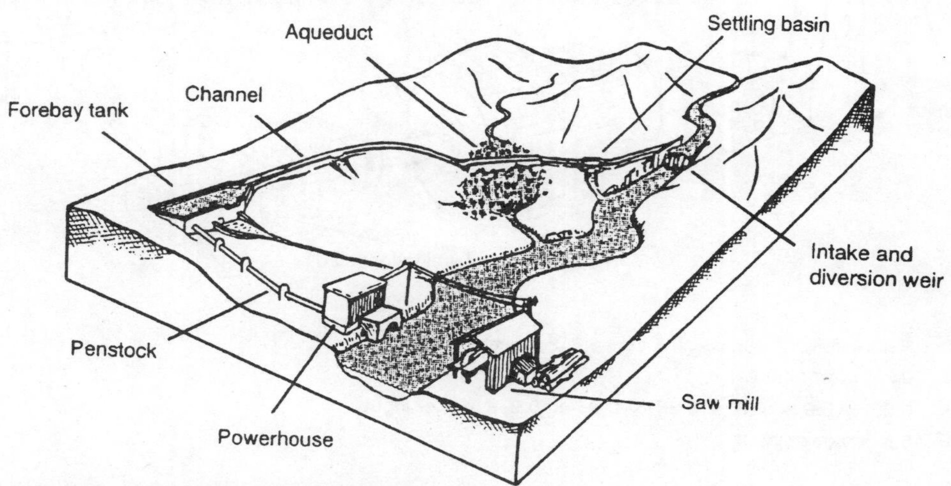

Fig 1.2.1 Major components of a micro-hydro scheme

1.3

Power from water

A hydro scheme requires both water flow and a drop in height (referred to as a 'head') to produce useful power. It is a power conversion system, absorbing power in the form of head and flow, and delivering power in the form of electricity or mechanical shaft power. No power conversion system can deliver as much useful power as it absorbs – some power is lost by the system itself in the form of friction, heating, noise etc.

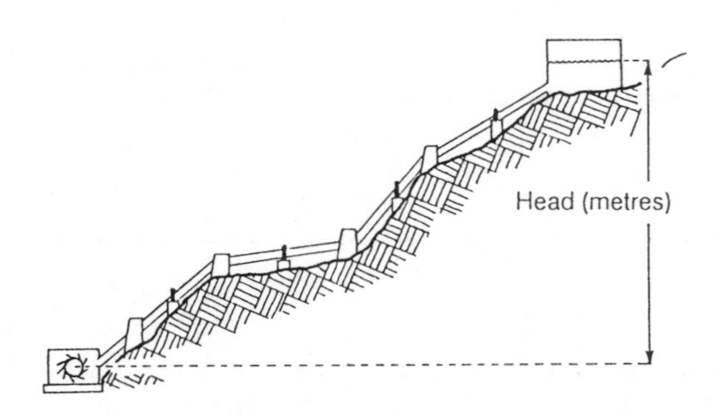

Fig 1.3.1 Head is the vertical height through which the water drops.

The conversion equation is:

Power input = power output + loss

or Power output = power input × conversion efficiency

For instance, if the scheme absorbs 200 kW and delivers 120 kW, then the loss is 80 kW. The efficiency is 60 per cent (120 = 200 × 60%).

The equation above is usually expressed slightly differently. The power input, or total power absorbed by the hydro scheme, is the gross power, P_{gross}. The power usefully delivered is the net power, P_{net}. The overall efficiency of the scheme (Fig 1.3.2) is termed e_o.

$$P_{net} = P_{gross} \times e_o \qquad \text{kW}$$

The gross power is gross head (h_{gross}) multiplied by flow (Q) and also multiplied by a factor of 10, so the fundamental equation of hydro power is:

$$P_{net} = h_{gross} \times Q \times 10 \times e_o \qquad \text{kW}$$

where head is in metres, and flow is in cubic metres per second. This simple equation should be memorized: it is at the heart of all hydro power design work. It is important to use correct units (see Example 1.3.1).

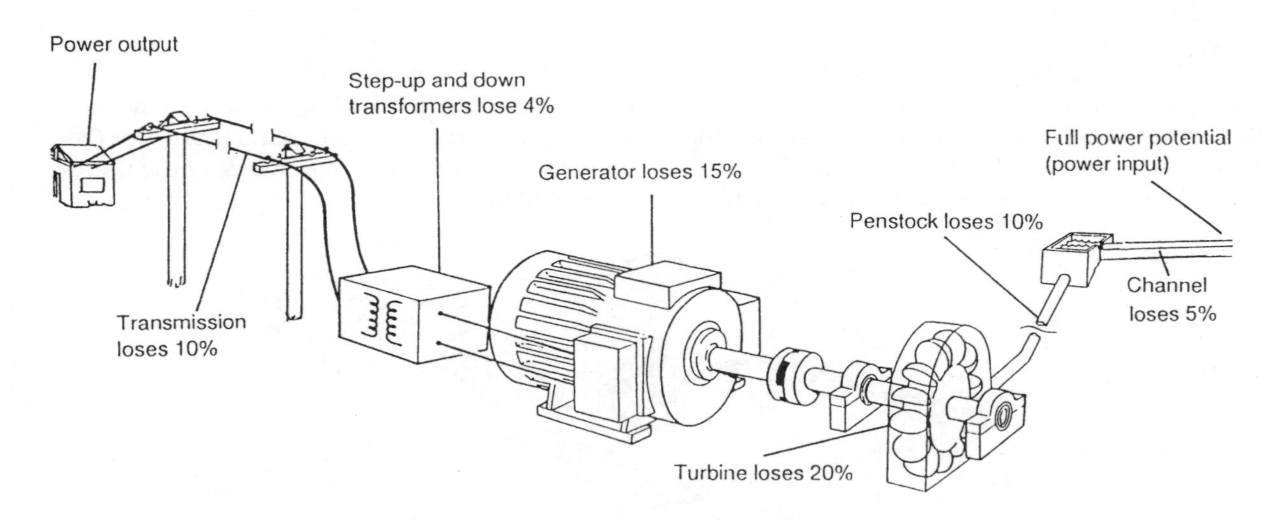

$$\text{Power output} = e_{civil\ works} \times e_{penstock} \times e_{turbine} \times e_{generator} \times e_{line} \times \text{power input}$$
$$= e_o \times \text{power input}$$
$$= 0.95 \times 0.9 \times 0.8 \times 0.85 \times 0.96 \times 0.9 \times \text{power input}$$
$$= 0.5 \times \text{power input}$$

Fig 1.3.2 Typical system efficiencies for a scheme running at full design flow. For part-flow efficiencies see Note 5.1.1.

The physics behind the power equation **Note 1.3.1**

The energy released by a falling body is its weight multiplied by the vertical distance over which it exerts its weight. The weight of falling water is the downward force it exerts. This force is a product of its mass (m) and the acceleration due to gravity (g). The vertical distance is the head (h_{gross}).

$$\text{Energy released} = m\ g\ h_{gross}\ \text{Joules}$$

The mass of the water is its density (ρ) × its volume (V) so that

$$\text{Energy released} = V\ \rho\ g\ h_{gross}\ \text{Joules}$$

Since the water enters the turbine at a certain volume flow rate of Q cubic metres each sec, the energy released can be expressed in terms of power (power is energy released per unit of time)

$$\text{Gross power } (P_{gross}) = \rho\ Q\ g\ h_{gross}\ \text{Joules/sec or Watts}$$

Water can be considered to have a density of 1,000 kg/m³, and g is 9.8 m/s². The power produced at the turbine will be much less than the gross power, because of friction losses in the penstock and in the turbine. The power output of the generator is less again, because of inefficiencies in the drive system and generator; further, transmission of the power involves losses, with the final result that the consumer receives only about half the gross power capacity of the site. The overall efficiency of the scheme (e_o) in fact tends to vary between 0.4 and 0.6. The power received by the consumer, P_{net}, is:

$$P_{net} = e_o\ \rho\ Q\ g\ h_{gross}\ \text{Watts}$$

$$P_{net} = e_o\ 1000\ Q\ 9.8\ h_{gross}\ \text{Watts}$$

$$P_{net} = e_o\ Q\ 9.8\ h_{gross}\ \text{kW}$$

The net power is often quickly estimated by taking e_o as 50% and rounding off:

$$P_{net\ (estimate)} = 0.5 \times Q \times 10 \times h_{gross}\ \text{kW}$$

Using the power equation **Example 1.3.1**

1 You have been asked to design a micro-hydro scheme to supply 50kW to a small rural factory. There is a waterfall nearby which is 20 metres high. How much flow is needed?

$$P_{net} = h_{gross} \times Q \times 10 \times e_o$$

To solve the problem the power equation must be changed so that Q is on the left hand side:

$$Q = \frac{P_{net}}{h_{gross} \times 10 \times e_o} = \frac{50}{20 \times 10 \times 0.5} = 0.5\ m^3/s$$

In small water power schemes flow is often measured in litres per second rather than cubic metres per second. It is important always to change to the standard unit , which is known as the 'SI' (System International) unit. All the equations in this book make use SI units as listed in Appendix 1. Always check that you know the correct units when using an equation in this book.

2 You are told that there is a flow of 150 litres per second in a small stream, and a head of 90 feet can be used for hydro. What is the estimated power output?

The unit given for flow in Appendix 1 is in m³/s. This is therefore the correct SI unit to use. Similarly, head is listed in units of metres. 90 feet is about 30 metres. 150 l/s is 0.15 m³/s.

$$P_{net} = 30 \times 0.15 \times 10 \times 0.5 = 22.5\ \text{kW}$$

1.4 Designing a scheme

The recommended approach to designing a scheme has four stages:

1 **Capability and demand survey:** It is essential to establish accurately how much energy is wanted for what purposes, when it is wanted, and where it is wanted. Can prospective consumers of the energy afford a new energy source, and how much are they willing to pay for it? This survey forms the basis of a financial study as described in Chapter 9.

It is also essential to assess the organizational capability of the users of the scheme. Micro-hydro is often planned for rural communities where most people do not use complex machines. The scheme will tend to involve large amounts of capital and some contributions in labour from local people, who will have high expectations of the benefits the new technology will bring to them. To avoid disappointment due to irregular maintenance and cash flow difficulties, it is best to ensure a sound management system before starting the project. The survey will recommend training in new skills, eg accounts, and a management system, eg a way of collecting tariffs and spending the funds. Most communities practise complex methods of organizing their affairs, such as irrigation water distribution, and the management of the hydro can build on such local skills and arrangements. Section 1.8 provides preliminary guidelines for a management capability and energy demand survey.

2 **Hydrology study and site survey:** This establishes the hydro power potential of the site. It shows how the water flow varies through the year, and where water must be taken for the cheapest and most effective scheme. It shows how much power will be available, and when it will be available. The study takes into account the various uses of water, for instance irrigation for agriculture, which will take precedence over hydro power. Chapter 2 describes some possible approaches to a hydrology study.

3 **Pre-feasibility study:** This is a quick cost study of a range of design options and rural energy sources. The designer of a hydro scheme will usually identify three or four different approaches to satisfying the consumer demand; there may for instance be two different designs of micro-hydro installation, and other possibilities, such as extension of the national grid lines, or use of a diesel generator. The pre-feasibility study compares these options and presents their main features. The consumers of power will want to know their options, and comparative costs, and so will the prospective funders.

The pre-feasibility will also compare the results of the energy demand survey with the results of the hydrology study. The demand survey tells us how the demand for power varies, while the hydrology study tells us how the supply of power varies. The pre-feasibility should make it clear how well supply and demand are matched. For example, it answers questions like "is the power needed for grain milling during the three dry months when there is little water, or are the villagers content without a milling service during these months?" These questions are answered by graphs and a 'plant factor' calculation as shown in Example 1.5.1.

The conclusions of the capability survey should be included here. These include recommendations for management structure, tariff structure, contingency plans and so on, as described in Section 1.8 following. The time-scale required for institution-building and management skills should be clearly stated. For instance, one year may be required. In this year it is also very important to monitor water flows and irrigation practices in order to correct the conclusions of the hydrology study against further findings.

In many cases it is useful in the pre-feasibility report to outline more than one engineering design option. For instance, a micro-hydro scheme could generate electricity which is transmitted to the centre of a village and used by a motor-driven grain mill. Or it could mill the grain with a direct mechanical drive from the turbine, but the villagers would have to walk some distance to the hydro turbine. The two options will have different advantages and disadvantages – these should be outlined in the pre-feasibility study. The report can be used as a basis for discussion.

To do the pre-feasibility study the chapters of this book can be followed through quickly, making guesses or estimates where possible.

4 **Full feasibility study:** If discussions following the pre-feasibility report indicate that one of the micro-hydro approaches is sensible, then proceed to detailed engineering calculations and costings. Also include a financial study, using the economic indicators described in this chapter and in Chapter 9. It is important also not to omit a full operation and maintenance (O+M) study. The golden rule of the feasibility study is:

"O+M first, economics and plant factor second, engineering design last"

It is essential to follow this rule, because the success of the scheme will depend in the end on correct operational procedures and effective management of the scheme when operating.

It is important to tailor the technical design to suit the level of operational and organizational resources (skills, finance, accessibility, repair workshop techniques and tools) in the region. Similarly, the technical design must be tailored to local economic conditions, such as the financial resources of the users, how much they can afford to pay for the hydro installation and how much time they can devote to managing it in future years given their other priorities.

Because of the golden rule, it is essential to make use of Chapter 10 on operation and maintenance in preparing a feasibility study, and also to make use of Chapters 1 and 9 on economic evaluation.

The feasibility will also set out in detail the tariff structure for the scheme, and how it will be implemented. This might include provisions for welfare, provisions for accumulation of funds for development of new end-uses of the hydro power, contingency plans in case of technical or managerial difficulties, or in case of problems with tariff collection, terms of reference for supervisory committees and so on. If the hydro is used both as a public service (domestic electricity) and a power source for a business, the allocation of priority rights over use of the power and the relative obligations of the different parties must be defined carefully in the form of contracts. In the same way relative obligations and priority rights must be set out for the different uses of the water supply, irrigation and hydro power, which will help resolve difficulties caused by unexpected changes in the water supply in future years, or changes in demand for either water or power.

Photo 1.4.1 Domestic lighting is now one of the major uses of micro-hydro power. (Sri Lanka)

Photo 1.4.2 An oil extractor driven by hydro in Nepal. (IT/Jeremy Hartley)

1.5
Plant factor

The best way to explain the plant factor concept is to take a simple example.

Let's say an engineer installs a hydro scheme to supply electrical lighting for a village. The scheme costs $10,000 to install. The total number of houses in the village is 50, but to begin with the village has only 25 houses wired for lighting, each consuming 200 watts of electricity. The other 25 villagers decide to wait for five years before they connect their houses. The installation is sized to provide 50 houses with 200 watts, so even on the first day it has a power capacity of $50 \times 200 = 10kW$. During the first five years only $25 \times 200 = 5kW$ is actually used. Consider the ratio of power used to power capacity:

$$\text{Power ratio} = \frac{\text{power used}}{\text{power installed}}$$

In this case, the power ratio in the first five years = 5kW/10kW = 0.5. In the second five years, it is 10kW/10kW = 1.

If the power ratio is expressed instead as an energy ratio, by multiplication by the time for which power is available or used, we arrive at the plant factor (also called 'capacity factor').

$$\text{Plant factor} = \frac{\text{power used} \times \text{time power used}}{\text{power installed} \times \text{period considered}} = \frac{\text{energy used}}{\text{energy available}}$$

In the first five years, the first 25 households are using 5kW for lighting. They require lighting in the evening hours only, 6pm to 12pm, that is, for 6 hours out of 24 hours. In this case the time power is used is 6 hours and the period considered is 24 hours. The plant factor is:

$$\frac{5\ \text{kW} \times 6\ \text{hours}}{10\ \text{kW} \times 24\ \text{hours}} = 0.125$$

In the second five years, the next 25 households are connected and the total power consumed in the evening hours is 10kW. The new plant factor is:

$$\frac{10\ \text{kW} \times 6\ \text{hours}}{10\ \text{kW} \times 24\ \text{hours}} = 0.25$$

These numbers, 0.125 or 0.25 plant factor, are a quick assessment of whether or not the hydro scheme is likely to be successful.

Suppose that the initial $10,000 capital cost was received as a loan from a bank. The bank charges interest and expects a repayment of $2000 each year for 10 years. In the first five years, only 25 villagers are connected, and the yearly cost per household of paying back the loan is $2000/25 or $80 per household per year, nearly $7 per month. On top of this there are maintenance costs to cope with, payment of the wages of the hydro plant operator, and so on. The village may be troubled by this high cost; it has to decide whether the 25 houses with lighting pay the full cost themselves or whether the other villagers, who are not receiving electrical lighting, must also pay something. The scheme is presenting the villagers with economic problems, as is indicated by the low plant factor of 0.125. In the second 5 year period the situation has improved because all the households are connected, and the shared loan repayment is half as much, just over $3 per month. The higher plant factor of 0.25 indicates a better scheme.

In practice the design engineer would be very reluctant to go ahead with such a scheme. A good design will aim for a predicted plant factor of above 0.4 even in the first years after installation and above 0.6 in later years. This is because a low factor implies costly power, and it indicates that some other sort of power (for instance a diesel generator for evening lighting) might well be better for the villagers.

A key design role for micro-hydro is therefore:

"Design for the highest possible plant factor "

Photo 1.5.1 The heat storage cooker under development by IT uses electricity generated during sleeping hours.

The importance of this rule is illustrated by Example 1.5.1, which shows how the final cost of the energy produced by the hydro depends very much on the plant factor. The example also shows how a high plant factor can be achieved: by careful seasonal and daily matching of water and power requirements, to water and power availability, and also good planning and financing of maintenance tasks. A careful energy survey will help achieve a high plant factor.

In the example it is found that the flow of 160 l/s is sufficient to meet demand, and that this flow is available all through the year. In many proposed hydro schemes the required flow is not available throughout the year. For instance, suppose 200 l/s were required from the same stream discussed in the example which has a flow pattern as shown in the hydrograph:

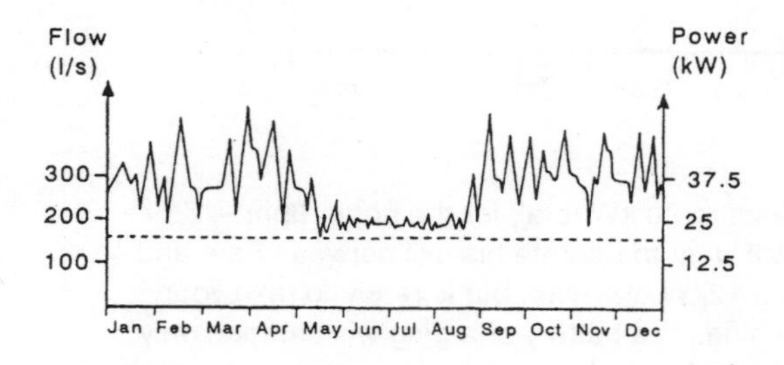

Fig 1.5.1 Simple hydrograph

In this case power is only available 9 months of the year. It would seem that the plant factor could never be higher than $9/12 = 0.75$ even if all the power generated were consumed. In Example 1.5.1 the highest achievable plant factor is 1, and it is found difficult to achieve an economically viable plant factor value of 0.6. It will be much more difficult to produce an economically viable plant factor if the plant is not operational 3 months of the year due to flow shortage. For instance, if a plant factor of 0.5 is achieved when the plant is working, non-operation for 3 months will reduce this to $9/12 \times 0.5 = 0.38$.

It is in fact possible to install a turbine which will continue to operate when the water flow reduces. This type of turbine is known as a 'variable flow' turbine, as opposed to a 'fixed flow' turbine. A variable flow turbine will usually be more expensive but often it is worthwhile because at least some power can be generated during lean flow periods of the year. (Examples of such turbines are multi-jet Peltons and crossflows fitted with regulating vanes; these are described in Chapter 5). In the example considered here, a turbine with good 'part-flow' efficiency might produce 25 kW at 200 l/s during the wet nine months and 20 kW at 160 l/s during the dry three months. As a result the maximum achievable plant factor would be $9/12 + (3/12 \times 20/25) = 0.95$. If the plant factor during the wet nine months is 0.5, then the overall plant factor may not reduce at all in this situation.

The hydrograph shows how flow varies through the year, and also from it we can see how many months of the year a certain flow is exceeded (eg 200 l/s is exceeded for nine months of the year in the example above). This same information is often presented in an 'exceedance curve' or 'flow duration curve' (FDC) for the stream. The hydrograph is converted to the FDC simply by taking all the flow records over many years and placing them with the highest figures on the left and the lower figures placed progressively over to the right.

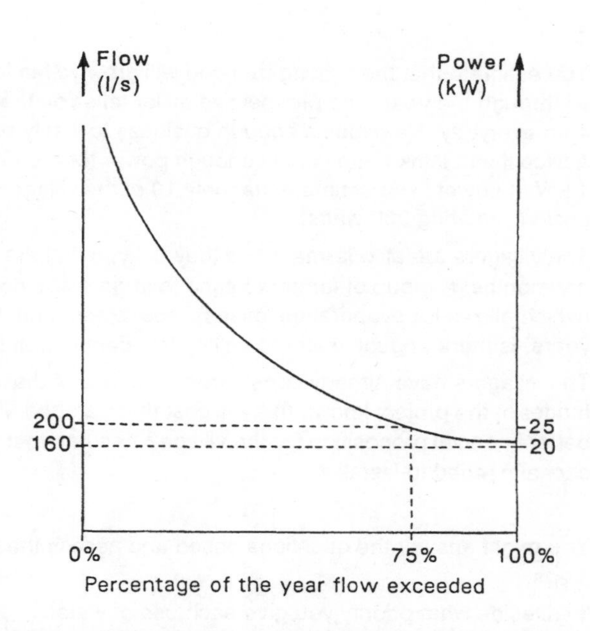

Fig 1.5.2 The flow duration curve (FDC) Note that the power derived from various flows can be shown on the same graph.

The FDC is useful because the power equivalent of the flow can be superimposed onto it, so that it is possible to read off the amount of time of the year that certain power levels can be obtained. This is a useful planning tool, allowing a choice of size of turbine to be made, together with an indication of required variable flow performance of turbine and an indication of the plant factor constraints which will result from any particular choice of turbine size.

Example 1.5.1 **Plant factor: Matching power supply and demand**

A proposed micro-hydro scheme will supply power to a village for electrical lighting and grain milling. There is also some interest in battery charging and heat storage cooking. The gross head is 25 metres. The flow in the river varies through the year as shown in this hydrograph:

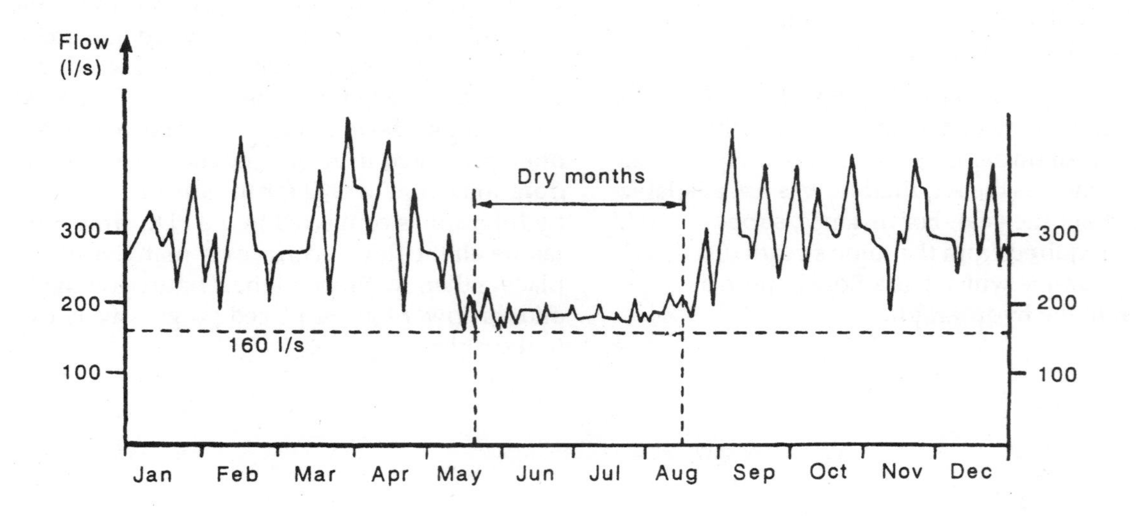

You estimate that the lighting demand will grow within four years to 20 kW total, for the hours 6pm – 12pm all through the year. The prospective miller tells you that he will want to operate his mill between 8am and 4pm everyday. He expects enough business to justify buying a 12kW machine, but is ready to take your advice if you think there is only enough power for a 6 kW machine. The battery charging will demand only 1 kW of power. You estimate that only 10 of the villagers are likely to purchase heat storage cookers, each cooker requiring 200 watts.

The villagers are also farmers and they tell you that the river water will be needed for irrigation during the 3 dry months. A group of farmers intend to irrigate 400 hectares, each hectare needing 5 m^3 of water per day (which allows for evaporation losses). You expect that the irrigation demand will double within about 4 years. Is there enough water to satisfy this demand, at the same time as the lighting and milling demands?

The villagers have already considered purchase of diesel powered mill and generator. The prospective funder of the project knows this will cost them around $0.08 per kWh of energy. Is the hydro scheme a better financial proposition for the village? Assume that the scheme will cost $4000 per year to pay off the capital needed to install it.

You might answer the questions posed and assess the feasibility of this proposal by taking the following steps:

1 Decide what priority you give each use of water.
2 Consider how the water demand variation through the year compares with water availability. To do this prepare a demand/supply graph for a typical year.
3 Prepare a demand/supply graph for a typical day.
4 Calculate the plant factor considering only the two primary loads, lighting and milling.
5 Calculate the plant factor with the addition of storage cooking and battery charging loads.
6 Consider possible down-time and modify the plant factor accordingly. Consider how to avoid the down-time.
7 To compare with the alternative option of a diesel generator/mill, calculate unit energy cost (see Section 1.7).

Continued opposite

Plant factor: Matching power supply and demand **Example 1.5.1** *continued*

Taking each step in turn:

1 **Priorities** Usually irrigation has first priority, since it is basic to agriculture and the economic security of the village. Both milling and lighting have the next highest priorities. Milling is usually more important, since it is a valued service and it brings in revenue which underpins the financial viability of the scheme. (The revenue also effectively reduces tariffs on lighting.) In this case you might decide battery-charging has the next priority after domestic lighting since it allows a wider population access to electricity. Storage cooking may be the bottom priority, since you estimate very few people will be able to afford electrical storage cookers.

2 **Yearly graph** The demand for water and the availability of water can be both shown on the hydrograph by converting it to a yearly demand/supply graph:

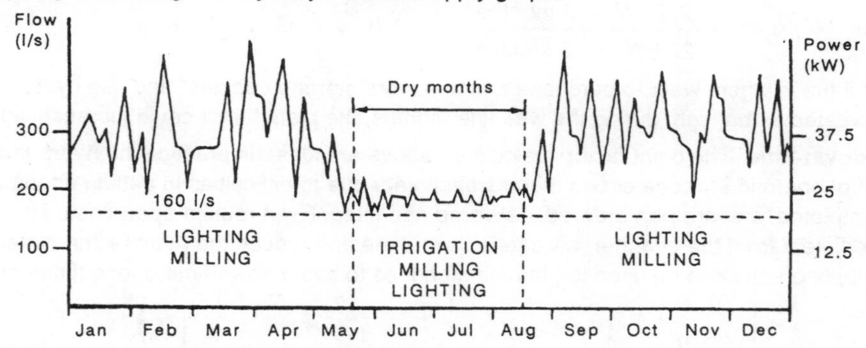

This graph is the conclusion of several calculations:

The hydrology study in this case consists of obtaining the average flow hydrograph, and noticing that the flow is unlikely in future years to drop below 160 litres/sec. The power equation gives:

$$P_{net} = h_{gross} \times Q \times 10 \times e_o = 25 \times 0.16 \times 10 \times 0.5 = 20 \text{ kW}$$

This shows that a 20 kW turbine could supply the 20 kW lighting demand throughout the year when lighting is needed, between 6pm and 12pm. If milling is going to take place between 8am and 4pm, the best time for irrigation will be the night, between 12pm and 8am, a period of 8 hours. Will there be enough water for irrigation during the night hours? Flow will be 0.16 m^3/sec, which over 8 hours is a total of $0.16 \times 60 \times 60 \times 8 = 4608$ m^3 of water. Daily water required for irrigation is $400 \times 5 \times 2$ (to allow for growth in irrigation demand) = 4000 m^3 per day.

3 **Daily graph** A demand/supply graph for a typical day in the dry season can now be drawn:

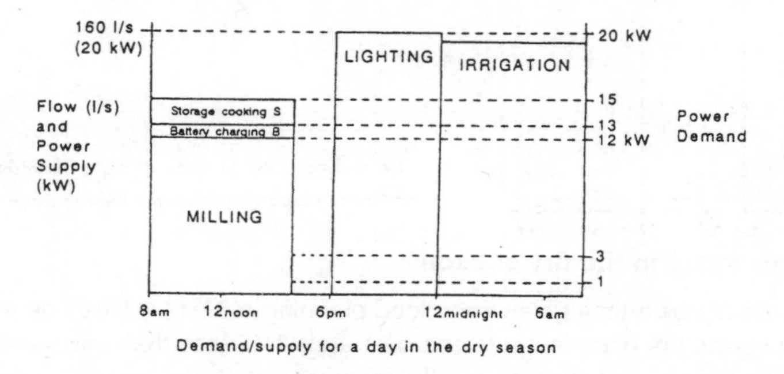

Demand/supply for a day in the dry season

The graph shows that the full irrigation demand is met at night. The miller can therefore install a 12 kW machine and the full lighting load can be met even in the dry season when water is needed for irrigation. The extra 3 kW needed for storage cooking and battery charging can be supplied during milling hours, but the turbine cannot supply 3 kW between 4 pm and 6 pm. This is because the part-power efficiency of the system will be reduced (see Example 5.1.1). It would be possible to switch in an artificial 'ballast' load (or possibly some of the lighting load) to enable the system to operate efficiently for small demands between 4 pm and 6 pm. This would apply also to battery charging and storage cooking loads during the night during the wet nine months of the year when irrigation is not needed. A similar day graph should be drawn for the wet season.

Continued overleaf

Example 1.5.1 *continued* **Plant Factor: Matching power supply and demand**

4 Calculate plant factor for primary loads:

$$= \frac{(8 \text{ hours} \times 12 \text{ kW milling}) + (6 \text{ hours} \times 20 \text{ kW lighting})}{20 \text{ kW} \times 24 \text{ hours}} = 0.45$$

5 Introduce secondary loads The plant factor in the 3 dry months will not be increased much by the secondary loads and they may not be used in conditions where time is restricted. Do not calculate plant factor on uncertain loads such as this. But for nine months of the year, if a suitable ballast is used to maintain system efficiency, storage cooking could consume 2 kW continuously and battery charging could consume an average of 0.5 kW, assuming that the charging unit is only used half the available time. Calculating plant factor on a yearly basis:

$$\text{Plant factor} = 0.45 + \frac{2.5 \text{ kW} \times 9 \text{ months}}{20 \text{ kW} \times 12 \text{ months}} = 0.45 + 0.09 = 0.54$$

Notice that if the villagers were to purchase and use more storage cookers, and use hydro energy otherwise wasted in the night during the wet nine months, the plant factor could increase significantly.

6 Allow for down-time The plant factors calculated above are ideal. In practice the hydro machinery may be out of operation for one or two months each year due to difficulties in getting a fault diagnosed, a repair completed, a spare part ordered, delivered and fitted. Even routine operations, such as cleaning sediment from the channel, will cause down-time and reduce the plant factor. Note that the demand/supply graphs can be used to plan maintenance to avoid down-time during times of peak power use:

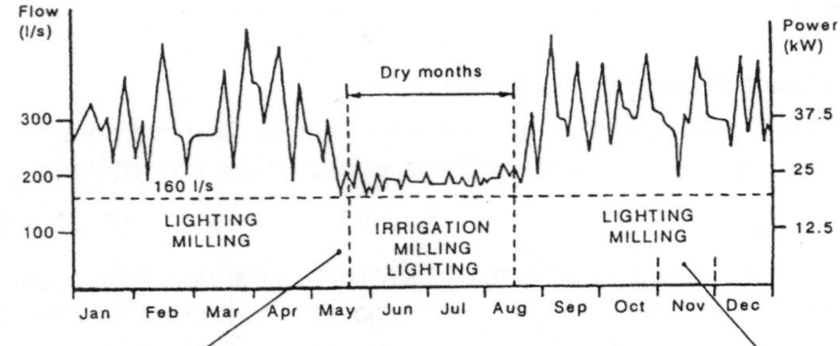

After seasonal floods repair weir and overhaul hydro.

Villagers not busy at this time of year. Good time for maintenance tasks, training for operators, spare part inventory, routine repairs.

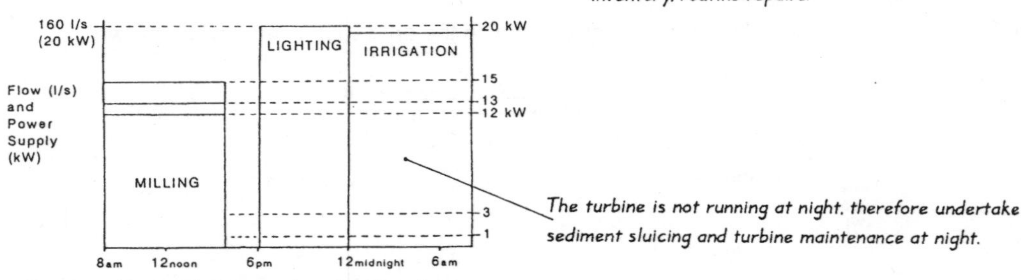

Demand/supply for a day in the dry season

An approximate estimate at down-time given very good planning at O+M will be one month per year. Since this is 1/12 of the year, the plant factor is approximately 1/12 less than calculated:

$$\text{Plant factor} = 0.54 \times \frac{11}{12} = 0.5$$

7 Unit energy cost (see Section 1.6) Allow 10% of the annual capital cost for operational and maintenance costs each year. The annual capital cost is \$4000:

$$\text{Unit cost} = \frac{C_{annual} + (O+M)}{P_{installed} \times 8760 \times PF} = \frac{4000 + (0.1 \times 4000)}{20 \times 8760 \times 0.5} = 0.05 \ \$/\text{kWh}$$

This provides a useful comparison with the diesel cost of 0.08 \$/kWh.

Load and plant factors

The term 'load factor' is sometimes mistakenly used to mean the same thing as plant factor as defined here. When planning a power supply system (and when devising a tariff system) you may find it useful to use the term load factor in its correct sense:

Load factor

$$= \frac{\text{total energy used by consumers}}{\text{total energy capacity connected to consumers}}$$

$$= \frac{\text{total energy consumed during period}}{(\text{maximum possible power demand in period}) \times (\text{hours in period})}$$

In the example given in the text, only 25 houses out of the total of 50 are connected in the first five years. If they all have all their lights switched on for six hours each day, then the total energy used each day is

$$25 \times 200\ \text{W} \times 6\ \text{hrs} = 30\ \text{kWh}$$

The total energy capacity connected to the households (to the 'load' which is electrical lighting) is:

$$25 \times 200\ \text{W} \times 24\ \text{hrs} = 120\ \text{kWh}$$

Therefore the load factor is 30/120 = 0.25 in the first five years, if all lights are always switched on for 6 hrs each day. (It reduces if some consumers switch their lights off for some of the time.)

Photo 1.5.2 This turbine drives a lineshaft which transfers power to a husker, an oil extractor and other devices including this small electrical generator. Electricity is generated only in the evening when the mills are out of use. (Nepal)

Note that we found the plant factor during this period to be half as much (0.125). This is because the plant factor is the ratio of energy used to machinery (or 'plant') installed, whereas the load factor is the ratio of energy used to load connected. (The 'load connected' in this sense is the energy consumed if the maximum consumer load was switched on during the entire period.) The load factor considers consumer behaviour patterns and average consumption of power by households. There is no need to use this conventional definition of load factor in micro-hydro planning, except sometimes when planning a tariff structure it can be a useful concept. A similar concept, the 'diversity factor' is also useful when sizing generators (see Chapter 8).

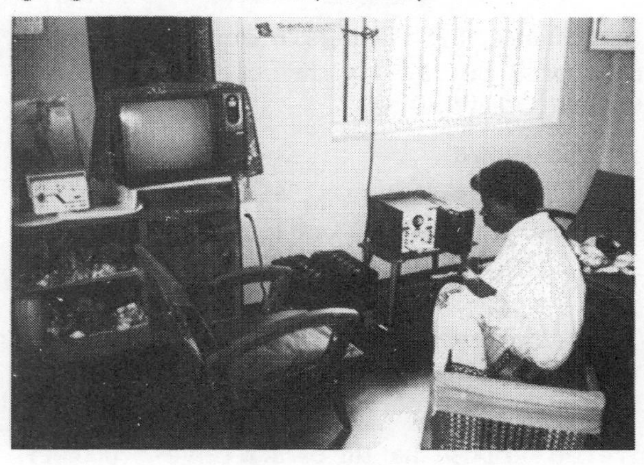

Photo 1.5.3 Often battery charging is the most practical way of distributing micro-hydro power to rural households for lighting and communications. (Sri Lanka)

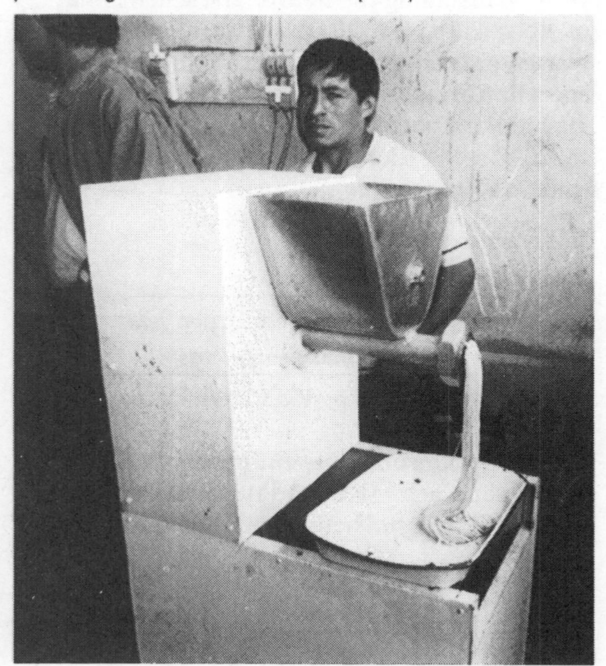

Photo 1.5.4 Micro-hydro power can be used for food processing industries in rural areas. (Peru)

1.6 Unit energy cost

The first thing we need to know about a hydro scheme is whether the electricity it produces is cheaper or more expensive than electricity produced in other ways. Buying electricity from the grid, for instance, may cost villagers $0.04 per kWh. This is the 'unit energy cost' of grid electricity. The kWh is a unit of energy and is simply calculated:

Energy

= power × period for which power used

kWh

= power in kW × hours for which power used

For instance, if 25 villagers consume 200 watts each for six hours a day, their energy consumption each year will be:

kWh consumed

= 6 (hours per day) × 365 (days per year) × 0.2 (kW per villager) × 25

= 10950 kWh per year

To find out whether the villagers are better or worse off with hydro, the unit energy cost for hydro can be calculated.

We can suppose that the bank receives $2000 per year in loan repayment for the capital borrowed. This is the annual cost (C_{annual}) of the capital spent on installing the hydro. The scheme may be expected to cost $200 per year in operators' salaries and repair costs; these are the annual 'operation and maintenance' (O+M) costs. The total cost each year is then C_{annual} + (O+M) = $2200 (see Section 9.6 to calculate annual cost from total start-up capital cost).

Unit energy cost

$$= \frac{\text{total annual cost}}{\text{energy consumed usefully per year}}$$

$$= \frac{\text{total annual cost}}{\text{power used} \times \text{period for which power used}}$$

The energy consumed usefully per year depends on the plant factor (PF), and the installed power capacity ($P_{installed}$). There are 8760 hrs in 1 year:

Unit energy cost

$$= \frac{C_{annual} + (O+M)}{P_{installed} \times 8760 \times PF}$$

If, in the above example, the predicted plant factor in the first five years of operation is 0.125 what is the unit energy cost?

Unit energy cost

$$= \frac{2000 + (200)}{10\text{ kW} \times 8760 \times 0.125} = 0.2\ \$/\text{kWh}$$

In the second five years the load factor rises to 0.25:

Unit energy cost

$$= \frac{2000 + (200)}{10\text{ kW} \times 8760 \times 0.25} = 0.1\ \$/\text{kWh}$$

This calculation shows that the micro-hydro scheme considered provides very expensive electricity compared to the grid. A much higher plant factor is needed to obtain comparable unit cost. In this case a plant factor of 0.6 would produce electricity at 0.04 $/kWh. One way to increase the plant factor is to introduce a daytime load, like a grain mill, or storage cookers.

The comparison in this case may be unfair since often the cost of grid electricity is subsidized.

Comparing unit energy cost is one way of assessing the financial viability of a hydro scheme. Chapter 9 surveys some other methods.

Photo 1.6.1 Electrical cooking is one application of hydro power in Nepal. These locally-made cookers use less than 200 W and can be used directly for cooking or for preheating water overnight.

1.7

Cost-benefit decisions

The costs of a micro-hydro scheme fall into two major categories, capital costs and running costs. A typical cost sheet is given in Example 1.7.1. The sheet includes a calculation of unit energy cost, because this is a way of including running costs and obtaining a comprehensive single financial indicator for the scheme.

The costs of a micro-hydro scheme will vary from location to location. Example 1.7.1 cannot be taken as representative of many schemes. The breakdown of costs within a scheme often follows a pattern in a particular region (Fig 1.7.1) but can also vary widely within a region.

The cost sheet shows remarks on cost-benefit sensitivity. Ask the question for each component, can a big difference be made to the success of the scheme with a small extra investment in cost or engineering design effort? To answer, you need some experience in a particular country and specific locations. Often a pattern will emerge. It is useful to include these remarks in order to discover the pattern. This helps you to concentrate your attention on the components which make the most difference to performance, benefits and economic success.

It is very clear, for example, that a small change in the plant factor will make a much bigger change to the final unit cost than that made by a small change in the cost of electrical transmission. As the design engineer, you may be presented with the option of spending a few days revising the transmission line design, or spending a few days planning for a higher plant factor. It is quite possible that doing the one job will not leave time for the other. It is important to attend to design tasks which give high returns. This is a cost-benefit decision.

We saw in the previous section that plant factor is reduced by equipment failure. Careful management of spare parts inventories and effective training of operators will avoid equipment failure and long down-times. The cost of O+M as indicated by Example 1.7.1 is relatively low (only 12% of the total cost) yet the contribution to benefit is very high. This means that this is an area which is very sensitive to improvement, and the design engineer should concentrate his time more in this area than in a less sensitive area.

For the same reasons the planning process and design studies are important. In this example they cost only 3% of the total capital cost, but the benefits resulting from a small extra investment, for instance a longer and more effective energy survey or hydrology study, can be very large.

Do not forget to allow finance for the period of institution-forming which is often necessary before installation of a scheme. It may be necessary to train local managers to implement a tariff system and an O+M system. There may be legal expenses also if the scheme is to be managed by a newly formed private company or collective.

In summary, cost-benefit analysis is an important part of the work of the micro-hydro engineer.

- Keep cost sheet records
- Include cost-benefit remarks
- Concentrate your work where cost reduction potential is greatest
- Concentrate your work where benefits are highest (O+M, plant factor)

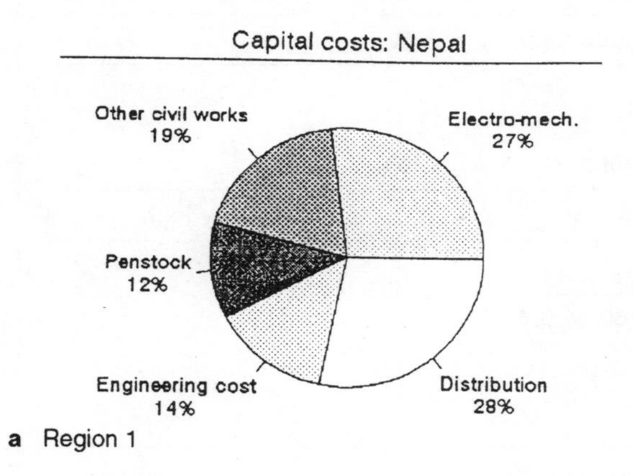

a Region 1

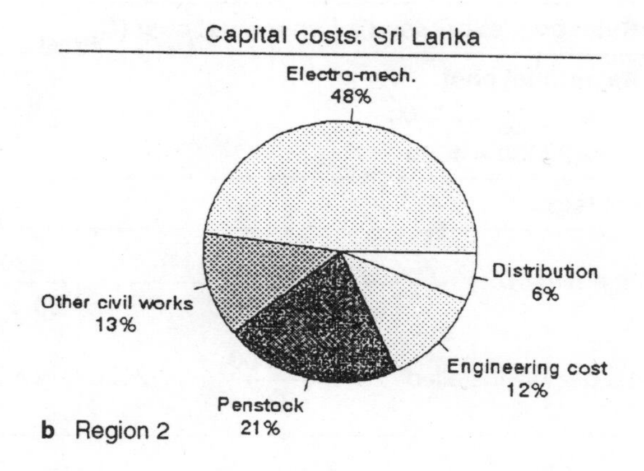

b Region 2

Fig 1.7.1 Cost breakdowns for different regions Opportunities for cost saving in distribution should be investigated in Region 2, and in electro-mechanical components in Region 1.

Example 1.7.1 **Cost sheet**

Capital costs (60kW scheme)	Cost $	Proportion of total cost	Contribution to benefits
1 **Planning/design** Engineering, energy survey, hydrology study, site survey, pre-feasibility report, feasibility report, supervision fees, commissioning fees, training manuals	4000	3%	High
2 **Management and finance** Institution formation, funding procurement, legal & insurance, training for management	2000	1%	High
3 **Penstock**	37000	27%	Medium
4 **Other civil works** Weir & intake, channel, powerhouse, site preparation/access roads, other	35000	25%	Medium
5 **Electro mechanical** Turbine, generator, switchgear, other	36000	26%	
6 **Distribution of electricity** Transmission lines, domestic connections	12000	9%	
7 **Appliances**	3000	2%	
8 **Contingency**	10000	7%	
Total capital cost	**139000**	**100%**	
Running costs			
1 **Fixed annual (O+M) costs** Labour for wages (O+M staff) Management committee (O+M) Specialist overhaul, maintenance, other	2000/year	6%	High
2 **Variable running costs** O+M staff recruitment, initial O+M training 5-yearly O+M training refresher, spare parts, tools, materials, specialist advice, replacement equipment, other	Allow 1000/year	3%	High
3 **Contingency**	Allow 1000/year	3%	
Estimated total yearly running cost (O+M)	**4000**	**12%**	**High**
Capital cost expressed as an annual cost (C_{annual})*	**28000/year**	**88%**	
Total annual cost $= C_{annual} + (O+M)$ $= 28000 + 4000$	**32000**	**100%**	
Plant factor	0.4		Very high

$$\text{Unit energy cost} = \frac{C_{annual} + (O+M)}{P_{installed} \times 8760 \times PF} = \frac{28000 + 4000}{60 \times 8760 \times 0.4} = 0.15\ \$/\text{kWh}$$

$$\text{Cost per kW installed} = \frac{139000}{60} = 2300\ \$/\text{kW}$$

* See Chapter 9 for an explanation of how C_{annual} is calculated. It is dependent on discount or interest rates. In this case the life of the scheme is assumed to be 10 years and the discount rate which applies is assumed to be 15%.

1.8

Capability and demand survey

We will assume in this section that a new system for energy supply is proposed for a rural community, such as a village of between ten and one hundred households. The guidance given here is only preliminary thought on the topic; much careful preparation is needed before embarking on the study.

A survey is needed to establish whether the scheme will be effectively managed over the next fifteen or so years. Effective management will depend on local capability and past experience in such things as organization of tariff collection, keeping of financial accounts, resolving conflict, distributing welfare benefits, etc. The major purpose of the survey is to assess this capability and from its results assess what assistance is required to raise capability to the required level.

Secondly the survey will establish what demand there is for a new energy source; how much is needed, where is it needed, what form is needed (mechanical effort, heat, or electricity?), and whether there is a genuine willingness and ability to pay for the proposed new supply. It will also investigate the methods by which the new energy supply could bring benefits to the less advantaged people in the village, and it should expose the disadvantages of the new system (for instance loss of jobs).

By the time the survey is finished, a report can be drawn up which covers the topics listed in Note 1.8.1. Some of the points in Note 1.8.1 are essential for technical planning as well as operational planning – for instance, a map or sketch of the village is needed to design electrical transmission lines, and the demand graphs are needed to size the turbine and generator, as shown in Section 1.5.

When collecting information, it may well be best not to draw up formal questionnaires. Instead prepare some checklists following the guidance given in Note 1.8.2. Use these checklists at the end of quite loose and informal discussions with as many of the villagers as possible. This is called 'open interviewing' (or qualitative interviewing) and can be more effective because it allows the person interviewed to express things in their own way. It is possible to guide an informal discussion to the topics you want to hear about, but do not expect to get back a particular answer, or one of a range of predetermined alternatives. Make very few notes during the conversation, and be prepared for it to take place at a time and place that suits the interviewee. (For instance, an unexpected conversation in the street may be the best way for women or children to talk unconstrained by household rules of seniority.) After the conversation make a written note of all the main points following your checklist.

Photo 1.8.1 Extensive discussions with villagers during early planning stages will provide a basis for successful management, operation and maintenance of a micro-hydro scheme. Fully formulated subsidy agreements, tariff structures and financial accounting systems must be introduced before construction starts. (Sri Lanka).

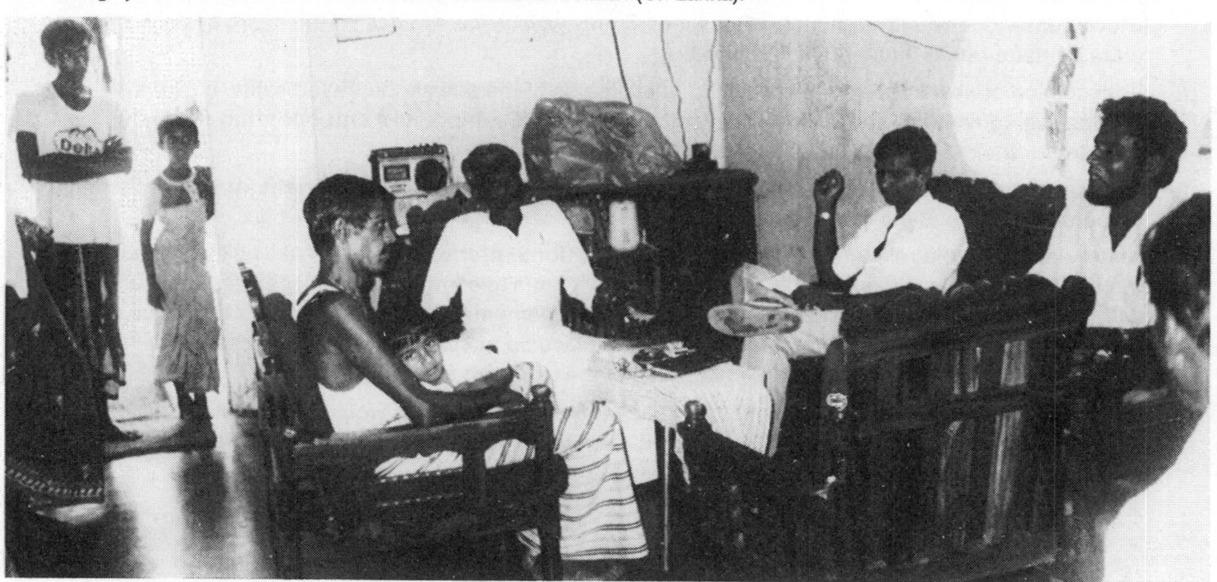

It is very easy to arrive at a biased view of energy demand, which reflects the interests of only one group of villagers. Often the most senior person in a household or village shop will talk with you, but his aspirations or needs may be different to those of his daughters, or his grandchildren. Interviews must be conducted with all types of people – that is why it is first important to identify the different types (point 2 of Note 1.8.1). Very often certain people will seem to be unavailable, or may be described as too sick or too old. These people may turn out to be the poorer members of the community, and your survey will be biased if you leave them out.

The most common bias is leaving out the interests and needs of women, which is usually because men will present themselves for interview. This bias will make your survey ineffective, even worthless, because in practice women undertake important tasks – childcare, agricultural processing, cooking, farming, and so on. They may very well have more accurate answers to many of your questions, about the cost of fuel, the problems associated with using different fuels, the needs of the children in five years time from now, and so on. Make the women your first priority.

Note 1.8.1 **Capability and demand survey**

The survey will result in a report covering the following topics:

1 A **map** or sketch of the village showing distances and positions of all houses and possible future productive or commercial activities (for instance a grain mill or battery charging unit).

2 A summary of the different **types of people** in the village with comments on how the proposed scheme might affect their economic security and opportunities in the future.

3 A summary of the different **institutions, organizations**, business leaders, or leading members of the village who may help organize the finance, maintenance, and operation of new schemes. This should include descriptions of their **past experience** in organizing finance and collective activities.

4 A full description of the current **irrigation system** and its management and **future plans** for irrigation, and a full report on how the villagers expect hydro power to affect their irrigation arrangements. Are there currently in existence methods of keeping accounts, maintaining machinery, etc?

5 An assessment of the **capability** of a local organization or individual to manage a complex scheme involving finance, welfare distribution, operation of machinery, and maintenance of machinery.

6 **Interview notes** from people and institutions with answers to questions such as those presented in Note 1.8.2. An assessment of different types of people's **willingness and ability** to pay for the proposed scheme (for instance through tariffs for electricity). Project this forward to future demand (five or ten years from now).

7 The quantity of energy required, what for (household domestic use and/or productive businesses such as a grain mill), when it is required. This should be presented in the form of **daily and yearly demand graphs** as shown in Example 1.5.1. It is essential that these such graphs are also drawn up for the **future demand** levels (five and ten years time). It may well be wise to design the hydro system for future demand rather than present demand.

8 A description of new **appliances** brought in to make use of the power source (eg lights, cookers, oil-expelling equipment) and how they would be **purchased, maintained and operated**, and how **revenue** for their use would be collected and used.

9 An assessment of the likelihood of effective and long lasting **distribution of benefit** from the scheme to poorer members of the community.

10 Where appropriate, a recommendation as to **organizational preconditions** for raising the capability to required levels (for instance, forming of a supervisory committee involving local agencies and bank staff, and provision of training in account-keeping, management and operation and maintenance of machinery). Such actions would be required to take place up to specified level of success, monitored through a smaller pilot project, before installation of the full hydro scheme.

11 A plan for a **management system**, explaining how revenues and tariffs would be collected and accumulated, how spent, how operation and maintenance would be organized, and **contingency plans** to accommodate possible future difficulties.

Interview checklist – some ideas – Capability and demand study **Note 1.8.2**

1 Types of people	Identify types of people and estimate their numbers. A list like this is used to check that interviews have been held with every type of villager, to avoid bias. • Men with farms, men without farms • Women with no cash income, women with cash incomes • Children, average per household • Elderly people • Disadvantaged people (disabled, chronically ill, lack of cash income, lack of work or source of food) • Members of religious community • People living in village only part of the year, and visitors • Professionals from outside (eg school teachers, government officers) • People with jobs outside the village • How will the distribution of types of people change over the next five or ten years? • Are new people coming to live in the area? Or are people leaving? • Will an electrical supply attract people, what type of people? • Others • What are the main activities of each type of person listed above, and their roles in village life? (eg teaching, carrying fuel supplies)
2 Institutions	Make up a list to check that you are investigating all the possible institutions which may have a capability to manage a hydro scheme. Note their activities and their membership. • Private businesses • Societies • Local bank • Government offices, extension workers • Voluntary organizations • Religious communities • Local manufacture of machinery etc • For each of the above, note proficiency and experience in particular skills (eg welfare, accountability, maintenance of machinery)
3 Energy sources	What types of fuel are used (wood, kerosene, dung, animals, etc)? For each main type of fuel: • Is the supply regular and easy, or are there shortages at times? • What does it cost in terms of labour input (eg walking and carrying by hand), inconvenience (time spent), cash payments • If it were replaced by another fuel like electricity what inconvenience would this cause (eg collecting the fuel is also a useful opportunity to do other things, talking with neighbours, grazing animals, collecting medicine plants, etc and using the fuel might have side-benefits eg wood smoke helps to clear insects) • What are the difficulties of using this fuel (health complaints, accident risk, etc) • What are the particular benefits of using this fuel (eg wood fires give high frying temperatures)
The future	Going back over these questions, will your answers be the same if you think five or ten years into the future? • What chance is there of a mains electricity grid connection, and how soon? • How much will the grid connection cost, and how will the village organize the finance?

Continued overleaf

Note 1.8.2 *continued* **Interview checklist – some ideas – Capability and demand study**

4 Village organizations	• Are there any organizations, for instance wider school activities, wider temple activities, a village welfare society, a childcare group, a farmers' association, a business, a government project? • Do villagers pay money into this or receive money from it? How are accounts kept? Are they open to inspection? • Could one of these organizations expand its activities to manage finance and waged labour for a village electrification scheme? Would the scheme be used to benefit everybody or only a few people? How would electricity tariffs be collected and how would accounts be kept? • How long has the most suitable organization been working? Have there been problems?
The future	• Is the organization likely to stop soon, or leave the area?
5 Households & individuals	• Size of household, number of occupants, ages, etc • Cash incomes, home-grown food, non-cash income • Amount of farm land, cattle, tools • Desire for new energy source (eg electricity supply) • Do you already own electrical appliances? Which ones? • If not, do you know their cost? Would finding enough money to buy them cause problems? • Would you wait some years before buying appliances? • How much could you afford to pay per month for an electricity supply (say for enough to power a radio/TV and two lights). How would you spend that money otherwise? What is the maximum amount you could pay? • Would carrying a battery for recharging be as good, if it cost less than a connection to your house? • Would you move your house if it was the only way of getting electricity? Is there anywhere you would be allowed to build it nearer the generator? • Do the women think differently? Do the elderly think differently? If so, why? • Is the idea of electric lights and TV good for women? Will you get light for the kitchen?
The future	• Will these things be different in five years or ten years? (For instance, the children will want electricity much more than we do now; there will be more cash available to pay a higher tariff). • Will the new scheme cause divisions and conflicts in the village, if for instance not everyone is connected to an electricity supply? Or is this quite normal in other fields of activity, for instance irrigation provision? What methods could be used to avoid difficulties?

Continued opposite

Interview checklist – some ideas – Capability and demand study **Note 1.8.2** *continued*

6 Entrepreneurs & officials	Consider past activities and present ones (eg a business or collective ownership of a water supply system, a tractor or an irrigation system): • Describe methods used to keep records of maintenance of machinery; and to keep records of finance • Are tariffs collected? Are there methods of including less wealthy people in the village? How are welfare benefits financed? • What problems have been encountered? How have conflicts been resolved in the past? • How would an electricity supply be organized here? Who would employ the operators and arrange training? Who would keep accounts, and ensure the spare parts stock was checked and updated? How would tariffs be collected? Would a committee of external officials supervise activities?) • Would an electricity supply be organized to ensure benefits for poorer villagers as well as successful farmers or professionals? How could this be done? Considering a business based on the hydro: • What business activity is proposed? (For instance, grain-milling, wood turning, rubber mill) • What is the market for the product, and how secure is it? Would a bank lend start-up finance? • How much energy (mechanical drive direct from the turbine, and/or electricity from the generator) would be needed, and how many hours each day, which seasons of the year? • How much would the business pay for this energy supply? • Would the business conflict with other interests? (For instance, some villagers would lose jobs, or a business in a neighbouring village would lose its market) • Would some of the less well-off villagers benefit, through new jobs, or through access to services and products? • Would the business generate enough revenue for a hydro fund so that electricity tariffs could be reduced to levels easily affordable by the less well off? • What supervisory committee could be set up to ensure equitable access to the benefits of the hydro?
The future	Would the proposed arrangements hold good over time? • How would equitable access be monitored? • What corrective actions could be taken if conflicts occurred or the benefits of hydro became progressively less accessible to most villagers, or if maintenance and operation procedures did not work well? • Going back to all the questions above, how would your answers differ in five or ten years time?
7 Other villages and model schemes	• Identify similar schemes in neighbouring villages. • Take note of studies made of demand and capability elsewhere • Find out how things have gone and whether there are lessons to be learned. Include this experience in the study.

You probably will not be able to talk to everyone, so you will need to use your interview results to estimate the energy needs of the rest of the village. Give each household on your sketch map a number and note on the map which houses gave interviews; include a few key points as to the future energy demand of these households, the level of tariff they can afford and what type of persons (gender, age, wealth level) interviewed at each household. Now check that you have spread your interviews evenly in a geographical way – you shouldn't have most interviews near the centre, or near the stream, or on the flat land, since the people living further away, away from water, or on the hillside, are also members of the community. A geographical bias can be very misleading.

Time is a very important factor to consider. Take note as to whether the village is growing or decreasing in population, and whether the new scheme will encourage growth in demand. If you are not sure how things will be in 5 or 10 years, then design for current demand and leave provision for possible growth – eg the penstock might be designed for a larger flow, while the turbine and generator are small, but at a later date they could be sold on to another village and replaced with a larger system.

Irrigation is a very important factor. It affects the supply of water, so is part of the hydrology study. But it requires organization also and the use of water for power will affect the organization of irrigation. Make sure that you fully understand the irrigation arrangements as they exist now, how they are managed, and make sure that the implications of water power use are fully understood by everyone involved in the village irrigation arrangements.

Questions which relate to organizational capability can be asked together with questions about energy demand. Much of the most useful information about capability can be learned from open interviews with villagers, so any part of the checklist of Note 1.8.2 could be used in any interview according to your own best judgement.

The information you collect on capability is the most important information of all. It must be used to assess the need for 'institution-building' – for instance, you may decide after doing the survey that the new energy scheme should not be installed for another year, or two, until certain financial, accounting or organizational principles are well understood and practised. A delay like this of one or two years may well make the difference in the long term between a successful or a failed scheme. It has the advantage of allowing time not only to organize, but also to do a careful check of the hydrology study and irrigation requirements. It is best always to explain at the very start of discussions that no scheme will be built for two years, then there is no disappointment if the delay is necessary.

1.9

Feasibility reports

The following list of headings is suitable for both a pre-feasibility and a full feasibility study. Each section should be brief; between half a page and 2 pages (all detailed information and calculations should be placed in the appendices, which will be longer). Always use tables, graphs, and sketches to save words wherever possible.

It is never wise to omit a heading simply because insufficient work has been done on the topic in question. For instance, under 'Management provisions', heading number 12b, it may be unclear what should be said, so it is tempting to remove the heading altogether. It is very much better to keep the heading in place and write "No conclusions have been reached as yet on provision for strengthening management capability, for instance through a pilot scheme, or through specialist training". The feasibility report then becomes an open document to which other people can contribute. In this example, the prospective funder may read this section and recommend a specialist to investigate. This would be a better outcome than if the scheme was funded and subsequently experienced problems stemming from management difficulties. The same approach should be taken to other headings (such as number 5 'Energy supply options'). Always include these headings even if you do not have at your immediate disposal the time, money, or access to specialists to cover the topic: just write whatever is the truth; "Insufficient data" or "Not fully investigated; approximate guess-estimates indicate that ...". There are certain topics which must never be left uninvestigated: for instance, ***topic 10 on O+M, Section 12a on Management structure and Section 7 on Hydro potential must always be fully investigated.***

There is a list of suggested headings in Note 1.9.1.

Feasibility reports – Headings **Note 1.9.1**

1a Summary	Briefly present all the major conclusions reached in the report. Include requests for grants or subsidy. State whether or not the financial requirement is typical or exceptional and whether it belongs to a general policy for finance of schemes in the region. Include economic comparison with other energy options.
1b Key illustrations	For instance a simple sketch map of village houses and transmissions, diagrams of layout of turbines and driven machinery, etc. Include simple energy supply/demand graphs (from 9 below).
1c Key data	Some of the key data can be presented on the diagrams/sketch maps. Ratings of turbine, generator, part flow arrangements, system efficiencies at part flow, table showing requirements for loans, connection charges, subsidy, and so on.
2 Energy demand	Summarize the results of your capability and demand study, concentrating on the daily and seasonal profile of energy demand. Include simple graphs. (Present the full results of your study as Supporting document 1.) Also estimate future demand trends over forthcoming 10 years and illustrate.
3 Water demand	Summarize the requirement for water from the hydro catchment for irrigation. Include any other uses of water, such as domestic or industrial, which may compete with hydro. Comment on possible multiple use of water. Estimate future demand over forthcoming 10 years. The full details should be in document 2 and some relevant data and sources in document 1.
4 Future demand trends	State whether the proposed scheme is intended to meet present or future demand for energy, and whether it allows for future growth in water use. Specify whether plans are for 5 years hence or 10 years hence, or as appropriate. Clearly specify whether conclusions of all sections below are projections into the future, and how growth in demand is accommodated.
5 Energy supply options	A brief survey and costing tables of various energy inputs including traditional fuels currently in use. Comment on future trends in supply of fuels, eg prospects for fuel wood replanting, kerosene price fluctuations. Comparative costing of energy sources which are alternatives to hydro (or could be used as auxiliary sources in combination with hydro) – eg solar photo-voltaic, solar thermal, wind, biogas, diesel; include hydro in comparative tables. Follow the same criteria as for hydro for socio-economic viability – life cycle cost, feasibility of management and maintenance, potential for accommodating growth in demand, welfare implications, convenience and sustainability, etc. Detailed information, source data references and calculations should be included in Supporting document 8.
6 Management capability	Briefly summarize the results of the capability and demand study (Supporting document 1) focusing this section on existing organizational arrangements, abilities and experience. Refer ahead to Section 11 for proposed future management arrangements.
7 Hydro potential	This section contains 2 key elements of the feasibility study: a hydrograph as in Fig 1.5.1 and an FDC as in Fig 1.5.2. The hydrograph should show irrigation and other non-hydro water demand. Both graphs should have axes marked to show the conversion of flow to hydro power. In cases where variable flow turbines are considered, an extended graph showing varying system efficiency may be necessary, as described in Example 5.1.1. An exceedence table as presented in Example 2.2.2 can be used to replace the FDC. All detailed information and hydrology data, sources, and site measurement data such as maps, should be placed in Supporting document 3, a full hydrology study for the site in question.

Continued overleaf

Note 1.9.1 *continued*	**Feasibility reports – Headings**
8 Hydro design	This section sub-divides into civil works components, penstock, turbine and generator, application of power and distribution of power. For each part present a design philosophy, eg "the channel is constructed from local materials in order to facilitate maintenance". State sources of materials (names and locations of manufacturers) and include sketches which present the dimensions and characteristics of each major component. Do not include calculations; these are in your full design study, Supporting document 4.
9 Plant factor: Matching of supply and demand	Comment on whether the power is available when it is needed, the effect of irrigation water demand. Present graphs as shown in Example 1.5.1. Calculate plant factor and discuss future trends in plant factor.
10a O+M structure	Summarize your plans for Operation and Maintenance of the installation and demonstrate that you are protecting against possible future difficulties: eg lack of motivation of operators. Present solutions such as bonus payments in return for continuous running of the hydro, exchange visits with neighbouring installations, etc. Summarize the operation schedules and arrangements for advance ordering of spare parts; full details to be in document 5.
10b O+M training	Describe requirements for training: eg translation of documents into local languages, visits by equipment manufacturers, refresher courses, future training of newly recruited operators.
10c O+M costs	All aspects of O+M should be costed (eg spare parts, wages, training and translations) and allow a contingency sum of 50%. If experimental or newly designed and manufactured equipment is used (eg locally made turbine) allow a full replacement cost. If batteries are used as part of the installation (as in domestic trickle-charge distribution) then allow for complete replacement every 3 years (or as appropriate). Allow for rising prices of spare parts and transport. Include management costs here, as calculated in 12b below.
11 Integrated water use	Describe how management of the hydro will be integrated with management of irrigation and industrial and domestic users of the available water supply. Detail the extent of involvement of farmers and other water-users in the planning, implementation, and management of the hydro scheme.
12a Management structure	How are the O+M procedures above, and the integrated water use procedures, going to be implemented? Who pays the operators and recruits new operators? How is the fund for O+M kept up, who keeps the accounts and visits the bank? How are conflicting interests in hydro and irrigation going to be resolved during drought periods? These problems are simplified in cases of private ownership but must be elaborated very carefully in cases of collective responsibility.
12b Management provisions	Refer this topic to an experienced specialist in rural development projects (eg water supply, agricultural assistance, other hydro schemes in the area). State which management skills may be lacking and how training could help. Consider the benefits of delaying start of construction for a year, while a management committee is formed, and procedures such as accounts, double-checking, tariff structure, O+M training, contingency planning, are established. Ideally build these skills around a pilot project, such as a diesel generator or a much smaller hydro installation (which may be portable and usable elsewhere later), or a diesel engine drive to a device intended eventually for hydro power.
12c Management costs	However the management is arranged, it will require finance; for instance, a pilot scheme, training costs, a wage for a full time manager, or a profit incentive for a private owner, or bonus schemes for the operators, managers, and supervisory committee members. Include these costs in your financial analysis under O+M.

Continued opposite

Feasibility reports – Headings **Note 1.9.1** ***continued***

13 Schedule of operations	Provide a time chart (divided by months and years) starting with planning and design approval stages, through to commissioning. Include the first year of operation during which monitoring and O+M training procedures will still be required. Do not omit management provisions and pilot schemes which may take a year or two before construction starts. O+M training will also appear on the chart. Validation of the hydrology study by further flow measurements (and rainfall measurements) can be shown on the chart.
14 Cost analysis	Provide a one-page cost sheet as in Example 1.7.1. Include running costs, contingency, plant factor, and unit energy cost (or other comparative economic indicators). More detailed breakdowns under each heading of the cost sheet can be provided on further sheets or in document 6 (Cost data).
15 Revenue	Comment on the various ways in which the hydro scheme will generate revenue, for instance, via sale of power to a mill-owner or other commercial enterprise, by offering of services for fee (battery charging, milling), by tariffs for fixed-wattage domestic electricity supplies. Specify how certain you are of this revenue in each case, and how this will change over time. Use Supporting document 7 (Revenues) to record data on business and market studies for potential hydro-drive enterprises.
16 Welfare	Comment on the potential of the hydro scheme in increasing the economic security of the village as a whole; in introducing new jobs and in bringing benefits to the less wealthy members of the community. Also comment on possible dangers, for instance the creation of dissent due to unequal distribution of advantages and opportunities offered by the hydro, and possible loss of existing jobs due to substitution of fuels. Propose methods of ensuring welfare benefits and protecting against disbenefits (for instance, some members of the community may receive electricity supply without payment of connection charge, due to lack of cash income; people due to lose jobs may be chosen for new jobs associated with the hydro installation).
17a Tariff structure	In village electrification schemes it is often possible to present a comprehensive financial analysis in the form of a tariff proposal. The tariffs are prices paid by householders and entrepreneurs for use of electricity. The amounts paid are calculated from outgoings such as loan repayments, O+M costs, and welfare funds. A request for a grant or subsidy can be made on the basis of such a tariff calculation, which is based on a cash flow analysis (Chapter 9). State in this section whether the tariff structure has been discussed with the villagers and whether agreement in principle exists. Will the tariff arrangement work in practice? – refer to experiences elsewhere. Indicate how well matched it is to the findings of the capability and demand study (Appendix 1). Does the tariff structure reflect the willingness and ability of all villagers to pay for the scheme? Does it include workable provisions for disadvantaged households and does it reflect the practical advice of most villagers?
17b Financial analysis	This section presents the financial future of the scheme, for example by a cash flow analysis (Chapter 9) and by presenting economic indicators. It answers the question, is the scheme economically viable or not?
17c Sources of finance	This section states the recommended basis for financing of the scheme. For example, it may set out a request for a subsidy or grant which pays for professional inputs and 25% of the capital costs, and show how a further 25% is covered by private investment by the users of the scheme (eg a connection charge or labour inputs) while the remaining 50% is covered by a loan at commercial lending rates. Note whether or not the recommendation matches existing policy on finance, and whether or not the same approach has been adopted elsewhere with success.

Continued overleaf

Note 1.9.1 *continued*	**Feasibility reports – Headings**
18 Socio-economic viability	Carefully draw together conclusions already made above on financial viability, management capability, O+M arrangements, and comment on factors which may threaten sustainability.
19 Monitoring contingency plans	Describe how the proposed structures will be monitored and what alternative ownership, management and O+M provisions could be made should these structures prove not effective in forthcoming years.
20 Supporting documents	**1** Capability and demand study **2** Water use and irrigation **3** Hydrology study **4** Technical design calculations **5** O+M arrangements **6** Cost data **7** Revenues **8** Energy supply options

Photo 1.9.1 Amongst rural industries textiles may benefit from hydro power. (Nepal)

Photo 1.9.2 An important source of revenue for micro-hydro plants is provision of battery charging services for domestic lighting. (Sri Lanka)

Hydrology and site survey

2

2.1

Introduction

This chapter is divided into the following sections:

- Flow prediction by area-rainfall method
- Flow prediction by correlation method
- Head measurement
- Flow measurement
- Geology study

The simplest hydrology study is the 'smallest flow' approach. This applies to any proposed hydro project where it is possible to say that the smallest flow in the stream (in the driest part of the year) is always going to be sufficient to meet the required power demand. All that is required is some certainty that next year, or in some future year after the scheme is installed, this 'smallest flow' won't turn out to be smaller than expected. If this happens then the turbine will produce little or no power at all. It will end up over-sized and therefore over-expensive.

If there is not much danger of this error then it is only necessary to visit the stream during the 'smallest flow' period of the year, and take flow measurements as described in Section 2.4 below. At the same time head measurements can be taken, and a geology study undertaken; these last two are always necessary for any scheme to ensure low overall costs and reliable performance.

If there is uncertainty as to the smallest flow over the forthcoming years, if it is known that there is insufficient flow for some part of the year, or if it is expected that a variable flow turbine will be needed, a prediction study must be undertaken. A flow prediction study will always involve a hydrograph, and an exceedence curve or flow duration curve (FDC). These have been introduced in Chapter 1, Section 1.6.

In some cases the FDC and hydrograph will be your own guess-estimates to begin with, in other cases they will be based on careful data collection. If the FDC is based on historical flow information going back many years into the past, it offers the best guidance available for prediction of flows many years into the future. In general, it is important to collect data on flow over as many years as possible in order to make reliable predictions.

There are a number of quite complicated ways of predicting river and stream flow. Only two fairly simple methods are presented here by way of an introduction. These are the *area-rainfall method* and the *flow correlation method*, as summarized in Notes 2.2.1 and 2.3.1 respectively. In both cases it is important to realize that a large part of the hydrology study is not done by visiting the site, but is done in town, visiting libraries and government offices. This study depends on getting hold of relevant data and documents. For instance, rainfall records and contour maps are needed for the area-rainfall method, and records of river flow are needed for the correlation method. Data on rainfall is usually found in irrigation departments or meteorology departments, weather centres, airport authorities and so on. Data on flows in major rivers and gauged streams will also be found in irrigation departments, or possibly in the offices of larger hydro power authorities. For the flow-correlation method this data must be complemented by your own flow or rainfall measurements on site. For the area-rainfall method, any field data would only act as verification checks since they are not the long term historical data which are needed for a hydrology study.

Fig 2.1.1 The hydrology study is best done in town

One purpose of the hydrology study is to ensure that the turbine is the correct size for the water flow, and has the correct variable flow characteristics, so that expense is not wasted on over- or under-capacity. The sizing of the channel, penstock, etc is also determined by the hydrology study.

The study ensures that the power supply potential of the site is known accurately, so that it can be carefully matched over the seasons of the year with the power demand.

Flow measurements on site have three purposes:

1 They may be made during the driest season in aid of a 'smallest flow' hydrology study.

2 They may be made to verify that the data obtained in documents in town is sensible and relates to the same stream as recorded in the documents.

3 They are necessary in the application of the flow correlation method described in Section 2.3 below.

The purpose of the geology study is to identify the best placement of installations such as penstock, channel and powerhouse. A small amount of effort on geology is easily repaid in cost savings made on channel construction costs, sound penstock and turbine foundations, and safety from channel collapse due to slope instabilities.

Head measurements are necessary to ascertain power availability and so are complementary to the flow prediction study.

Fig 2.1.2 Site visit Remember to bring a compass with you as well as a map.

Photo 2.1.1 Community participation in site preparation and surveying in Peru. The long-term success of a project depends on local initiative rather than import of equipment and expertise.

2.2

Flow prediction by area-rainfall method

A careful examination of a map shows that a number of possibilities exist in the design of a hydro scheme. Consider Fig 2.2.1, which shows a section of a map. Two possible micro-hydro schemes have been drawn on the map, both making use of a turbine placed at point C. Scheme A draws water at point A, then channels it along a contour line to point A1. The small drop needed for the channel is ignored since this is a preliminary study. A1 represents a proposed forebay tank feeding a penstock which drops 150 metres to the turbine at C. There is an alternative proposal for an intake site at B. The channel feeding water to the penstock at B1 (giving a turbine head of 75 metres) is shorter than channel A, and covers ground which is less steep.

Would you draw water for your scheme from point A or point B?

Since obtaining the maximum head is very important in a hydro scheme (high head turbines are usually cheaper than low head turbines), an intake sited at point A could be preferable to one sited at B. On the other hand, the penstock for scheme A is longer and therefore more expensive. The length of channel is greater, and because it traverses a steep slope it may be expensive to build and maintain. The geological survey will establish whether problems with landslips or storm runoff will make channel A unfeasible.

It is evident that a greater flow of water can be obtained at B, which may mean that more power is available from scheme B. To calculate the relative flows, first study the contours and consider the area in which rainfall collects and runs towards points A and B. Draw in the perimeters of these areas, as shown in Fig 2.2.1 with dashed lines; the two areas form the rainfall catchment area for point B, whereas the smaller single area above A is the catchment for A.

Notice that catchment B includes catchment A.

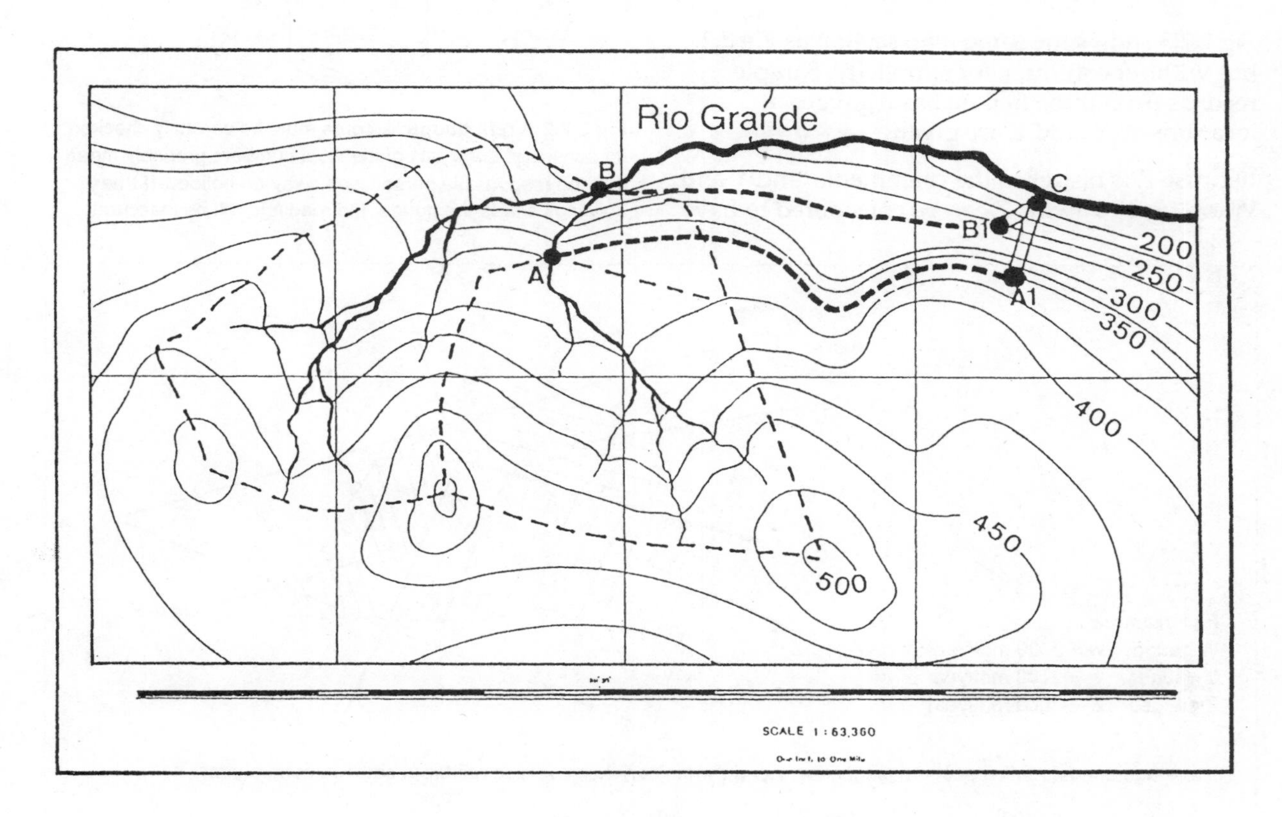

Fig 2.2.1 Portion of a map showing prospective schemes. Scheme A has an intake site at point A, a channel leading to point A1, and a small catchment area compared to scheme B. Notice that the contours indicate a gross head of 150 metres for scheme A and 75 metres for scheme B. The catchment areas are shown marked out with dotted lines. The lines are drawn by identifying watershed or ridge lines (thinking whether or not the water will flow into the chosen stream above the proposed intake) and by joining peaks also along watershed lines.

The area-rainfall approach to flow prediction is summarized in Note 2.2.1. It is recommended that you refer to this Note while reading the text, and later use it as a guide when undertaking similar calculations.

Finding ADF (annual average daily flow)

An estimate of average yearly flow, or discharge, at points A and B can be made from existing records of rainfall, such as information collected at nearby rain gauges. A rain gauge (Fig 2.2.2) measures the depth of water collected from rainfall over the year.

We can compare catchments A and B by doing a simple numerical example, supposing that three rain gauges are situated somewhere near the catchments of interest. Such rain gauges should be located on the map; if not it is necessary to mark them on the map yourself.

Any number of rain gauges can be used in the manner described below; three gauges are taken only as an example.

Fig 2.2.3 shows the same map section as Fig 2.2.1, but without contours, for simplicity. Sample records from three fictional rain gauges at locations W,Y, and Z are given.

Because Z is nearer to the catchments, and Y and W are further away, Z can be considered to have more influence on the average rainfall in the catchment. One way of weighting the average is to draw the polygon shown, linking the gauge locations, and then draw a perpendicular bisector through each link, so referring a part of the catchment to each rain gauge. If we label the area of catchment referring to Z as 'Area Z' and the total catchment area 'Area total' then the average rainfall for the whole catchment is given by the expression shown in Fig 2.2.3. (This method of estimating rainfall is known as the Thiessen method. It is useful in many hilly areas but not suitable for mountainous areas, where advice should be sought from a specialist.)

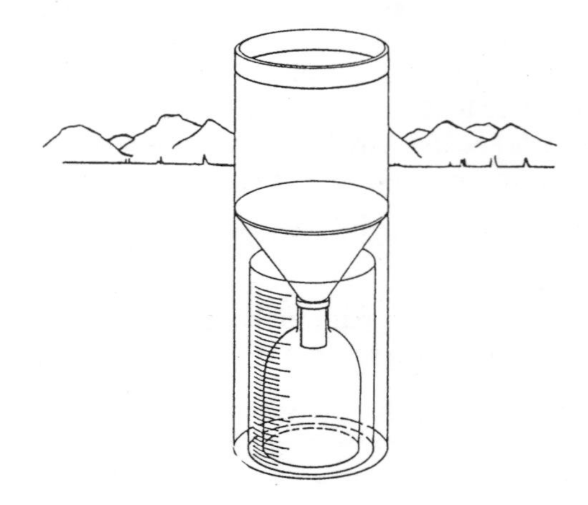

Fig 2.2.2 A rain gauge Records must be carefully checked for accuracy. Long lists of identical numbers probably mean that the responsible person was away on holiday. If trees or buildings shade the gauge, the readings will be inaccurate.

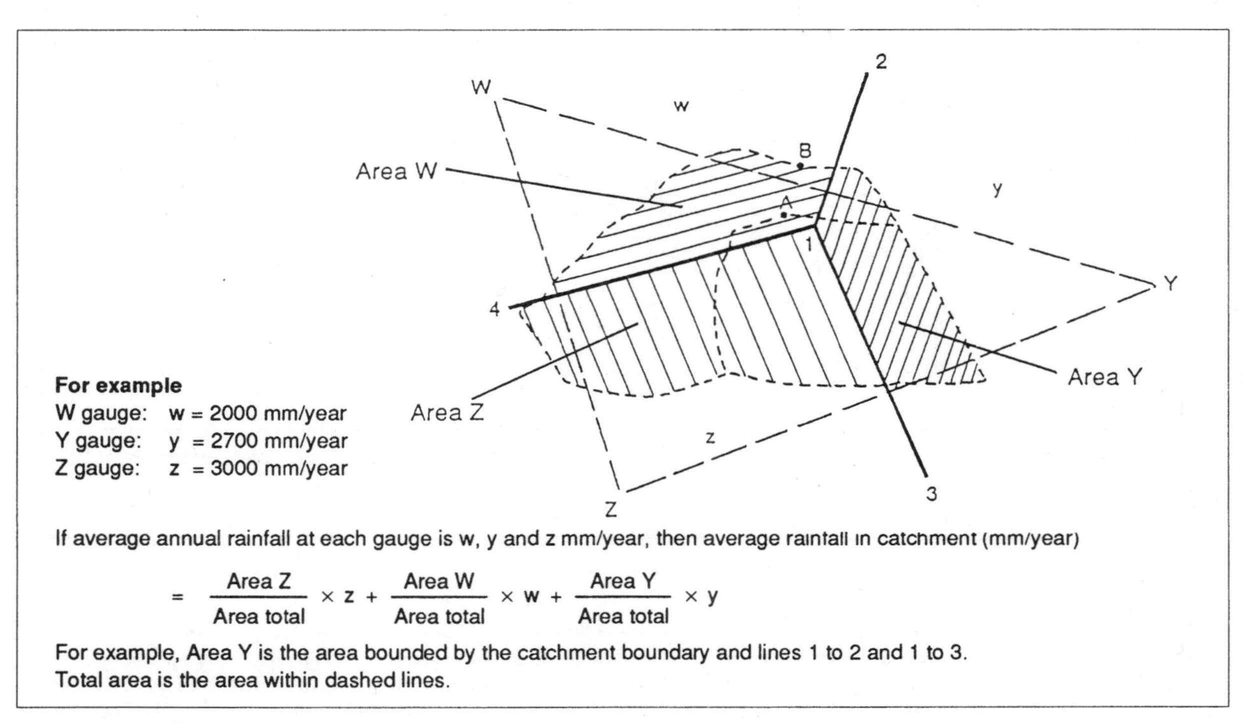

W
w
2
B
A
y
Area W
1
Y
4
Area Y
Area Z
z
3
Z
For example
W gauge: w = 2000 mm/year
Y gauge: y = 2700 mm/year
Z gauge: z = 3000 mm/year
If average annual rainfall at each gauge is w, y and z mm/year, then average rainfall in catchment (mm/year)
= Area Z / Area total × z + Area W / Area total × w + Area Y / Area total × y
For example, Area Y is the area bounded by the catchment boundary and lines 1 to 2 and 1 to 3.
Total area is the area within dashed lines.

$$= \frac{\text{Area Z}}{\text{Area total}} \times z + \frac{\text{Area W}}{\text{Area total}} \times w + \frac{\text{Area Y}}{\text{Area total}} \times y$$

Fig 2.2.3 Estimating average rainfall

Fig 2.2.4 Estimating catchment area

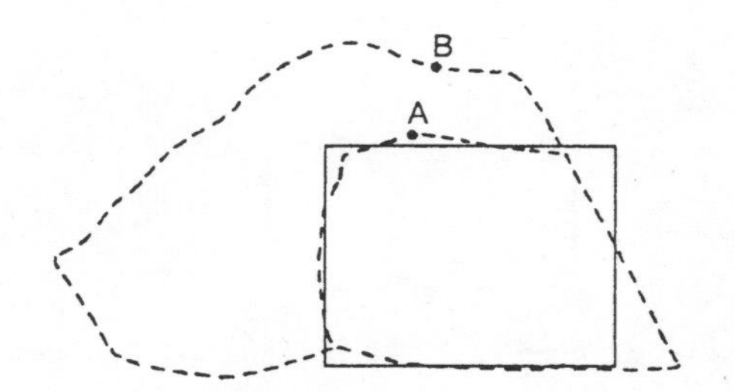

a 'Blocking' This method consists of drawing a rectangle (block) which seems to the eye to have the same area as the catchment outline. Once the block is drawn, simply measure its sides and calculate its area.

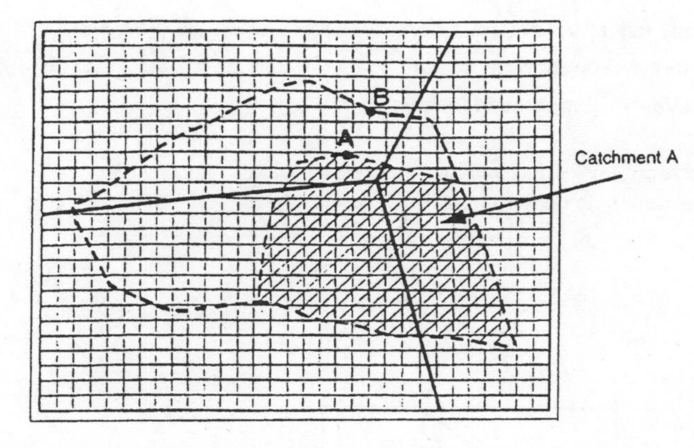

b Counting squares As an example, catchment A is made up of 135 squares. The scale of the map is one inch to one mile, and the squares are 3 mm × 3 mm. See Example 2.2.1: the area is calculated to be 4.88 square kilometres, effectively 5 km² since hydrology calculations can never be very precise.

It now only remains to find a technique for estimating areas from the map. The simplest method is to estimate by 'blocking' as shown in Fig 2.2.4. Alternatively, use squared tracing paper, which can be made by tracing or photocopying graph paper onto tracing paper.

If you place the squared tracing paper onto the map you can trace the perimeter of the catchment area, and then construct the perpendicular bisectors as shown in Fig 2.2.4b on the tracing paper. It is now possible to count and sum the total catchment areas and the individual areas referenced to each gauge. The exercise can be done for both catchments.

The total volume of rain falling each year on the catchment is now known, as shown in Example 2.2.1. Some of this water evaporates, and some flows underground; a simple representation of the hydrological cycle is shown in Fig 2.2.5. It is sufficiently accurate in most cases to calculate the average yearly streamflow (called the 'runoff') at points A and B as:

$$\text{Runoff} = \text{Rainfall} - \text{Evaporation}$$

Strictly, the runoff is also reduced by loss of rainfall seeping into the ground and not recovered later in the form of sub-surface flow.
In practice it is sufficiently accurate to ignore this effect.

A great deal of water evaporates if the air is dry and the sky is clear. Wind also increases evaporation, as does the length of the day, and air

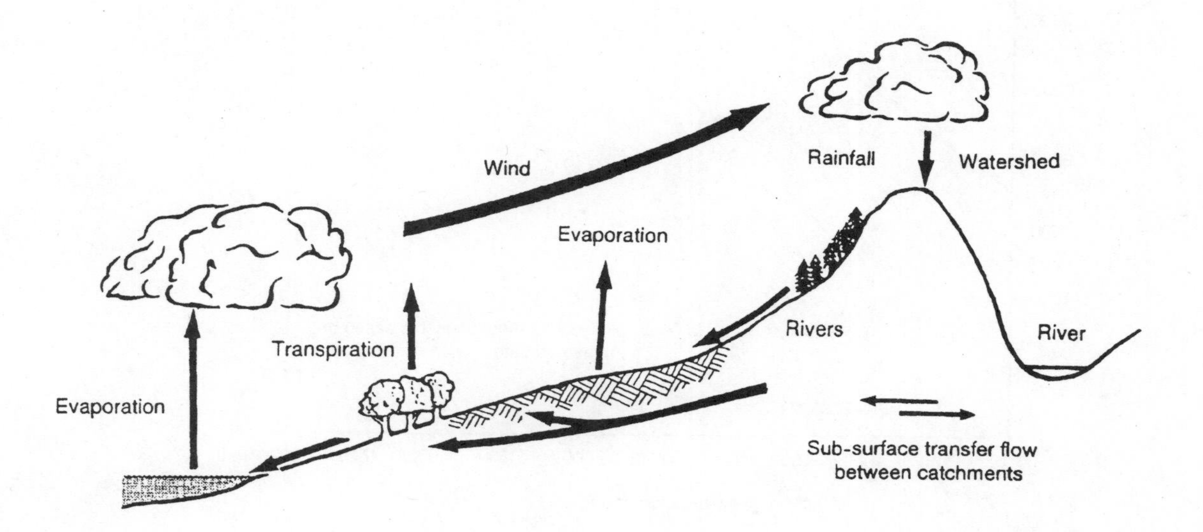

Fig 2.2.5 The hydrological cycle Notice that sub-surface flow may increase the river flow or reduce it, if the catchment is below higher watershed areas. The rainfall/runoff graph of Fig 2.2.7 takes into account such effects because it is based on measurements. This chapter assumes that net transfer flow is zero; in practice this assumption should be checked.

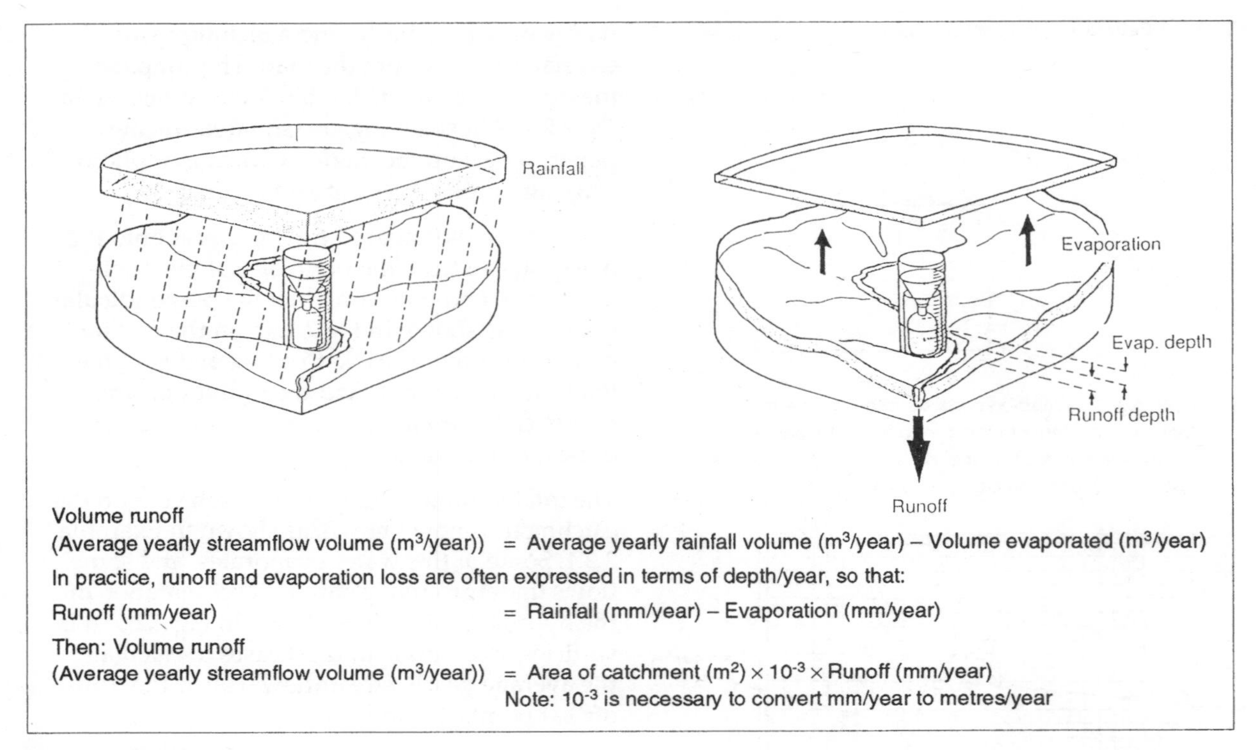

Rainfall
Evaporation
Evap. depth
Runoff depth
Runoff
Volume runoff
(Average yearly streamflow volume (m³/year)) = Average yearly rainfall volume (m³/year) – Volume evaporated (m³/year)
In practice, runoff and evaporation loss are often expressed in terms of depth/year, so that:
Runoff (mm/year) = Rainfall (mm/year) – Evaporation (mm/year)
Then: Volume runoff
(Average yearly streamflow volume (m³/year)) = Area of catchment (m²) × 10⁻³ × Runoff (mm/year)
Note: 10⁻³ is necessary to convert mm/year to metres/year

Fig 2.2.6 Runoff

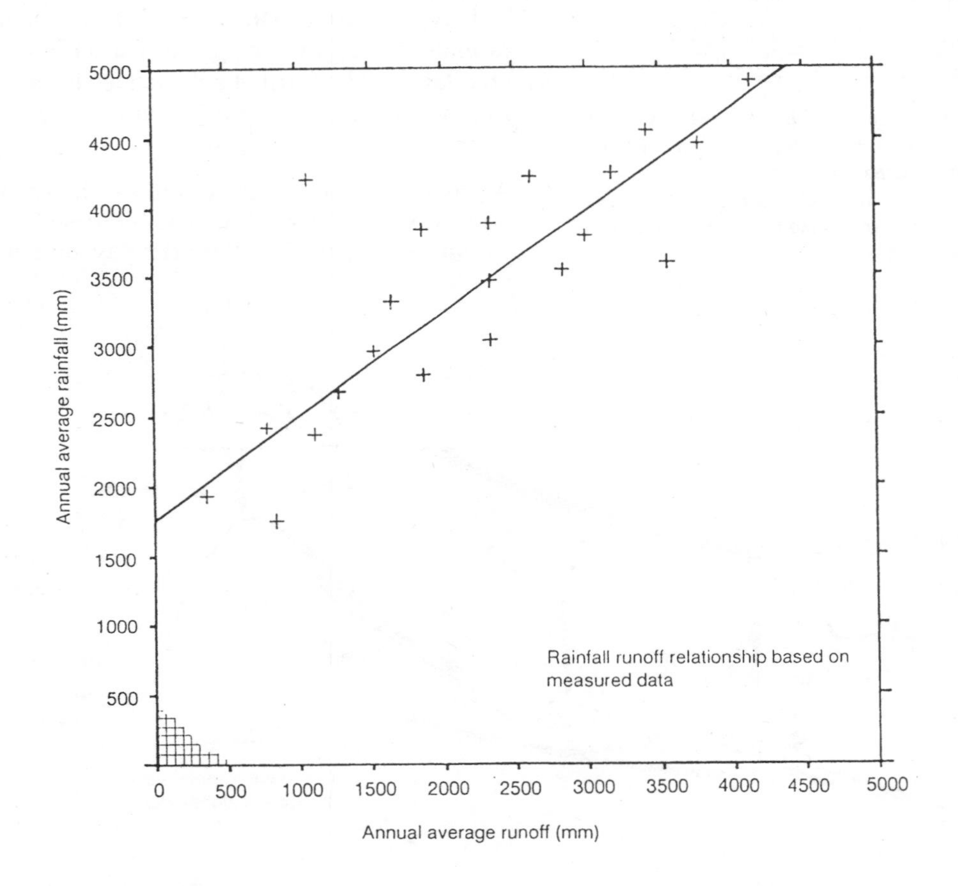

Fig 2.2.7 A rainfall-runoff graph The graph is created by entering measured values of runoff and rainfall at particular catchments in the region. If the dots cluster as they do in this example, a straight line can be drawn to fit the cluster. This line can be used to make a prediction of runoff at new sites. *The graph shown here is based on data from SW Sri Lanka – do not use it in other regions.*

temperature. These factors are taken together to calculate evaporation by the Penman and NRECA methods which are described in other texts. Such calculations can be side-stepped if you are fortunate enough to be in a country where 'rainfall-runoff' graphs are available.

An example is Fig 2.2.7. Often the relationship between runoff and rainfall is very similar in different valleys in the same region of the country, so you can find a graph amongst government records which applies to your catchment area.

Area-rainfall method **Note 2.2.1**

1 **Area**
Identify a proposed intake site and find its area of rain catchment (Fig 2.2.1, Example 2.2.1).

2 **Rainfall**
Find average rainfall per year for the catchment (Figs 2.2.3, 2.2.4, Example 2.2.1).

3 **Evaporation**
Convert rainfall to runoff using a rainfall-runoff graph for your locality or the NRECA or Penman methods which are described in other texts. Fig 2.2.7 is an example which cannot be used anywhere except South West Sri Lanka.

4 **ADF**
Average annual daily flow. Calculate volume of water flowing past the intake site each year (Example 2.2.1), and convert to an ADF figure expressed in m^3/sec.

5 **Validity of FDC**
Obtain a flow duration curve (FDC) which is valid for the catchment (Fig 2.2.12). Check its validity; it should be based on daily data.

The site should have characteristics (soil type, vegetation cover, and extent of exposed rock) which correspond to the terrain described by the FDC.

A visit to the site is necessary. Also a site flow measurement is useful (Section 2.5 below), and conversations with local people using the stream may provide a check as to flow levels during the driest part of the year.

6 **Net flow**
Consult local farmers about present and future use of water. Also consider future industrial uses – a brick factory for instance may need water. Draw daily and yearly demand/supply graphs as in Example 1.5.1, to see how the water flow can be shared during the day and night, and during the year. Alternatively a preliminary study can make allowance for irrigation by simply subtracting an estimated allowance from the ADF to arrive at a figure for ADF_{net} (see Example 2.2.2).

7 **Exceedence table**
Use the FDC to draw up a table showing the net flows which are exceeded for various proportions of the year, as in Example 2.2.2. Convert each ADF_{net} figure to a power output figure. Use this table to plan the hydro in consultation with manufacturers and the users.

8 **Demand/supply graphs**
In addition to the table, or instead of it, modify the FDC or the hydrograph to show other water uses such as irrigation, and to show power demand. Draw a yearly and daily demand profile. See Example 1.5.1 and Example 5.1.1, which discuss the use of variable flow turbines.

9 **Cost**
Draw up a quick economic study for the intake site proposed, with channel and penstock costs.

10 **Optimization**
Repeat steps 1 to 8 for other catchment areas and intake sites. Compare the results of steps 7 and 8 in order to choose the best intake site.

Example 2.2.1 **Calculation of catchment area, average annual rainfall and ADF**

Examples of a map and tracings are given in Figs 2.2.1 and 2.2.4. Rainfall figures are given in Fig 2.2.3. The scale of the map is 1 km to 10 mm, and you are using 1 mm squares. Refer to summary of method, Note 2.2.1.

- What is the total volume of water collected in the year in catchment A?
- How do you take into account the volume of water lost each year in evaporation?

1 and 2 Area and rainfall

Total number of squares in A = Squares (total) = 135

Number of squares referred to gauge Y = Squares (Y) = 58

Number of squares referred to gauge Z = Squares (Z) = 69

Number of squares referred to gauge W = Squares (W) = 8

Double check: does (Squares (W) + Squares (Z) + Squares (Y)) add up to Squares (total)?

Rainfall in catchment A

$$= \frac{\text{Squares (Y)}}{\text{Squares (total)}} \times y + \frac{\text{Squares (W)}}{\text{Squares (total)}} \times w + \frac{\text{Squares (Z)}}{\text{Squares (total)}} \times z$$

$$= \frac{58}{135} \times 2700 + \frac{8}{135} \times 2000 + \frac{69}{135} \times 3000 = 2812 \text{ mm/year} = 2.81 \text{ m/year}$$

Scale of map: 1 inch to 1 mile (1 to 63360)

Suppose that the squares are 3 mm across representing 3×63.36 metres

Therefore area of one square

$$= (3 \times 63.36)^2 \text{ m}^2$$

Therefore total catchment area A

$$= \text{Squares(total)} \times \text{Area of one square} = 4.88 \times 10^6 \text{ m}^2$$

3 Evaporation

Refer to Fig 2.2.6, and enter the rainfall on catchment A expressed as a depth in mm.
Find the net runoff once evaporation and subsurface flow are accounted for (Fig 2.2.7):

Runoff from catchment A

$$= 1420 \text{ mm/year} = 1.42 \text{ m/year}$$

4 Find ADF

Runoff from A expressed as volume flow per year:

Total catchment area × runoff expressed as a depth

$$= 4.88 \times 10^6 \times 1.42 = 6.93 \times 10^6 \text{ m}^3\text{/year}$$

Convert this figure to a volume flow per second. This nominal average flow can, in some circumstances, be called the ADF (annual average daily flow):

$$\text{ADF} = \frac{\text{Volume flow per year}}{\text{Number of seconds in year}} = \frac{6.93 \times 10^6}{365 \times 24 \times 60 \times 60} = 0.22 \text{ m}^3\text{/sec}$$

Scheme B can be assessed in the same way. What is the final value of ADF for scheme B? Check that you arrive at a value of 0.38 m^3/s for scheme B.

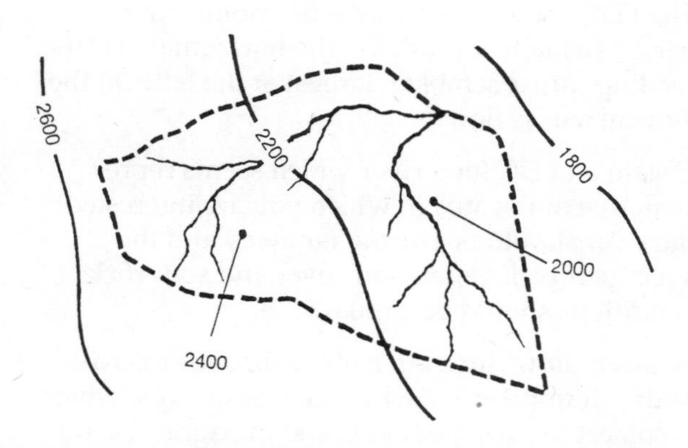

Fig 2.2.8 **Isohyetal map;** interpolation and averaging will give a very approximate indication of rainfall.

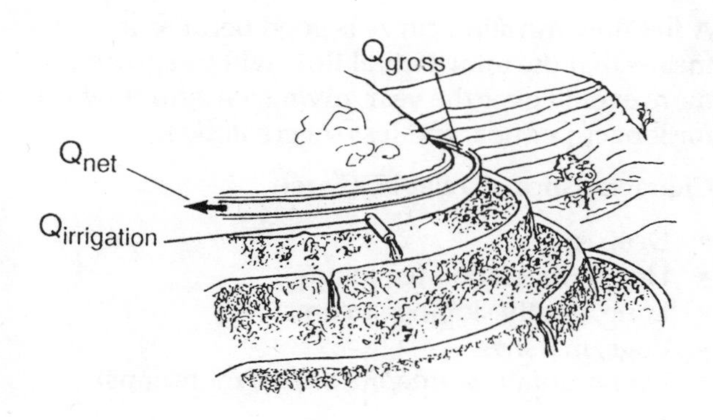

Fig 2.2.9 A small pipe set into the channel wall can easily be blocked and unblocked with a small piece of turf.

Fig 2.2.10 **Collective water use planning** Farmers can use the hydro channel for irrigation.Seasonal flow rates and maintenance schedules must be planned before designing the hydro scheme.

Absence of rain gauges

If no rain gauge records are available, the following methods may help:

- If you have one or two years to wait for planning and finance clearance, immediately install a flow measuring device such as a notched weir into the stream, and monitor as frequently as possible.
- Set up and monitor at least one rain gauge in the region of interest.
- Do not use short-term records on their own, as two years' data can be misleading (fifteen years' data are required) but correlate them with other data.
- Consult a professional hydrologist.
- Use the flow correlation method as described in Section 2.3.
- Often data in the form of isohyetal maps are available. These show lines of constant rainfall. They should never be used as a single indication of rainfall, but are sometimes useful as a check on other indications. On the whole their use should be avoided, since in micro-hydro applications the catchments are too small for sufficient accuracy using isohyets.

Net flow:
Irrigation planning and seepage loss

An important use for the water flow is irrigation. An essential aspect of hydro planning is to include the farmers of the region in the hydro planning process. Both agricultural and industrial water use is likely to reduce the amount of water available for hydro power (Fig 2.2.10).

It may be possible to divert the flow for irrigation during the day or the night and use it for hydro in the evening. Yearly and daily demand/supply graphs are needed, see Example 1.5.1. It may be that irrigation water is only needed during part of the year. The next section, *flow variation,* discusses flow duration curves and annual hydrographs – both of these can be modified to take account of non-hydro water use and so show the net flow available to hydro.

If you are uncertain, it is best simply to subtract a portion of flow from the ADF to allow for estimated irrigation and other water uses, treating this diversion as a bulk yearly quantity, and so calculate a figure for net flow (ADF_{net}) available for the hydro. It is also necessary to allow for water lost in seepage from the channel.

$$ADF_{net} = ADF - Q_{irrigation} - Q_{seepage}$$

Since $Q_{irrigation}$ may vary considerably over the time the scheme is running, it is necessary to predict these variations, and to estimate the maximum future value of $Q_{irrigation}$. It may be for instance that the number of irrigated fields will double over the next ten years; if this is a possibility, use double the present value for $Q_{irrigation}$ when calculating ADF_{net}.

Flow variation

The river flow varies during the year. There are two ways of expressing this: the annual hydrograph (Fig 2.2.11), and the flow duration curve or FDC (Fig 2.2.12). Both of these are drawn up from data collected by Government hydrologists over many years.

Ideally, they need to be based on records taken every day for at least fifteen years. But an FDC or hydrograph based on as little as one or two years' data can be useful if checked against rainfall figures for a longer period (to assess whether these were fairly typical years or not). It is important that you do not use FDC curves based on monthly measurements. In the example here, the assumption is made that all data is based on a daily or hourly measurement; this allows the average flow to be called an 'annual average daily flow' or ADF.

The FDC shows how flow is distributed over a period (usually a year). On the horizontal axis the readings are assembled, largest at the left. On the vertical axis is flow.

Obtain an FDC for a river which seems representative of the area in which you are interested; the river should not be too far away and the average mix of vegetation cover and soil/rock conditions should be similar.

A 'steep' flow duration curve is bad for micro-hydro. It implies a 'flashy' catchment – one which is subject to extreme floods and droughts. Factors which cause a catchment to be 'flashy' are:

- Rocky, shallow soil
- Lack of vegetation cover
- Steep, short streams
- Uneven rainfall
  (frequent storms, long dry periods)

A flat flow duration curve is good because it means that the total annual flow will be spread more evenly over the year, giving a useful flow for longer periods, and less severe floods.

Characteristics of a flat FDC are:

- Deep soil
- Heavy vegetation (eg jungle)
- Long, gently sloping streams
- Bogs, marshes
- Even rainfall (temperate or two monsoons)

Example 2.2.2 shows how the FDC is used to calculate an exceedence table, indicating the power which can be supplied for different proportions of the year.

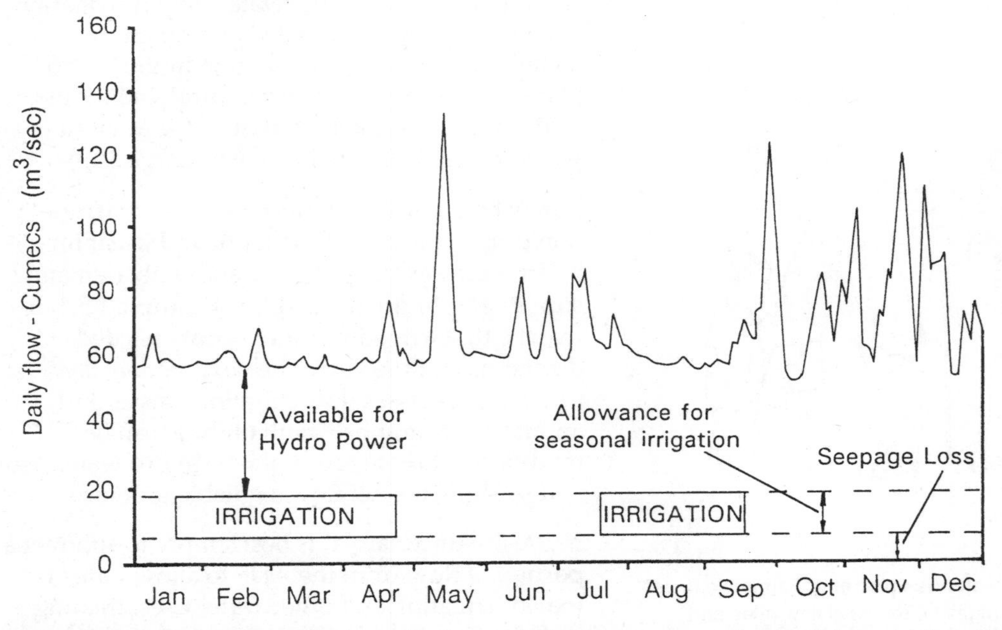

Fig 2.2.11 The annual hydrograph shows flow in a particular river or average flow in all rivers monitored in a region. The irrigation or industrial demand for water can be included on the hydrograph.

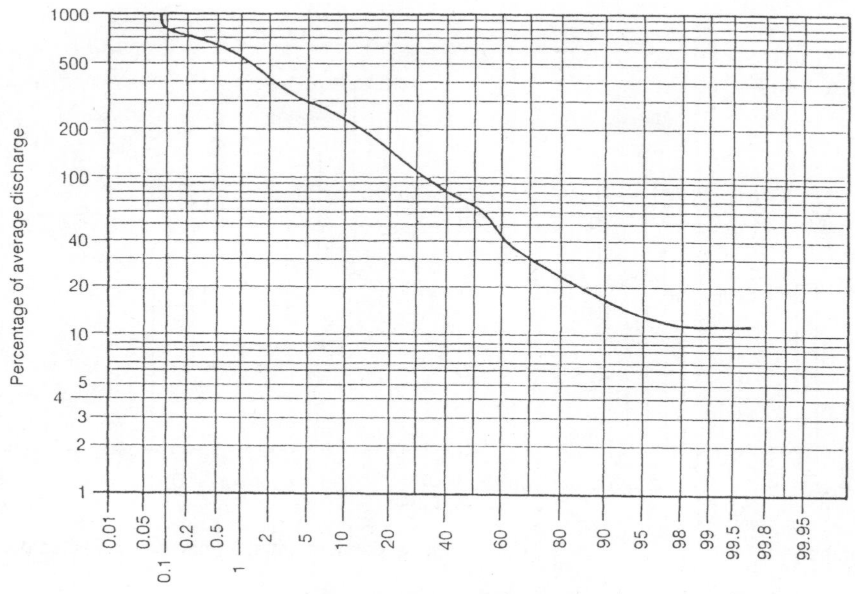

Fig 2.2.12 The FDC (Flow duration curve) represents flow variation based on sample measurements from a river. This one is for the Way Ganga in South West Sri Lanka. The same FDC can be applied to other rivers or streams in the area if soil type and vegetation cover do not differ between the catchments. *Do not use this graph except in some parts of SW Sri Lanka.*

Photo 2.2.1 Planning water use for irrigation in Peru. This is an important part of the planning process for micro-hydro power.

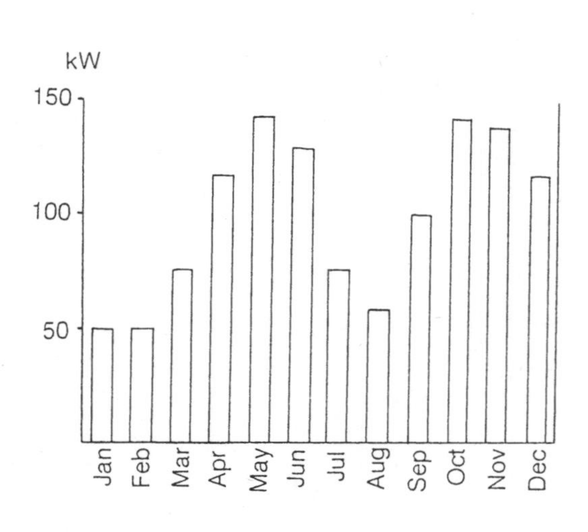

Power demand graph for the year

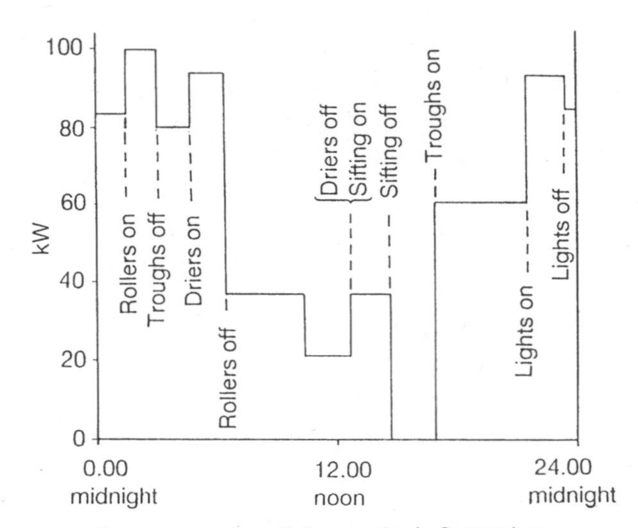

Power demand graph for one day in September

Fig 2.2.13 Annual and daily demand profiles in a Sri Lankan tea factory

Matching power demand and supply

The power users will usually have requirements which vary during the year. This is the case in a Sri Lankan tea factory, where an annual demand profile might be as illustrated in Fig 2.2.13. A typical daily variation in load is also shown. It may be that the hydro scheme is proposed as a secondary source of power which partially replaces an existing source such as an electricity grid or diesel generator. In this case the sizing of the turbine (and selection of Q_{number}) will depend on comparing the cost of each different hydro option with the savings made in existing power costs by each hydro option.

If there is no existing source of power then there are four approaches which must be compared and which are often used in combination with each other:

1 Choose the turbine and generator which will provide for the highest power demand. If this requires more than the smallest flow, then choose a variable flow type of turbine. See Example 5.1.1 to make sure part-flow performance will be adequate to provide for the lowest power demand. In Example 2.2.2 this approach could mean selection of a 165 kW turbine and generator. The FDC shows that this could run for 3 months of the year, but it is first necessary to superimpose the yearly demand profile on the hydrograph, before it is possible to say that supply will match demand through the year. Consider option **3** below at the same time.

2 Choose a turbine/generator size which matches only part of the demand, and purchase an auxiliary power source such as a diesel generator to provide for peak power demand. The saving in cost of hydro equipment may easily pay for the auxiliary equipment and the overall economics of this option may well be better than **1** because of the improved plant factor of the hydro.

3 Consider methods of storing hydro energy which is surplus at low demand periods, and then using it to meet peak demand.

4 Consider changing the yearly and daily demand pattern to suit the power supply pattern.

Photo 2.2.2 A tea factory in Sri Lanka with a micro-hydro electricity supply.

Exceedence tables — Example 2.2.2

Example 2.2.1 continues as follows. The calculation steps are summarized in Note 2.2.1.

5 Validity of FDC

We can suppose that the FDC in Fig 2.2.12 has been drawn up from measurements on rivers flowing out of a large catchment area, and that catchments A and B lie in this area, and have deep soil, good cover, and no exposed rocks. The conditions are representative of the basin described by the FDC. It is also confirmed that the FDC is based on daily data. So we decide that it is valid for our application.

6 Net flow

Discussions with local farmers reveal that a flow of 15 l/s will be required for irrigation. This is not all year round but as it is needed in the driest month. It can be treated in a preliminary study as a year-round figure. Discussions also reveal that no industrial users are likely but more land may be irrigated near the channel in future years, and there may be informal abstractions of channel water. You estimate therefore that irrigation will require 30 l/s. Calculations as in Section 3.6 reveal that seepage loss will be 10 l/s. The ADF calculated in Example 2.2.1 was 0.22 m^3/s

$$ADF_{net} = ADF - Q_{irrigation} - Q_{seepage} = 0.22 - 0.03 - 0.01 = 0.18 \quad m^3/s$$

Note that there may be other ways of sharing water between hydro and irrigation which will preserve a higher value of ADF_{net} – see Example 1.5.1 where irrigation is undertaken at night while the hydro is not running.

7 Exceedence table

Consult the FDC (Fig 2.2.12). Identify where 100% of ADF (which is 0.22 m^3/s) falls on the FDC. We find that for 33% of the year, or four months, there will be more runoff at point A than 0.22 m^3/s. Using the FDC, list the flows obtained for 140%, 120%, 100%, 80%, 60%, 40%, 30%, 20%, of ADF. Make up a list of ADF_{net} values for each row on a table as shown below. Each flow rate of water represents a different turbine power output. Assuming overall efficiency (e_o) is 0.5, tabulate the respective power delivery values.

$$\text{Site power output} = P_{net} = e_o \times ADF_{net} \times 9.8 \times h_{gross} \quad kW$$

The gross head for scheme A is found from Fig 2.2.1 to be approximately 150 metres.

Power delivered by catchment A = $0.5 \times ADF_{net} \times 9.8 \times 150$ kW. (Note Example 5.1.1)

Table 2.2.1 Exceedences for Scheme A

%	ADF (m^3/s)	ADF_{net} (m^3/s)	P_{net} (kW)	Exceedence	Q number	Expressed as months per year
140	0.308	0.268	197	22%	Q22	3 months
120	0.264	0.224	165	28%	Q28	3 months
100	0.220	0.180	132	33%	Q33	4 months
80	0.176	0.136	100	42%	Q42	5 months
60	0.132	0.092	68	57%	Q57	7 months
40	0.088	0.048	35	64%	Q64	8 months
30	0.066	0.026	19	70%	Q70	9 months
20	0.044	0.004	3	84%	Q84	10 months

What size turbine and alternator will you install for scheme A? A 132 kW turbine will operate for four months of the year (the Q33 scheme). It could also be run, at reduced power, for a further period if the turbine is specified to take reduced flow. But a large turbine will have a heavy capital cost, and will be out of action or underused for much of the year. A smaller turbine might be more economic as it is used more heavily – a 68 kW turbine (the Q57 scheme) would be used at full power for 7 months in the year, and part-power for some of the remaining months. If part-flow/part-power is considered, give careful consideration to the part-load efficiency of both turbine and generator (see Example 5.1.1).

Example 2.2.2 *continued* **Exceedence tables**

8 Matching delivery and demand for power

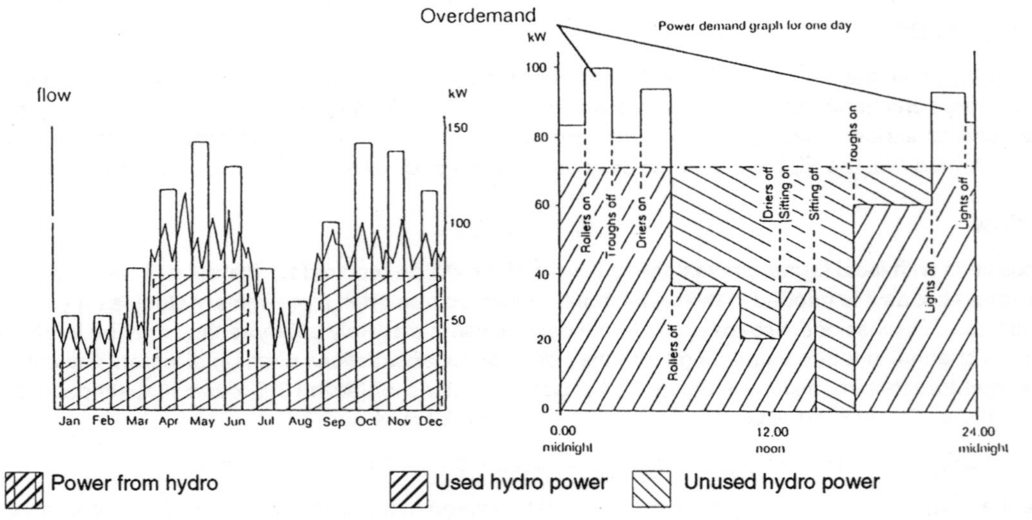

Consider a Q57 scheme for catchment A. An annual hydrograph indicates the seasons during which peak power is available. The 68 kW supply can now be shown on the annual demand profile. It can also be shown on the daily profile assuming this is representative of wet season activity. Notice that during some of the day the hydro scheme is producing less power than is needed. When the scheme is producing more power than needed, a governor or load controller/ballast will be needed.

9 Cost

Suppose that an approximate design is produced for Scheme A which results in a total cost figure of $140,000. This includes some expensive channel sections and a long penstock. It is important to cost approximately each option, because then a comparison can be made with other intake site options (eg Scheme B).

10 Optimization

Alternative catchments can now be considered. Repeating the above calculations for catchment B, for instance, results in the following table:

Table 2.2.2 Exceedences for Scheme B

%	ADF (m^3/s)	ADF_{net} (m^3/s)	P_{net} (kW)	Exceedence	Q number	expressed as months/year
140	0.530	0.490	180	22%	Q22	2 months
120	0.460	0.420	154	28%	Q28	3 months
100	0.380	0.340	125	33%	Q33	4 months
80	0.300	0.260	96	42%	Q42	5 months
60	0.230	0.190	70	57%	Q57	7 months
40	0.150	0.110	40	64%	Q64	8 months
30	0.110	0.070	26	70%	Q70	9 months
20	0.080	0.040	15	84%	Q84	10 months

Now compare the cost and performance of each catchment option. Schemes A and B both provide approximately the same kW performance. The channel of Scheme A is more expensive because it is built on steeper slopes and the penstock is more expensive because it is longer. If in this case the turbine and generator costs are about the same, the overall cost of B is less than $140,000 while providing the same power to the factory. Other factors such as integration into local irrigation planning also affect the choice. For instance, scheme A may be found to be better because although it is more expensive, it has a channel higher up on the hillside and so offers greater irrigation advantage to local farmers.

2.3

Flow prediction by correlation method

In many cases there are insufficient rainfall records for the proposed site. Instead, it may be possible to find flow measurement records for a river in the region. If these exist, it is very likely that an FDC will also be available for this gauged river. If not, the data could be used to prepare an FDC. The correlation method (summarized in Note 2.3.1) will then allow you to draw up an FDC for the stream in which you are interested with surprising ease.

The first step is to make some on-site measurements of flow at the site of interest. This is the 'ungauged site'. Only about ten or twelve measurements are needed, taken at different times during the year, at random intervals. Quite often it will be known in advance that the flow is more than sufficient for the proposed hydro scheme during the rainy part of the year, and the only critical period needing careful flow prediction is the dry season. In this case you could take six to ten measurements spread out at random intervals over the dry period only.

Having obtained two sets of readings, one for each river or stream, find the flow at the gauged site on each of the days that you obtained a flow reading from your proposed hydro site. You now have a list of corresponding flows. Example 2.3.1 illustrates this.

On a graph of streamflow at the ungauged site vs streamflow at the gauged site, plot the pairs of data points. Then, looking at the points, does there seem to be a correlation between most of these points? Can a straight line or smooth curve be drawn through this maze of points so that most points are relatively near this curve?

Often in the dry season there is little variation in flow readings between the catchments, whereas wet season flows can be affected by sudden rainfalls and significant day-to-day variations which are different for each catchment. Consequently, in fitting a curve, one would expect the curve sometimes to have a closer fit with data points associated with low flows than with those associated with high flows. The closer the fit, the more accurate the results of this method.

From the FDC of the gauged site, select the flow at several exceedence values, find the corresponding flow in the ungauged site from the previous curve, and plot an FDC for the ungauged site.

Photo 2.3.1 One suitable method of flow measurement for correlation is the salt dilution method. This requires a conductivity meter and a weighed quantity of salt.

The flow correlation method **Note 2.3.1**

1. Identify a proposed intake site, and consult government offices for hydrological records of gauged flows of rivers in the same region. Obtain these records and an FDC for the gauged river.
2. Take a series of measurements at the ungauged site. About 10 or 12 measurements are enough in a year. (If the dry season only is of critical interest, as few as 6 measurements taken at random intervals will be sufficient).
3. Find the corresponding flow on the gauged river for each of the flows recorded at your site of interest. Plot a graph of streamflow at the gauged site versus streamflow at the ungauged site.
4. Draw a correlation line through these points – do the points cluster sufficiently to indicate a reasonable correlation?
5. If so, use the FDC to select flows at certain exceedence values, convert these to corresponding flows for the ungauged site and plot an FDC for the ungauged site.

Example 2.3.1 — The flow correlation method

Fig 2.3.1 shows a map of north western Thailand showing two catchment areas.

A micro-hydro station is proposed at point B, which is fed by a catchment area of 51 km².

This area is part of the larger catchment of 2600 km² feeding site E.

There is a government river flow gauging station at site E.

Prepare a flow duration curve (FDC) for the stream at site B.

1 Obtain government flow records for site E and also an FDC (Fig 2.3.2).

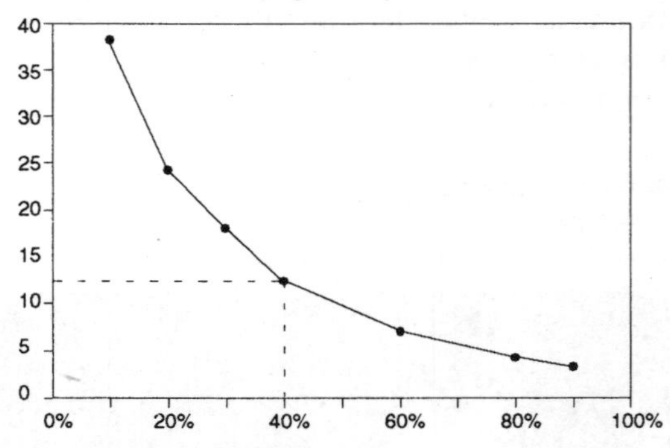

Fig 2.3.2 FDC for site E

2 Take about ten measurements of the flow at site B at random dates as shown in Table 2.3.1 (lefthand column). List the corresponding flows at E for those dates.

Date	Flow Q_B	Q_E
5 Feb 87	0.48	9.55
22 Mar 87	0.34	4.80
3 May 87	0.27	4.40
5 Jul 87	0.27	3.40
25 Jul 87	1.23	15.30
6 Sep 87	1.09	20.70
14 Sep 87	2.28	26.90
25 Sep 87	1.16	28.90
13 Oct 87	1.42	26.60
28 Oct 87	1.30	22.10

3 Plot the corresponding flows on a graph of flow at E vs flow at B. If possible, draw a straight line through the scatter of points (Fig 2.3.3). The closer the fit of the line to the clustering of points, the more accurate the method will be for your purposes.

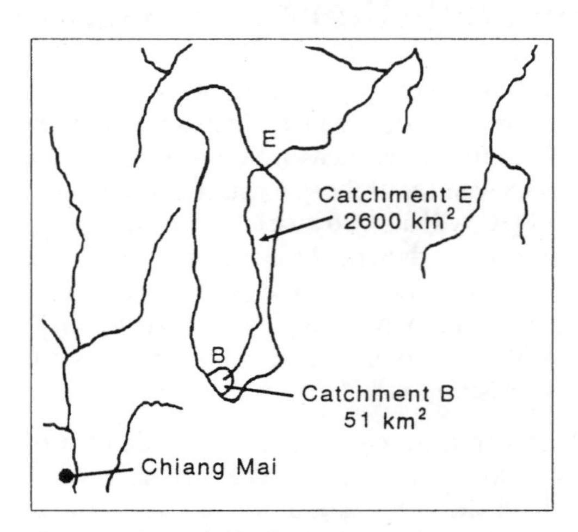

Fig 2.3.1 Catchments B and E

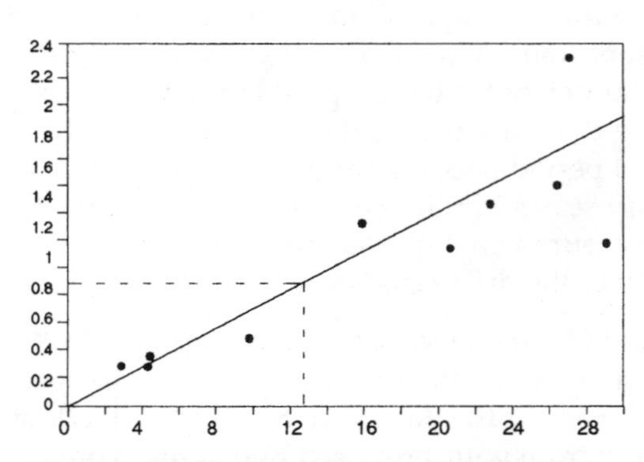

Fig 2.3.3 Correlation of flows at sites B and E

4 Use the FDC of the gauged site to select a flow at a specific exceedence value (eg 40% as shown in Fig 2.3.2). Find the corresponding flow at the ungauged site from Fig 2.3.3 and plot the value on a new FDC graph for site B. Repeat this procedure for other exceedence values and draw a curve through the points as shown in Fig 2.3.4.

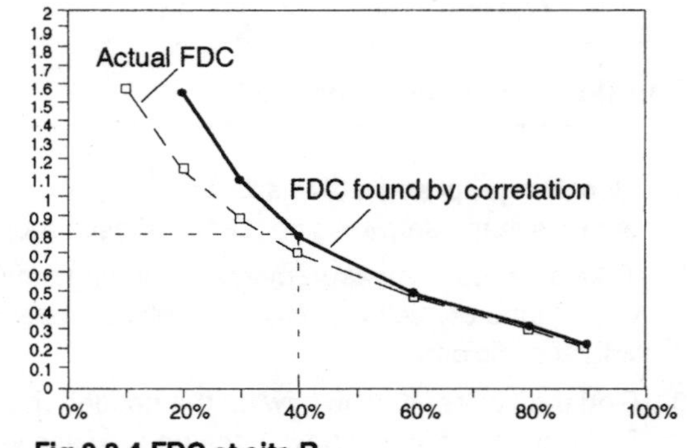

Fig 2.3.4 FDC at site B

The FDC obtained by correlation is compared to the actual FDC for site B, in order to show that this method can give results to acceptable accuracy.

2.4
Head measurements

In Section 2.2 it was suggested that a map could be used to estimate the heads at various sites. This represents a quick first look at various possibilities in the region, but does not represent a careful measurement of head. Because head is the most important factor in hydro design and costing, it must be measured accurately at the sites being considered. An accuracy of ± 3% is required.

One measurement of head is no use; it is in fact very dangerous, because it will almost certainly lead to expensive mistakes. At least three separate measurements should be taken at each site. Always plan for enough time to allow on-site comparison of survey results. Never leave the site before analysing results as usually mistakes are easy to rectify at site, but impossible to rectify once you have returned home.

If the three measurements do not agree closely with each other, you will need to take further measurements until you are satisfied you have enough agreement. Of the three, at least two should involve quite different methods of measurement.

There are a number of methods of head measurement; some are listed in Table 2.4.1. The table indicates the accuracy of each method but this really depends on the skill with which the method is used.

Some methods are more suitable on low head sites, but are inaccurate on high heads; some are only suitable on high head sites. Always choose the most accurate method given the equipment available.

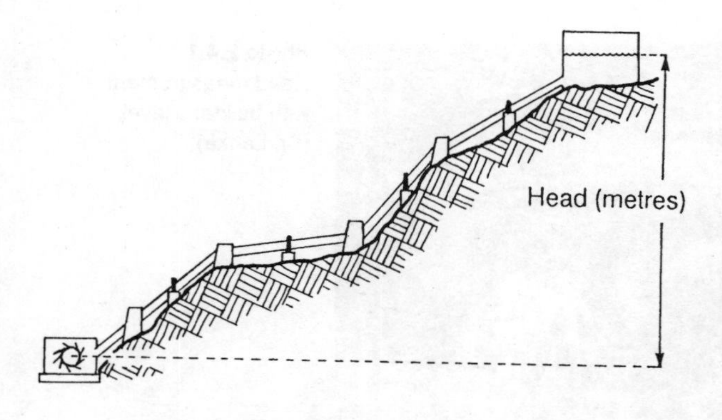

Fig 2.4.1 The concept of 'head'

The methods described here are:

- Water-filled tube (with rods or person)
- Water-filled tube and pressure gauge
- Spirit level and plank (or string)
- Altimeter
- Sighting meters
- Sighting with spirit level
- Builder's levels
- Map

Water-filled tube

Fig 2.4.2 shows this method. The method is useful for low head sites, since it is cheap and reasonably accurate. A person can act as a reference height if necessary. Two or three separate attempts must be made, so that you can make sure that your final results are correct and reliable. In addition the results should be cross-checked against measurements made by another method, for instance the water-filled tube and pressure gauge method.

The accuracy of this method can be quite surprising even when a person is used as a reference height. Villagers in a Colombian village measured a head as 'forty-eight and a half Loises' (Lois was the man who conducted the survey), which worked out at 81.6 metres. Later surveys, made at great expense, gave 82.2 metres, less than 1% difference.

Equipment

Nylon hose

Either transparent or with transparent ends. A hose diameter of between 4 and 10 mm is convenient. Fill with water before ascending.

Height markers

You can use a person as a reference height, as shown in Fig 2.4.2 a.

If graded rods are used (Fig 2.4.2 b) decimetre or centimetre markings are sufficient. A steel tape measure could be taped to a wooden rod.

Only one graded rod is necessary.

Prepared record sheet

As shown in Fig 2.4.2 c

Table 2.4.1 **Comparison of techniques**

Method	Comments	Advantages and limitations	Accuracy	Precautions
Water-filled tube and rods (or person)	Weight: Light Expense: Low	Long-winded for high heads	Approx 5%	Repeat measurements
Water-filled tube and pressure gauge	Calibration chart must be drawn up Weight: Light Expense: Low	Fast, quite foolproof, can measure penstock length at same time	Good (<5%) if gauge calibrated	Recalibrate the gauge. Repeat measurements
Spirit level and plank (or string)	Weight: Light Expense: Low	Unsuitable for long gentle slopes Slow to use. Best done with two people	Approx 5% on steep slopes 10-20% accuracy on gentle slopes (1:10)	Repeat measurements
Altimeter	Weight: Can be heavy, some are light. Expense: High	Useful on medium and high heads (>40 m) Can be fast, but more reliable if continuous monitoring undertaken	Gross errors (30%) possible. Used with skill, 2% at high heads	Experience and skill needed. Must be calibrated and temperature corrected
Sighting meters (clinometers or Abney levels)	Weight: Light Expense: Moderate	Fast in clear ground; two people required	Good 5%	Experience, skill, and calibration needed
Sighting with a spirit level	Weight: Reasonable Expense: Low	Useful on all sites especially where other methods are too slow	2-5% on steep slopes 10-20% on gentle slopes (10°)	
Builders' levels (Dumpy) theodolite	Weight: Heavy Expense: Can be hired, since in common use	Not good on wooded sites Fast	Very good	Liable to error Calculations can introduce errors
Map	Map-reading skills needed Weight: Light Expense: Low	High heads only Wrong site may be identified	Depends on quality and scale of map	Map may be incorrect Check correct site identified

Photo 2.4.1
Head measurement with builder's level. (Sri Lanka)

Fig 2.4.2 Water-filled tube

a *Using person height*

i Person Y matches water level in the tube to his eyes.

Person X keeps the water level at the other end of tube to the expected forebay surface level.

b *Using graded rods*

i Person Y measures height A1 at expected forebay surface level

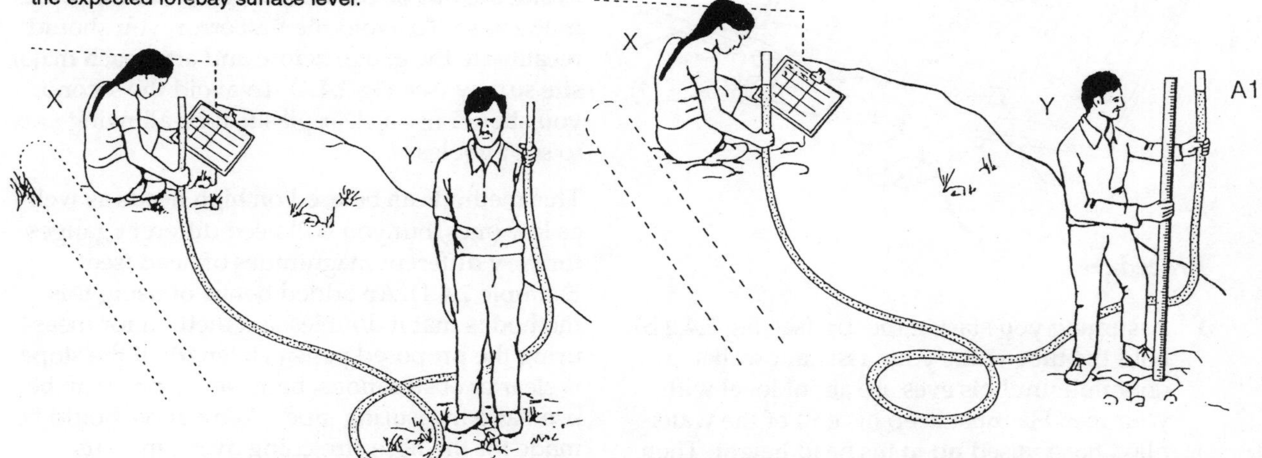

ii Count the number of Y heights

ii Person Y stays in the same place and measures B1, just as person X has moved downhill and measured A2.

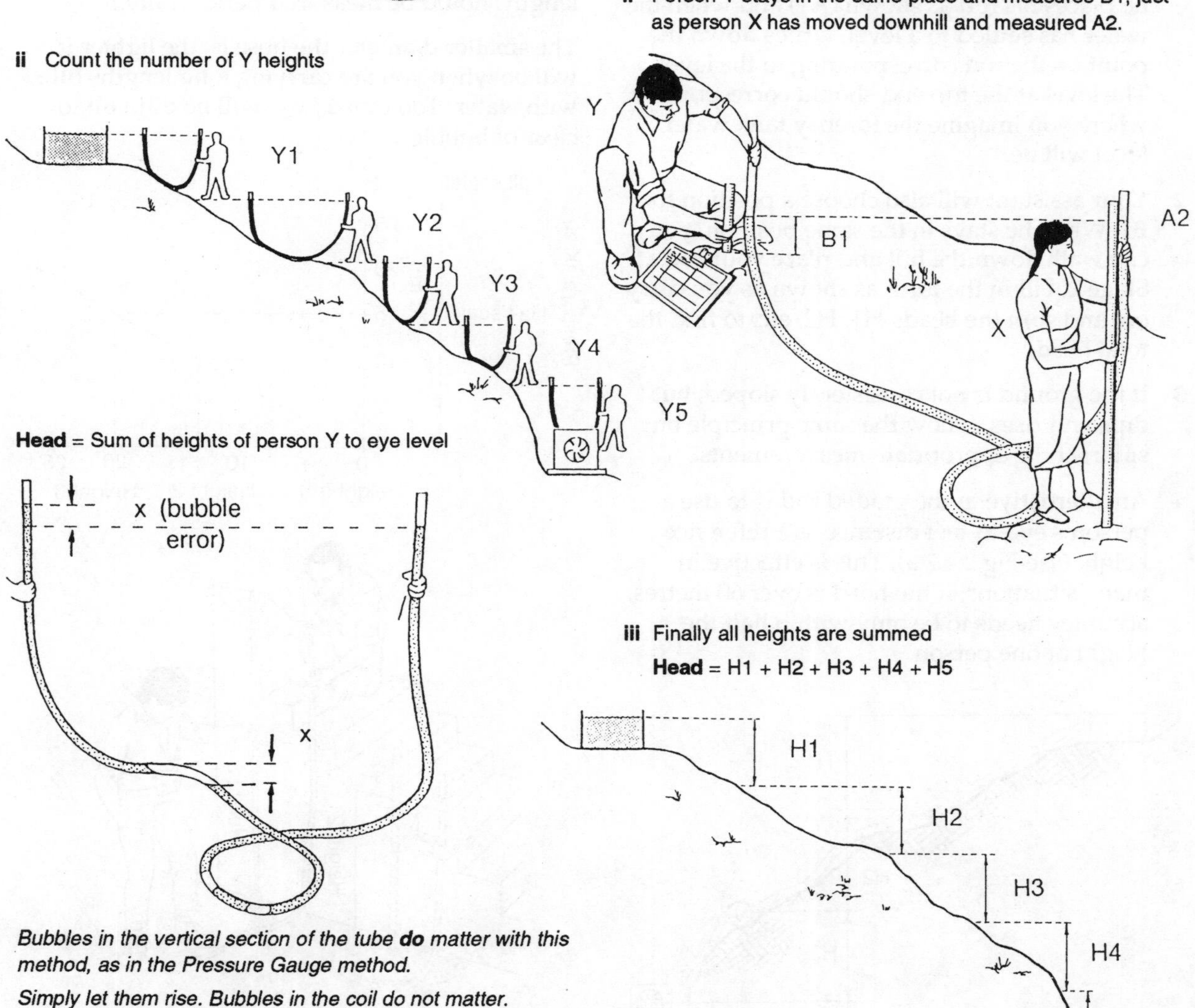

Head = Sum of heights of person Y to eye level

iii Finally all heights are summed

Head = H1 + H2 + H3 + H4 + H5

Bubbles in the vertical section of the tube ***do*** *matter with this method, as in the Pressure Gauge method.*

Simply let them rise. Bubbles in the coil do not matter.

If the vertical height of the bubbles can be seen easily, then the correction can be made while taking the reading, as indicated in the diagram above.

Fig 2.4.2c Example of charting measurements and summing for gross head

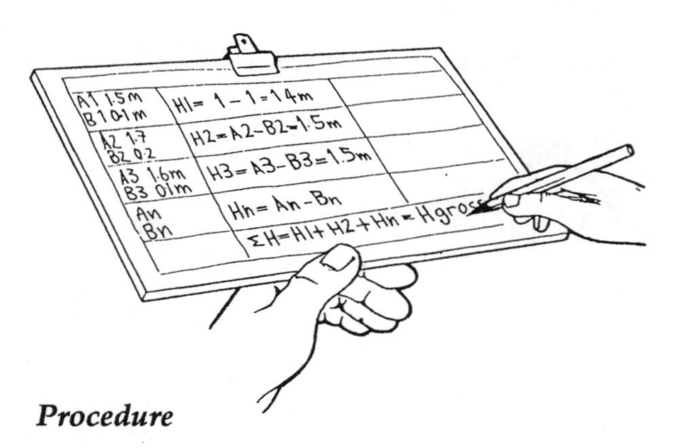

Procedure

1 Assuming you start at the top (see Fig 2.4.2 b), hold the tube while your assistant walks downhill until his eyes are about level with your feet. He must keep his end of the water-filled hose raised up to his head height. Then he places his rod as shown (A1) and when the water has settled to a level, writes down the point on the rod corresponding to the level. The level at the top end should correspond to where you imagine the forebay tank water level will be.

2 Your assistant will also choose a position for B1. While he stays in the same position, you can walk down the hill and place your rod at Stage 2. Fill in the form as shown as you carry on, and sum the heads H1, H2, etc. to find the total head.

3 If the ground is not consistently sloped, but dips and rises, follow the same principle but subtract the appropriate measurements.

4 An alternative to the graded rod is to use a person's eye-to-feet distance as a reference height (see Fig 2.4.2 a). This is effective in many situations; if the head is over 60 metres, accuracy needs to be only within half the height of one person.

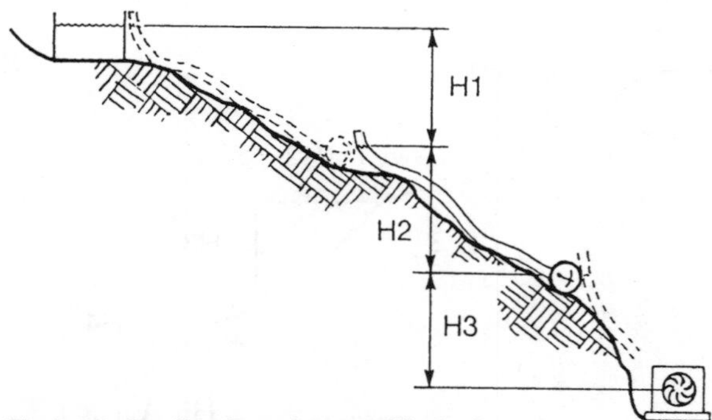

Fig 2.4.3 Application of water-filled tube and pressure gauge

Water-filled tube and pressure gauge

This is probably the best of the simple methods available, but it does have its pitfalls. The two sources of error, which you must take care to avoid, are out-of-calibration gauges and bubbles in the hose. To avoid the first error, you should recalibrate the gauge before and after each major site survey (see Fig 2.4.4). To avoid the second, you should use a clear plastic tube allowing you to see bubbles.

This method can be used on high heads as well as low ones, but you will need different gauges for very different magnitudes of head (see Example 2.4.1). An added bonus of using this method is that it doubles as a method for measuring the proposed penstock length, if the slope is clear of obstructions, because the hose can be used as a measuring tape. (Allowance should be made for the hose stretching over time – its length should be measured periodically.)

The smaller diameter the hose is, the lighter it will be when you are carrying long lengths filled with water. Too thin a hose will be difficult to clear of bubbles.

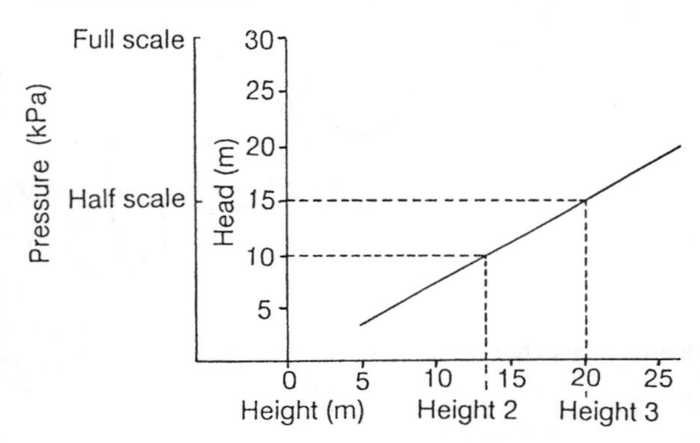

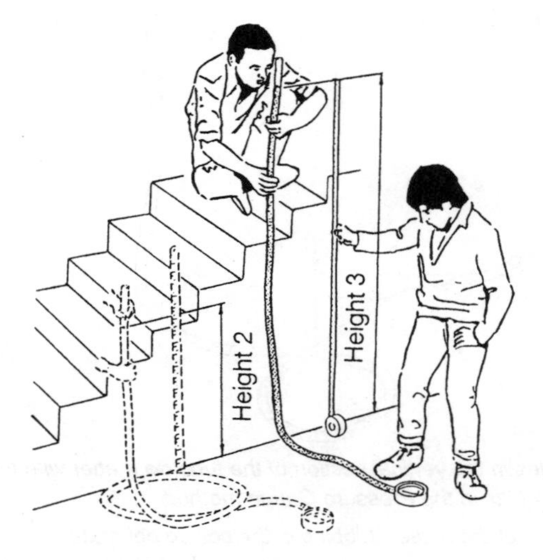

Fig 2.4.4 Calibrating a pressure gauge
Take about five readings, ranging from low on the gauge scale to full scale.

Find about 20 metres of petrol pipe hose, which has a diameter of around 6 to 8 mm, fill it at base, and take care not to empty it. It is worth making a good job of connecting it to the pressure gauge while you are still at base.

Alternatively, the plastic tube can be filled at site by immersing it in the stream. Water should be allowed to flow through the tube to drive all the bubbles out.

Equipment

Clear plastic hose

Calibration chart
Follow procedure shown in Fig 2.4.4

Pressure gauge
For selection of gauge, see Example 2.4.1

Prepared record sheet
Similar to that shown in Fig 2.4.2

Procedure

First calibrate the gauge following Fig 2.4.4.

When on-site note each pressure reading on a prepared sheet and convert the table to a head. From Example 2.4.1 you will see that a gauge reading in kPa or psi gives you head in metres by the equations:

$$h\ (m) = \frac{p(kPa)}{9.8}$$

$$h\ (m) = \frac{p(psi)}{9.8} \times \frac{100}{14.5} = 0.704 \times p(psi)$$

Spirit level and plank (or string)

This method is identical in principle to the water-filled tube and rod method. The difference is that a horizontal sighting is established not by water levels but by a carpenter's spirit level placed on a reliably straight plank of wood. Fig 2.4.5 shows the principle. On gentle slopes the method is very slow, but on steep slopes it is useful. Mark one end of the plank and each reading turn the plank and level round so that the marked end is alternately near the ground and away from the ground. This cancels out some of the possible errors.

The spirit level and string method involves three people. For this you must use a 'line level' which is attached or hung from a piece of string. One person holds one end of the string at foot level while a second person walks downhill until he can hold the string taut at eye-level. The third person checks the level, ensuring the string is horizontal. Then the uphill person walks downhill while the downhill person stays on the same spot. The head is the average of the heights of the two string-holders, multiplied by the number of measurements taken.

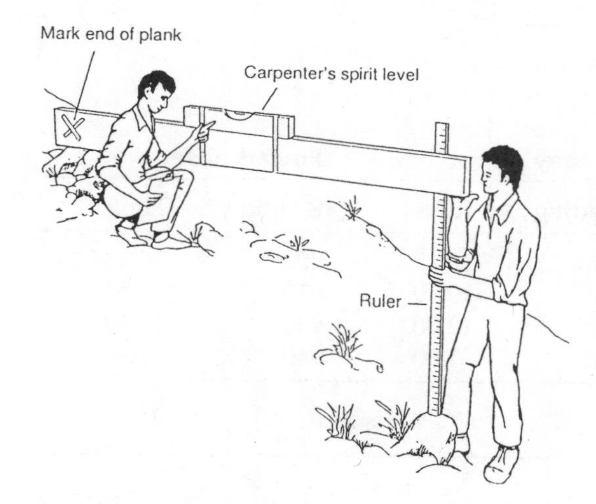

Fig 2.4.5 Spirit level method

Altimeter

Conventional altimeters are difficult to use effectively; errors of ± 30% can occur in inexperienced hands. New digital altimeters are also available which are easier to use and increasingly safe in inexperienced hands.

In general, it is best to think of the altimeter as a device which a professional surveyor, who knows the equipment well, might use. Whoever uses the altimeter, the results should always be checked by measuring head with another, quite different method.

The principle of the altimeter is that it measures atmospheric pressure. Atmospheric pressure changes by 9 mm head of mercury for every 100 metre change in elevation.

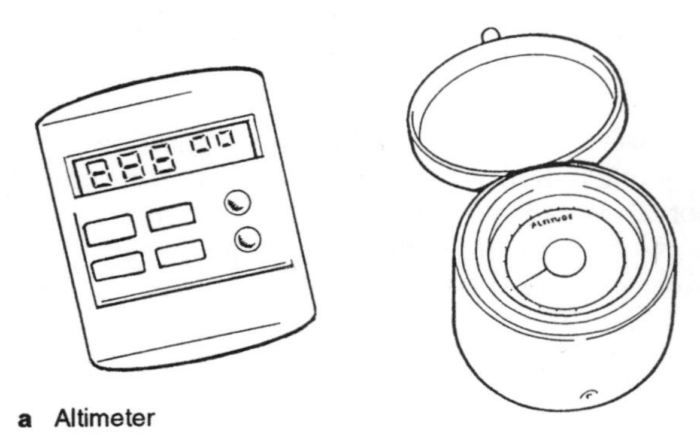

a Altimeter

Forebay		Powerhouse	
Reading	Time	Reading	Time
1000	10.15	900	10.20
1010	10.50	915	10.55
1015	12.00	930	12.30
1015	1.00	940	1.30

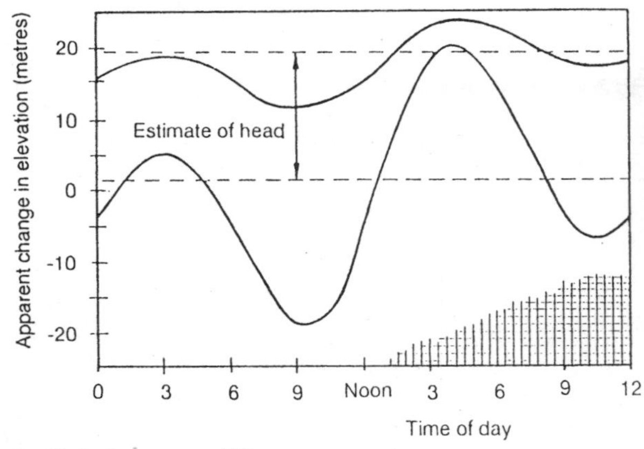

b Plot of pressure drift

Fig 2.4.6 Use of altimeters

Altimeters require skill to obtain accurate results because their readings are affected by changes during the day in temperature, atmospheric pressure and humidity. If a single altimeter is used, these conditions will change between a base measurement and elevated measurement. The changes must also be carefully measured and used to correct the altimeter readings. Two altimeters can be used simultaneously, but experience shows that this tends to produce errors rather than eliminate them. The best plan is to take alternate measurements at the powerhouse and forebay proposed sites, and record the time at which measurement is taken. The interval between measurements should be short. The pressure drift on both sets of readings should then be plotted as shown in Fig 2.4.6 and an estimate made of head. On the whole, use of altimeters is not recommended.

Photo 2.4.2 A pressure gauge attached to the end of a long nylon hose is an effective low cost instrument for head measurement.

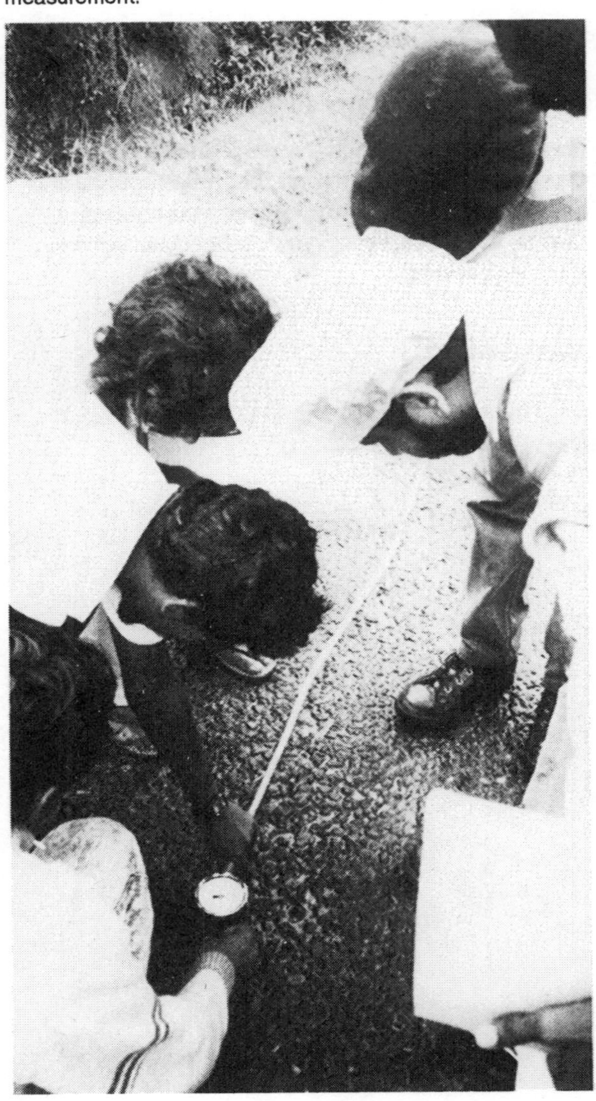

Pressure expressed as 'head' **Note 2.4.1**

Pressure is defined as the force exerted per unit area, and so is measured in Newtons per square metre (N/m^2) which are known as Pascals or Pa. Force is defined by Newton's equation as mass (kg) × acceleration (m/s^2). The downwards force exerted by a column of water is therefore its mass × the acceleration due to gravity ($g = 9.8\ m/s^2$) since if the water were free to move it would accelerate at this rate. The mass of the column can be expressed as its volume × its density ($m^3 \times kg/m^3 = kg$). Volume can be expressed as the horizontal cross-sectional area of the column × its height.

In summary:

$$\text{Force} = \text{mass} \times \text{acceleration} = \text{volume} \times \text{density} \times \text{acceleration}$$

$$\text{Pressure (p)} = \frac{\text{Force}}{\text{area}} = \frac{\text{volume} \times \text{density} \times \text{acceleration}}{\text{cross-sectional area}}$$

$$p = \frac{V \times \rho \times g}{A}$$ but since volume is height (h) × area (A):

$$= \frac{h \times A \times \rho \times g}{A} = \rho \times g \times h$$

To express a pressure as a head it is only necessary to turn this equation around, taking great care to keep the units coherent: see Example 2.4.1.

$$\text{head} = \frac{p}{\rho \times g}$$

Sighting meters

Hand-held sighting meters measure angle of inclination of a slope (they are often called clinometers or Abney levels). They can be very accurate if used by an experienced person. They are small and compact, and sometimes include range-finders which save the trouble of measuring linear distance. Since the method demands that the linear distance along the slope is recorded, it can have the advantage of doubling as a measure of the length of penstock pipe.

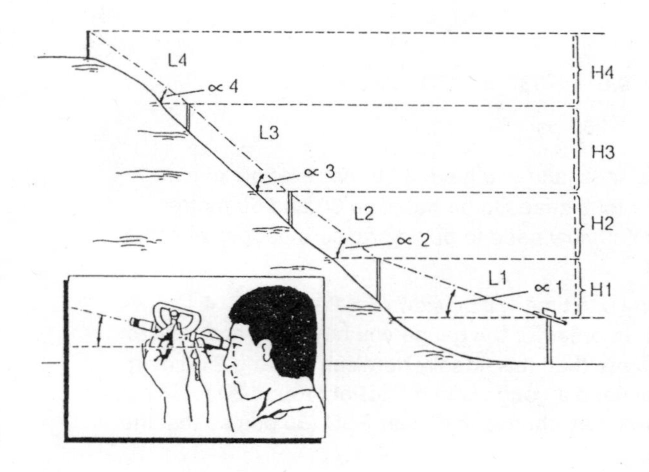

Fig 2.4.7 Use of sighting meters

Equipment

Measuring tape

The tape should, as far as possible, measure the hypotenuse distance. A range-finder may be integrated into the sighting meter and the readout could be directly of head.

An assistant

The method is difficult with one person, unless a range-finder is used.

Record chart

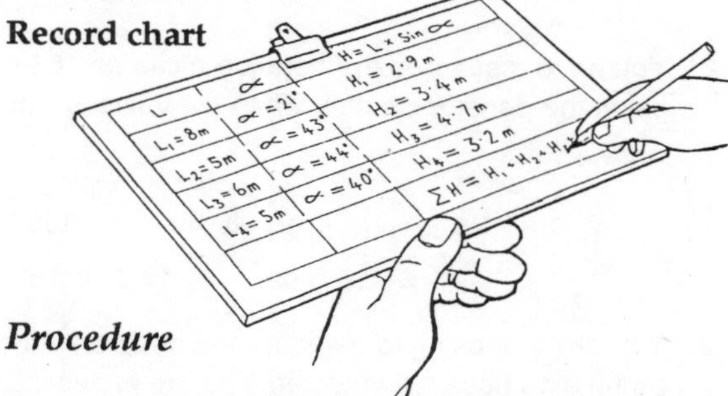

Procedure

Fig 2.4.7 shows the method. First place two equal-length stakes on the slope, within visible range of each other. The distance between the stakes can be as long as the measuring tape or range-finder will allow. Place the angle-sighting meter on top of one stake, and read and record the angle made to the top of the next stake. Measure and record the distance between the stakes. Use the trigonometric function $H = L \times \sin \alpha$ to calculate head between the stakes.

Example 2.4.1 **Pressure expressed as 'head'**

1 In the funnel-shaped flask shown, what is the pressure at (1) and at (2)? (Density of water (ρ) is 1000 kg/m³, and 'g' is 9.8 m/s²). Express your answer in Pascals, in kPa, in bar, and in psi. Finally express your answer as a 'head' in metres.

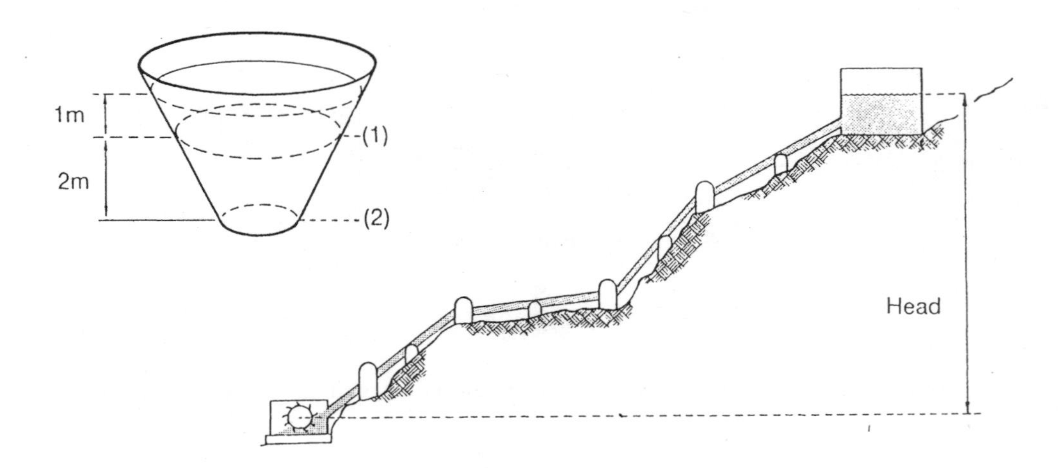

Fig 2.4.8 Measuring head and converting to pressure

Answer:

at (1) $p = \rho \times g \times h = 1000 \times 9.8 \times 1 = 9800$ Pa $= 9.8$ kPa
$= 0.1$ bar $= 14.5 \times 0.1 = 1.45$ psi $=$ 1 metre head

at (2) $p = \rho \times g \times h = 1000 \times 9.8 \times 3 = 29400$ Pa $= 29.4$ kPa
$= 0.29$ bar $= 14.5 \times 0.29 = 4.2$ psi $=$ 3 metres head

Notice you knew the answer in metres head immediately on looking at the diagram.

2 You have measured the head on a site as 163 metres, as shown. What is the pressure in bar and psi at the penstock outlet, when the valve is closed?

Answer:

$p = \rho \times g \times h = 1000 \times 9.8 \times 163 = 1597$ kPa
$= 16.0$ bar $= 14.5 \times 16 = 232$ psi

3 You are proposing to measure the head on various sites, and you have a 20 metre length of clear petrol pipe hose full of water. You are expecting the total heads to be between 30 and 60 metres. The slope is quite steep; you estimate it to be 45°. You now need to buy a pressure gauge: what full-scale deflection, in psi, will you buy?

Answer: the maximum head you will measure at any one time will be sine 45×20, so $h = 14.1$ m. This is a pressure of 138 kPa, or 1.4 bar, or 20 psi. In order for the gauge you buy to be accurate for your purposes, the full-scale deflection should be more than this, ideally between 1.5 and 2.5 times larger than the expected measurement. So you will need a gauge with a FSD of around 30 to 50 psi, or between 210 and 350 kPa. If you have a choice, you may choose the lower FSD (30 psi) so that the gauge is also useful to measure smaller heads.

Dumpy (builders') levels and theodolites

The Dumpy level is capable of 1 mm accuracy but is expensive, heavy and slow. Skilled operators are needed and mistakes are often made in the long strings of calculations required. This is a common survey method however, so availability is good. Distance can be measured simultaneously. Not suitable for steep, wooded sites.

The theodolite is expensive and heavy, but capable of fast and accurate work where the ground is fairly clear. It is very slow on wooded sites and there is plenty of opportunity for operator error and for machine error with this method, but again availability is often good.

Both devices require calibration and periodic re-calibration.

Beware of borrowed instruments – always calibrate before use.

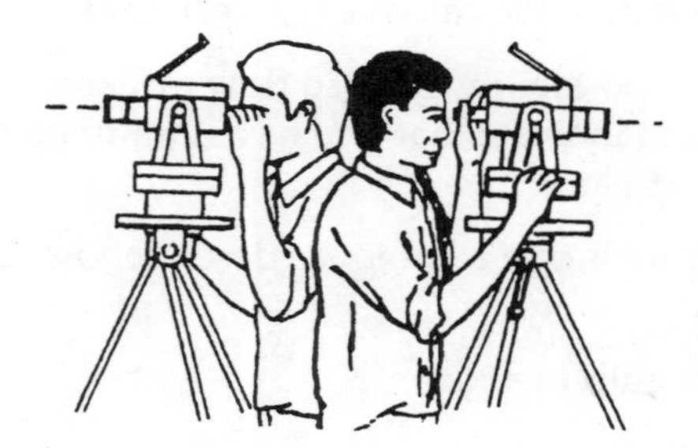

Fig 2.4.9 Use of levels

2.5

Site measurement of flow

The purpose of the hydrology study is to predict flow, as it varies through the year. The results of the hydrology study will always be the proper record of flow at the site. The hydrology study should be based on many years of daily records; a flow measurement is a single record and is therefore of less use. However, it acts as a check that the hydrological analysis is not misleading as a result of mistakes or changes to the catchment or abstraction from the river.

There are other reasons for taking site measurements during the design phase. These are:

- Flow measurements, taken daily over a complete year, are part of a hydrology study, and are useful to cross-check other data used in the study. Weekly and monthly readings of flow are also useful, but less so.
- When using the flow correlation method (Section 2.3).
- A few measurements taken over a few days are likely to be misleading, whether or not they correlate with the hydrograph you are using (see Fig 2.5.1). Nevertheless, if all the measurements taken fit into the 'envelope' formed by the hydrograph during high flow then there is some indication of the accuracy of the hydrograph. During low flow periods the envelope is much narrower and measurements outside it are an indication that the hydrograph may be wrong. It is useful to gather information locally on the river flow variation (exceedence) and also compare this with your measurements and hydrograph.

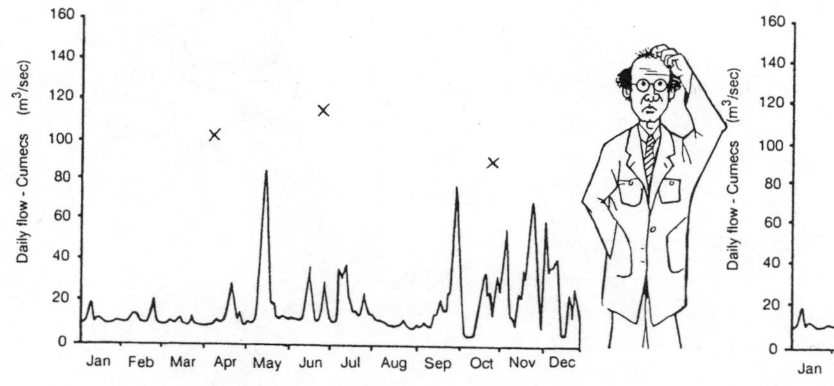

a Measured flows do not agree with predictions

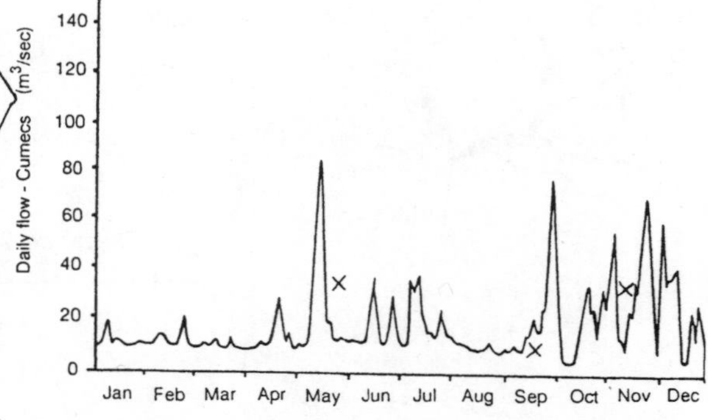

b Measured flows show correlation with predictions

Fig 2.5.1 Comparing measurements with hydrograph

- You may not have found any map for the area or the map you used to identify the stream may be inadequate. You are forced in this situation to carry out some mapping yourself and measure flows in order to develop an impression of the catchments upstream.
- Another possibility is that a flow measurement will indicate that you have identified the wrong stream.

The flow measuring techniques described here are:

- the salt gulp method;
- the bucket method;
- the float method;
- propeller devices
- stage/control methods, including weir methods.

It is necessary to study the distinctive features of each of these in order to find a suitable method for any particular site.

The salt gulp method

The 'salt gulp' or salt dilution method of flow measurement is described first because it has proved easy to accomplish, accurate (< ±7%), and reliable in a wide range of stream types.

Using this method stream flow can be measured in less than 10 minutes and very little equipment is needed. The main device is a conductivity meter. The calculations following take a little longer if done manually, as described here; alternatively they can be done automatically if you purchase an integrating meter.The disadvantage of this method is the cost of the meter which is in the order of US $50 - $400.

The probe should be immersed, close to the bed of the stream (not near the surface) or ideally at mid-depth. If the probe is in a fast-flowing section, bubbles can occur inside the probe shield and upset the measurement. To avoid this, place the probe in a slower flowing section, for instance behind a rock in the stream.

A bucket of heavily salted water should be thrown into the stream. The 'cloud' of salty water in the stream starts to spread itself out while travelling downstream. At a certain point downstream it will have filled the width of the stream. The cloud will have a leading part which is weak in salt, a middle part which is strong in salt, and a lagging part which is weak. Salt conducts electricity, so the saltiness of the water can be measured with an electrical conductivity meter. If the stream is flowing slowly, and has a small volume of water passing the meter each second, it will take a long time for the cloud to pass. Since not much water is present, it will not dilute the salt very much, so the electrical conductivity (which is greater the saltier the water) will be high. It is apparent, then, that low flows are indicated by high conductivity measured in the cloud, and long cloud passing times.

Conversely, high flows are indicated by weak dilutions (low salinity or conductivity) and short passing times. Flow will therefore be inversely proportional to both cloud passing time and degree of conductivity. This is all you need to know to intuitively understand the method.

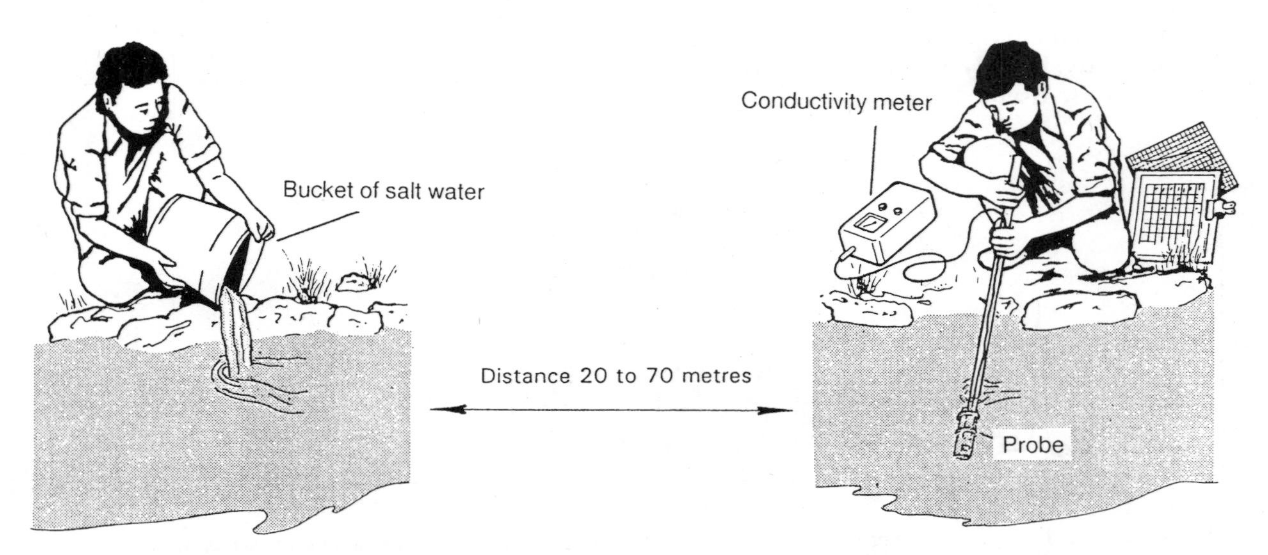

Fig 2.5.2 Salt dilution gauging The intervening distance should be 20 metres (if the stream is turbulent and will quickly mix in the salt) or 70 metres (if otherwise).

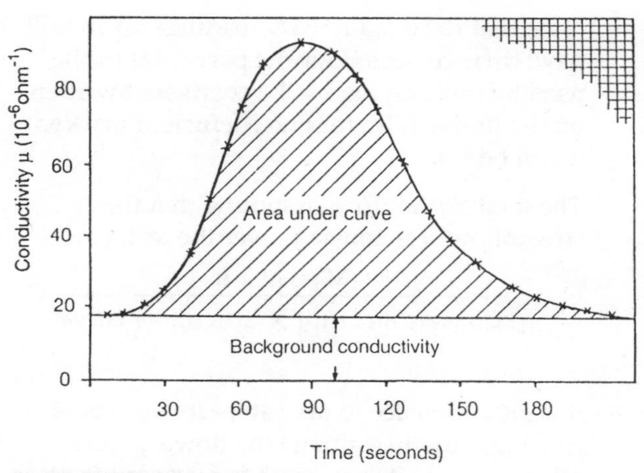

Fig 2.5.3 Plotting your measurements

Equipment

Thermometer

Range 0° - 40°C

Conductivity meter

Range 0 - 1000 mS (Siemens (S) are identical to ohm^{-1}). Many conductivity meters are temperature corrected in which case a separate thermometer is unnecessary.

Integrator (optional)

An integrator will spare you the trouble of taking five second readings and plotting them out, but it can introduce errors.

Pure table salt

A useful guide to quantity is 100 grams for each 0.1 m^3/s of expected streamflow.

Bucket

For mixing salt and water.

Procedure

1 Decide on the amount of salt, and the distance between injecting the salt and monitoring conductivity (Typically, for 0.1 m^3/s, use 100 g of salt and 50 metres distance.). Dry the salt, since the principle of this method depends entirely on knowing a mass of salt rather than a mass of a salt/water mixture. Record the mass of salt, in case you forget later.

2 Take a temperature reading of the stream, and record it.

3 Mix the salt in the bucket with stream water, being careful not to spill any.

4 Choosing a length of stream with no large stagnant pools, and no significant inflows or outflows, place the probe in the stream at a fairly fast flowing point, keeping it in position with a rock. Read the base conductivity and record it.

5 Get an assistant to go the decided distance upstream and pour the salt in at a fast-moving point.

6 As soon as the meter reading starts to increase take conductivity readings every five seconds. Two people may be needed for this. The reading will rise, reach a peak, then fall back to the base level. If the reading is jumpy, the salt has not mixed – repeat with a longer distance. If the peak reading is less than twice the base reading, repeat with more salt. For examples of good and bad curve shapes see Fig 2.5.4.

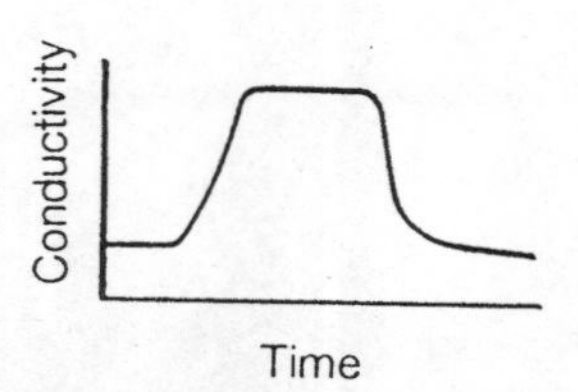

a Meter saturated. Change scale or use less salt.

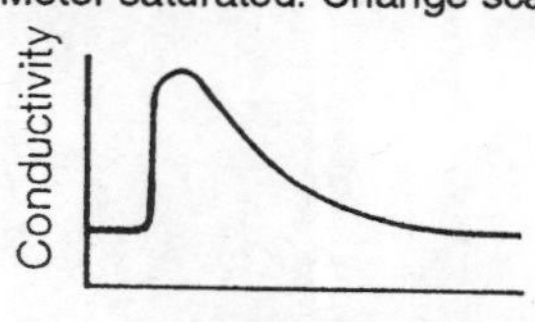

b Badly skewed curve. Use longer distance.

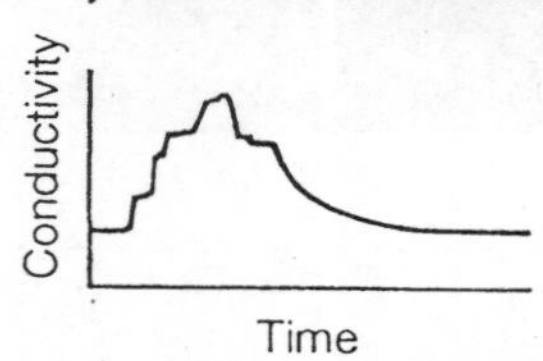

c Uneven reading. Salt not mixed. Use longer distance.

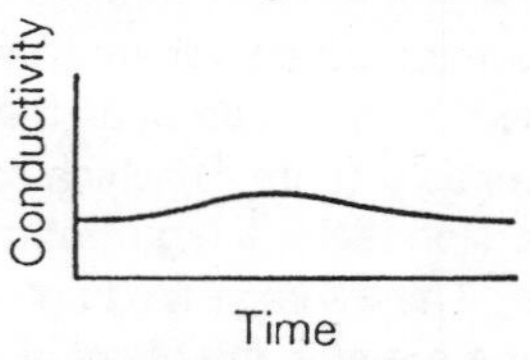

d Insufficient response compared to base level. Use more salt

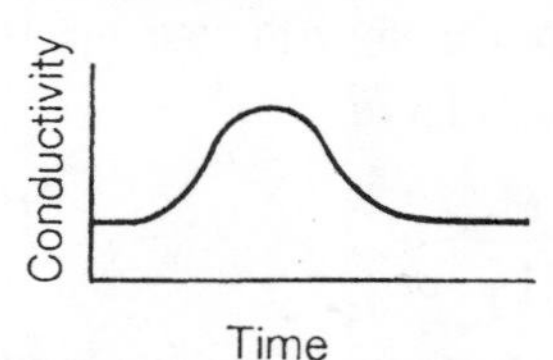

e Perfect

Fig 2.5.4 Examples of good and bad readings

7 Plot the readings as a graph of conductivity against time (Fig 2.5.3) and find the area under the curve. This can be done by plotting on graph paper and counting squares. Alternatively, the area can be calculated with the help of a pocket computer program. A calculator containing an integrator will give the area directly. This is the area under the curve with the background conductivity level taken as zero (see Fig 2.5.3). This area in units of $ohm^{-1} \times 10^{-6}$ seconds (micro Siemens × seconds) is converted to concentration by multiplying by a conversion factor for the temperature concerned ie at about 22°C the reciprocal of the conversion factor is 2.04 $ohm^{-1} \times 10^{-6}$ per mg/litre.

Note that the conductivity readings taken will have to be corrected for the particular probe used by multiplying by the coefficient written on the probe. If there is no coefficient marked, it will be 1.0.

The total streamflow, assuming that the streamflow was constant over the test, is

$$Q = \frac{\text{Mass of salt}}{\text{Conversion factor} \times \text{area under curve}}$$

8 It is good practice to take at least two sets of readings, and to estimate the flow by some other method. It is sensible to work out the test results (approximately) before leaving the site.

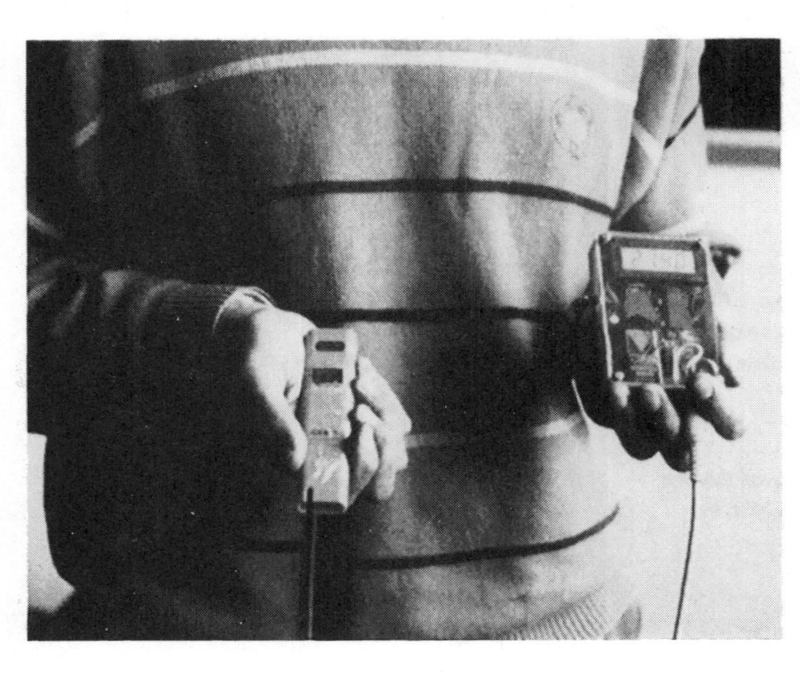

Photo 2.5.1 A conductivity meter combined with a microprocessor can be used to give direct readings of flow (per unit mass of salt injected into the stream). (UK)

Note 2.5.1 **Salt dilution – the mathematics**

There is no need to understand the mathematics of the method, but here is a quick exposition for those who are interested. In a short unit of time, *dt*, the average conductivity of the water while the cloud passes is measured by the meter as μ (ohm^{-1}), which can be converted to an average salt concentration by multiplying by the conversion factor k (kg/m^3ohm^{-1}). In the same time period *dt* the volume of water passing is Q*dt* ($m^3/s \times s$). The average mass of salt passing in *dt* is therefore $\mu k Q dt$. All these masses summed together must be equal to the mass of salt (M_{salt}) originally placed in the stream:

$$M_{salt} = \Sigma \mu k Q \, dt$$

Since μ is the only factor varying with time, this is expressed

$$M_{salt} = k Q \int \mu \, dt$$

or

$$Q = \frac{M_{salt}}{k \int \mu \, dt}$$

where the integral is evaluated over the whole period that the cloud passes. The conversion factor 'k' of course varies with water temperature, so must be based on a measurement of temperature during the site visit.

Example 2.5.1 — Salt dilution

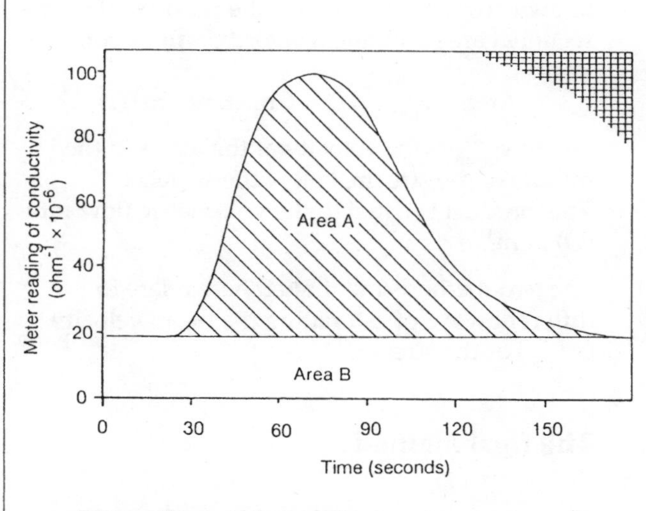

a Plot readings on blank graph paper

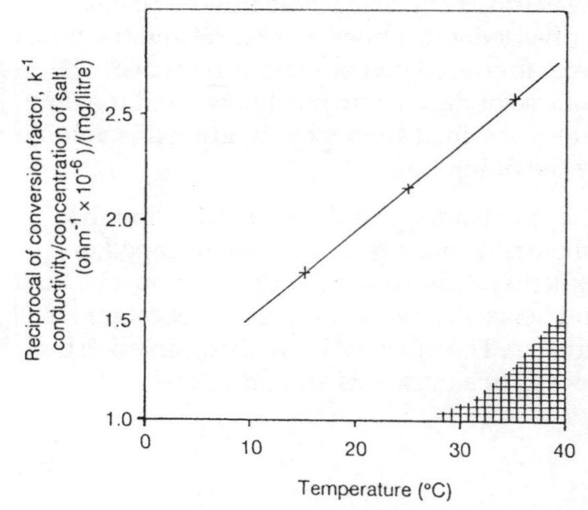

b Note here that the graph gives 1/k (the reciprocal, written as k^{-1})

Fig 2.5.5 Example graph and conversion chart

Amount of salt: 100 g
Temperature of water: 22°C

Area A is in units of [conductivity × time] i. e. [$ohm^{-1} \times 10^{-6} \times s$]

Measurements are plotted on a graph as in **a**, the choice of scale being your decision. Do not forget to allow for the scales when area is calculated. If each square is 5s × 5 $ohm^{-1} \times 10^{-6}$ and there are 145 squares:

$$A = 145 \times 5 \times 5 = 3625 \; [s \times 10^{-6} \times ohm^{-1}]$$

The units on the conversion factor graph are not standard SI but can be converted, remembering that a litre (l) is a thousandth of a cubic metre (m^3). Grams (g) are of course thousandths of a kilogram. These units will be encountered often and this example therefore serves as guidance on their manipulation.

From **b**: At 22°C the reciprocal of the conversion factor k^{-1} is 2.04×10^{-6} [$ohm^{-1}/mg\, l^{-1}$] The calibration factor of the conductivity probe can be assumed to be 1.0. The conversion factor k is $1/2.04 \times 10^{-6}$, but there is no need to calculate this separately, simply amend the flow equation instead with k^{-1} on top instead of k below. (This form is encountered in practice.) The equation as given below operates non-SI units and gives flow in litres per second rather than m^3 per second.

$$Q = \frac{M \times k^{-1}}{A}$$

$$= \frac{100 \times 10^3 \times 2.04}{3625} \left[\frac{mg \times ohm^{-1} \times 10^{-6}}{mg \times l^{-1} \times s \times ohm^{-1} \times 10^{-6}} \right]$$

$$= 56.27 \;\; l/s$$

The bucket method

The bucket method is a very simple way of measuring flow. The whole flow to be measured is diverted into a bucket or barrel and the time it takes for the container to fill is recorded. The volume of the container is known and the flow rate is obtained simply by dividing this volume by the filling time.

The disadvantage of this method is that the whole flow must be channelled or piped to the container. Often a temporary dam must be built and hence the method is only practical for small streams. The practical limit, using an oil drum instead of a bucket, is around 20 l/s.

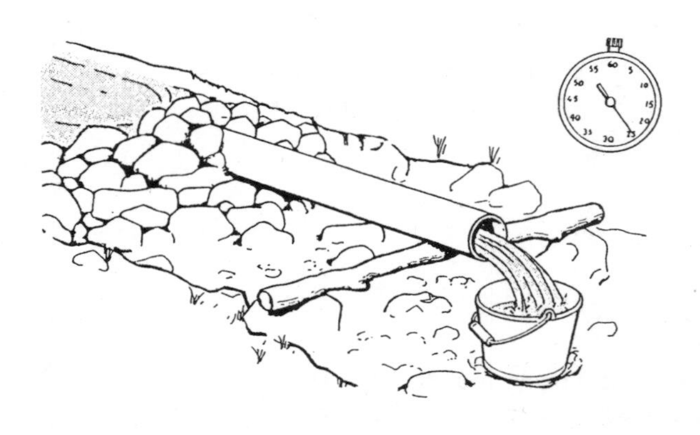

Fig 2.5.6 Using the bucket method

Velocity-area methods

For a fluid of constant density flowing through a known cross-sectional area, the product of cross-sectional area and mean velocity will be constant:

$$\text{Area} \times v_{mean} = Q = \text{Constant} \ (m^3/s)$$

where v_{mean} is the velocity of the water in the middle of the stream, below the surface.
This product is equal to the volumetric flow rate (Q) in m^3/s.

The two methods mentioned here relate to different ways of estimating the mean velocity (v_{mean}) of the stream.

The float method

The cross-sectional profile of a streambed is charted and an average cross-section established for a known length of the stream (Fig 2.5.7). A series of floats, perhaps convenient pieces of wood, are then timed over a measured length of stream. Results are averaged and a flow velocity is obtained. This velocity must then be reduced by a correction factor which estimates the true mean velocity as opposed to the surface velocity. The factor depends on stream depth (see Note 2.5.2). By multiplying the average cross-sectional area by the averaged and corrected flow velocity, an estimate of volume flow rate can be made.

Note 2.5.2 **Float method: Correction factors for stream velocity**

Using the float method total errors can be 100%, especially in shallow, rocky streams. Reduce surface velocity by the following factors to find average velocity:

Large, slow, clear stream	0.75
Concrete channel, rectangular section, smooth	0.85
Small regular stream, smooth bed	0.65
Shallow (0.5 m) turbulent stream	0.45
Very shallow, rocky stream	0.25

For greater accuracy in lined channels, submerged floats (half-filled bottles) can be used in conjunction with tables (see Appendix 3, Bibliography). Points to watch are as follows:

- Choose as regular a section as possible with no pools, waterfalls, outflows or in-flows.
- Allow floats to accelerate before the 'start'.
- Use different floats (sticks, leaves, bottles) and ignore any very slow ones.

Photo 2.5.2 A triangular or V-notch weir permanently installed in a forebay tank. This is very useful for a hydrology study. (UK)

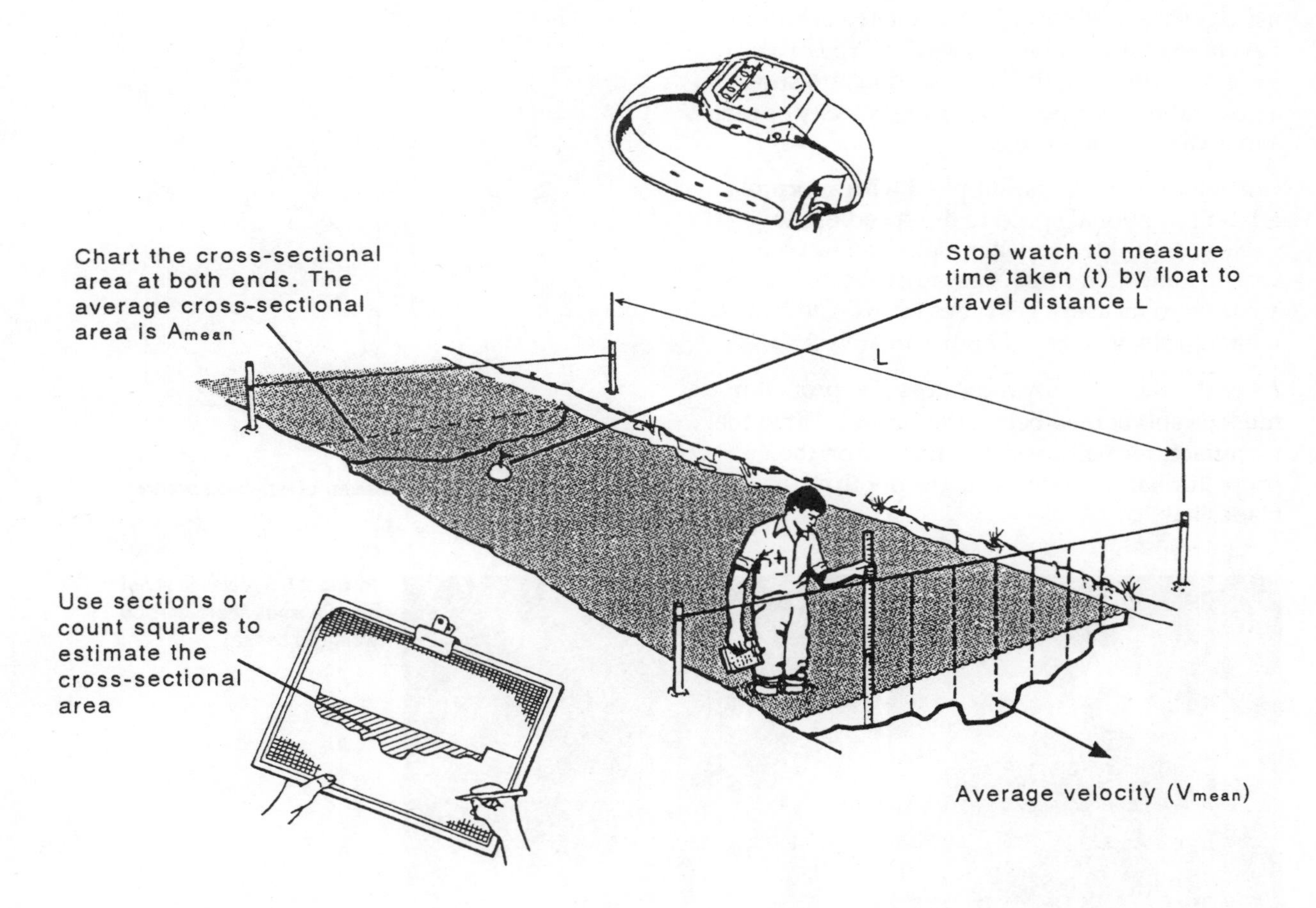
Chart the cross-sectional area at both ends. The average cross-sectional area is A_{mean}
Stop watch to measure time taken (t) by float to travel distance L
L
Use sections or count squares to estimate the cross-sectional area
Average velocity (V_{mean})

$$\text{Flow (Q)} = A_{mean} \times v_{mean} = A_{mean} \times \frac{L}{t} \times \text{correction factor}$$

Fig 2.5.7 Charting the cross-sectional area of a stream

The inaccuracies of this method are obvious. Unless a smooth regular channel is considered, obtaining an accurate figure for the cross-sectional area of the stream will be very difficult and tedious. The average velocity obtained will not be the mean velocity of the stream since the float is on top of the water, and the correction factor can only be an approximate adjustment.

In general, choose the longest possible length of parallel-sided stream with unchanging cross-sectional area along this length. A rough-bottomed stream with obstructions to flow, such as large rocks, will tend to give unrealistic results.

Propeller devices

Often called current meters, these consist of a shaft with a propeller or cups connected to the end. The propeller is free to rotate and the speed of rotation is of course related to the stream velocity. A simple mechanical counter records the number of revolutions of a propeller placed at a desired depth. More sophisticated devices use electrical pulses. Procedures have been developed for calculating the discharge of a stream by dividing the cross-section into sub-areas and measuring the average velocity of each sub-area (see references).

Current meters will be supplied with a formula relating rotational speed to the speed of the stream. A simple propeller meter can be constructed and calibrated. Generally these devices are used to measure velocities from 0.2 to 5 m/s with a probable error of approximately 2%.

As with other velocity indicators, the propellor must be submerged below the surface. Often the manufacturer will provide a marker on the propellor handle to indicate the depth of the blades.

Stage-control method, natural sections

This is similar to the weir method (next section), except that a physical feature of the stream is used to control the relation between stage and discharge. The term stage refers to a measured depth of water. A control section is established where for a given change in discharge (flow rate) a relatively large, measurable change in stage can be measured. A broad control section should be avoided because changes in flow will result in very small changes in stage.

If debris obstructs the control section or erosion causes its slope to change, then readings will no longer remain valid. The gauge, typically a graded staff, should be sited where it is easily accessible for reading and not exposed to damage from debris.

Note that this method is valid for comparing one flow to another, but that a reference flow must be known and related to the graduated staff in order to obtain a quantitative estimation of flow rate.

Fig 2.5.8 Stage marker; upstream of a natural control section

Photo 2.5.3 A plastic sheet seals a small sharp-crested weir. (Sri Lanka)

Stage-control method, common sharp-crested weirs

A weir is a structure such as a low wall across a stream. A flow measurement weir has a notch through which all the water in the stream flows. Volumetric discharge can be determined from a single reading of the difference in height between the upstream water level and the bottom of the notch. For reliable results the crest of the notch must be kept 'sharp' and sediment must be kept from accumulating behind the weir. Crests are normally kept sharp and durable by forming from sheet metal strip.

After the weir is constructed, a method for measuring the head 'h' should be included. This measurement should be made far enough upstream of the weir to prevent the reading from being affected by the downward curve of the water heading through the notch. A distance at least four times the head is usually recommended. One method for determining 'h' is to drive a stake into the streambed until it is precisely level with the lower edge of the notch, as shown in Fig 2.5.9. To measure head 'h', a scale is placed vertically on top of the stake and the level of the stream is read directly off the scale. Numerous variations of this method can be used.

Fig 2.5.9 shows a rectangular weir, suitable for large flows. Small flows with wide variation are better measured by a triangular weir (Fig 2.5.10).

Weirs can be timber, concrete or metal and are always oriented at right angles to the stream surface. Siting of the weir should be at a point where the stream is straight and free from eddies. Upstream, the distance between the streambed and the crest of the weir should be at least twice the maximum head to be measured. There should be no obstructions to flow near the notch and the weir must be perfectly sealed against leakage. Plastic sheet can be used to do this. The crest of the weir should be high enough to allow water to flow freely leaving an air space under the outflowing sheet of water.

The crest of triangular and rectangular weirs must be level. Triangular weirs can be used with a range of notch angles (a 90° notch angle is often used). Equations for most sharp-crested weirs are usually not accurate for very small heads less than about 5 cm.

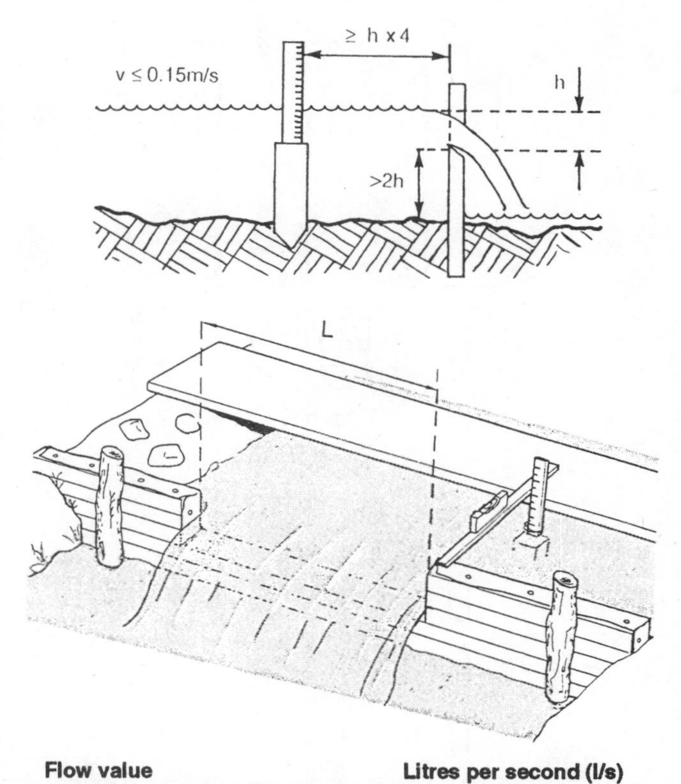

Flow value — **Litres per second (l/s)**

$Q = 1.8 (L - 0.2 h) h^{1.5}$ m^3/s

Overflow height	Weir width L			
	0.5 m	1.0 m	1.5 m	2.0 m
5 cm	10	20	30	40
10 cm	27	56	84	113
20 cm	74	155	235	316

Fig 2.5.9 Measuring flow from a rectangular weir

Photo 2.5.4 A rectangular weir constructed from wood. It uses a thin metal strip to form a sharp crest. (Sri Lanka)

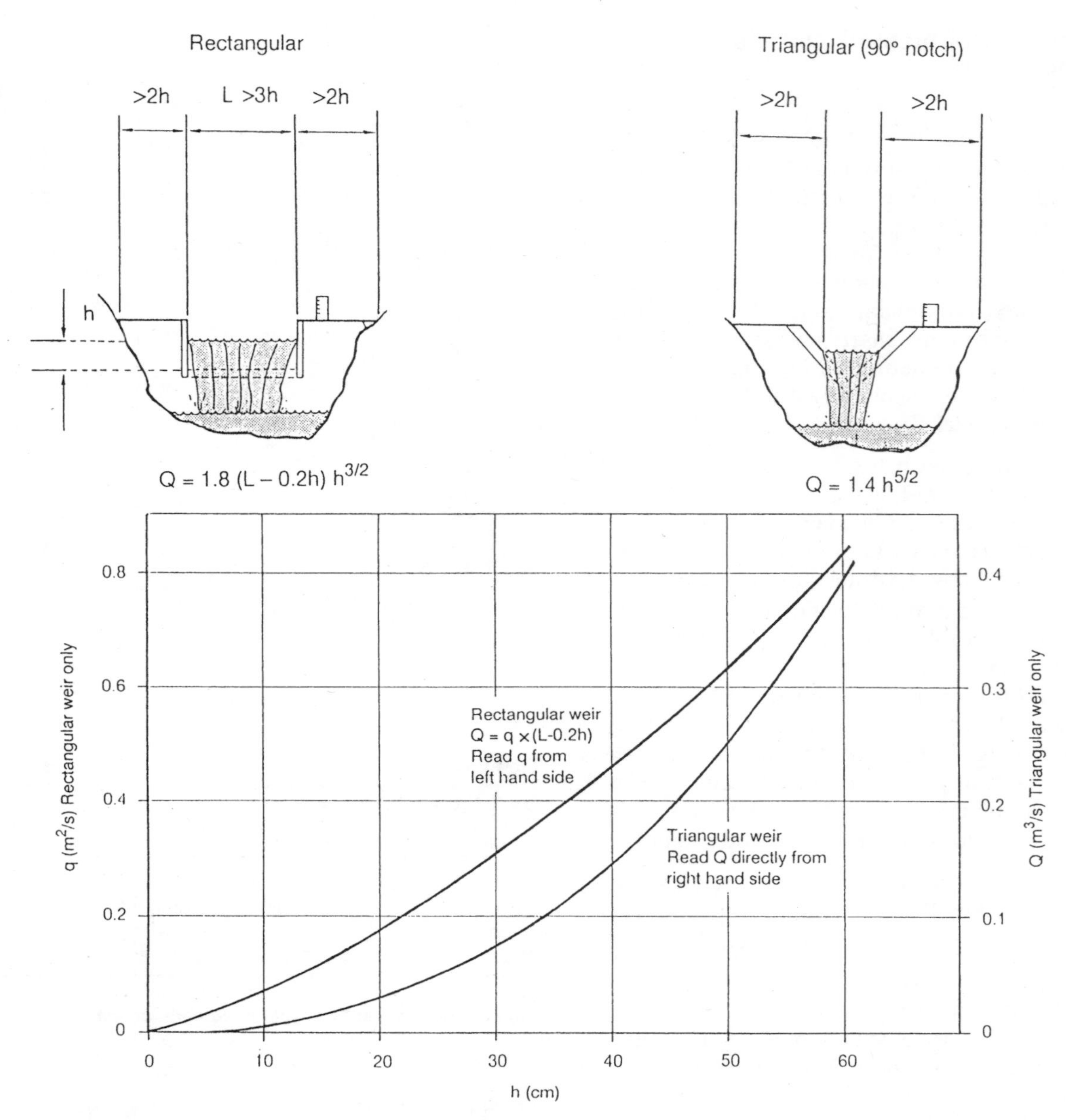

Fig 2.5.10 Measuring flows with rectangular and triangular weirs

A triangular weir can measure a greater range of flows than the other weirs. The notch of a weir must be wide enough to accommodate the largest expected discharge, hence some knowledge of probable flow rates is necessary before a weir is selected or designed. If stream velocities above 0.15 m/s are encountered it will be necessary to correct for the effect of the stream approach velocity.

Disadvantages of weirs include:

- If the notch is too wide or deep the formulae tend to underestimate the discharge.
- If the approach velocity is too high discharge is again underestimated.
- If the weir becomes submerged due to insufficient room for the downstream free fall of water, application of the formula will overestimate discharge.
- Temporary weirs are arduous to install and leakage is difficult to prevent.
- Annual or flash flooding may destroy temporary weirs.

Another method of flow measurement that is worth considering is the slope/area method: this uses friction coefficients (given in Table 3.6.4 in Chapter 3) and relates the flow to loss of height (or head) in the river.

2.6

Geological study

Your visit to the proposed site should include a geological survey. Aim to return home with some idea of the following:

- **future surface movements;** for example, loose rock slopes that may be disturbed by construction work or by heavy rainfall, dry mud indicating mud flows, storm gulleys that may take torrents and rock flows during heavy rainfall, signs of flood behaviour at valley base level;
- **future sub-surface movements;** for example, landslip and subsidence;
- **soil and rock types;** information is needed in order to design the foundations of civil works, to decide which materials to use in channel construction, and to assess which building materials are available on-site.

Just as in the hydrological study, the on-site survey must be preceded by careful use of a map, if one is available. Try also to obtain aerial photographs of the proposed micro-hydro area. Use the map and photographs to sketch out the basic geological characteristics of the area.

When you visit the site, bring your sketch along and add to it your observations.

An example sketch is shown in Fig 2.6.1. Remember to survey slopes high above the proposed hydro installation, as activities such as rockfalls far above are likely to have knock-on effects causing geological activity near the installations. In a similar way, extend your survey to the land below the likely sites of civil works, as ground movements below have repercussions above.

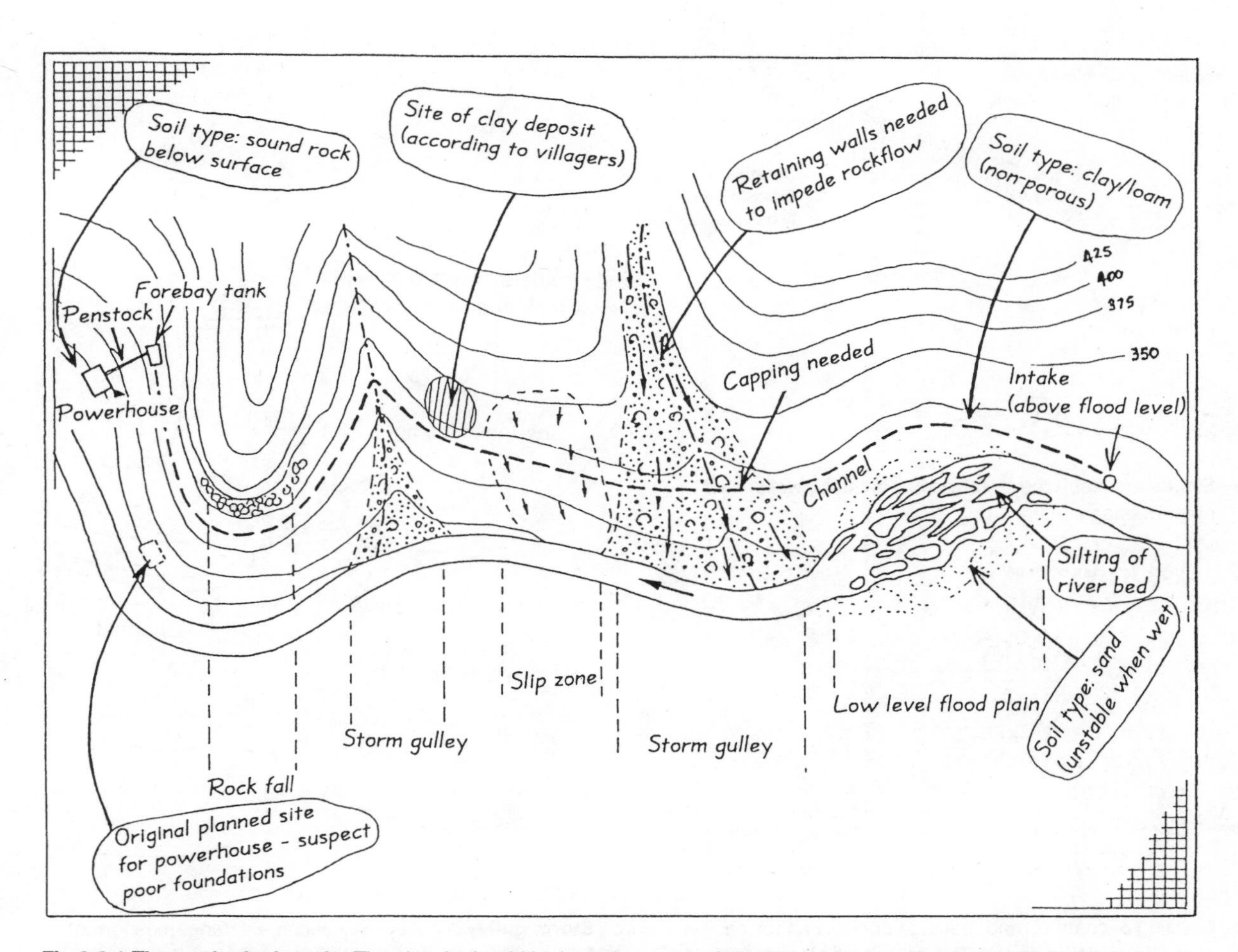

Fig 2.6.1 The geological study. The sketch should be based on maps, interviews, photographs and your own site observations. Use squared paper when visiting the site to help you draw to scale – costing of the channel and comparing the costs of different channel and works locations is then much easier.

It is very possible that the construction work required for the hydro installation may cause disturbance which activates surface and sub-surface movement. The installation itself over time could promote disturbance, for instance by water leakage from channels and tanks into the hillside below. Care should be taken to avoid these effects.

The main purpose of the geological survey is to assess the best locations for proposed civil works, and to estimate their construction costs and future maintenance costs.

Note 2.6.2 lists a few items that are useful to bring with you on a geological survey. An essential source of information is local knowledge. Tour the area and then speak to people who have known the area in question for a long time. Return for a second look to follow up what they have been saying, and to check their estimations of flood level, size of landslide, and so on, against your own measurements and assessments.

Future surface movements

Fig 2.6.1 shows a river from which water is drawn at an intake site. Contours are shown, and the proposed line for a channel conveying water to a tank (the forebay) at the top of a delivery pipe (the penstock). The figure could be your sketch for a proposed location of civil works, like scheme A in Fig 2.2.2. Remember that there will be other possible intake sites (like scheme B in Fig 2.2.2) which should be included for comparison.

Loose slopes

A loose slope is usually self-evident or indicated by debris collecting at its base. Civil works must be protected by stabilization of the slopes above and below the works. The best stabilization method is planting or seeding of local grasses and bushes.

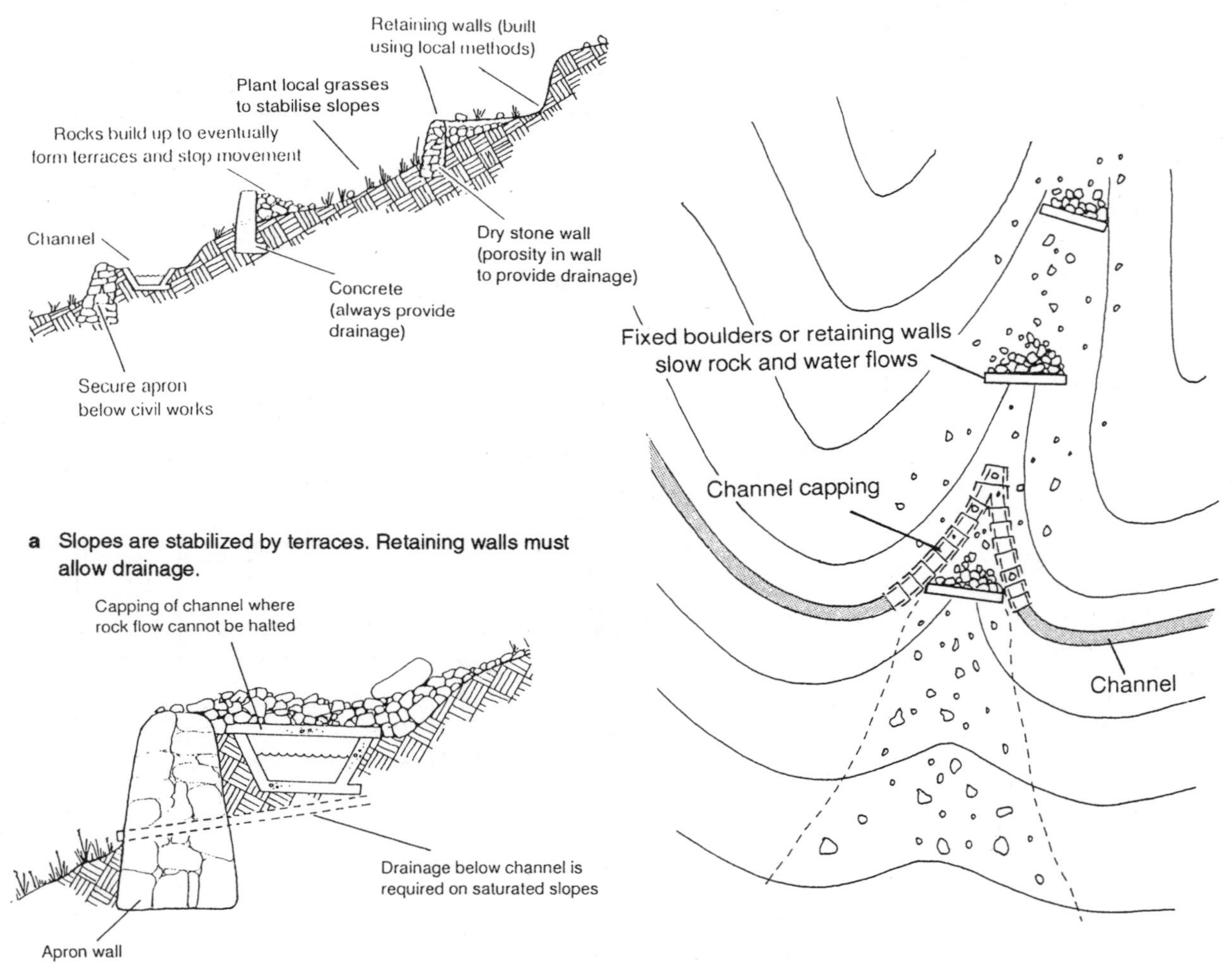

a Slopes are stabilized by terraces. Retaining walls must allow drainage.

b Capping of channel, and drainage under channel, can be cost-effective techniques. The apron wall also protects the channel from surface movements below.

c **Storm gulley** A gulley may become a dangerous torrent full of rolling rocks only once in ten years. Ask local farmers how they protect their irrigation channels.

Fig 2.6.2 Slope stabilization

Checklist for geology study — Note 2.6.1

Desk study

Sketch from map and aerial photographs

Interviews

- Road building, irrigation, civil contractors with knowledge of the area.
- Villagers for land movement history, flood levels, irrigation channel protection, stabilization methods.

Useful tools

- Notebook, pencil, calculator, squared paper, sketches from map, reference tables (eg Table 2.6.1).
- Map. Compass. Measuring tape.
- Hammer (to test weathering of rock).
- Pickaxe/spade for inspection trenches.
- Measuring cylinder to settle out clay, loam, sand constituents of soil.
- Soil permeability measuring kit (see Note 3.6.3).
- Head and flow measuring equipment (see Sections 2.4 and 2.5).

Site survey

Geological phenomena	Identification	Cost implications
Loose slope, subject to movement under heavy rainfall	Debris, dry mud lack of vegetation	Plant grasses, terracing, retaining walls, drainage, apron. Capping on channel.
Storm gulleys	Debris	As above and bridges either for gulley or for channel.
Flood plains	As described in text. Local residents' knowledge	Avoid location of civil works.
Landslip fault ('Rotation')	Semi-circular crack or step in hillside (Dipping of paths, broken walls)	Avoid location. Bridging, catenary wires etc. Sealing from leaks. Sealing of spillway drains.
Layer faulting	Debris, layers visible	Avoid location. Gabions, pins.
Subsidence		
Soluble rock	White/yellow limestone, pitted, cavernous rock.	Avoid. Deep foundations or piles.
Sand below surface	Identification trench. Near river bank, surface sand, sandstone rock area	Avoid. Deep foundations, piles.
Soil types	Table 2.6.1	Cost of importing gravel for concrete, clay/cement for sealing, aggregate for gabions.

It is worthwhile thinning out such vegetation nearby and transplanting to the unstable slope. This must always be done on slopes which have been stripped of vegetation during construction work, and on soils and rubbles which have been moved (known as 'made-up ground').

Never build on made-up ground. Another method of slope stabilization is terracing or use of retaining walls, as illustrated in Fig 2.6.2. In all cases of terrace construction water drainage must be provided, to avoid collapse of terraces. Gentle traverse sloping of the terraces can provide protection of penstocks and powerhouses from slope erosion and eventual rock/ mud slides (Fig 2.6.3).

The slopes below the civil works are as important as the slopes above. The same techniques of grass-planting and terracing must be applied to stabilize lower slopes. This is because movement below can cause collapse above.

A secure retaining wall or 'apron' can be built immediately below any civil construction (see Fig 2.6.2) if there is reason to suspect lower slope instability and planting/ terracing methods are not helpful.

Storm gulleys

Unusually heavy rainfall or snowmelt will release torrents of water down hillside slopes. Sudden release can be caused by overtopping of upland lakes. Very often the water will erode gashes or gulleys in the hillside which are easily detectable. Loose rocks and boulders are usually swept along with the torrent (boulders become lighter in water and are easily moved). The torrent is therefore capable of destroying anything in its path and many wrongly placed micro-hydro installations have been lost in this way. This danger can be seen by looking for collections of rock debris at the foot of dry stream beds and eroded sections of the hillside.

Take care when surveying to identify small gulleys, which may look too small to worry about, but may nevertheless become forceful torrents in a bad ten-year storm. The presence of vegetation on such a gulley does not mean it is harmless; vegetation can spring up within two years but the gulley may only release a torrent every ten years. Usually there will be loose rock deposits along the gulley bed.

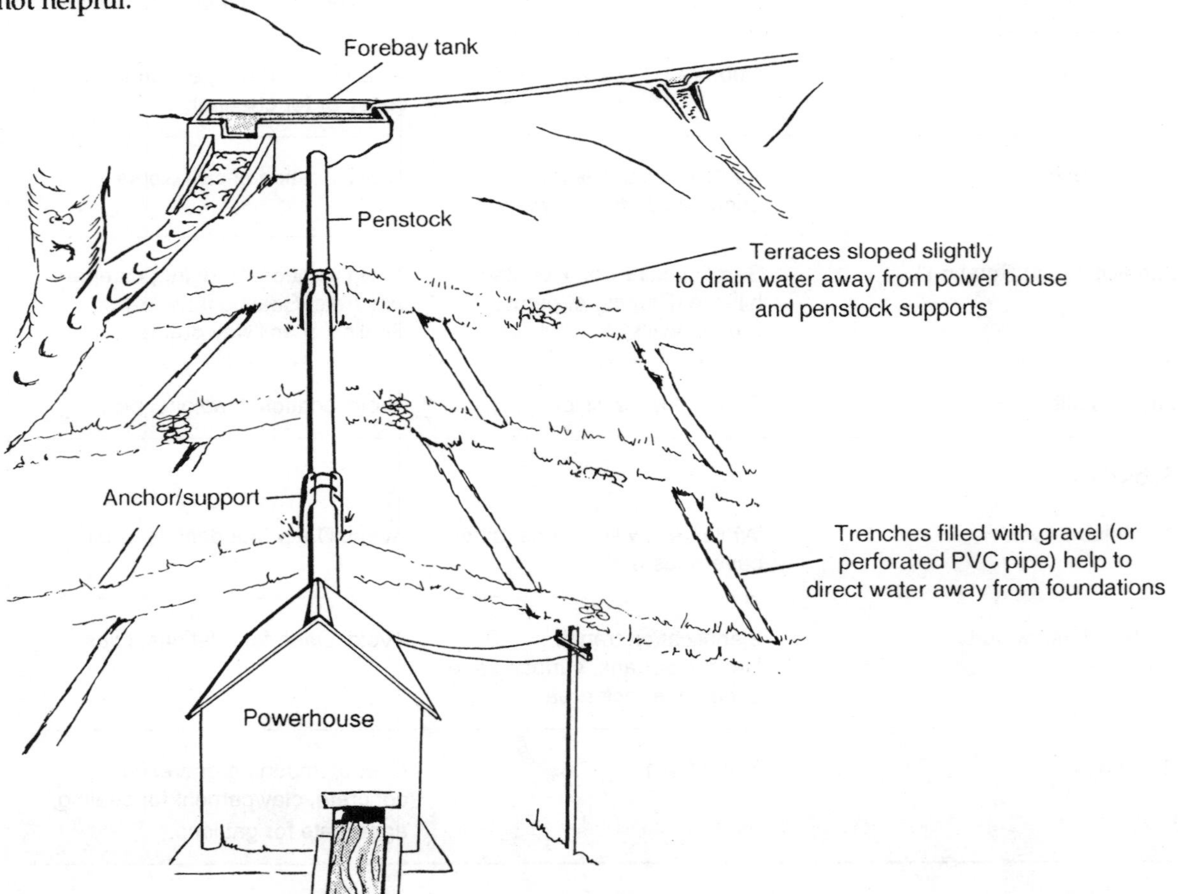

Fig 2.6.3 The powerhouse and penstock could be destroyed within a few years by the action of heavy rains eroding the foundations of the penstock. Careful traverse sloping of terraces can direct water away from the powerhouse and penstock support blocks.

Civil works should not be located in a gulley or below one. Where a traverse by a channel is unavoidable, precautions should be taken as shown in Fig 2.6.2 c. Beware especially of attempting to conduct water in a small gulley underneath a channel in a culvert. It is much more sensible to direct the water channel across a bridge or 'channel crossing' (see Fig 2.6.4).

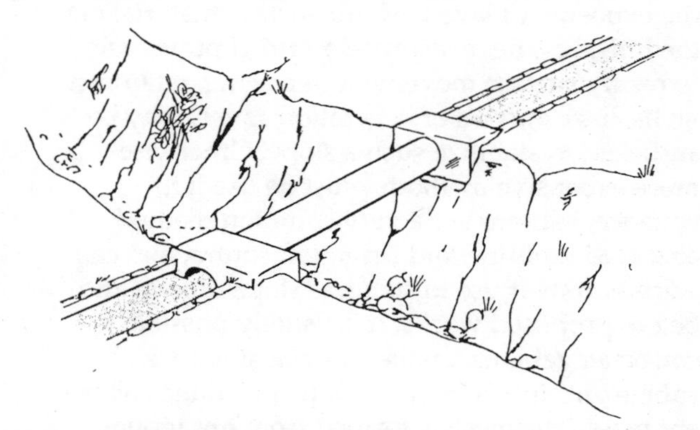

a A pipe bridge or aqueduct over a 10 year storm gulley. The bridge is suitable only if higher than suspected rock flow.

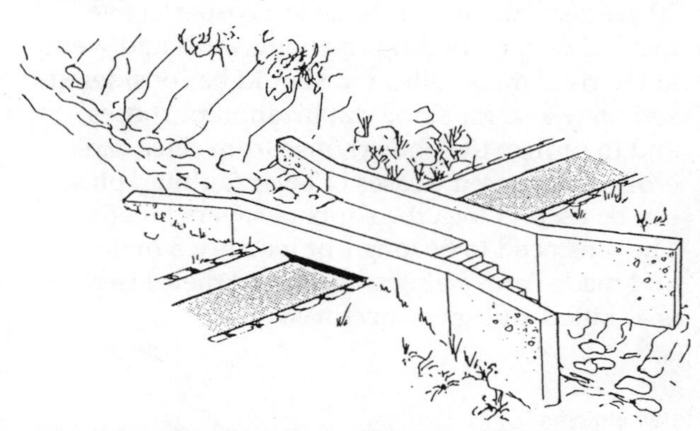

b A storm gulley directed over the channel Often this is safer than building a culvert under the channel, since a small culvert is easily blocked in a storm.

Fig 2.6.4 Bridge and storm gulley

Flood plains

A map will often indicate a low-level flood plain by showing an area without contours. Signs of periodic flooding will be seen on site, and information on floods will be obtainable from people who have lived in the area a long time. Typical signs of flooding are: low ground near the river, ponds, suspected water borne debris on the river banks or caught in bushes, wet or dry secondary water courses, sand gravel and dried mud deposits (the latter often showing shrinkage cracks). Flood plains pose various risks: the river path might change, sediments build up, sandy ground may be unsuitable for placing foundations. Powerhouses and intakes must be built above flood plain levels, although sometimes use of wing walls can protect intakes from flooding (see Chapter 3).

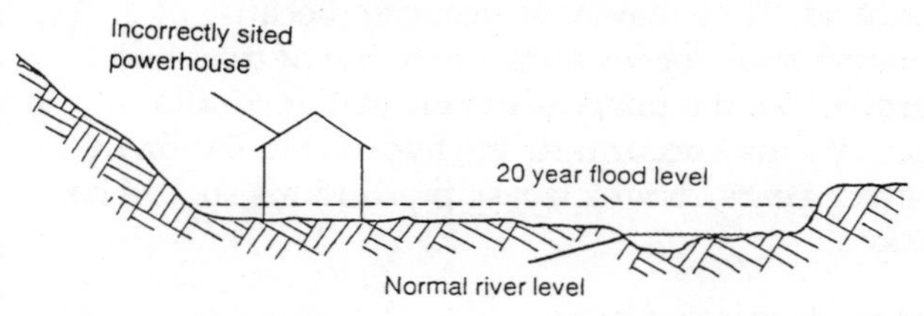

Fig 2.6.5 Powerhouse. Always build the powerhouse above the 20-year flood level. To find the level speak to local residents and look for evidence of flood debris (sand, gravel, dry mud, flotsam).

Future sub-surface movements

Landslip

A landslip zone can be represented on the geological sketch (Fig 2.6.1) as shown. The same effect viewed from the side is shown in Fig 2.6.6. Slip zones can be detected by carefully surveying for sudden steps in the hillside, or cracks. A step is found where the fault line meets the surface. Even steps which have become vegetated and rounded may indicate fault lines which could be freshly activated. Sometimes a sudden dip or rise in a path indicates a fault, or a broken section of wall. An oddly shaped tree trunk can be indicative. Weeping of water from cracks on the slope can indicate the lower exposed end of a fault line.

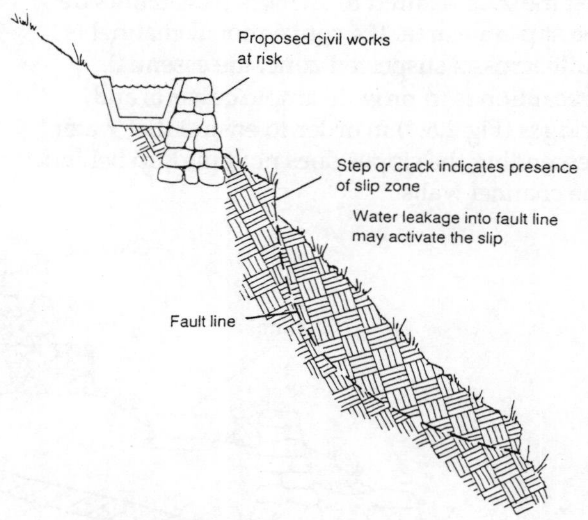

Fig 2.6.6 Landslip or 'rotation'. Note that disturbances caused by construction may cause the slip to start. A slip below the civil works will prompt collapse above eventually. Water leaking from a channel or forebay, or from a spillway, may enter the fault line and activate it. Careful direction of channel and spillway drains is therefore very important.

It is important to study the larger area surrounding the site. Landslip occurrences some distance away on similar ground indicate danger nearer at hand. They may have occurred because of disturbance, for instance, stripping of trees and grasses for the purpose of cultivation. Similar activity may occur near the hydro installation in future years, in which case precautions should be taken.

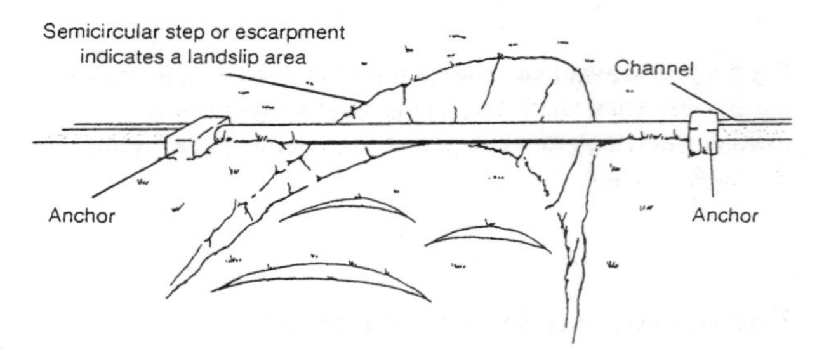

Fig 2.6.7 A slip zone traversed by an aqueduct. The aqueduct or pipe must be capable of supporting the weight of water within it and its own weight. Long traverses are therefore impossible, and catenary or suspension wires, or a pipe bridge are necessary.

The general rule is to avoid location of works near landslip faults. In the rare case where it is decided to traverse a channel across a suspected slip zone then a pipe or an open aqueduct can be used, but only if the traverse is short (Fig 2.6.7). The traverse length is limited by the length of pipe or aqueduct that can support its own weight and the weight of water inside it. For longer lengths a pipe bridge must be built, or catenary wires used to support the pipe. These last must be secured to anchors placed outside the slip zone area. If a conventional channel is built across a suspected zone, the essential precaution is to provide ample culverts and bridges (Fig 2.6.4) in order to ensure that water descending the slopes does not build up behind the channel walls.

Layer faulting

A second type of sub-surface fault pattern can cause severe slope instability. This is 'layer faulting' and consists of parallel faults sloping downwards in the direction of the slope (Fig 2.6.8). Layer faulting is detected by the presence of debris collecting at the foot of the slope and the exposure of layers of sub-surface material on the hillside. The measures described previously to rectify surface movement (terracing, retaining walls, introduction of vegetation cover) may be sufficient to stabilize such a slope. Otherwise more expensive methods must be used; for instance, gabions are a very common method and road building and irrigation contractors can advise on their use to stabilize slopes above and below proposed works. It is usually possible to construct gabions on-site by enclosing rock rubble within a wire mesh to form a huge coherent mass. The mesh is formed from hexagonal sections of galvanized steel wire of 2.5 mm to 3.0 mm diameter (giving a corrosion-proof life of 15-20 years). Gabions can be a very cost-effective method of constructing retaining walls and weirs at the river intake site, and should be considered seriously also for slope stabilization purposes, and to protect the powerhouse from river bank erosion. In extreme cases of layer faulting, pins can be used to lock the slippage layers together. The pins need to be long (for instance 5 metres) and made from stainless steel, and their insertion is a skilled and specialized task.

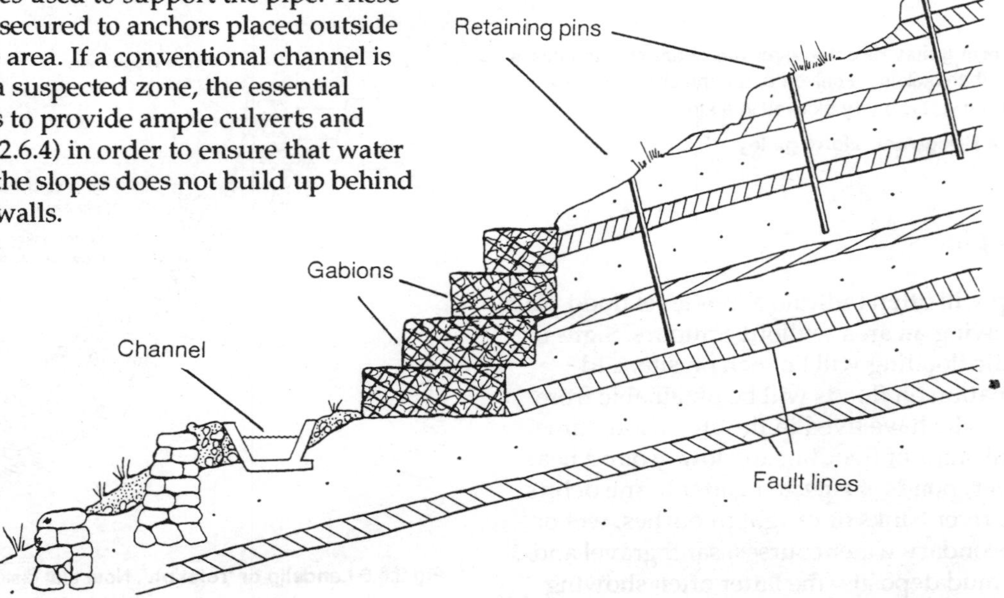

Fig 2.6.8 Instability is sometimes caused by fault lines sloping in the direction of the hillside. As the rock weathers it loses cohesiveness and crumbles away. Slippage along the fault lines contributes to surface movement. Attempts to stabilize such a slope are expensive and difficult. Gabions (wire cages enclosing a mass of stones) can be used to anchor the slope, as shown. Retaining pins are more expensive to purchase and install. Specialist advice should be sought in cases like these.

Subsidence

A third form of sub-surface movement is subsidence. This is caused by the presence of soluble rocks such as limestone and the action of water and chemicals underground. Limestone outcrops (white or yellowish rock often pitted with small holes) are the first immediate indicator of this danger. Underground water courses and dissolving rock are indicated by the sudden appearance of streams and their sudden disappearance into the surface. A clear indication is the presence of 'sink-holes'. These are small, circular (2 metres to 10 metres diameter) depressions in the surface, cone-shaped, which may have a small hole at their deepest point into which water trickles, or may be dry. The sink-holes are formed when limestone below dissolves away to the point at which the soil above collapses.

It is important not to place civil works such as penstock anchors in such an area to avoid collapse of foundations. This is especially true if newly introduced water courses, such as the spill from a forebay tank, are not carefully diverted as they may accelerate subsidence activity below ground. Another example of possible risk is the presence of unusually acidic water in rivers or streams, the effluent perhaps of a mining operation upstream.

Sandy soils, or soils containing layers of sand below the surface, pose dangers of sub-surface movement. The presence of additional mass above them, and seepage of water into them, can promote movement and then subsidence. Care should be taken to inspect soil types below the surface and avoid building on sandy soils, or drive piles deep below the weak layers.

Photo 2.6.1 Grass growth will help stabilize a hillside.

Soil and rock types

Your survey and sketch of the proposed installation area should include careful notes as to soil and rock types. It is worthwhile to dig inspection trenches at various points around the surveyed area.

The purpose of the analysis is to assess:

- Suitability of soil/rock for placement of foundations. Likelihood of movement of foundations.
- Suitability of soils and rock for use as construction materials.

It is well known that foundations must be laid on firm soil or rock. Where these are not found at a reasonable depth from the surface, or found only partially, piles may be driven deeper underground until firm material is found. All parts of a foundation must be laid on an equally firm base, to avoid differential settling of the building.

Table 2.6.1 lists techniques for identification of the various soil types, and their usefulness in construction and for foundations.

Note 2.6.1 provides a checklist which summarizes this section.

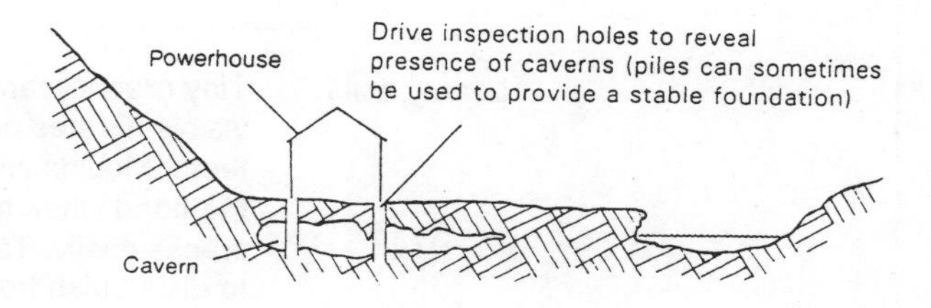

Fig 2.6.9 Subsidence. Limestone foundations can be dissolved slowly by water and eventually collapse. In one case a powerhouse collapsed because mining activities up-river released acidic waters which accelerated the process.

Photo 2.6.2 A powerhouse placed too close to the river.

Table 2.6.1 **Rock and soil characteristics**

	Code	Identification	Usefulness
Rock: Weathering and decomposition	Solid	Survives hammer blow. Bright clean. Cracks more than 30 cm apart.	Good base for foundations. Boulders can be used as ballast in cement weirs and anchor blocks.
	Weak	Breaks under hammer. Discoloured. Cracks 5 cm to 30 cm apart.	Reasonable base for foundations. Loose or crumbly material must be removed. Good for aggregate in cement work.
	Ghost	Keeps form of rock but crumbles easily. Equivalent to soil.	Unsuitable for foundations. Drive piles through or dig for deep concrete work.
Soil: Firmness and cohesion properties when wet and when dry	Gravel/ Gravelly soil	Small to medium stones. Crunches under foot.	If gravel content very high, forms good base for foundations. Gravel itself very useful as aggregate in concrete. Gravelly soil drains well. Can be used to fill drainage trenches.
	Sand/Sandy soil	Small grains, sugar grain sized. Each grain separable and visible. Cannot be moulded by hand when wet or dry.	Not suitable for foundations since sand can flow and compress when wet and dry. Important construction material – mix with cement.
	Loamy soils	Tiny grains scarcely visible. Grates between teeth. Moulds slightly in the hand when moist but cracks easily. Take care to distinguish from clay or sand.	Not suitable for foundations. Avoid partial presence of loamy or sandy soils as a foundation to prevent differential settling. Drive piles or dig for deep concrete work. Can be deceptively stable when dry but very mobile when wet. Weakened by frost action.
	Clay/Clay soils	Invisible particles. Moulds well when moist, can be formed into a ribbon between the fingers. Difficult to break by hand when dry. Impermeable. Expands when wet and shrinks and cracks when dry.	Very useful as an impermeable sealing material for channels. Medium load-bearing capacity. Suitable as foundation base for light construction work if well-designed concrete footings are inserted into clay bed.
	Peat/Organic soils	Brown or black. Can have rotting smell. Found in deposits in marshy areas.	Not useful for construction. Must be removed. Highly compressible and unreliable. Can be used as a combustion fuel and for cultivation when oxygenated and limed.

Civil works

3

3.1 Introduction

Various possibilities exist for the general layout of a hydro scheme. Fig 3.1.1 shows some of these. The discussion here focuses on schemes which utilize a long penstock or pressure pipe; that is, high head schemes.

A decision must be made with such schemes as to the relative lengths of the penstock and channel sections, and how to route them. Section 3.2 discusses the water supply route.

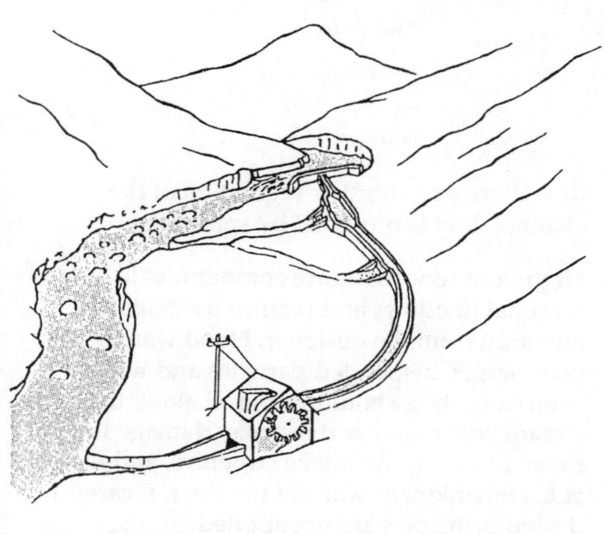

Low head with channel

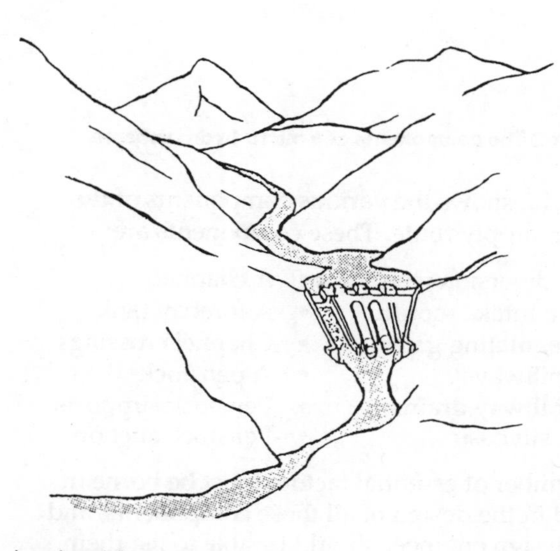

Low head river barrage

High head with no channel

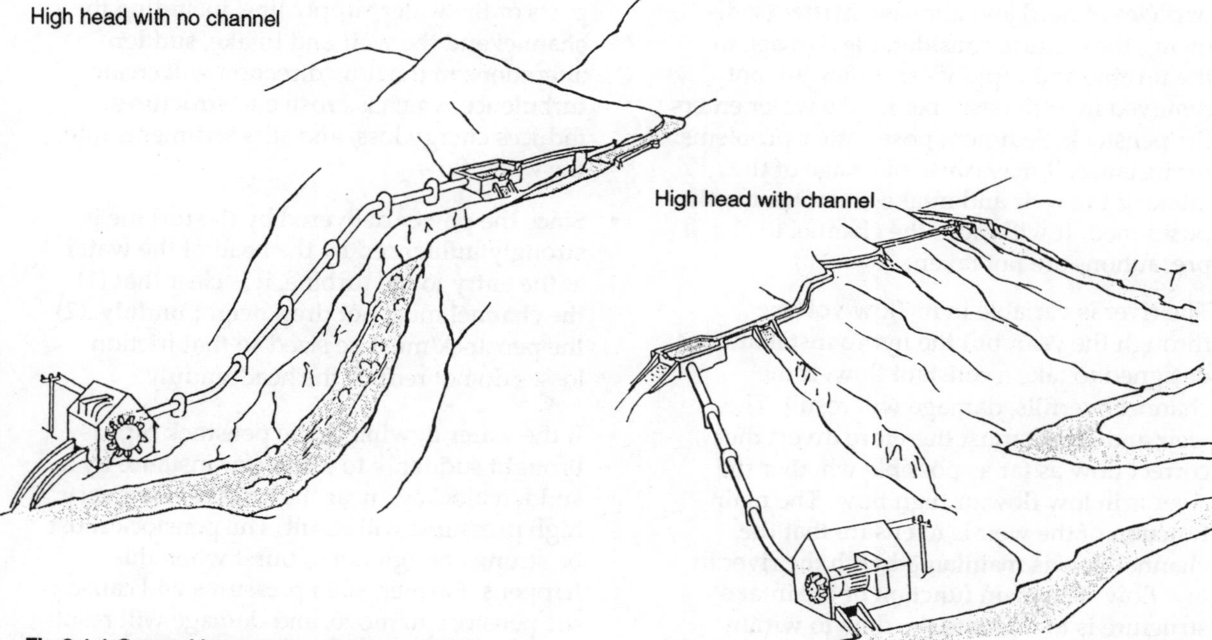

Fig 3.1.1 General layouts for a hydro scheme

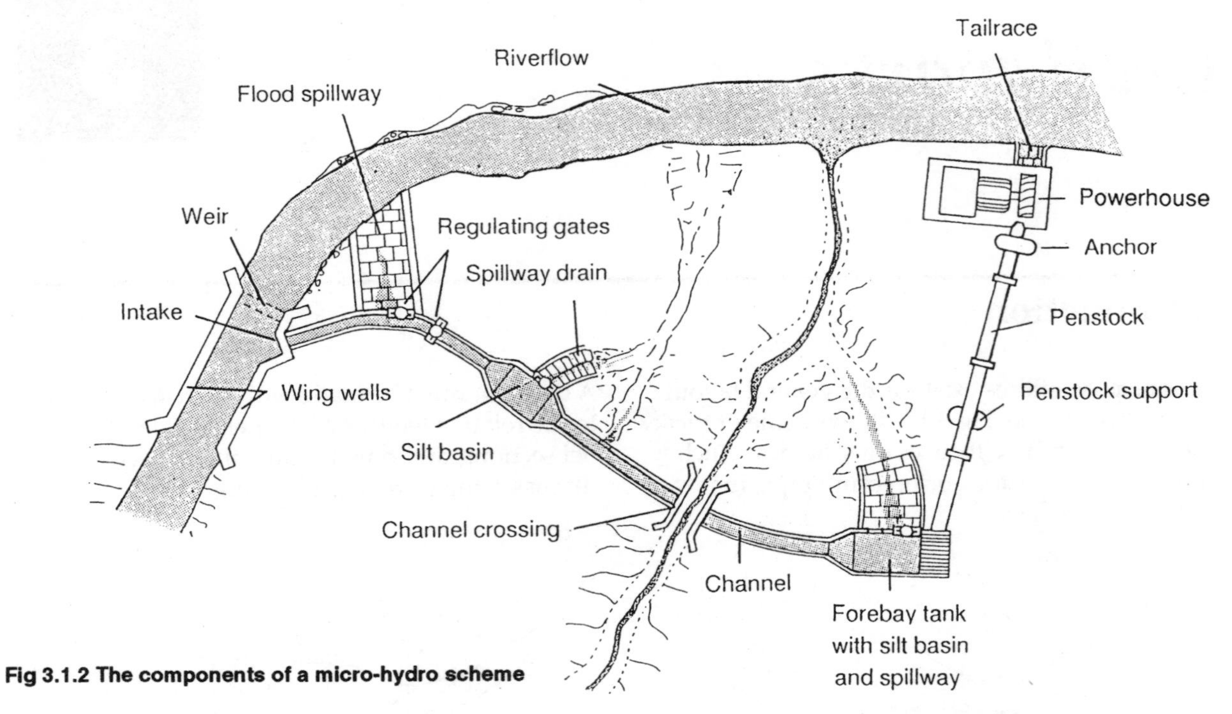

Fig 3.1.2 The components of a micro-hydro scheme

Fig 3.1.2 shows the various components of the water supply route. These components are:

- A diversion weir
- An intake mouth
- Regulating gates
- Spillways
- Spillway drains
- A silt basin
- A channel
- A forebay tank
- Channel crossings
- A penstock
- Penstock supports
- Penstock anchors

A number of essential factors must be borne in mind in the design of all these components, and the design engineer should be able to list them. For example:

- Flowing water in the river will carry small particles of hard and abrasive matter (sediment); these cause considerable damage to the turbine and rapid wear if they are not removed in a silt basin before the water enters the penstock. Sediment poses other problems: for instance, it may cause blockage of the intake if the weir and intake are not correctly positioned. It will cause the channel to clog if precautions are not taken.
- The river is variable in its flow volume through the year, but the hydro installation is designed to take a constant flow. If the channel overfills, damage will result. The weir and intake must therefore divert the correct flow as far as possible whether the river is in low flow or high flow. The main function of the weir is to ensure that the channel flow is maintained with the river in low flow. The main function of the intake structure is to regulate the flow to within reasonable limits when the river is in high flow. Further control or regulation of the channel flow is provided by spillways.
- High flow conditions are common, as in seasonal flooding, and require particular attention from the designer. Flood waters will carry larger suspended particles and will even cause large boulders to roll along the stream bed; together these may damage the diversion weir, the intake structure, and the side embankment walls of the river, if careful design principles are not applied.
- Another factor demanding attention is the effect of turbulence in the water flow; in all parts of the water supply line, including the channel and the weir and intake, sudden alterations to the flow direction will create turbulence which is erosive to structures, induces energy loss, and stirs sediments into the water flow.
- Since the power delivered by the turbine is strongly influenced by the head of the water at the entry to the turbine, it is clear that (1) the channel must not drop height unduly; (2) the penstock must be sized so that friction losses do not reduce the head unduly.
- If the water flowing in the penstock pipe is brought suddenly to a halt (for instance by a sudden blockage near the turbine) then very high pressures will result. The penstock must be strong enough not to burst when this happens. Further, such pressures will cause the penstock to move, and damage will result if it is not retained firmly with anchors.

3.2 System layout

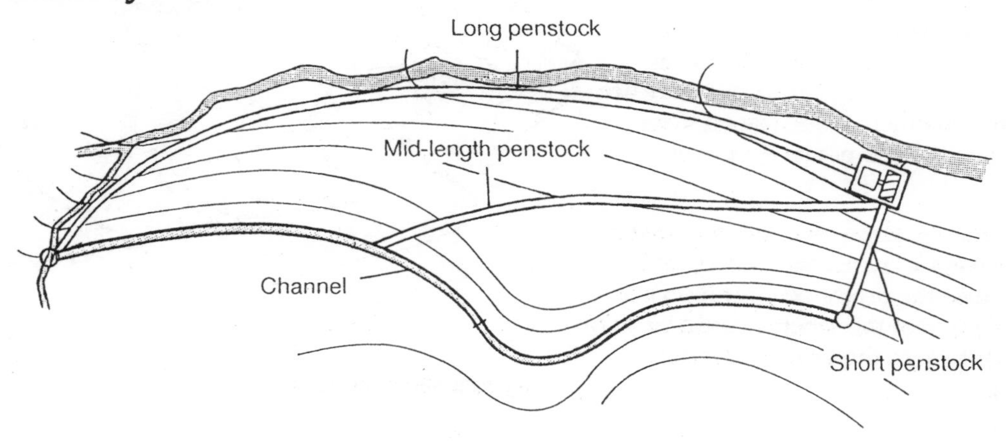

Fig 3.2.1 Channel and penstock options Notice that the channel can be shortened to avoid the risks and expense of construction across a steep slope.

Penstock pipework is considerably more expensive than open channel work. The basic rule in laying out a system is:

Keep the penstock as short as possible.

Fig 3.2.1 reproduces the map showing scheme A as considered in Chapter 2. Three possible penstock routes are shown. The 'short penstock' option will in most cases be the one leading to the most economic scheme, but this is not necessarily the case.

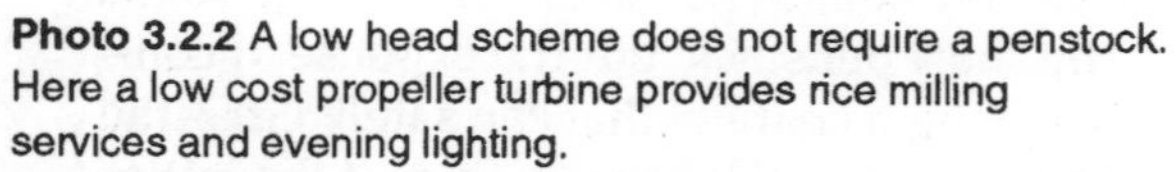

Photo 3.2.1 A straight penstock can keep costs to a minimum.

Photo 3.2.2 A low head scheme does not require a penstock. Here a low cost propeller turbine provides rice milling services and evening lighting.

Considering each option in turn:

- **Short penstock**

  Here the penstock is short but the channel is long. The long channel is exposed to the greater risk of blockage, or of collapse or deterioration as a result of poor maintenance. Installing the channel across a steep slope may be difficult and expensive, or even impossible.

  The risk of a steep slope eroding may make the short penstock layout an unacceptable option, because the projected operation and maintenance cost of the scheme could be very great, and outweigh the benefit of initial purchase cost.

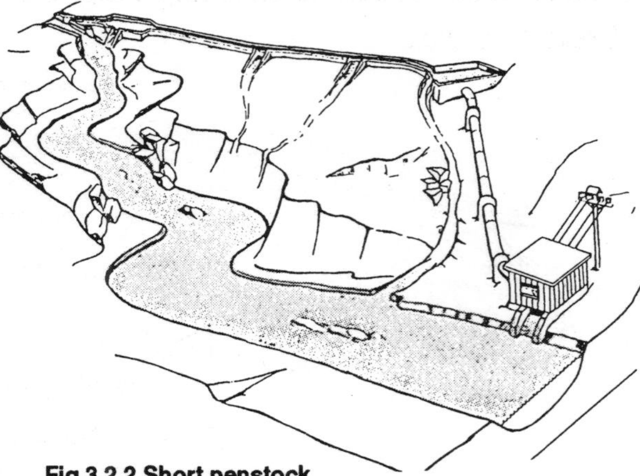

Fig 3.2.2 Short penstock

- **Long penstock**

  In this case the penstock follows the river. If this layout is necessary because of an impossible terrain on which to construct a channel, certain precautions must be taken. The most important one is to ensure that seasonal flooding of the river will not damage or deteriorate the penstock. It is always important to calculate the most economic diameter of penstock; in the case of a long penstock, this is particularly important, since the cost will be particularly high.

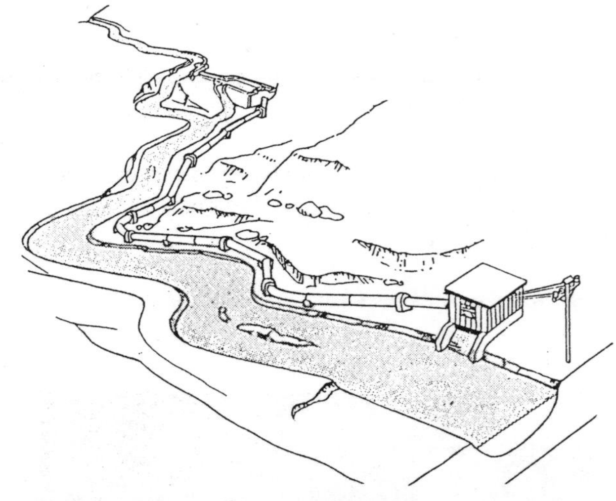

Fig 3.2.3 Long penstock following river

- **Mid-length penstock**

  The mid-length penstock will cost more than the short penstock, but the expense of constructing a channel that can safely cross the steep slope will be saved. Even if the initial purchase and construction costs are greater in this case, this option may be preferable if there are signs of instability in the steep slope.

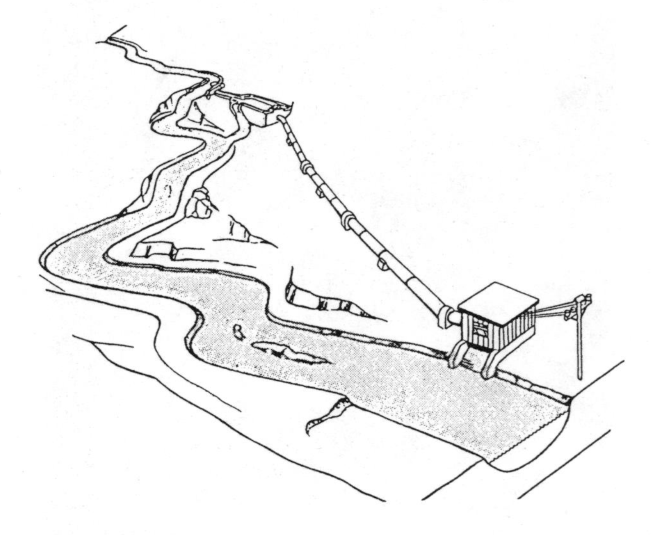

Fig 3.2.4 Mid-length penstock

In some cases the soil may be particularly sandy and permeable and leak water away from the channel. A shorter channel may be a wise option in this case, to limit water losses by seepage.

Various other features may raise the cost of the channel, as discussed in sections 2.6 and 3.6. One point to consider is the cost of stream crossings over the channel. Where floods are expected, the cost of providing such crossings will increase the channel cost. Some of these difficulties can be overcome by the use of closed pipes as channel sections, or by covering and lining channels. Although the costs of improved channels are high, they will usually be less than penstock costs.

In order to make a choice of system layout, some rapid costing is necessary. Section 3.11 describes a method of estimating penstock cost. This cost must be applied to each of the route options and combined with estimates of the associated channel costs.

3.3
Weir and intake

A hydro scheme must extract water from the river in a reliable and controllable way. The water flowing in the channel must be regulated during high river flow and low flow conditions. Figs 3.3.3 and 3.3.6 describe how this is done by weir and intake structures. Sometimes it is possible to avoid the expense of building weirs, instead using the natural features of the river. A natural permanent pool in the river may provide the same function as a weir.

Another consideration in siting an intake is to protect it from damage – this is shown in Fig 3.3.1. A great many variations in intake and weir design are possible, to suit a wide range of natural conditions. Only very few of the possible options are discussed here, in order to introduce a simple design approach. Once the fundamental principles of this design approach are understood, it will be possible to solve design problems presented by new conditions.

Fig 3.3.2 shows how the intake passes water into a headrace (a fast-flow channel), which then continues past a spillway to a silt basin. Because the intake is sited close to the river, it is often not easily accessible. During flood flow conditions, for instance, it may be difficult or dangerous to

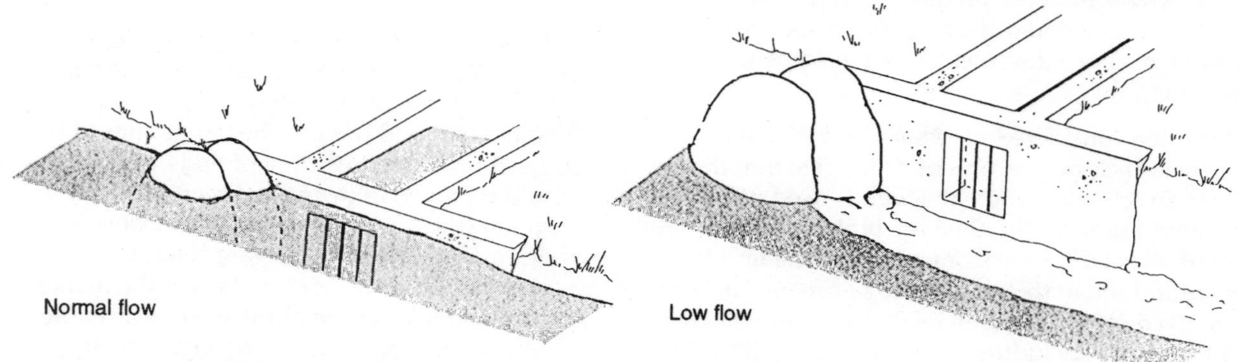

a The intake shown here has been built flush into the side of the stream to allow debris to roll or float past.

b The intake shown here has been left high and dry because over a period of some years the river has changed its course. This effect may occur only during the low flow season. A weir is necessary to ensure submergence of the intake mouth.

Fig 3.3.1 Intake protection An intake is subject to damage by moving debris during floods. One method to avoid this is to site it behind permanent boulders or outcrops of bed rock. Remember that boulders will move in flood waters and are best allowed to roll downstream past the intake.

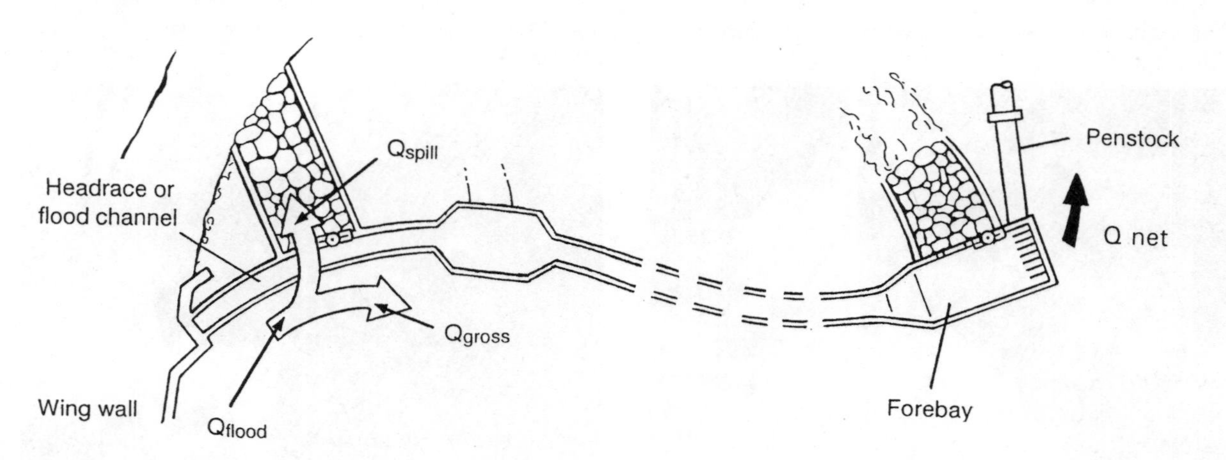

Fig 3.3.2 A sufficiently high water velocity in the headrace channel will allow sediment and debris entering the intake to be flushed down to the silt basin. No regulation of flow or screening is provided at the intake mouth, since access can sometimes be difficult or dangerous. Notice the distinction between flows Q_{flood}, Q_{gross} and Q_{net}. Q_{net} is the ADF_{net} flow calculated in Section 2.2.

attempt any control operations at the intake mouth. The headrace allows control functions and the silt basin to be placed where access is safe, away from the river.

A stream will often carry sand with it, sometimes in large quantities when heavy rains wash topsoils into the waterways. This sand may be deposited as sediment behind a weir and threaten to create a blockage around the mouth of the intake. This is unlikely to cause problems with water intakes in many mountainous streams, since the sand can be drawn into the intake and swept through to the silt basin where it is removed.

Nevertheless, it is wise to study the length of the river carefully to find a position for the intake which is least likely to be a point of deposition of sediments. Guidance on this is given in the section below under the heading 'Avoiding sediment concentrations'. Also see Appendix 3, Bibliography.

It is often considered necessary by hydro engineers to include a sediment sluice gate into the weir. In schemes where the weir is used partly as a dam to store water, then the build up of sediment will reduce the storage volume behind the weir or dam. In these cases the sediment can be removed either by manual labour during the dry season or by including large sluice gates into the weir.

It is generally recommended that micro-hydro schemes do not incorporate storage reservoirs upstream of the weir and that therefore the expense of sluice gates and sluicing operations can be saved. However it is necessary to ensure that sediments do not clog the intake mouth itself. This is best done in the following ways:

1 Place the intake mouth some distance behind the weir in order to allow sediment and debris to build up downstream of the mouth. They can be removed manually during low flow periods. (Fig 3.3.6 b)

2 Assuming that you have first chosen an intake mouth position which is least vulnerable to silt accumulation, allow the intake mouth to draw all silt and entrained debris into the headrace channel and then allow it to settle out in the silt basin where it can be removed. The direct intake (Fig 3.3.6 c) can be a better design than the side intake to ensure this happens. Do not use a fine screen, instead use a widely spaced 'rack'.

3 If **1** and **2** are not sufficient to avoid clogging, place in the weir a pipe of large cross-section which is closed off at the downstream end by a bolted disc and flange. The flange should be accessible during low flow periods when the disc is removed allowing sediment to wash through (Fig 3.3.6 b). Alternatively, consider redesigning a small sluice gate consisting of vertically placed wooden rods near the intake mouth and to one side of the weir, near to the intake, which can easily be removed during low flow periods. This will allow accumulations to be washed downstream.

4 Consider design variations which encourage high velocity flow near to the intake. These may for instance consist of appropriate changes in the crest level of the weir.

Photo 3.3.1 A concrete diversion weir under construction. (Sri Lanka)

Photo 3.3.2 An intake mouth controls flood flow. (Sri Lanka)

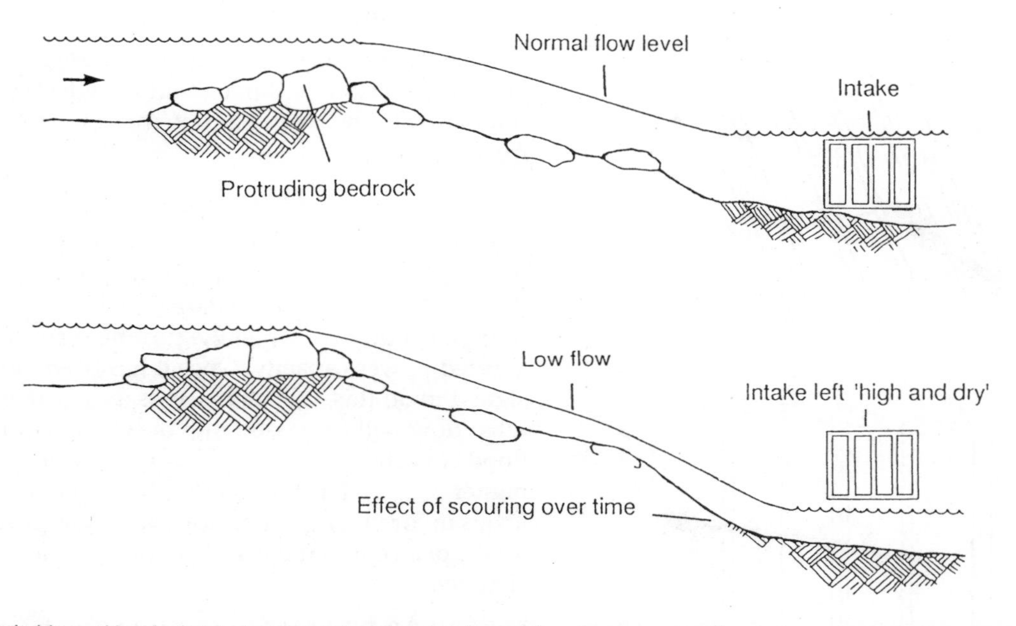

a Unsuitable position for intake because of surface level reduction due to seasonal variations of flow or scouring of river-bed over time. Fig 3.3.1 shows the effect of seasonal variation in flow and scouring on river position.

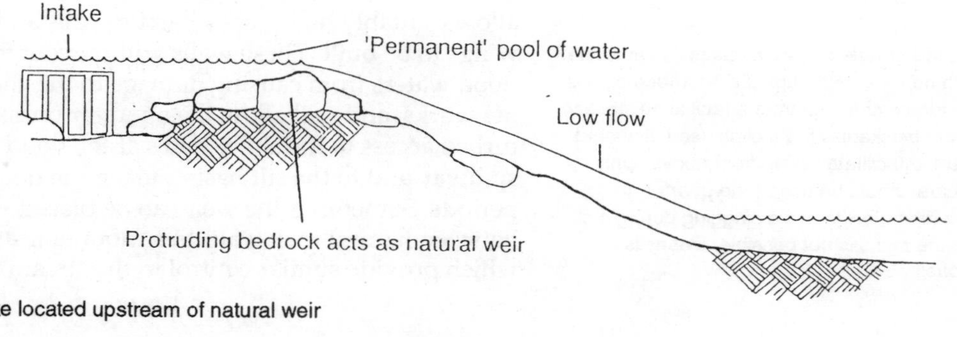

b Intake located upstream of natural weir

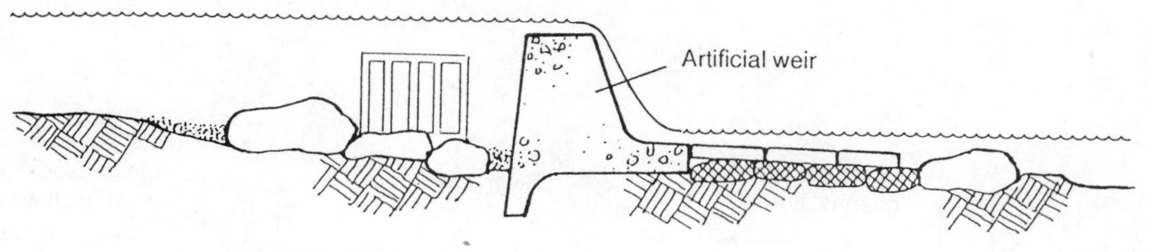

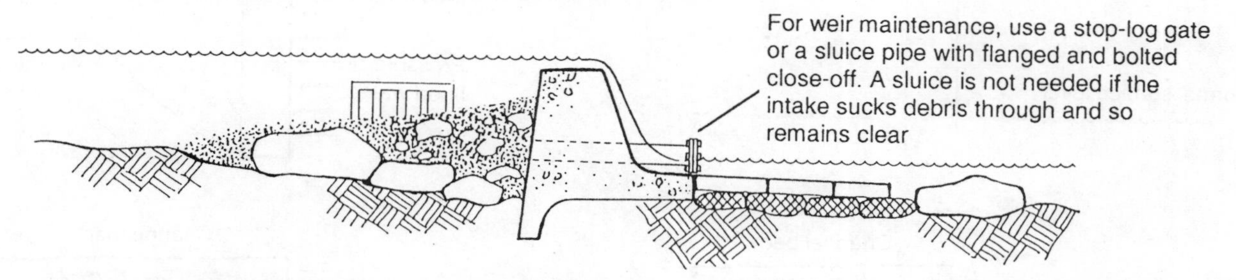

c Side intake located upstream of artificial weir. Sediment and stones build up, which may eventually clog the intake. Careful positioning of the intake behind large rocks may avoid this, or design such as the direct intake shown in Note 3.3.3 which sucks debris into the headrace channel. The sluice pipe shown is therefore not needed for sediment removal as such, but only for draining of the weir for maintenance, eg checking for cracks and for repair. If included it should be positioned near the intake mouth area. Sediment collecting at the weir as shown does not obstruct the function of the weir and can be left in place; it actually helps to seal the weir.

Fig 3.3.3 The function of a weir Every micro-hydro scheme must make use of a natural or an artificial weir. The function of the weir is to maintain a permanent water level above the intake mouth during both high and low flow seasons.

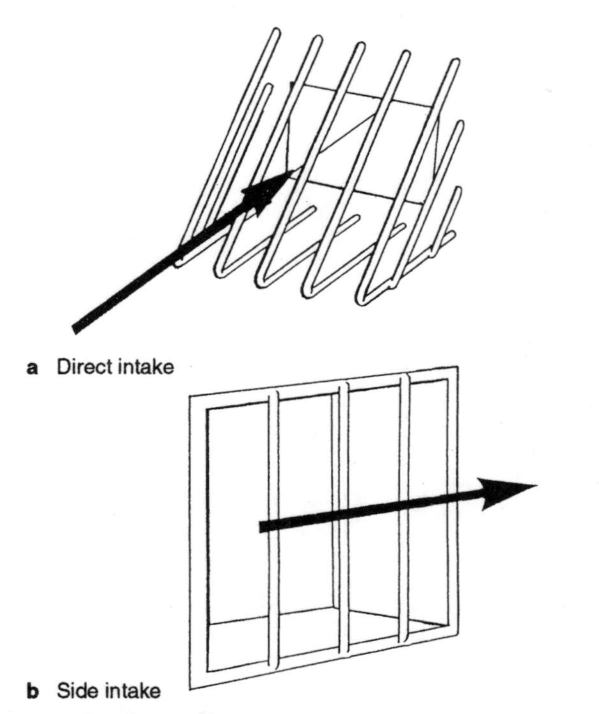

Fig 3.3.4 Direct and side intakes In both cases it is possible to design intakes with no rack. (Although if a headrace pipe is used instead of a headrace channel, then a rack is necessary to protect the pipe from blockage.) If a rack is used, it should consist of vertical bars to facilitate raking from above. Only very widely spaced bars should be used (one quarter of channel width or pipe diameter) to avoid clogging during high flows when raking of the racks is not possible. Debris is removed by the silt basin.

In all aspects of intake and channel design, it is always important to think about the effects of a flood flow in the river. Keep in mind the fundamental principle:

"Think flood"

In flood conditions the river flows faster and the water surface is higher. The intake mouth is designed to pass the full channel flow (Q_{gross}) in normal conditions, when the turbine is running at rated power capacity. Once it has been designed to do this, it is then necessary to calculate what flow will be passed into the channel during flood conditions when the water surface level is higher (Q_{flood}). Fig 3.3.5 shows the effect of flood levels in forcing a greater flow rate through the intake orifice into the fast-flow, or headrace, channel.

Example 3.3.1 describes a method for estimating the size of the largest flood likely to occur in the lifetime of the hydro scheme. This estimate allows suitably high intake barrier walls and wings to be built. These walls will prevent the flood waters from causing damage to the channel works, and will allow the operator undisturbed access to the sluice gates at the flood spillway and to the silt basin during the flood periods. Sometimes the weir can be placed between natural rock walls in a mountain stream which provide similar control to the stream in

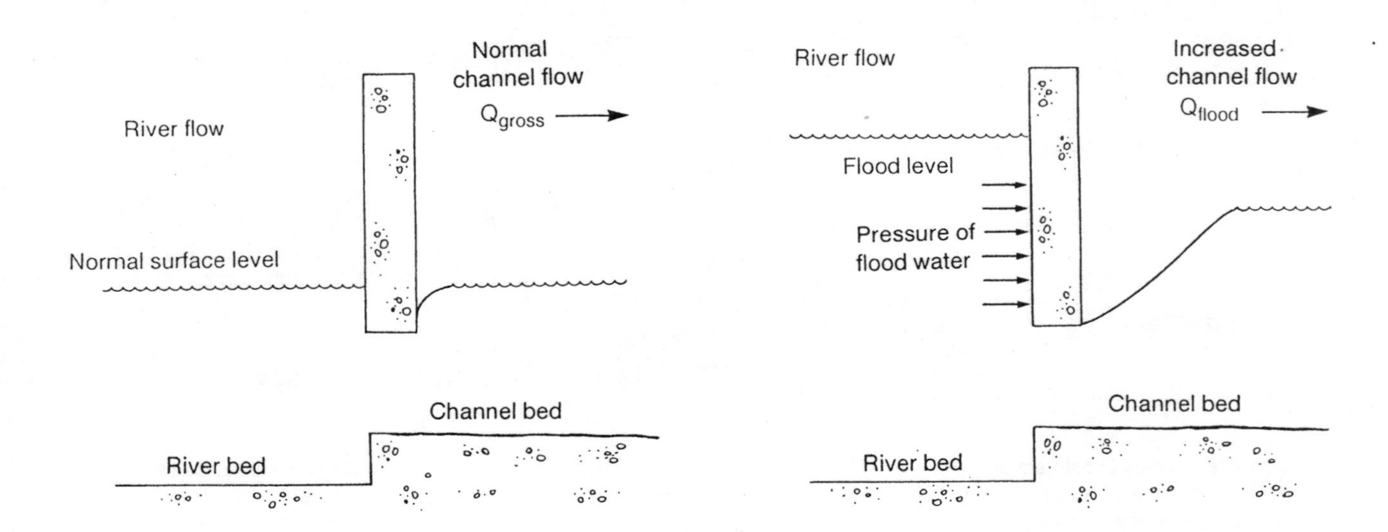

Fig 3.3.5 Cross-section of intake showing flood flows (Q_{flood}) and normal (Q_{gross}) flows in the fast-flow channel. Notice that the intake barrier wall must be strong enough to take the pressure of flood waters.

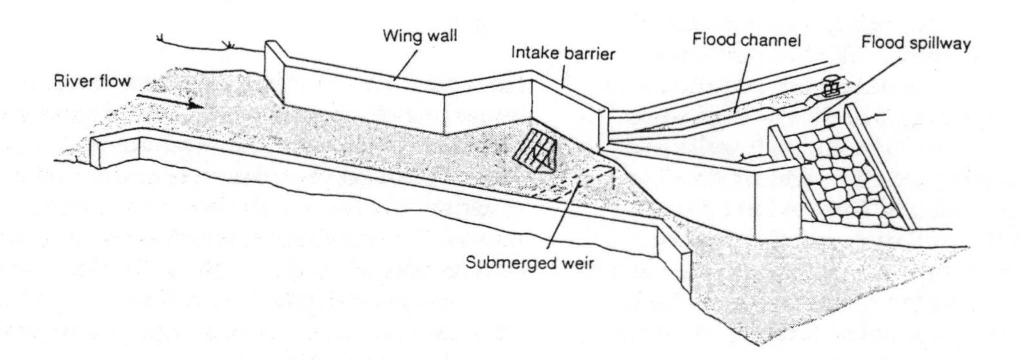

a Notice that flood spillway is accessible during floods while the weir and intake mouth are not. The rack is not essential and can be omitted from the design. Some designers will be forced by site conditions to replace the headrace channel with a headrace pipe; it is important then to include the rack to protect the pipe from blockage.

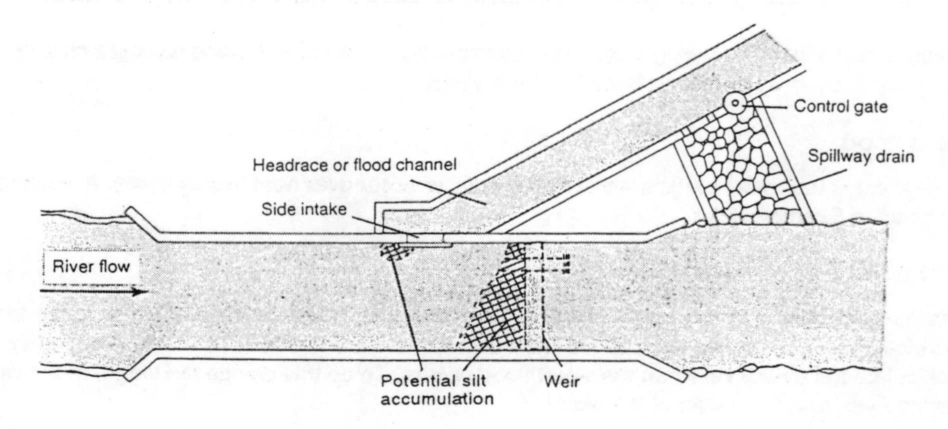

b Sediment build-up downstream of this side intake can be removed manually instead during low flow periods when access is feasible. In many sites the side intake will remain clear of blockage by acting in the same way as a direct intake, that is, sucking all sediments through to the silt basin.

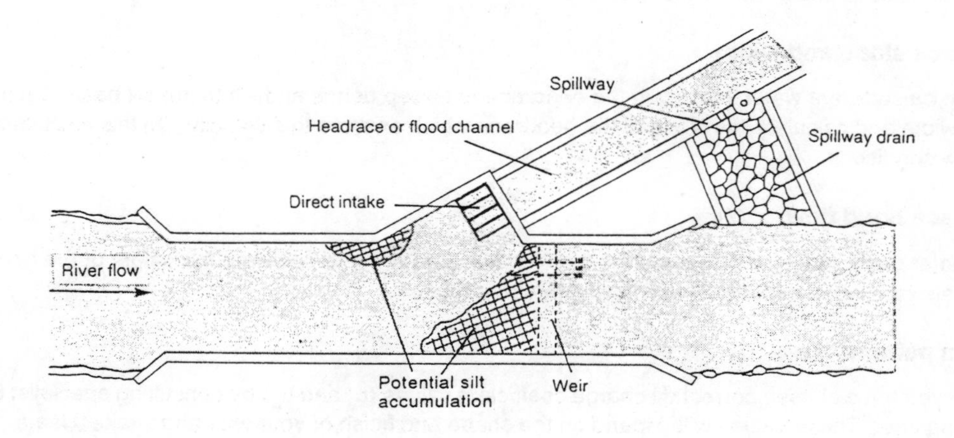

c A direct intake will automatically stay clear of blockage by encouraging the debris to flow through the intake rather than collect at the intake mouth.

Fig 3.3.6 Intake system Notice that both side and direct intakes are possible. Intake racks should always have wide spacings to avoid clogging, since access to clear a clogged rack during floods may be impossible. Notice that the debris and sediment collected behind the weir does not pose any difficulties. Sediment behind the weir can actually be a benefit, helping to seal the weir from leakage. The debris entering the intake is later flushed from the silt basin.

flood periods. The walls can be built from gabions (wire cages filled with boulders), or concrete, or rock and mortar to a substantial thickness. In Example 3.3.1, the calculation results in a recommendation for extended weir length and for approximately 4 m high walls. Although expensive to build, such walls will provide valuable protection against possible flood damage which is likely to have a cost greater than the wall cost, and also against expensive down-time during floods when access difficulties become too severe to correct faults resulting in stoppage of the turbines. In some areas the likely worst flood size will be less than assumed in the example, leading to smaller barrier wall heights and shorter weir crest length.

Example 3.3.1 illustrates a method of calculating intake orifice size and weir height, together with headrace dimensions and headrace bed slope. This calculation provides an estimate of the water surface level in the headrace during normal flow conditions, which is useful in the next section when the height of the flood spillway crest is established. To follow the method of calculation of headrace bed slope you must first refer ahead to Section 3.6.

Note 3.3.1 **Summary of intake calculations**

The calculation is taken in the following steps (see Example 3.3.1). If you are using a programmable calculator, the summary of equations in Note 3.3.2 will assist.

1 The worst flood

Estimate largest expected river flow (the worst flood) likely to occur over next twenty years. A method of doing this is given in Example 3.3.1.

2 Height of flood barrier walls

Flood waters will overfill the channel and damage it. Careful use of natural features such as rocky embankments can prevent this, but sometimes artificial walls and barriers are needed. To find how high they should be, calculate the river level when the worst flood occurs. To do this decide the length of the weir crest across the river, and the height of the weir.

3 Intake dimensions

During normal (non-flood) conditions a certain flow of water is required in the channel. By setting various design parameters, such as the water depth in the headrace, an intake size can be calculated which ensures that this channel flow is achieved.

4 Headrace slope and width

There must be sufficient water velocity in the headrace to sweep debris and silt to the silt basin. Choose a headrace width and calculate the slope of the headrace which ensures this velocity. Do this calculation for normal flow conditions.

5 Headrace flood flow

Find the water depth in the headrace during the worst flood in order to calculate the height of the headrace wall. Find also the worst expected flood flow in the headrace.

6 Design parameters

Make sure you have chosen correct discharge coefficient values (c_d and c_w) by consulting specialist books or a civil engineer. These values will depend on the shape and finish of your weir and intake. Use a programmable calculator to allow you to quickly try out different values for dimensions such as the weir crest length, weir height etc. See the summary in Note 3.3.2.

7 Costing

Calculate the cost of the resulting structure and run through the calculation steps again varying sizes such as the length of the weir until the most effective design is achieved.

The examples provide a simple theoretical methodology for design which must be developed and modified by each engineer in the light of experience and by reference to other texts such as Lauterjung (see Bibliography). Nevertheless these calculations are a useful starting point and framework for successful design work.

Example 3.3.1 describes a method for calculating intake barrier and wing wall heights. It continues the design example of Chapter 2, concentrating on the proposal for a scheme fed by catchment A, as illustrated in Fig 2.2.1. The example also introduces one aspect of weir design, the choice of crest length. In some cases, the river has natural bedrock protrusions which form suitable foundations for the weir and result in longer overall crest length (Fig 3.3.7). This can be an advantage, since a longer weir crest will have the effect of reducing flood levels and the necessary wing wall and barrier height.

The calculations here assume that the river level downstream of the weir is below the weir crest level. In some cases the weir becomes submerged, in which case some modification of the weir equation given here is necessary – see a specialist text.

Intake calculations with a programmable calculator **Note 3.3.2**

If you wish to speed up the calculation by using a programmable calculator (or pocket computer, see Appendix 2) the following summary of equations will help (see Example 3.3.1 for the full equations):

Estimate worst flood

Set L_{weir} and h_{weir} (the weir crest length and the height of the weir)

Set c_w (the weir discharge coefficient)

Find $h_{over\text{-}top}$ from c_w, L_{weir}, $Q_{river\ flood}$

Find $h_{barrier}$ from h_{weir}, $h_{over\text{-}top}$

Set c_d (the intake discharge coefficient)

Set d (the height of the intake mouth)

Set $h_{h\ (normal)}$ (the headrace water depth during normal flow conditions)

Find w (the intake width) from Q_{gross}, c_d, h_{weir}, $h_{h\ (normal)}$, d

Check v_i is not erosive

Set v_h (the normal velocity in the headrace) to a high value, 2.0 m/s or above

Find w_h and S (width and slope of headrace)

Find $h_{h\ (flood)}$ by guessing values until the Manning and Orifice Discharge equations both give the same figure for Q_{flood}

Adopt Q_{flood} as design value for spillway

Find the wall height of the headrace channel from worst depth

Estimate cost of resulting design

Return to the start of the calculation, adjusting the set design values (eg L_{weir}, h_{weir}, d, v_h, $h_{h\ (normal)}$ etc) to find the most cost-effective design.

Weir coeffient c_w **Table 3.3.1**

Profile of crest of weir	c_w	Profile of crest of weir	c_w
broad; sharp edges	1.5	sharp-edged	1.9
broad; round edges	1.6	rounded	2.2
round overfall	2.1	roof-shaped	2.3

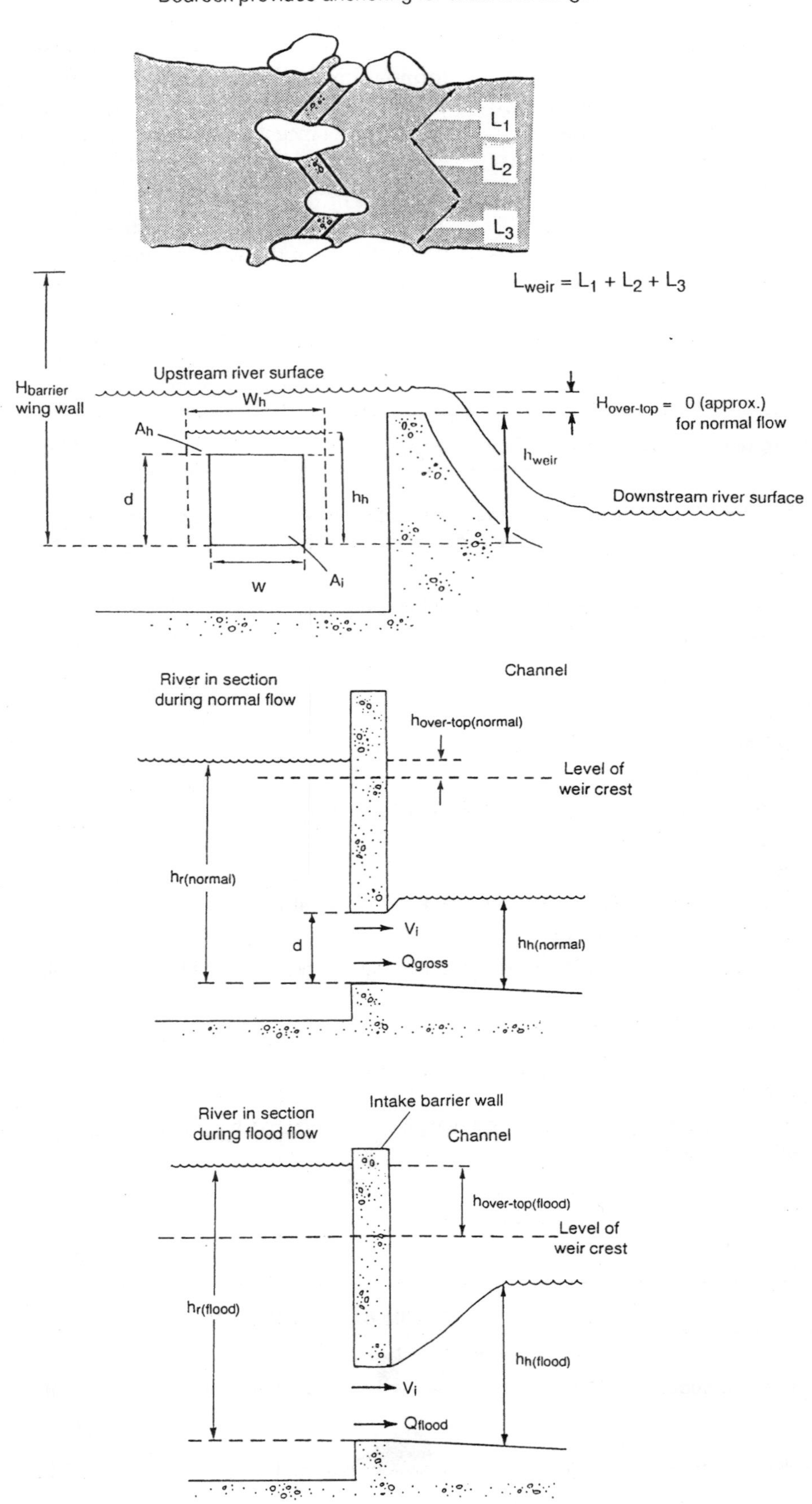

Fig 3.3.7 Weir and intake dimensions

Flood containment, intake and headrace — Example 3.3.1

In Example 2.2.2 a hydro scheme is proposed which delivers 68 kW for 7 months of the year. These are the months when rainfall is high, and the possibility of severe floods must be considered. The flow through the turbine (Q_{net} in Fig 3.3.2) is 92 litres/sec, while the flow into the channel (Q_{gross}) is 132 litres/sec – the difference is the allowance made for water losses from the channel to irrigation, informal domestic abstractions, and soil seepage or leaks. Design an intake and headrace structure. Fig 3.3.8 shows the findings of the calculations here.

1 The worst flood

Often when planning micro-hydro schemes the FDC is only your own estimate based on limited data from only one or two years. This means that the largest flow shown on the curve is not necessarily the largest flow that will occur in the next fifteen years. Even if the FDC is based on daily measurements over many years there is always a possibility that a flash flood lasting only an hour will be a larger flow than shown. For both these reasons it is wise to make a conservative estimate of the worst flood likely to be encountered in the future, by multiplying the largest flow on the FDC by a factor of between 20 and 50. To emphasize this point the following example applies the factor 50 to the FDC of Fig 2.2.12. The largest flow shown is 10 × ADF; our flood estimate is therefore 50 × 10 × ADF = 50 × 10 × 0.22 m^3/sec = 110 m^3/sec.

2 Height of flood barrier walls

Fig 3.3.7 shows the weir and intake mouth. The surface level of the river is only slightly higher than the weir crest, since water is only trickling over the top of the weir. This additional height is termed '$h_{over\text{-}top}$' and can be taken as zero during normal flow conditions.

The height of the intake barrier ($h_{barrier}$) is the height to which water is likely to rise in the worst flood conditions. This is shown in Fig 3.3.7. The length of the weir crest can be varied by constructing obliquely across the river, as shown in Fig 3.3.7.

The wing walls and intake barrier are built to contain floods, and so must be at least as high as the worst flood surface level. The flood level ($h_{over\text{-}top}$) can be simply related to the river flood flow ($Q_{river\ flood}$) and the weir crest length (L_{weir}) by the standard weir equation:

$$Q_{river\ flood} = c_w \times L_{weir} \times \left(h_{over\text{-}top}\right)^{1.5}$$

where c_w is the coefficient of discharge for the weir (see Table 3.3.1). Therefore:

$$h_{over\text{-}top} = \left(\frac{Q_{river\ flood}}{c_w \times L_{weir}}\right)^{0.667}$$

To investigate the effect of an extension of the weir length, find the wing wall and intake height needed for both a straight weir of 10 metres length and an oblique one of 15 metres.

The coefficient c_w varies with the shape of the crest of the weir. Consult Table 3.3.1; for simply built square-topped weirs with round edges a value of 1.6 is suitable.

Assuming a square-topped weir of length 10 metres:

$$h_{over\text{-}top} = \left(\frac{50 \times 10 \times 0.22}{1.6 \times 10}\right)^{0.667} = 3.6 \text{ m}$$

If the weir crest length is 15 metres:

$$h_{over\text{-}top} = \left(\frac{50 \times 15 \times 0.22}{1.6 \times 15}\right)^{0.667} = 2.8 \text{ m}$$

The lower value of $h_{over\text{-}top}$ is preferable; the longer weir crest length controls the river better. The total height of the structures is, given a crest length of 15 metres:

$$h_{barrier,\ wing\ wall} = h_{over\text{-}top} + h_{weir} = 2.8 + h_{weir}$$

continued overleaf

Example 3.3.1 *continued* **Flood containment, intake and headrace**

The height of the weir crest above the intake base line can now be chosen. The river surface level will be held constant behind the weir, so too high a weir may incur cost in extra wing wall length extending upriver, and will also cost more in weir construction, materials and maintenance. The weir should not be so low that the intake mouth cannot be sited above the river bed and below the weir crest level. A value of 1.0 metre is chosen as a starting point in this example for h_{weir}; the designer can later return to this point to adopt a revised value if appropriate, once the effect of the weir height on other dimensions becomes apparent.

$$h_{barrier} = h_{over\text{-}top} + h_{weir} = 2.8 + 1 = 3.8\text{m}$$

Fig 3.3.7 shows the intake mouth as a rectangle of depth d and width w. The figure also shows the water flow through the intake during normal flow (Q_{gross}) and during flood flows (Q_{flood}). During floods the amount of water over-topping the weir is considerable, and the extra head of river water ($h_{r\,(flood)}$) acts to drive more flow through the intake, so that Q_{flood} is larger than Q_{gross}.

3 Intake dimensions, normal flow conditions

The intake behaves according to the orifice discharge equation:

$$Q = A_i v_i = A_i c_d \sqrt{2g\,(h_r - h_h)}$$

where v_i is the velocity of the water passing through the intake, and c_d is the coefficient of discharge of the intake orifice. A_i is the cross-sectional area of the intake, h_r is the head of river water upstream of the weir while h_h is the depth of water in the headrace channel. The equation is used to calculate velocity in the intake and the size of the intake mouth. First a suitable value for c_d must be chosen.

The coefficient of velocity (c_d) has a value which decreases with the amount of turbulence induced by the intake. A sharp-edged and roughly finished intake will typically have a c_d as low as 0.6 whereas a carefully finished aperture will have a c_d of 0.8. In this example we adopt 0.6 for c_d; in practice it is necessary to consult other books or a specialist adviser, to choose a value suitable for the shape and finish of the intake mouth you are proposing to build.

Next choose a suitable value for the water depth in the headrace ($h_{h\,(normal)}$) during normal flow conditions. If we would like to see the surface of the water in the headrace at the same level as the upper edge of the intake orifice, then $h_{h\,(normal)}$ = d, the depth of the intake mouth. Let's say we will set d to be 0.2 m, therefore $h_{h\,(normal)}$ is also 0.2 m.

Finally, calculate $h_{r\,(normal)}$. The first guess for h_{weir} is 1 m, and during normal conditions $h_{over\text{-}top}$ is approximately zero.

$$h_{r\,(normal)} = h_{weir} + h_{over\text{-}top} = 1.0 + 0 = 1.0\text{ m}$$

$$v_i = c_d \sqrt{2\,g(h_r - h_h)} = 0.6\sqrt{2 \times 9.8 \times (1.0 - 0.2)} = 2.4\text{ m/s}$$

Check that this velocity is not excessive; continuously maintained velocities greater than 4 m/s will erode the edges of the intake mouth. To reduce velocity lower the weir crest, or raise h_h, or raise the position of the intake to reduce h_r. Remembering that d is set to 0.2 m:

$$Q_{gross} = A_i v_i = d \times w \times v_i$$

$$w = \frac{Q_{gross}}{d \times v_i} = \frac{0.132}{0.2 \times 2.4} = 0.28\text{ m}$$

4 Headrace slope and width, normal flow

The velocity of the water in the headrace must be sufficiently high to sweep debris and sediment through to the spillway and silt basin. Choose a suitable value for headrace velocity (v_h), say 2.0 m/s. The headrace must slope downwards sufficiently to maintain this velocity while the water depth ($h_{h\,(normal)}$) stays at the value chosen, in this case 0.2 m. If the width of the headrace channel is w_h and the cross-sectional area is A_h, we can write:

$$Q_{gross} = v_h A_h = v_h h_{h\,(normal)} w_h$$

$$w_h = \frac{Q_{gross}}{v_h h_{h\,(normal)}} = \frac{0.132}{2.0 \times 0.2} = 0.33\text{ m}$$

continued opposite

Flood containment, intake and headrace **Example 3.3.1** *continued*

The slope of the headrace is found using the following equation (Manning's equation) which is introduced in Section 3.6:

$$S = \left(\frac{n \times v_h}{R^{0.667}}\right)^2 \qquad \text{where } R = \frac{w_h\, h_h}{w_h + 2h_h}$$

'n' is a 'roughness' value for the material of the headrace; the value for concrete is 0.015.

$$R = 0.090 \qquad S = \frac{0.015 \times 2.0}{0.090^{0.667}} = 0.022 \text{ m/m}$$

This value for slope tells us that if the silt basin is placed 15 metres distant from the intake, it should be dropped below the level of the intake by a height of 0.022 × 15 = 0.3 metres.

5 Headrace flood flow

Flood conditions will increase the flow in the headrace. The water depth ($h_{h\,(flood)}$) will usually increase also, although an increase in headrace velocity during floods will offset this.

To calculate $h_{h\,(flood)}$ it is necessary to take several guesses for $h_{h\,(flood)}$ and each time calculate the flood flow in the channel using the two equations below. (The first one is the discharge orifice equation and the second is Manning's equation.) The correct value of $h_{h\,(flood)}$ is known when both equations give approximately the same result.

equation 1 $$Q_{flood} = d \times w \times c_d\sqrt{2g\left(h_{r\,(flood)} - h_{h\,(flood)}\right)}$$

equation 2 $$Q_{flood} = h_{h\,(flood)} \times w_h \times \frac{S^{0.5}}{n} \times \left(\frac{w_h\, h_{h\,(flood)}}{w_h + 2h_{h\,(flood)}}\right)^{0.667}$$

For instance we could try 0.4 m as a first guess for headrace water depth. Remember in Example 3.3.1 we found the river height in flood to be 3.8 m.

equation 1 $$h_{r\,(flood)} = h_{weir\,(flood)} + h_{over\text{-}top\,(flood)} = 1.0 + 2.8 = 3.8 \text{ m}$$

$$Q_{flood} = 0.20 \times 0.28 \times 0.6\sqrt{2 \times 9.8 \times (3.8 - 0.4)} = 0.27 \text{ m}^3/\text{s}$$

equation 2 $$Q_{flood} = 0.4 \times 0.33 \times \frac{0.022^{0.5}}{0.015}\left(\frac{0.33 \times 0.4}{0.33 + 2 \times 0.4}\right)^{0.667} = 0.31 \text{ m}^3/\text{s}$$

As a second guess, try $h_{h\,(flood)} = 0.35$ m:

equation 1 $Q_{flood} = 0.28 \text{ m}^3/\text{s}$

equation 2 $Q_{flood} = 0.28 \text{ m}^3/\text{s}$

Since both equations agree we can make two conclusions:

- The wall height of the headrace upstream of the spillway must be about 0.4 m (see Fig 3.3.8) in order to contain the flood flow of 0.35 m depth. Note that the water depth under normal flow conditions should also be checked against wall height.
- The flow in the channel in the worst flood will be around 0.28 m³/s, about twice the design flow. The spillway (see Example 3.4.1) will have to remove the excess flow, which is $Q_{flood} - Q_{gross}$.

6 Design parameters

Repeat the above steps with revised values for L_{weir}, h_{weir}, d, $h_{h\,(normal)}$, v_h, in order to achieve a satisfactory design. Cost your design and revise the dimensions to reduce overall cost.

continued overleaf

Example 3.3.1 *continued* **Flood containment, intake and headrace**

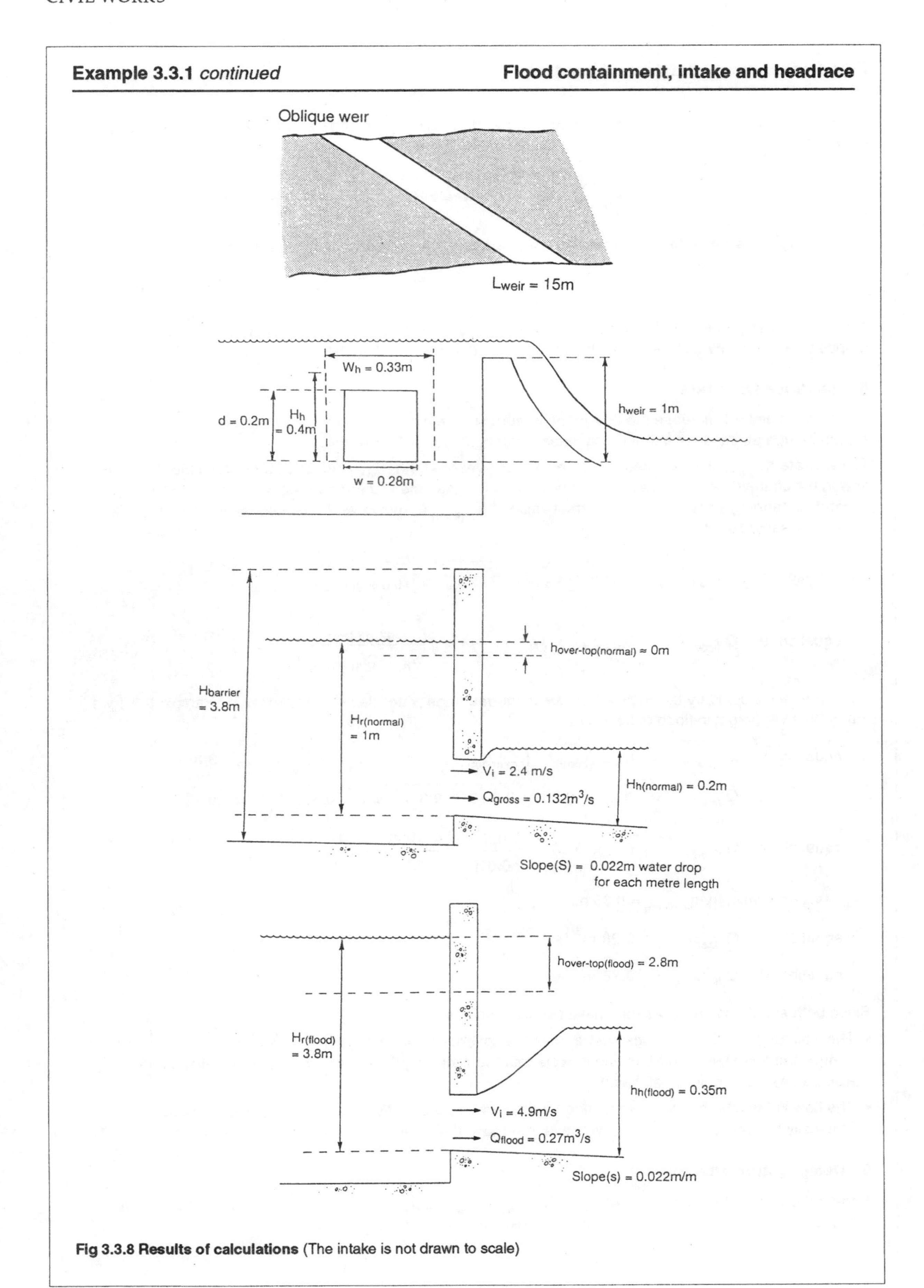

Fig 3.3.8 Results of calculations (The intake is not drawn to scale)

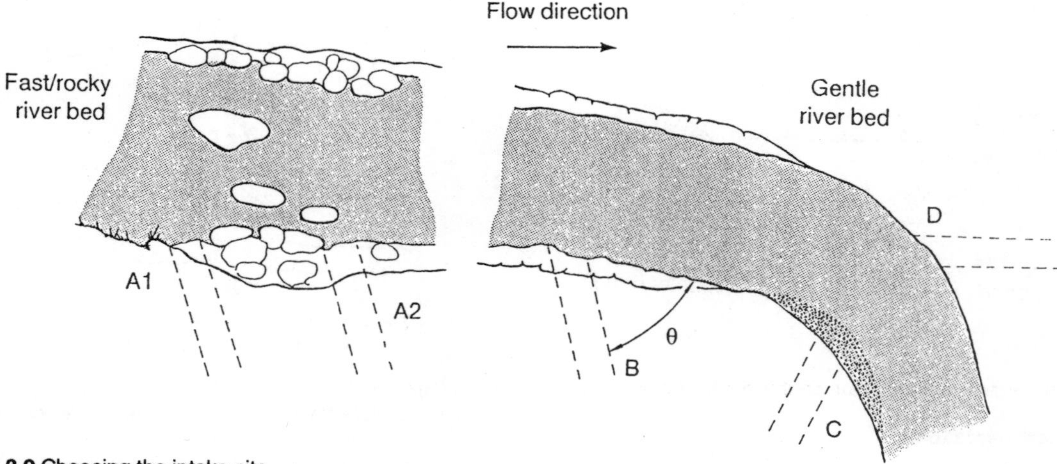

Fig 3.3.9 Choosing the intake site

Avoiding sediment concentrations

Silt passing through the intake can be removed by silt basins in the channel. Nevertheless, rivers carrying a heavy silt load may overburden the silt basins and clog the intake mouth. The intake and weir should be positioned to avoid this.

Fig 3.3.9 shows a river in aerial view. A mountainous and rocky section is shown, followed by a gentler section and a bend. Suppose that intake sites must be chosen somewhere along these stretches of river.

An intake sited in the bend at point C is unsuitable, since silt tends to deposit on the inside of a bend. This is due to a spiral flow effect under the surface of the water, as shown in Fig 3.3.10.

An intake in the straight section at B draws water from the river, so creating a bend in the flow itself. This will also set up a spiral flow which may also cause silt clogging of the intake (Fig 3.3.10 a). One method of avoiding this effect is to deflect the flow on the other side of the river with a 'deflecting groyne', as shown in Fig 3.3.10 b. This can be added at a later stage once it is established whether silting of the intake is a problem. It may be sufficient to ensure that the angle θ is either very small, to avoid spiralling, or alternatively more than 80°, so that the side intake principle is used.

Site D is not recommended since it is liable to damage from boulders rolled downstream by flood flows, and it will collect floating debris very rapidly. Nevertheless it could be a suitable site from the point of view of avoiding sediment (refer to the study by Hildebrandt listed in Appendix 3, Bibiliography).

If the bed is loose and sandy at B, a trench type weir (Fig 3.3.12) could be designed and costed before taking a decision. If the bed is sound, a permanent or a temporary weir could be constructed. Temporary weirs are discussed below.

Sites A1 and A2 may pose difficulties with respect to access and works on a steep embankment. Even when an intake is built, operation and maintenance could be problematic in high flow conditions. A1 is suitable for the placement of a weir anchored on boulders, assuming there is no danger of their movement in a flood. A2 offers the advantage of natural protection of the intake mouth.

Temporary weirs

Usually it is sensible to adopt traditional water management techniques known to local people, and possibly methods of constructing temporary weirs are among these. The principle of the temporary weir is to construct a simple structure at low cost using local labour, skills, and materials. It is expected to be destroyed by annual or bi-annual flooding. Advanced planning is made for rebuilding of the weir whenever necessary. Fig 3.3.11 shows an example of construction methods in placement of temporary weirs.

Skimmers

An aid to a hydro scheme is a skimmer which protects the intake from floating debris. This is useful during low flow seasons when the surface of the water may drop below the intake mouth roof. The skimmer is usually a wooden pole which itself floats on the water surface.

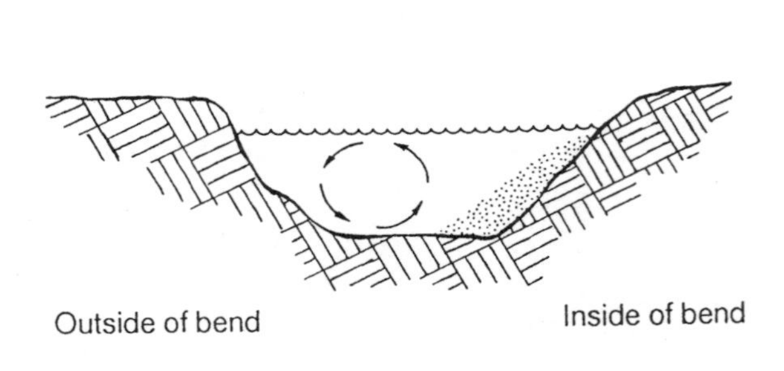

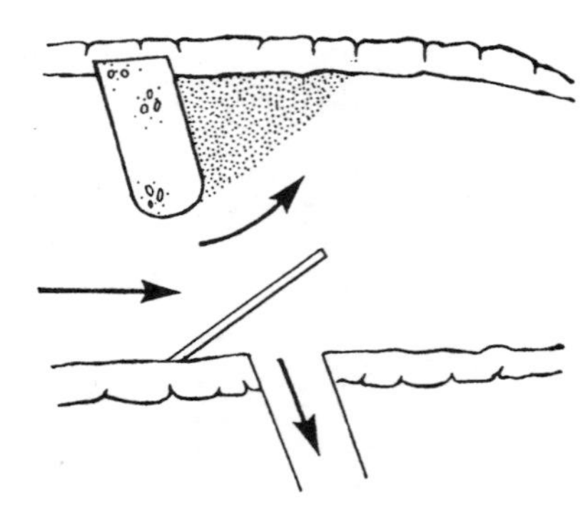

a Spiral flow causes sedimentation on the inside of a bend

b Deflecting groyne constructed from concrete or gabions. The skimmer deflects floating debris from the intake area.

Fig 3.3.10 Sedimentation

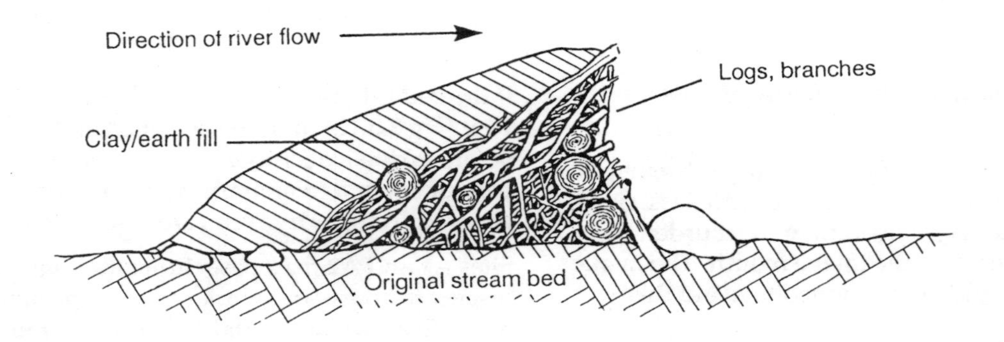

Fig 3.3.11 Example of a temporary weir. Such weirs are usually destroyed by floods and then rebuilt. Refer to local irrigation practice.

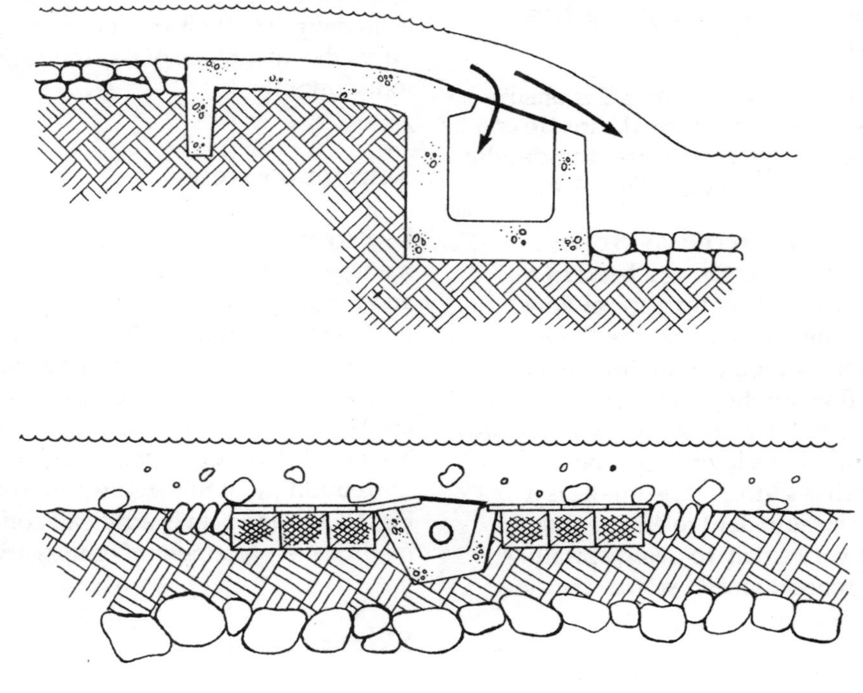

Fig 3.3.12 Trench intake. Fast flowing mountain streams are often laden with excessive silt loads and larger debris such as stones, grasses and branches. The intake rack shown here allows these objects to flow over it, the water velocity itself acting to keep the rack clean. Any debris which collects on the rack during low flow periods will be washed off in high flow periods. If such intakes are designed carefully, with correct mesh size, slope angle, evenness of slope, width, etc, they can operate in difficult conditions for years on end without attention.

3.4 Spillways

Fig 3.4.1 shows the flood spillway in greater detail together with flow control and channel emptying gates. In Example 3.3.1 it was shown that, when the river is in flood, flows in the channel (Q_{flood}) can be twice the normal channel flow (Q_{gross}). The spillway must be large enough to remove the excess flow ($Q_{spillway}$):

$$Q_{spillway} = Q_{flood} - Q_{gross}$$

The spillway is a channel flow regulator. In addition, it is combined with control gates to provide a means of channel emptying.

In certain circumstances it is essential to stop the channel flow quickly, for instance if a break in the channel wall has occurred downstream causing progressive channel collapse and foundation erosion.

Often emergencies will occur when the river is in flood. The control gates must be placed at a distance from the stream, to allow easy access when the river is in flood. It is sensible not to rely on one channel-emptying mechanism, but to have two, since a gate can fail to operate when needed, especially if not often used.

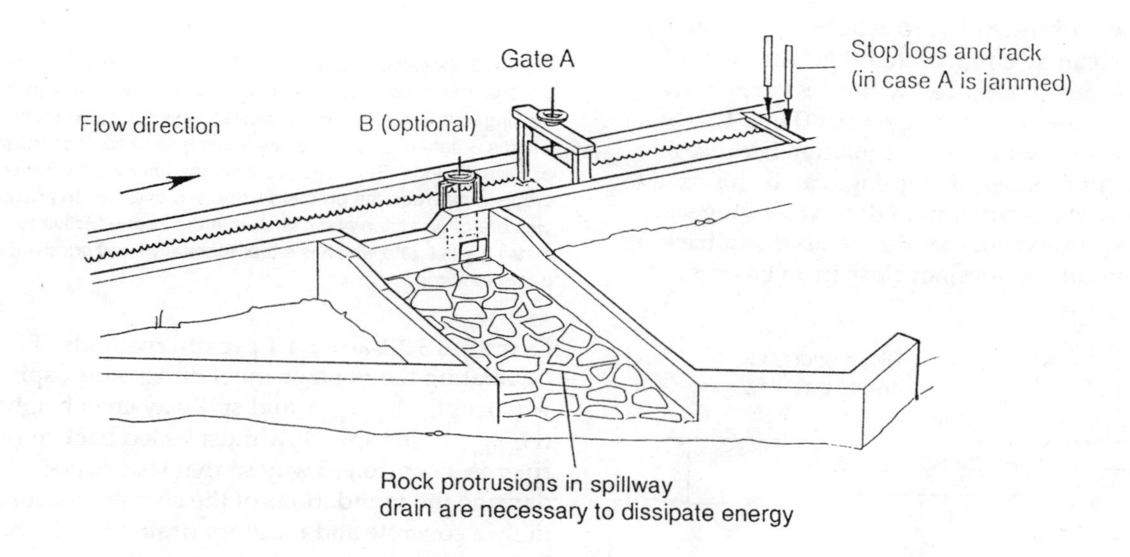

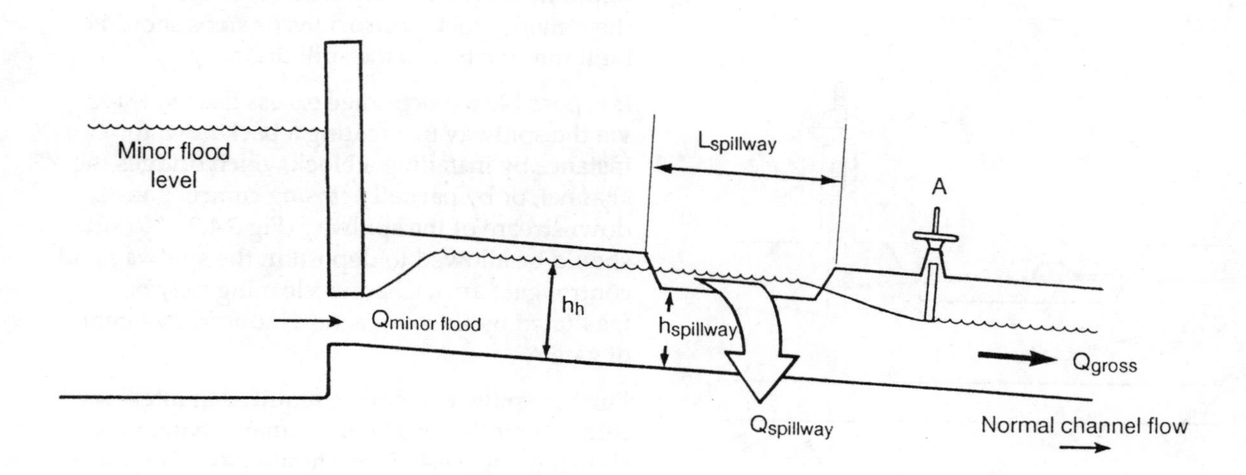

Fig 3.4.1 Flood spillway and flood control gates. Always place at a distance from flood activity. Gate B is an optional extra gate to reduce channel flow in case of failure of other methods during damaging flood flows. 'Stop-log' methods can be used to reduce costs.

In general, gates are more easily opened than closed, so gate B in the figure could be more reliable than gate A. Gate B will not stop flow but only reduce it, especially if flow obstruction is placed just downstream (such as the stop logs shown). Gate B only needs to be fitted in circumstances where reliable channel closure is considered a high priority – this includes cases where expensive channel works downstream are prone to damage from excess channel flows. If gate A is used to stop channel flow, the immediate effect will be for the full flood flow (Q_{flood}) to pass over the spillway. The flood channel walls and width must be sufficient to accommodate the full back flow in this case. Since A is a variable orifice size gate, it is also a useful mechanism for regulating the flow as shown in Fig 3.4.2 and also to achieve low flows for turbine part-flow testing.

In low-cost micro-hydro schemes gates such as shown can be eliminated altogether. Instead, gate A can be replaced with a less expensive stop-log and rack arrangement (Fig 3.4.3). Note that this provides flow regulation because a different number of stop-logs can be inserted to introduce a restriction to flow of varying size. This arrangement can also be used as a back-up in case gate A does not close in an emergency.

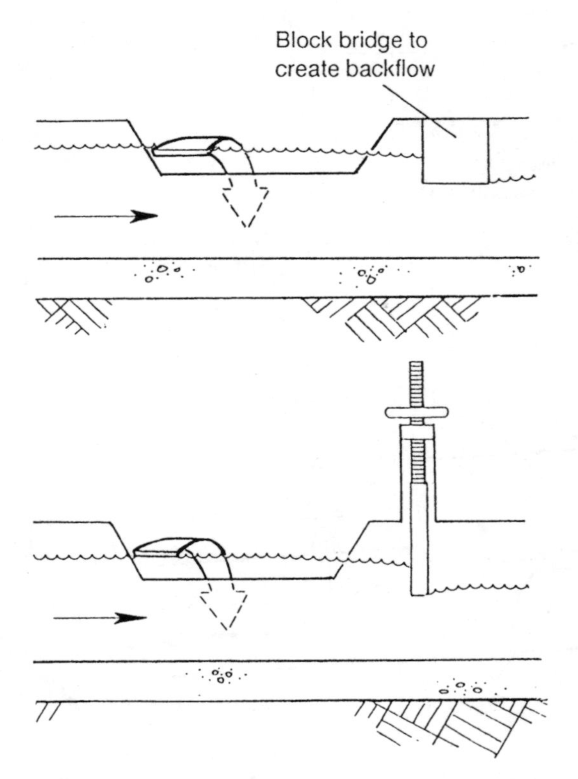

Fig 3.4.2 Regulating channel flow. Both these methods ensure that the channel flow is not more than Q_{gross} and so protect it from excessive loading. Partial closure of the channel with vertical stop-logs can also be used to achieve the same effect as partial closure of a gate.

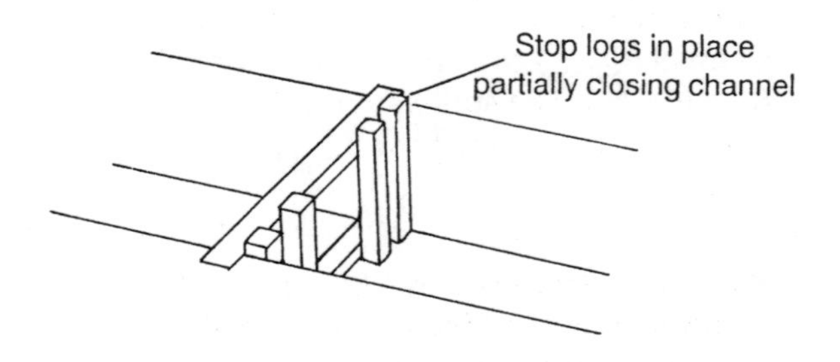

Fig 3.4.3 Stop-log and frame These can either fully close the channel or partially close it to regulate flow. The logs can be hinged or wired to the top rack to protect against theft. Chains are then attached to the lower end of the logs to lower or raise the logs. Stop logs are also often horizontal timbers dropped between the grooved sides of a channel. In practice logs are often stolen and it is often more cost-effective to install a steel gate which is either manually lifted or raised on a threaded rod.

Examples 3.3.1 and 3.4.1 present methods of calculating the flood channel dimensions, spillway length ($L_{spillway}$), and spillway crest height ($H_{spillway}$). The spill flow must be led back to the river in a controlled way so that it does not damage the foundations of the channel. Accordingly a concrete and masonry drain must be built to provide a resistant platform for the spill flow. Rapid flows can be very erosive; to dissipate their energy rock protrusions or steps should be built into the bed of the spill drain.

It is possible to encourage excess flow to leave via the spillway by creating a back flow: for instance by installing a block which bridges the channel, or by partially closing control gate C, downstream of the spillway (Fig 3.4.2). No silt should be allowed to deposit in the spillway and control gate area. Channel cleaning may be facilitated by flushing action induced by opening of gate A.

Further spillways may be required at intervals throughout the length of the main channel as shown in Fig 3.4.4. They should have durable drains constructed to lead spilt water away from the channel foundations. The cost of inclusion of these spillways and drains is considerably less than the cost of slope erosion and channel damage caused by an overflowing channel.

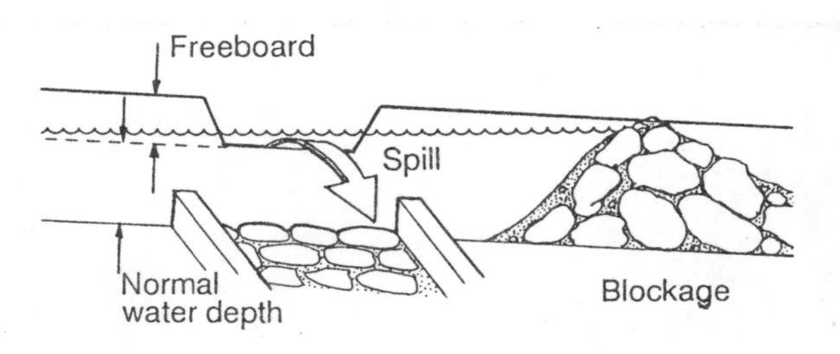

Fig 3.4.4 Spillways along the channel. Identify channel sections which are vulnerable to blockage. Identify areas where unwanted spills will cause damage to the hillside or to the channel walls. Protect these areas with emergency spillways, with additional freeboard or with 'capping' (Fig 3.6.1) to prevent any debris from entering the channel.

Photo 3.4.1 A block bridge and channel spillway.

Flood spillway dimensions — Example 3.4.1

Following Example 3.3.1, the flood spillway can be sized on the basis of the headrace water depth of 0.2 m at normal flows.

With reference to Fig 3.4.1, the height of the spill crest ($h_{spillway}$) should be aligned to the normal flow surface level or water depth.

$$h_{spillway} = h_{h\,(normal)} = 0.2\text{ m}$$

The spillway length ($L_{spillway}$) is found from the standard weir equation. It must be long enough to pass the spill flow (Q_{spill}) with the available excess head of water behind it. The spillway length is not found from worst flood conditions but from minor floods, when the headrace flow has only risen by 15% above its normal value. (This is because a longer spill length is needed when there is a smaller head driving water over the spill crest.)

Standard weir equation:

$$Q = c_w \times L_{weir} \times \left(h_{over\text{-}top}\right)^{1.5}$$

In this case:

$$Q_{spill} = Q_{minor\,flood} - Q_{gross} = c_w \times L_{spillway} \times \left(h_{minor\,flood} - h_{spillway}\right)^{1.5}$$

Adopt a value for the weir coefficient for spillways (c_w) of 1.6:

$$L_{spillway} = \frac{Q_{minor\,flood} - Q_{gross}}{1.6\left(h_{minor\,flood} - h_{spillway}\right)^{1.5}}$$

For a minor flood flow of 10% above normal flow:

$$Q_{minor\,flood} = 1.15 \times Q_{gross} = 1.15 \times 0.132 = 0.152\text{ m}^3/\text{s}$$

If we assume that the minor flood water travels at the same velocity as normal flows, the water depth for this increased flow will also be increased by the same ratio:

$$h_{minor\,flood} = 1.15 \times h_{spillway} = 1.15 \times 0.2 = 0.23\text{ m}$$

$$L_{spillway} = \frac{0.152 - 0.132}{1.6\,(0.23 - 0.20)^{1.5}} = 2.4\text{ m}$$

The calculation is very approximate since the water surface level behind the spillway will vary. For this reason and to ensure operation if partially blocked, a factor of safety of 1.5 is used, giving a length of 3.6 m. It is wise to fit a backflow device such as shown in Fig 3.4.3. This will allow more careful regulation of the channel flow than the spillway on its own. The bibliography (Appendix 3) lists references (eg Chadwick) which give a fuller explanation of flow over side weirs.

3.5

Silt basins and forebay tanks

The water drawn from the river and fed to the turbine will carry a suspension of small particles of solid matter. This 'silt load' will be composed of hard abrasive materials such as sand and will cause expensive damage and rapid wear to the turbine runners. To remove this material the water flow must be slowed in silt basins so that the silt particles settle on the basin floor where the deposits can be periodically flushed out to make room for further deposition. It is necessary to settle out the sediment both at the channel entry and at the penstock entry, or forebay tank (Fig 3.1.2).

Fig 3.5.1 shows a simple design for a silt basin at the channel entry, and Fig 3.5.2 shows a design for a forebay basin. Both basins must follow five important principles:

1 They must have length and width dimensions which are large enough to cause settling of the sediments but not so large that the basins are over-expensive and bulky.

2 They must allow for easy flushing out of deposits, undertaken at sufficiently frequent intervals.

3 Water removed from the flushing exit must be led carefully away from the installation. This avoids erosion of the soil surrounding and supporting the basin and penstock foundations. A walled and paved surface similar to a spillway drain will do this.

4 They must avoid flow turbulence caused by introduction of sharp area changes or bends, and they must avoid flow separation.

5 Sufficient capacity must be allowed for collection of sediment.

The designs shown are not necessarily the most suitable for your purpose. Many variations are possible, but they must all satisfy the design principles.

Separated flow and turbulence

Two effects must be avoided in the design of silt basins. These are turbulence and flow separation. Fig 3.5.3 a shows an incorrect design which encourages both effects. Turbulence must be avoided since it stirs up the silt bed load and maintains silt in suspension.

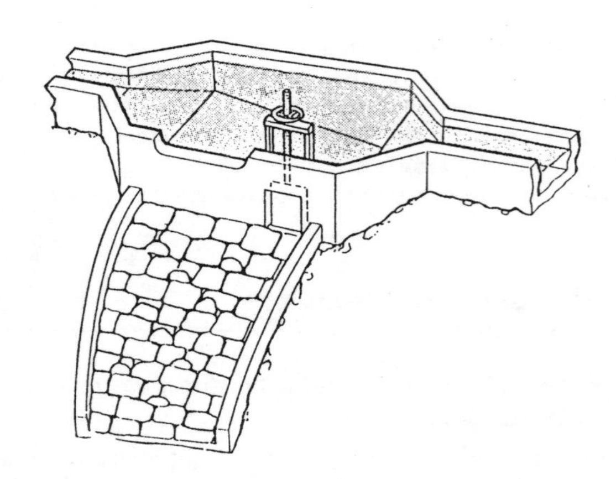

a The sluice gate is opened for periodic flushing of silt accumulation. Less costly alternatives to the sluice gate are shown in Fig 3.5.9. Notice that the opportunity is taken of incorporating a regulating spillway into the silt basin.

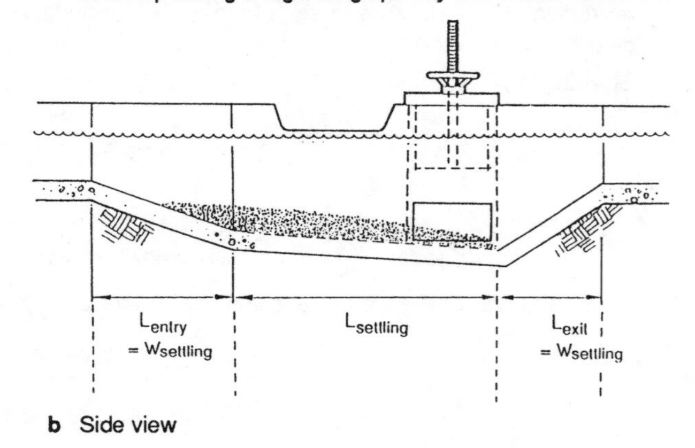

b Side view

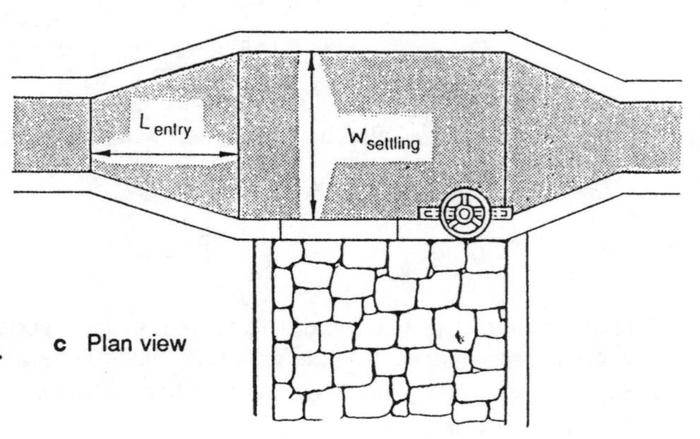

c Plan view

Fig 3.5.1 Silt basin at channel entry

Flow separation (sometimes loosely known as 'slipstreaming') is the tendency for a body of water to move quickly through the basin from entry to exit, carrying a sediment load with it. Fig 3.5.3 **b** shows the entry and exit profiles needed to avoid this, and Fig 3.5.4 shows the same principle with application to the depth of the silt basin.

a Notice that the paved drain is directed away from penstock supports. Note that the sluice gate can be replaced with a flushing pipe.

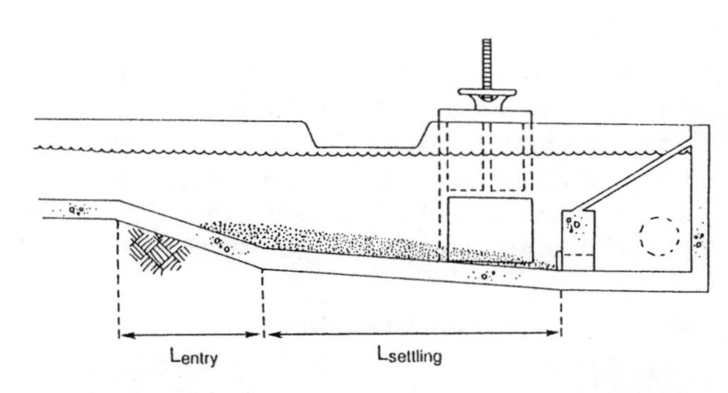

b Side view

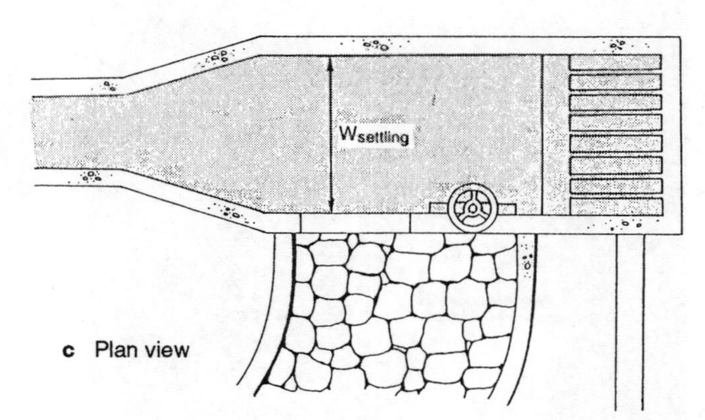

c Plan view

Fig 3.5.2 Forebay tank consisting of silt basin and trashrack, and spillway

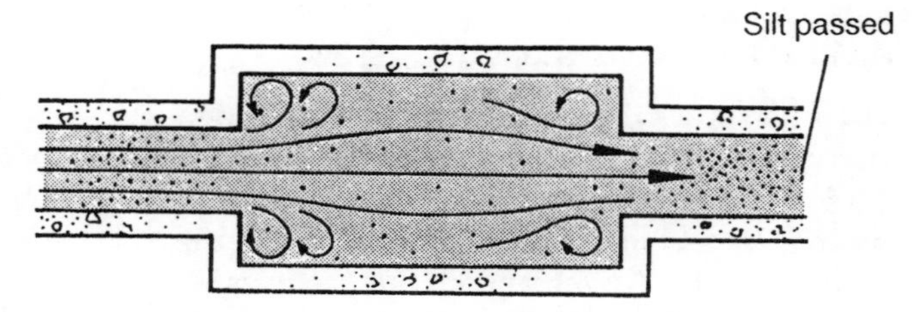

a Incorrect: high velocity in centre stream and turbulence in corners

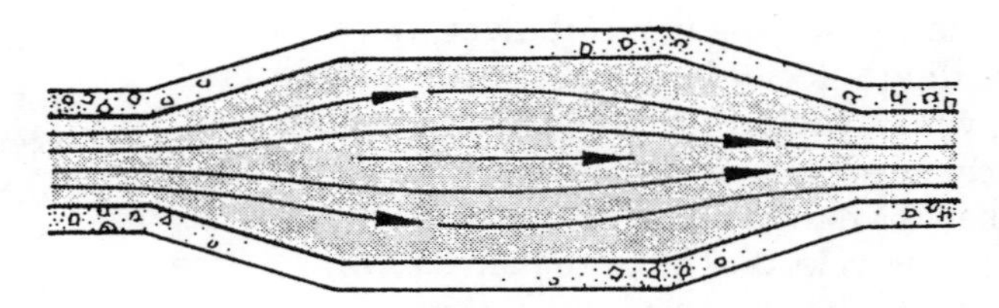

b Correct: low velocity throughout width, no turbulence. Entry and exit lengths should be at least as long as the basin is wide.

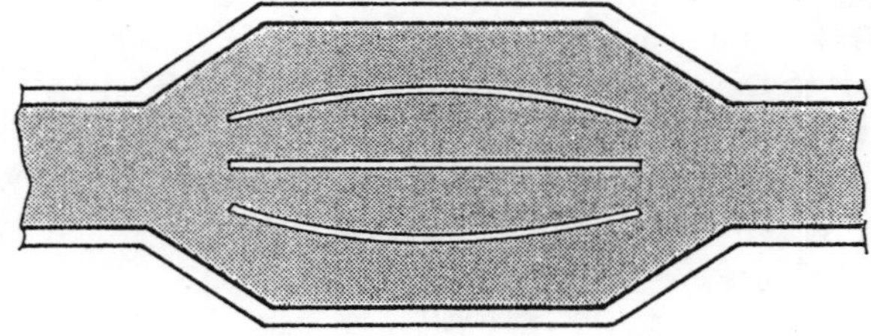

c One method of avoiding slipstreaming is to install flow distribution walls

Fig 3.5.3 Separated flow and turbulence

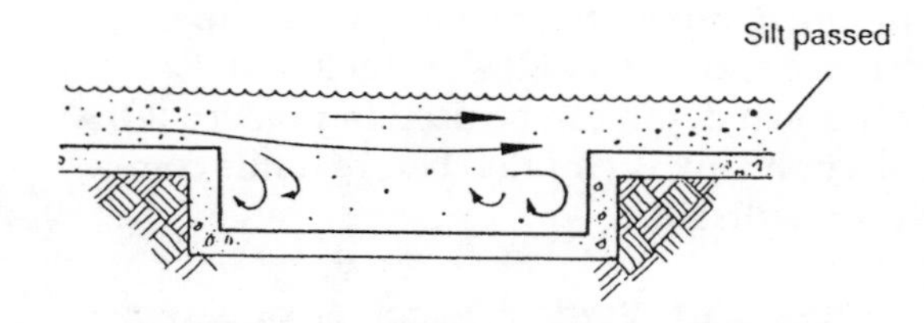

a Incorrect: high surface velocity and turbulence in corners.

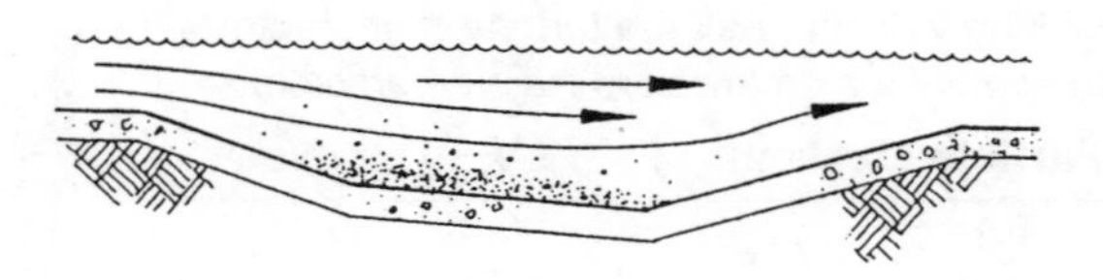

b Correct: low velocity throughout

Fig 3.5.4 Flow separation and turbulence

Width and length of settling areas

Fig 3.5.5 shows three side views of the basins.

The complete length of the basin is divided into three portions, L_{entry}, $L_{settling}$ and L_{exit}. The central portion is the settling area. The settling length ($L_{settling}$) and width ($W_{settling}$) is also shown on Figs 3.5.1 and 3.5.2. The depth of the basin is divided into two portions, $D_{settling}$ and $D_{collection}$.

Fig 3.5.5 shows the paths taken by suspended particles. While the collection tank is filling (Fig 3.5.5 **a**), the effective cross-sectional area of the basin is larger, so slowing the flow. When the tank is full (Fig 3.5.5 **b** and **c**), the particles will travel forwards faster. The speed at which the particles drop vertically depends on their size, shape, density and the extent of turbulence in the water. Assuming non-turbulent water, the settling velocity ($v_{vertical}$) of small particles is given in Table 3.5.1. In most micro-hydro schemes it is sufficient to remove particles bigger than 0.3 mm in diameter, which have nominal settling velocities of more than 0.03 m/s. Nevertheless, always consult the turbine manufacturer.

To design a silt basin, simply choose any width ($W_{settling}$) which will be practical to build and which is between 5 and 15 times the width of the channel. It is usually sufficient for $D_{settling}$ to be equal to the channel water depth (this is explained in Note 3.5.1). The length of the basin ($L_{settling}$) is then found from this equation:

$$L_{settling} = \frac{Q}{W_{settling} \times v_{vertical}}$$

If the length which results is inconvenient to build, then adjust the value of $W_{settling}$ until both width and length are convenient. Your design should carefully avoid turbulence and slipstreaming. The design approach given here is simplified, but is effective so long as tapered entrance and exit lengths are included. Additional depth can be used in practice if the collection tank is emptied before it fills completely with silt.

Table 3.5.1 Vertical velocities of particles

Use this table to find the relevant vertical sinking velocity. Ask the turbine manufacturer to specify maximum permissible particle size.

Particle size mm	$V_{vertical}$ m/s
0.1	0.02
0.3	0.03
0.5	0.1
1.0	0.4

Collection tank

Suspended matter content in a river section varies with season and with the type of terrain upstream. The variation over the year can be very large; for instance, the pruning of the tea bushes in a Sri Lanka tea estate causes heavy silting of the streams for a few weeks each year. Seasonal floods cause high velocities and turbulence, so raising silt load enormously. The silt basins should be designed to work during periods of high silt load or 'turbidity' (see Example 3.5.2). Turbidity is expressed as the weight of silt carried per m^3 of water.

A simple method of measuring silt load is to fill a bucket from the stream a number of times, each time in a different place or at a different depth, letting the water stand until silt is settled out, and then separating and weighing solid matter, and recording the volume of water processed. This is a very uncertain process unless there is some confidence that the worst part of the year is chosen. Discussions with local farmers, and samples taken from irrigation channels, will be very helpful. If other hydro schemes function nearby, it is essential that their operators are questioned about the frequency at which basins are emptied. Design work can then be based on observation of existing basin design and performance.

Tests should also be made on the silt content of tail water from turbines, and on the water entering the forebay tank. This is easily done by filling a graduated flask and then waiting for the silt content to settle out. These tests will help monitor the effectiveness of the silt basins.

Photo 3.5.1 A partitioned silt basin. (Nepal)

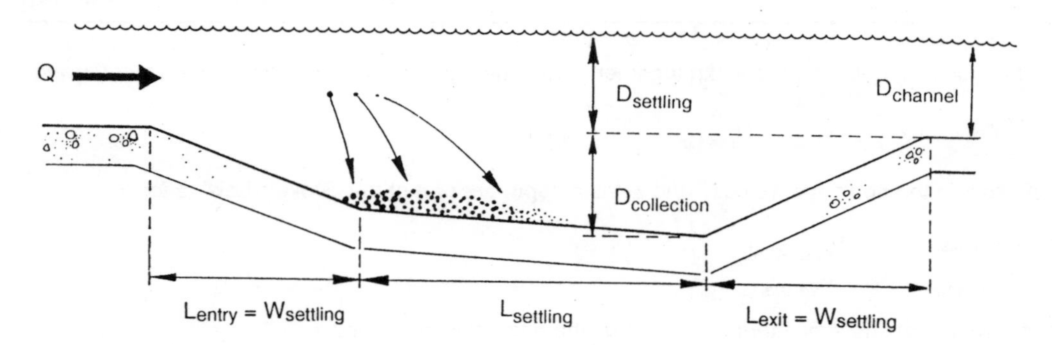

a The collection tank is empty just after the basin is sluiced. The silt particles begin to collect again, the lightest ones falling at the end of the basin.

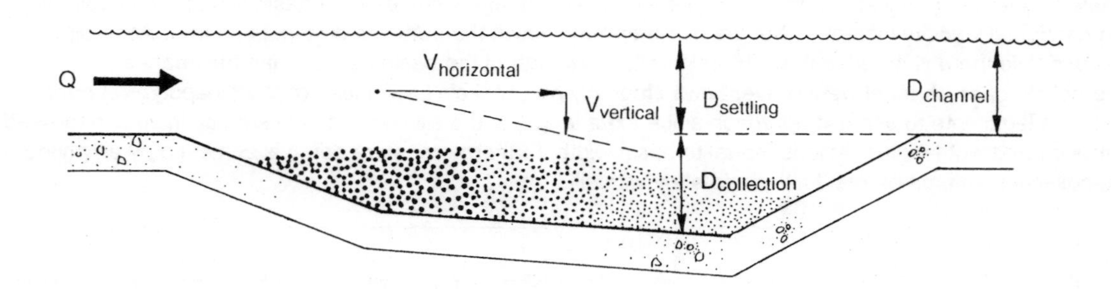

b After a period of time the collection tank is full of sediment. The lightest particles still fall at the end of the basin.

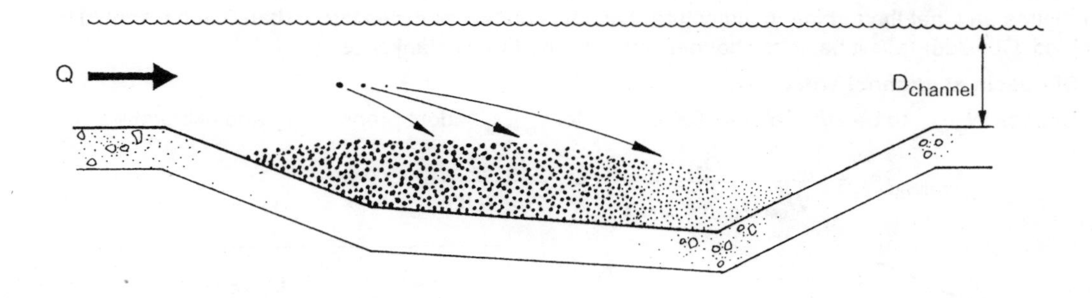

c The speed at which the lightest particle drops is $v_{vertical}$.

Note the ratio $\frac{L_{settling}}{D_{settling}} = \frac{v_{horizontal}}{v_{vertical}}$

The design equation for both length and width is

$$L_{settling} = \frac{Q}{W_{settling} \times v_{vertical}}$$

but it is also essential to use tapered entrance and exit lengths as shown. These remove turbulence which would make the basin ineffective. They also provide additional settling length, so giving a safety margin to ensure silt removal.

Fig 3.5.5 Sizing of silt basin

Note 3.5.1 **The silt basin design equation**

The depth of the settling basin and its width together determine the horizontal velocity of the water flowing through it.

$$Q = v_{horizontal} W_{settling} D_{settling}$$

Fig 3.5.5 shows that the horizontal velocity and settling depth are part of the 'Stoke's Law' ratio:

$$\frac{L_{settling}}{D_{settling}} = \frac{v_{horizontal}}{v_{vertical}}$$

Solving these two equations simultaneously gives us the basin design equation

$$L_{settling} = \frac{Q}{W_{settling} \times v_{vertical}}$$

Following this principle (Stoke's Law), the lightest silt particle will always arrive at the floor of the basin whatever the depth of water in the basin. As the collection tank volume of the basin fills up with sediment, the horizontal velocity increases but the depth through which the particles drop gets shorter. According to this principle there is no advantage in making the flow path of the basin deeper than the channel.
In practice, open channel water never flows smoothly and Stoke's Law does not strictly apply. Nevertheless, it is legitimate to use this approach since extra length in the silt basin is achieved by including tapered entrance and exit lengths, at least equal to basin width; furthermore, extra depth is achieved by emptying the collection volume before it fills completely with silt.

Example 3.5.1 **Sizing of silt basin**

Continuing Examples 3.3.1 and 3.4.1, find suitable silt basin sizes. Assume that the channel water depth is 0.5 metres and that the turbine manufacturer has told you that particles larger than 0.3 mm should be avoided. Consider (a) silt basin at channel entry, (b) the forebay tank size.

a **Silt basin at channel entry:** $Q_{gross} = 0.132 \text{ m}^3/\text{s}$

Choose $W_{settling}$ to be 2m. Refer to Table 3.5.1 for $v_{vertical}$. Adopt proper entry and exit profiles.

$$L_{settling} = \frac{Q}{W_{settling} \times v_{vertical}} = \frac{0.132}{2 \times 0.03} = 2.2 \text{ m}$$

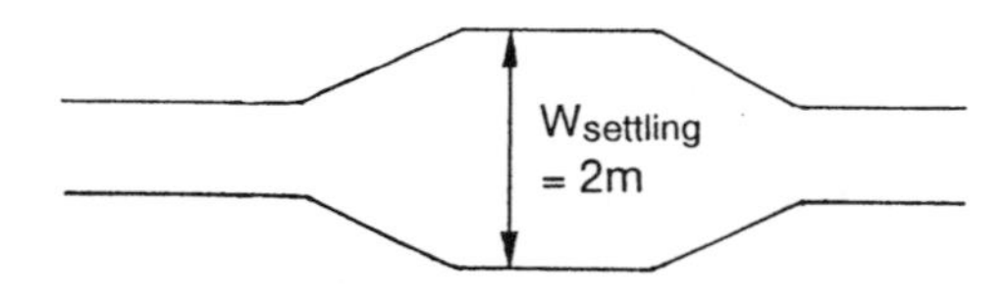

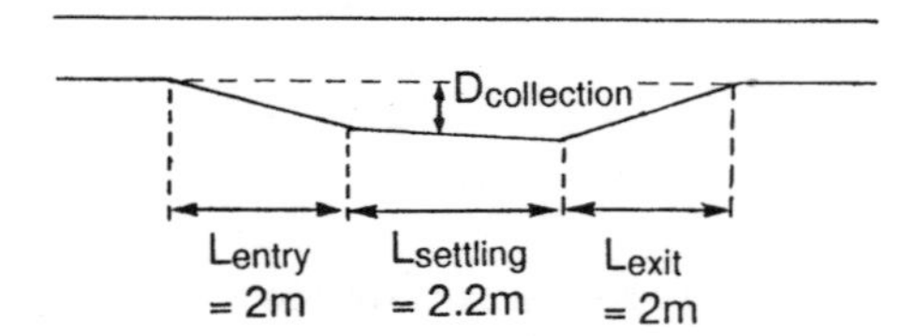

Note that $D_{settling}$ is equal to the channel depth but so far we have not designed the full depth of the basin which includes $D_{collection}$. This is done in Example 3.5.2. Note also that in order to avoid turbulence tapered entrance and exit lengths are needed. The design rule for these is to make them each equivalent in length to one basin width. The design will therefore have the dimensions shown here, with the value of $D_{collection}$ still unknown.

If the terrain is steep it may be more convenient to choose a narrower width, say 1.5 metres. This would result in a settling length of 2.9 m.

b **Forebay tank:** the flow through the forebay tank is the net flow, or turbine flow, which is less than Q_{gross} if irrigation draw off or seepage loss in the channel is significant. Nevertheless, it is safer to design the forebay tank for conditions where flow reduction does not occur, and adopt the value of Q_{gross}. Dimensions are therefore as above, although when we design the collection depth in Example 3.5.2 we may decide that less capacity is needed for the forebay, if silt entry into the channel is not expected to occur.

Collection tank capacity and basin emptying frequency **Example 3.5.2**

Continuing Example 3.5.1, specify the collection tank depth on the basis of a reasonable emptying frequency. Assume that discussions and measurements indicate a river carrying loads of 0.5 to 2 kg/m³ throughout the year.

A possible emptying frequency (T hours) during most of the year, when the carrying load (S) is 0.5 kg/m³, would be twice daily. In twelve hours the intake will suck in :

$$\text{Silt load} = Q \times T \times S = 0.132 \times 12 \times 3600 \times 0.5 = 2850 \text{ kg}$$

$$\text{Density of sand} = 2600 \text{ kg/m}^3\text{. Assume a packing density of 50\%}$$

$$\text{Volume capacity of silt} = \frac{1}{0.5} \times \frac{1}{2600} = 0.77 \times 10^{-3} \text{ m}^3/\text{kg}$$

$$\text{Required collection tank capacity} = 0.77 \times 10^{-3} \times 2850 \text{ kg} = 2.2 \text{ m}^3$$

Given basin dimensions of 2.0m ($W_{settling}$) and 2.2m ($L_{settling}$) this implies a collection tank depth ($D_{collection}$) of:

$$D_{collection} = \frac{\text{Tank capacity}}{W_{settling} \times L_{settling}} = \frac{2.2}{2.0 \times 2.2} = 0.5 \text{ m}$$

In practice, some extra volume is available due to tapering which allows a safety factor in this design calculation.

During some months of the year the carrying load may rise to 2 kg/m³, so demanding emptying of a 2.2 m³ tank 8 times each day, or every 3 hours. An alternative would be to increase the collection depth to 1 metre, but this may be unreasonable in practical terms. Another option is to double the basin width and length, if this is possible. The result will be an oversized basin in terms of the point of deposition of sediment, but a correctly sized basin in terms of feasible emptying procedures. If a large increase in width and length is not practical, a compromise should be reached: emptying frequency could be increased to 4 times daily, and surface area only doubled. During periods of exceptionally high turbidity above 2 kg/m³, very frequent emptying will be needed.

In practice, frequency of emptying will depend on observations made of collection tank filling time.

The forebay tank may, in theory, require a smaller collection capacity than the entry basin since the silt load of the channel should be low. In practice, silting of channel water from entry of debris into the channel can easily happen, or can result from failure of the entry basin. For this reason the forebay tank should be sized as large as the entry basin and should be emptied as frequently as necessary by observation.

Usually the silt basin can be emptied at a time when the turbine is not used. If the turbine is required to run continuously, a two-part silt basin can be devised (Fig 3.5.6). This allows channel flow to continue while one part of the silt basin is emptied. Each part can be sized to accommodate half the collection volume of silt and each part will normally take half the full flow. Care must be taken to ensure that the occasional passage of full flow does not disturb existing silt deposits – tapering is especially important.

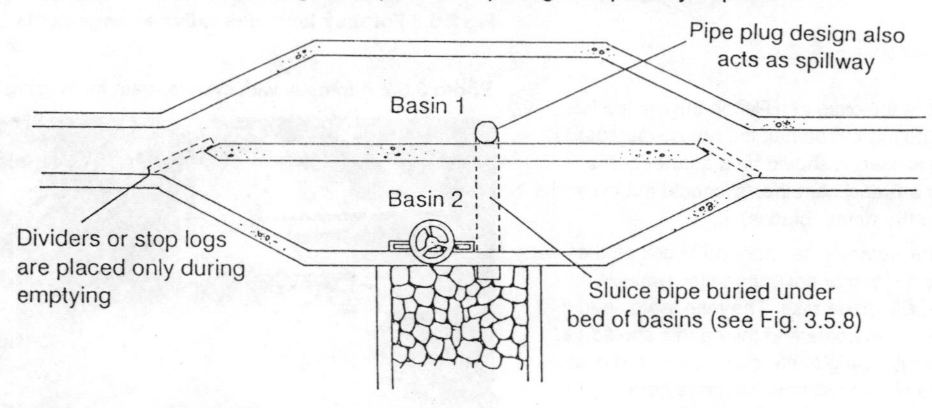

Fig 3.5.6 Partitioned silt basin to allow continuous flow to turbine while silt basin is emptied. Usually not necessary, since basins can be emptied with the turbine stopped.

Forebay tank

Fig 3.5.2 shows the forebay tank. The design of this tank follows exactly the procedure outlined above for silt basins in general, except that the exit portion is replaced by a trashrack and penstock entrance area. The dividing wall must be about 20 cm higher than the collection depth.

While designing a forebay tank it is important to keep the entrance to the penstock fully submerged. A rule of thumb is to submerge the penstock mouth by a distance below the water level of more than 4 penstock diameters. This is to prevent air being drawn into the pipe because of a vortex which can be formed on the surface if the penstock is closer to the surface.

It is necessary to install an air vent at the point where the penstock is joined to the forebay tank. The air vent will prevent damage to the penstock if for some reason the penstock mouth is blocked. If this happens, a low pressure is created inside the pipe which can make the penstock collapse inwards. An air vent will prevent this. The other reason for installing an air vent is to help remove air from inside the penstock during start-up. (In some forebay tank designs these bubbles can disturb silt deposits.) An air vent which is too small will be ineffective: if in doubt on this consult the Sourcebook (listed in Appendix 3, Bibliography).

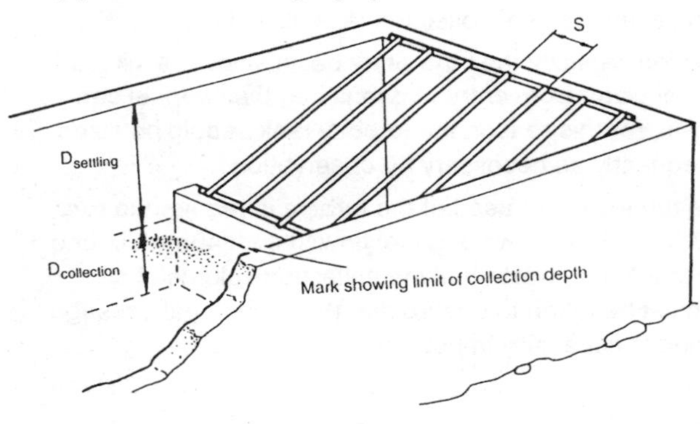

Trashrack spacing: in the case of a Pelton turbine the bar space 's' should not be more than half the nozzle diameter; where a spear valve is used, it should be a quarter of the nozzle diameter. For a Francis turbine, 's' should not exceed the distance between the runner blades.

Cleaning racks: either remove the rack and knock off the debris (two racks used together ensure one is always in place), or use a specially made rake. The rake teeth should be spaced to fit between the bars and the handle should be long enough to allow full raking of the rack. Racks are best placed at an angle as shown (60 to 80 degrees from horizontal) to help raking but also to allow gravity and buoyancy to keep the rack clear.

Fig 3.5.7 Forebay tank trashrack spacing and wall height

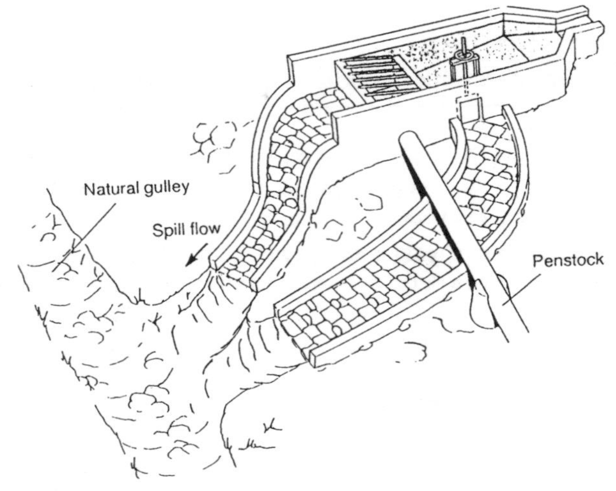

a Forebay tank showing alternative position of spillway in order to utilize a natural gulley as a drain. Sluice leakage is carefully directed away from penstock supports. Note that a side spillway next to the sluice gate could have been used as in Fig 3.5.2.

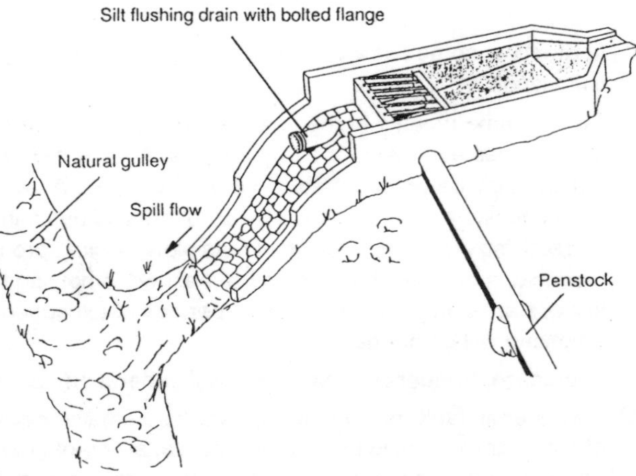

b A pipe can be used to flush silt forwards in order to avoid the risks of poor drainage affecting penstock support or slope stability. It may be a less costly option than **a**. The pipe can be blocked when not needed by water pressure on a front cap or board placed at its mouth, or as shown by an end cap.

Fig 3.5.8 Forebay tank alternative arrangements

Photo 3.5.2 A forebay with cylinder gate for sluicing. (Nepal)

Emptying of silt basins and power shut-down

Usually emptying is a fairly tedious process. The sluice gate is opened and then sediment is shovelled along the basin floor to direct it through the gate. The basin floor is slightly sloped to facilitate this.

During manual emptying the intake control gate must be closed so that the water flow stops. Since this means shutting down the turbine, it is necessary to empty at periods when power from the turbine is not needed. Usually it is possible to clear out the basins while the turbine is shut down for other reasons.

In some cases, for instance where very frequent emptying is needed (see Example 3.5.2), the power shut-down can become unacceptable. In these cases there are two recommended options – rapid emptying designs, and partitioning designs.

Rapid emptying consists of arranging the sluice gate so that when opened the outward rush of water sucks the bulk of the silt deposit with it. Fig 3.5.9 shows rapid emptying designs (both are proven and effective). Notice that the pipe plug design is especially convenient because it provides both a spillway and a sluice passage at the same time.

In both cases it is convenient to place a 'flush gate' immediately in advance of the silt basin. Once the basin has been flushed once, remaining deposits can be flushed again by closing this gate, allowing water to build up behind it, and then opening it quickly to wash through. An extra spillway is needed immediately before the 'flush gate'. (The gate should operate by quick lifting rather than by slow rotation of screw thread.)

These methods still require shut-down of power, even if only for a few minutes. To avoid this, it is possible to build a diversion channel around the silt basin which can carry flow while the basin is emptied. If there is a silt basin further downstream, for instance the forebay tank, then the passage of silt for a few minutes will not cause problems. The disadvantage of this approach is that the diversion channel may accidentally be left in operation for long periods, so running the risk of eroding the turbine. A safer option is a partitioning basin, as shown in Fig 3.5.6. The operation of such a basin is explained in Example 3.5.2.

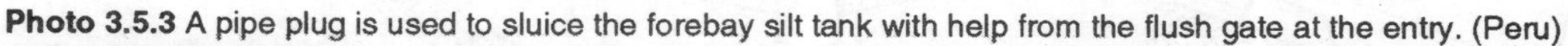

Photo 3.5.3 A pipe plug is used to sluice the forebay silt tank with help from the flush gate at the entry. (Peru)

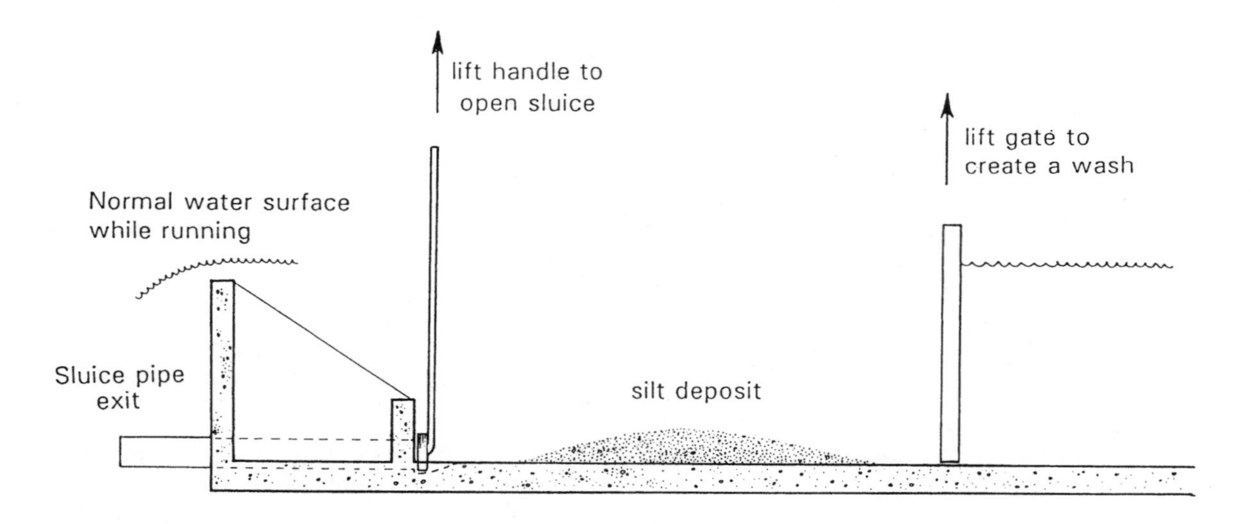

a A simple but effective and proven method of sluicing.

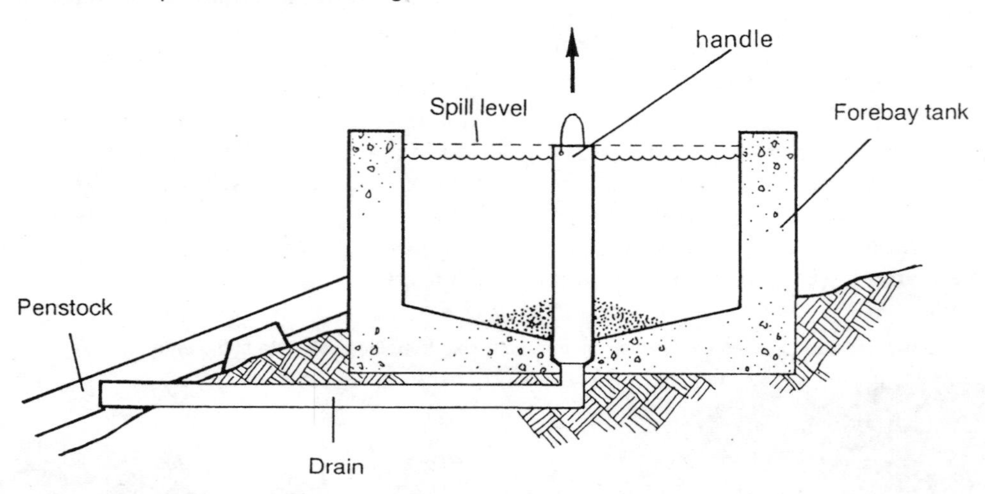

b Pipe-plug design for sluicing. Note the plug also acts as a spillway.

Fig 3.5.9 Rapid-emptying silt basins

Baffled tanks

It is often difficult to find space to build silt basins and the temptation is to reduce their size by incorporation of baffles. Fig 3.5.10 shows an incorrect baffle tank design. If baffles are narrowly spaced the cross-sectional area of flow will be small and high velocities will result, so that silt fails to settle. Sharp bends create turbulence and therefore further suspension of particles, and also can create slipstreaming effects. Baffled tanks are also often harder to sluice clear of silt deposits. Because of these dangers it is best to consult a specialist with experience of baffle tank designs which have been proven in practice to be effective.

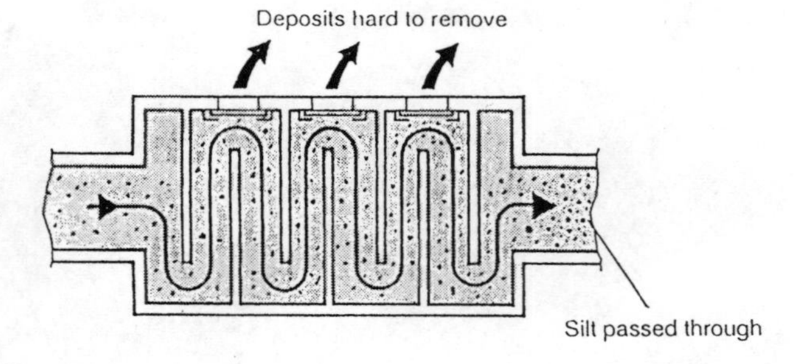

Fig 3.5.10 Incorrect baffle tank design The sharp turns encourage turbulence, making it impossible for the silt to settle out. The extra length of travel does not help because the width of each flow passage is narrow, so the velocity of water is high.

3.6

Channels

Fig 3.6.1 shows the various types of channel section which may be suitable for a particular installation. The various types considered here are:

- Simple earth excavation, no seal or lining.
- Earth excavation with a seal of either cement slurry or clay. ('Sealing' is the application of a thin layer of material with no structural strength; it serves only to reduce friction and leakage. Polythene sheet is also a sealant.)
- Masonry lining or concrete channels. ('Lining' is any method of adding structural strength to the channel walls.)
- Flumes or aqueducts made from galvanized steel sheet, wood, pipes, or pipes cut in half to form troughs, etc.

The choice of the correct type of channel for each part of the route is very important. Some guidance on this is given below. Once the channel type and associated lining or sealing material is chosen for each section, it is possible to calculate suitable dimensions, and the correct drop needed between the entry point and exit point (see Fig 3.6.2). The channel type and the material of channel determine the three basic factors which govern channel dimensions and head. These are described below.

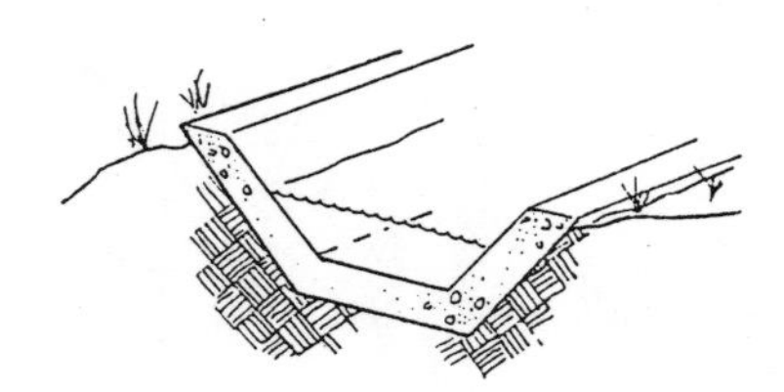

Open concrete

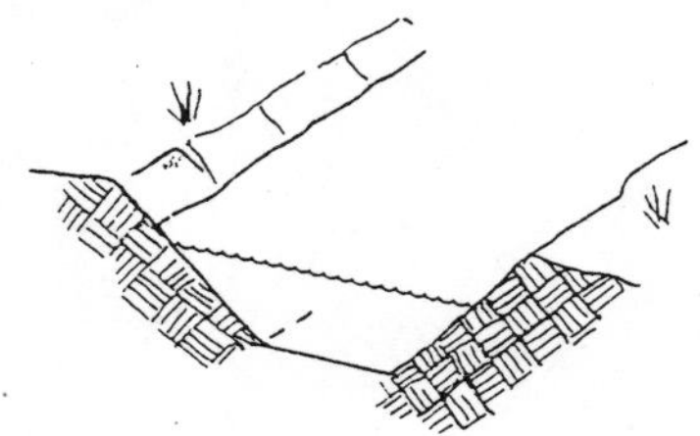

Simple earth channel

Aqueduct (suspended pipe)

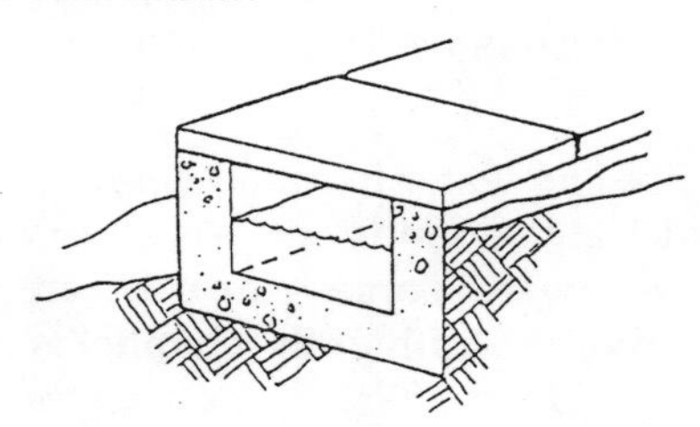

Capped concrete protects against blockage by debris

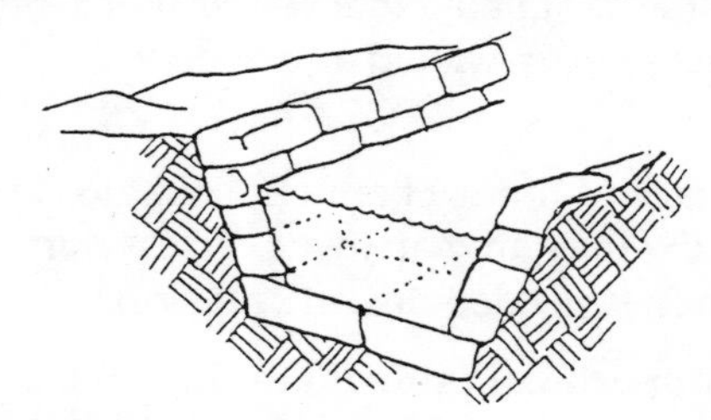

Lined channel (stone and mortar)

Aqueduct (galvanized iron or wooden flume)

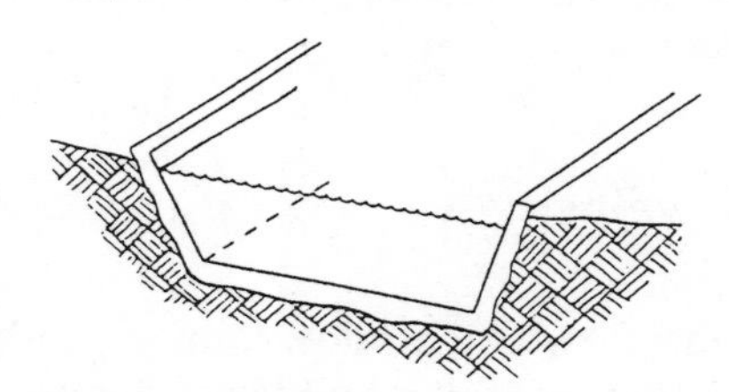

Sealed channel. Earth can sealed with clay or cement slurry, or at greater expense, with UV-resistant polythene sheet.

Fig 3.6.1 Various types of channel

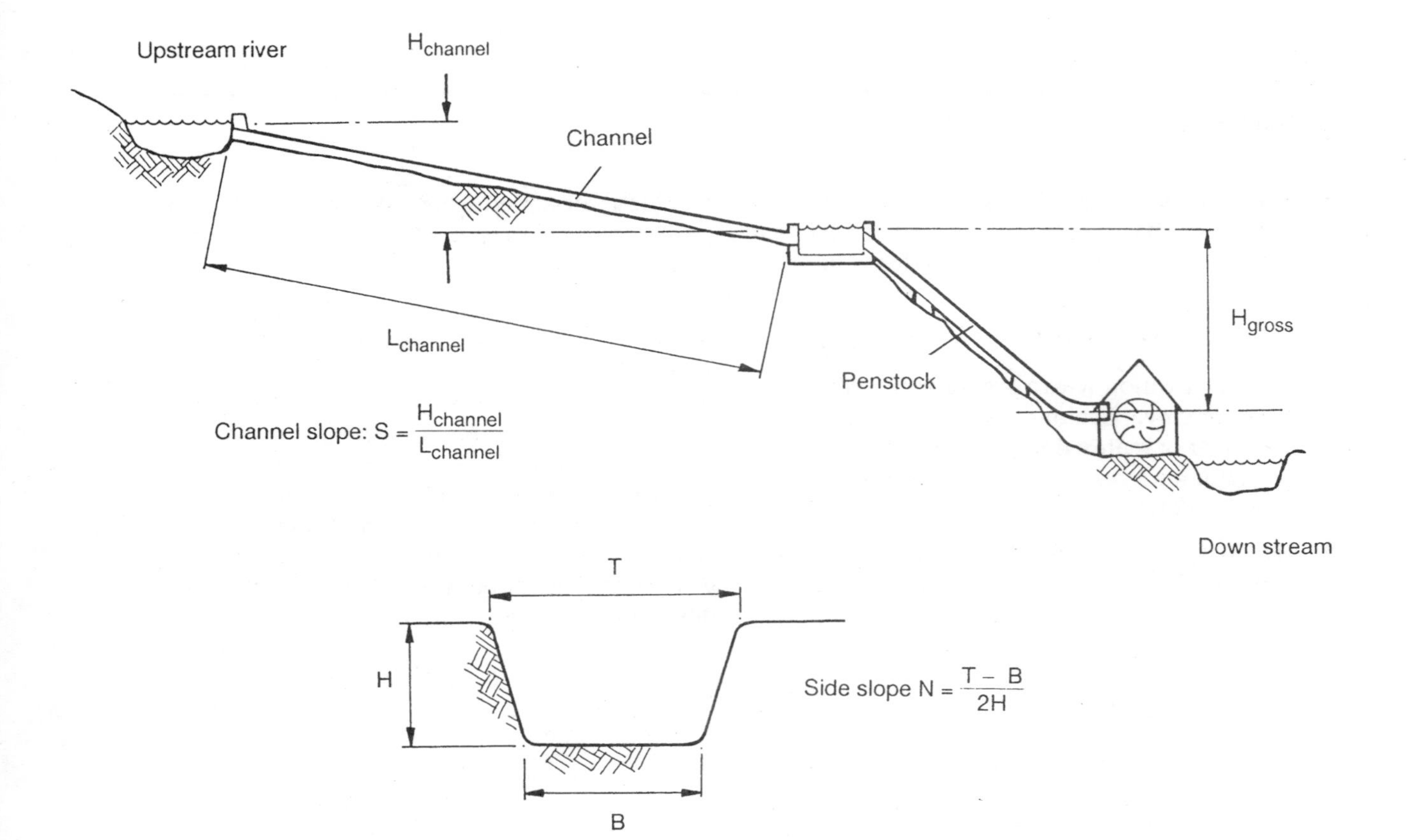

Fig 3.6.2 Channel dimensions and drop

Side slope and cross-sectional profile

The action of flowing water in a channel made from loose material like sandy soil will cause the walls to collapse inwards, unless the sides are sloped gently and the width of the channel is large relative to its depth. Fig 3.6.2 defines the side slope ('N') of a channel. The figures for N given in Table 3.6.1 show that the stronger the material, the steeper-walled and deeper the channel can be.

The advantage of lining channels is that to carry the same flow they are narrower, so less horizontal excavation is needed on difficult slopes.

Normally trapezoidal profiles are chosen, except in the case of wood, steel, cement or masonry/cement channels where rectangular ones are sometimes more easily built.

Water flow velocity 'v'

Excessively fast flow will erode a channel, and excessively slow flow will result in silt deposition and clogging of the channel. Table 3.6.2 gives recommended velocities for different materials.

Photo 3.6.1 An aqueduct (Sri Lanka)

Roughness 'n'

As water flows in the channel, it loses energy in the process of sliding past the walls and bed material. The rougher the material, the more frictional loss, and the greater the head drop or slope (see Fig 3.6.2) needed between entry to the channel and exit, for a given velocity.

A method for designing the channel is given in Note 3.6.1. Example 3.6.1 shows how this method is used in practice to size each channel section and calculate a slope for each section. The total channel head is the sum of all the individual section heads. The calculation is iterative. It starts with a guess as to a suitable flow velocity; the final result is that the channel dimensions are known and the required slope and head loss are known. If the head loss seems excessive then the calculation can be repeated for a smaller velocity, giving rise to a larger channel cross-sectional area and larger dimensions. If the head loss is very small, you can repeat the calculation for a faster velocity (if it is still safe from erosion), and make the channel more compact.

The optimum design will be governed by the five key principles, all of which must be encompassed in a successful design:

- The velocity of the water must be high enough to ensure that suspended solids (sediments) do not settle on the bed of the channel or the pipe.
- The velocity of the water in the channel must be low enough to ensure that its side walls are not eroded by the water flow. If this is impossible, without conflict with the previous principle, the use of improved lining of the channel wall must be considered.
- The head loss, or loss of vertical height, due to the channel must be minimized (this also implies minimum flow velocity). Table 3.6.4 gives the values of roughness factor 'n' for various materials. If an earth channel is sealed, the roughness of the sealant will determine the friction loss.
- The channel must be durable and reliable, not only free from sedimentation, but also protected from destruction by storm runoff, rockfalls crossing its path, or by landslip. It must also cope with unusually high flows along its length, if the intake structure does not already adequately protect it from this. Unusually high channel flows are regulated by the use of spillways. Channel crossings protect it from storm runoff.
- It must have the minimum possible material cost, construction cost and maintenance cost. The availability of necessary skills in construction and labour cost must be assessed.

The method shown in Note 3.6.1 refers to several new terms: the freeboard allowance (F), the wetted perimeter (P) and the hydraulic radius (R).

The 'freeboard allowance' is the amount the channel is oversized to allow for higher flows than the design flow. The channel must not spill when it is carrying excessive water, because damage will quickly result to its walls and the hillside on which it is built.

A common freeboard allowance is 20%, meaning that the channel can accommodate 1.2 times the design flow. It is important to increase this to 30% when building small unlined channels.

Note that the 'freeboard height' is the height of the channel sides above the water level. A freeboard height of 0.15 m is a minimum value for unlined channels.

The wetted perimeter (P) is the distance over which the cross-section wets the channel bed and sides during normal flow. In Fig 3.6.3, $P = P_1 + P_2 + P_3$.

The hydraulic radius ($R = A/P$) is a nominal quantity describing the channel efficiency. If the channel has a large cross-sectional area and relatively small wetted perimeter, then for a standard freeboard this implies it is efficient and can develop the required velocity with relatively little head loss. The most efficient profile for the cross-section of the flow is a semi-circle. The trapezoidal section is the best practical approximation to this.

Example 3.6.1 calculates dimensions for sections of a channel. Notice that the water depth changes from section to section. This is potentially dangerous where the change of depth occurs in an earth channel, as the turbulence created can be erosive. In such cases protect the transition area with a lining.

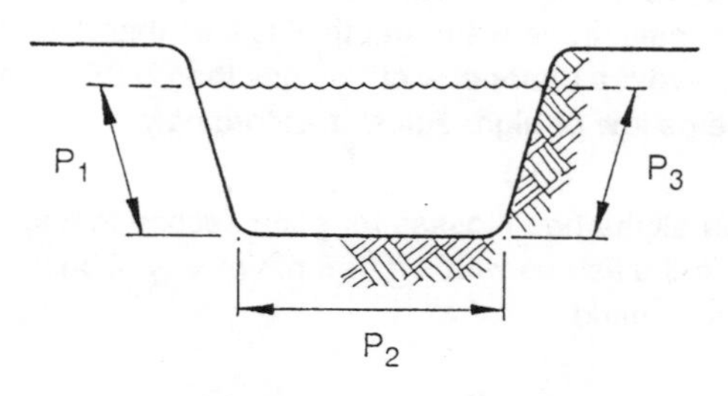

Fig 3.6.3 Wetted perimeter $P = P_1 + P_2 + P_3$

Note 3.6.1 **Summary of method; calculation of channel dimensions and head**

Before starting to calculate the dimensions of any particular section of the channel, you will need to first decide on its length (L), and the material with which it is made, lined, or sealed. (Remember 'lined' means that strength is added, whereas sealing only reduces friction.)

Calculation steps:

1 Choose a suitable velocity (v). Do not exceed the maximum or minimum velocities given on Table 3.6.2.

2 From Table 3.6.1 find side slope (N). For rectangular channels N = 0. From Table 3.6.4 estimate roughness of the wet surfaces (n). Choose a suitable freeboard allowance (F): F = 1.3 is usually suitable.

Calculate cross-sectional area (A) from equation: $A = Q \times \frac{F}{v}$

3 Calculate channel height (H), channel bed width (B), and the channel top width (T). (X is an index which expresses the chosen side slope N in terms of a semi-circle. This reduces friction loss to a minimum.)

$$X = 2\sqrt{(1 + N^2)} - 2 \times N \qquad H = \sqrt{\frac{A}{(X \times N)}}$$

$$B = H \times X \qquad T = B + (2 \times H \times N)$$

Notice that for rectangular channels N = 0 and X = 2, so:

$$H = \sqrt{\left(\frac{A}{2}\right)} \qquad T = B = 2 \times H$$

ie for a rectangular channel the width is twice the height. For stable uniform flow in a long channel, it is best to keep the velocity below the 'critical' limit, $v_c = \sqrt{(A \times g/T)}$.
Calculate v_c and reset v in Step 1 following the rule $v < 0.9\ v_c$.

4 From Table 3.6.3 calculate the wetted perimeter (P). Calculate the hydraulic mean radius (R) from the equation:

$$R = \frac{A}{P}$$

The slope (S) can now be found from Manning's equation:

$$S = \left(n \times \frac{v}{R^{0.667}}\right)^2$$

You are now able to construct the channel section with the required slope and cross-sectional dimensions. First take a note of the head loss (L is the length of the channel section): Head loss = L × S

5 The channel is sized for a flow of Q × F. The water depth for a flow Q will be less than H by the freeboard height, which is H × (F – l). Check that freeboard height is not less than 0.15 m for unlined channels. If the water depth H is less than 1 metre, use Note 3.6.2 for a possible revision of the value 'n'. Avoid a channel width of less than 0.25 m as narrow channels can easily be blocked. If the velocity seems low or high, adjust it accordingly.

6 Add all the head losses for each section to find the total head loss. If this is too great or too small, repeat all steps with a different velocity. Consider using a different lining or sealant, but keep overall cost in mind.

Channel sizing *following method of Note 3.6.1* **Example 3.6.1**

Channel flow (Q_{gross})	132 l/s
Total channel length (L)	1600 m
Length of unlined section (L_U) soil type: sandy loam	1400 m
Length of lined section (L_L) lining type: coarse concrete	150 m
Length of aqueduct (L_A) material: wood	50 m

	Unlined section	Lined section	Aqueduct
Calculation step 1			
Choose velocity from Table 3.6.2	v = 0.5 m/s	v = 1.2 m/s	v = 1.3 m/s
Calculation step 2			
Side slope N from Table 3.6.1	N = 2	N = 0.58	N = 0
Roughness coefficient from Table 3.6.4	n = 0.03	n = 0.02	n = 0.015
Freeboard allowance:	F = 1.3	F = 1.3	F = 1.3
Cross-sectional area: $A = \frac{(Q \times F)}{v}$	A = 0.343 m^2	A = 0.143 m^2	A = 0.132 m^2
Calculation step 3			
$X = 2 \times \sqrt{(1 + N^2)} - (2 \times N)$	X = 0.472	X = 1.155	X = 2
Height $H = \sqrt{\frac{A}{(X + N)}}$	H = 0.37 m	H = 0.287 m	H = 0.257 m
Bed width B = H × X	B = 0.18 m	B = 0.332 m	B = 2 × H = 0.514 m
Top width T = B + (2 × H × N)	T = 1.66 m	T = 0.63 m	T = B = 0.514 m
Wetted perimeter $P = B + \left[2 \times H\sqrt{(1 + N^2)}\right]$	P = 1.83 m	P = 0.996 m	P = B + 2 × H=1.028m
Calculation step 4			
Hydraulic mean radius $R = \frac{A}{P}$	R = 0.187 m	R = 0.144 m	R = 0.128 m
Slope $S = \left(\frac{n \times v}{R^{0.667}}\right)^2$	$S = 2.10 \times 10^{-3}$	$S = 7.64 \times 10^{-3}$	$S = 5.88 \times 10^{-3}$
Head loss HL = L × S	HL_U = 2.9 m	HL_L = 1.15 m	HL_A = 0.29 m

Calculation step 5

Check freeboard heights for unlined sections and check the roughness n. For the unlined section here the freeboard height is H × (F – l) = 0.11 which is too little; therefore increase F. Roughness n should be changed as Note 3.6.2.

Calculation step 6

Total head loss $HL_T = HL_U + HL_L + HL_A$ = 4.3 m

If the scheme you are considering has a turbine head in the order of 50 metres, this represents a channel head loss of about 9%. If the head loss is excessive repeat steps 1-4, choosing lower velocities. Take care not to choose velocities so low that sediment will settle in the channel. Note that a high head loss has the advantage that high velocities reduce the channel dimensions, allowing it to be compact and free from sedimentation. For a more compact channel, choose higher velocities.

Table 3.6.1 Side slopes

	Material	Side slope 'N'
a	**Trapezoidal sections**	
	Sandy loam	2
	Loam	1.5
	Clay loam	1
	Clay	0.58
	Concrete	0.58
b	**Rectangular sections**	0

These figures are for 'lining' (strengthening) with the material, but not for sealing. For sealed channels, use the N value for the surrounding soil and not the material.

Table 3.6.2 Flow velocities

Material	Maximum velocity to avoid erosion: Less than 0.3m deep	Less than 1.0m deep
Sandy loam	0.4 m/s	0.5 m/s
Loam	0.5 m/s	0.6 m/s
Clay loam	0.6 m/s	0.7 m/s
Clay	0.8 m/s	1.8 m/s
Masonry	1.5 m/s	2.0 m/s
Concrete	1.5 m/s	2.0 m/s

Minimum velocity

In order to avoid silt clogging of the channel the flow should not be too slow. If the water is always clear, this is not a problem. Silty water should not move at less than 0.3 m/s.

Photo 3.6.2 Wooden formers are very useful when constructing a channel.

Photo 3.6.3 A channel crossing offers better protection than a culvert

Table 3.6.3 Characteristics of cross-sections

Type of cross-section	Wetted perimeter (P)	Top width (T)
Rectangle	$B + 2 \times H$	B
Trapezoid	$B + 2 \times H\sqrt{(1 + N^2)}$	$B + 2 \times H \times N$
Triangle	$2 \times H\sqrt{(1 + N^2)}$	$2 \times H \times N$

Roughness coefficient 'n' — Table 3.6.4

See Note 3.6.2 for advice on channels of less than one metre depth.
(Approximate design values for small channels after ageing are shown in the left-hand column.)

		n
Earth channels	Clay, with stones and sand, after ageing	0.020
	Gravelly or sandy loam channels, maintained with minimum vegetation	0.030
	Lined with coarse stones, maintained with minimum vegetation	0.040
	For channels less than 1 metre deep, use the equation in Note 3.6.2. eg: Vegetated (useful to stabilize soil); water depth 0.7 metres water depth 0.3 metres	0.050 0.070
	Heavily overgrown, water depth 0.3 metres	0.150
Rock channels	Medium coarse rock muck	0.037
(0.040)	Rock muck from careful blasting	0.045
	Very coarse rock muck, great irregularities	0.059
Masonry channels	Brickwork, bricks, also clinker, well pointed	0.015
(0.030)	Normal masonry	0.017
	Coarse rubble masonry, stones only coarsely hewn	0.020
Concrete channels	Smooth cement finish	0.010
(0.020)	Concrete when wood formwork is used, unplastered	0.015
	Tamped concrete with smooth surface	0.016
	Coarse concrete lining	0.018
	Irregular concrete surfaces	0.020
Wooden channels	Planed, well-joined boards	0.011
(0.015)	Unplaned boards	0.012
	Older wooden channels	0.015
Metal channels (0.020)		0.020
Natural water courses	Natural river bed with solid bottom, without irregularities	0.024
	Natural river bed, weedy	0.031
	Natural river bed with rubble and irregularities	0.033
	Torrent with coarse rubble (head-sized stones), bed load at rest	0.038
	Torrent with coarse rubble, bed load in motion	0.050

Roughness effect for shallow channels — Note 3.6.2

Research at Wageningen University in the Netherlands demonstrated that the roughness effect is worse for channels under 1 metre in depth, because of the turbulence created by the side walls and bed surface. The research showed that the following equations can be used to find the roughness coefficient. H is the depth of water.

Well maintained channels with little vegetation: $n = 0.03/\sqrt{H}$ $H < 1\,m$

Channels with short vegetation: $n = 0.04/\sqrt{H}$ $H < 1\,m$

Heavily overgrown channels: $n = 0.08/\sqrt{H}$ $H < 1\,m$

In practice it is sensible to maintain short vegetation in order to protect the walls of channels:

'n' can therefore be found from the equation $n = 0.04/\sqrt{H}$ $H < 1\,m$

Seepage loss and choice of channel type

It is usually necessary to examine the soil along the route of the proposed channel. If the soil is very sandy it may be necessary to seal the channel against water loss by leakage. Either a sealant or a lining could be used for this purpose.

A soil porosity test can be made with the technique illustrated in Note 3.6.3. A cylinder of about 300 mm diameter can be quickly formed from any available piece of sheet metal which is then pushed into the ground and filled with water to a clearly marked level. Use a lid to prevent rainwater entry or evaporation. Seepage into the soil reduces the level as time goes by. Each day or each hour, refill the cylinder from a graduated flask so that the exact quantity added can be recorded. Several of these cylinders can be used along the route. The records taken then provide an estimate of soil porosity. Note 3.6.3 shows how the porosity indicates the soil type.

This test will also provide a direct measure of the water lost should you decide to use this soil for a simple excavated channel without lining or sealing. It is important that the infiltration rate observed in the cylinder is observed to be steady, while measurements are taken, since transient effects will take place while refilling. Very gentle refilling will overcome such effects.

To calculate the seepage flow ($Q_{seepage}$) read the relevant factor (s) from the table. Find the wetted perimeter value (P) from the channel sizing calculation (Note 3.6.1), and channel length (L).

$$Q_{seepage} = s \times P \times L \quad m^3/s$$

Unlined channels or channel sections are considerably cheaper to construct since they do not require so much labour or difficulty of transporting materials to the site. A typical excavation rate is 4 m^3/day/person but this depends on soil type. A proper survey of the route is necessary to detect where a lined channel is needed; this will be for instance where :

- the ground is excessively porous;
- rocky ground prevents excavation;
- the ground is steep and the soil unstable;

A full survey of the channel route is desirable but may be costly. It is very useful to inspect other channels in the area and consult local farmers with experience of irrigation channels.

If the route passes through sandy soil where the seepage will be excessive, a length of pipe or a lined section of channel can be used. If the soil is waterlogged, the area can be drained, using contour drains and lined main drains, or a pipe passed through the area. Rocks which obstruct the route can be removed, or the route diverted around them. Blasting or heating by fire is possible to shatter the rocks, but this can give rise to unexpected delays.

One important reason for using lined sections of a channel is where there is a risk of the channel wall being broken. Sometimes this can happen as a result of a small diversion being made when the channel flow is low by someone wishing to use the water. The danger is that extensive damage to the channel wall will occur by erosion when the channel flow increases. This kind of danger exists especially where you have built up the ground to maintain channel slope across undulating ground; in this case a lined section of channel is advisable.

Note 3.6.3 **Measuring seepage loss**

Soil type	Basic infiltration rate (mm/hour)	Seepage losses (s) (m^3/s per m^2 of wetted surface)
Sand	30+	8.3×10^{-6} and above
Sand loam	20-30	$5.6\text{-}8.3 \times 10^{-6}$
Silt loam	10-20	$2.8\text{-}5.6 \times 10^{-6}$
Clay loam	5-10	$1.4\text{-}2.8 \times 10^{-6}$
Clay	1-5	$0.3\text{-}1.4 \times 10^{-6}$

Fig 3.6.4 Measuring seepage loss. Avoid incorrect readings due to loss through evaporation by covering the cylinder. Sideways seepage can be restricted by sinking a second cylinder of larger diameter deeper into the ground around the first cylinder. The seepage loss is $s \times P \times L$ m^3/s where P is the wetted perimeter and L is the channel length.

Lined channels allow water to be carried at greater velocity since their walls are more resistant to erosion. This usually means that lined channels can be more compact, which is a useful feature where space is limited. It should be remembered that three disadvantages are associated with water travelling at low velocity, as in the case of unlined channels:

- risk of plant growth blocking the flow (although limited growth should be encouraged to stabilize the channel walls);
- risk of silt settling out;
- channel is larger, uses more space.

A very important feature of an open channel is its vulnerability to damage from such sources as landslip and rockfalls, and from stormwater runoff crossing its path. The cost of protection from these eventualities and their associated repair costs must be included in an estimation of the channel cost. It may be the case that the use of a low-pressure pipe to convey water instead of an open channel may be more expensive to purchase but the cheaper option in the long term because of savings in protection and maintenance. The use of a low-pressure pipe rather than an open channel will also save labour costs during construction; where marshy ground and ravine crossings are a particular problem, using pipe can save the cost of aqueduct construction.

Nevertheless, it is usually the case that the cost of a pipe greatly exceeds the cost of an open channel. A pipe also has a serious disadvantage compared to an open channel, in that access to its inside is considerably more difficult; if it is not designed for a sufficiently high velocity, internal silting will occur and will be hard to detect and correct. Further, a pipe is prone to blocking at its mouth.

Channel crossings

Where small streams or rivulets cross the path of the channel, very great care must be taken to protect the channel. Again, the golden rule is :

"think flood"

During storms, the rivulet will become a torrent easily capable of washing away the channel. Provision of a culvert (a small drain running the rivulet under the channel) is usually not adequate protection. It will tend to block just when most needed. In the long term it is economic to build a full crossing, as shown in Fig 3.6.5. This is sized to accommodate a flow of around 1000 times the usual wet season rivulet flow.

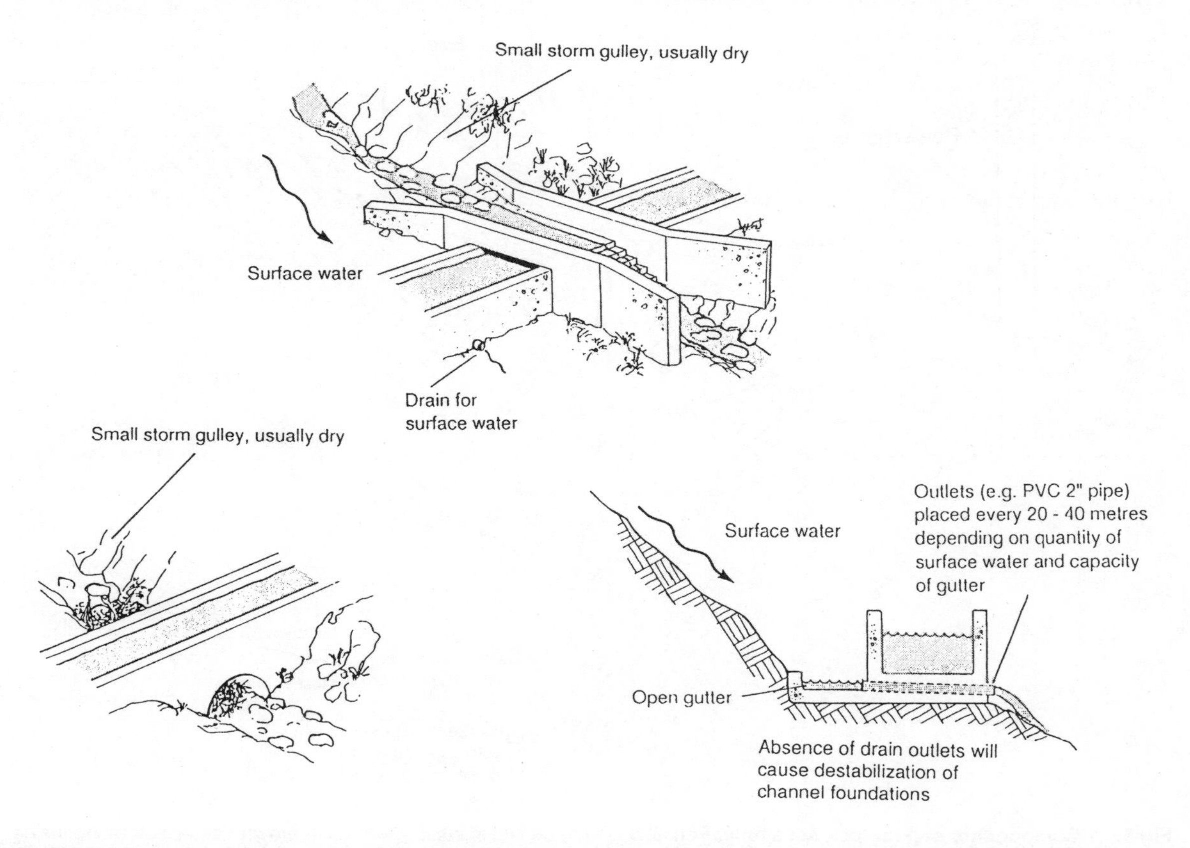

Fig 3.6.5. Channel crossing These are usually a better long-term economy than simple culverts. Simple culverts, as shown, will tend to block with debris during the rainy season. The rains will then destroy the channel. Drains are nevertheless needed under concrete channels to avoid trapping of surface water behind the channel wall.

3.7
Penstocks

The penstock is the pipe which conveys water under pressure to the turbine. The major components of the penstock assembly are shown in Fig 3.7.1.

The penstock constitutes a major expense in the total micro-hydro budget. It is therefore worthwhile to optimize the penstock design to minimize both lifetime running costs and initial purchase cost. To ensure low maintenance cost care should be taken to place the penstock anchors and supports on stable slopes and to find firm foundations.

There should be no danger of erosion from storm runoff on the slopes, and there should be safe access for repair and maintenance jobs (such as repainting).

In projecting the cost of the penstock it is easy to underestimate the expense of peripheral items such as joints and coats of paint. The choice between one penstock pipe material and another can make significant differences in overall cost if all these factors are included. For instance, plastic penstock piping may be cheap but the joints may be expensive or unreliable in some regions. A general method for selecting a penstock is given in Section 3.11.

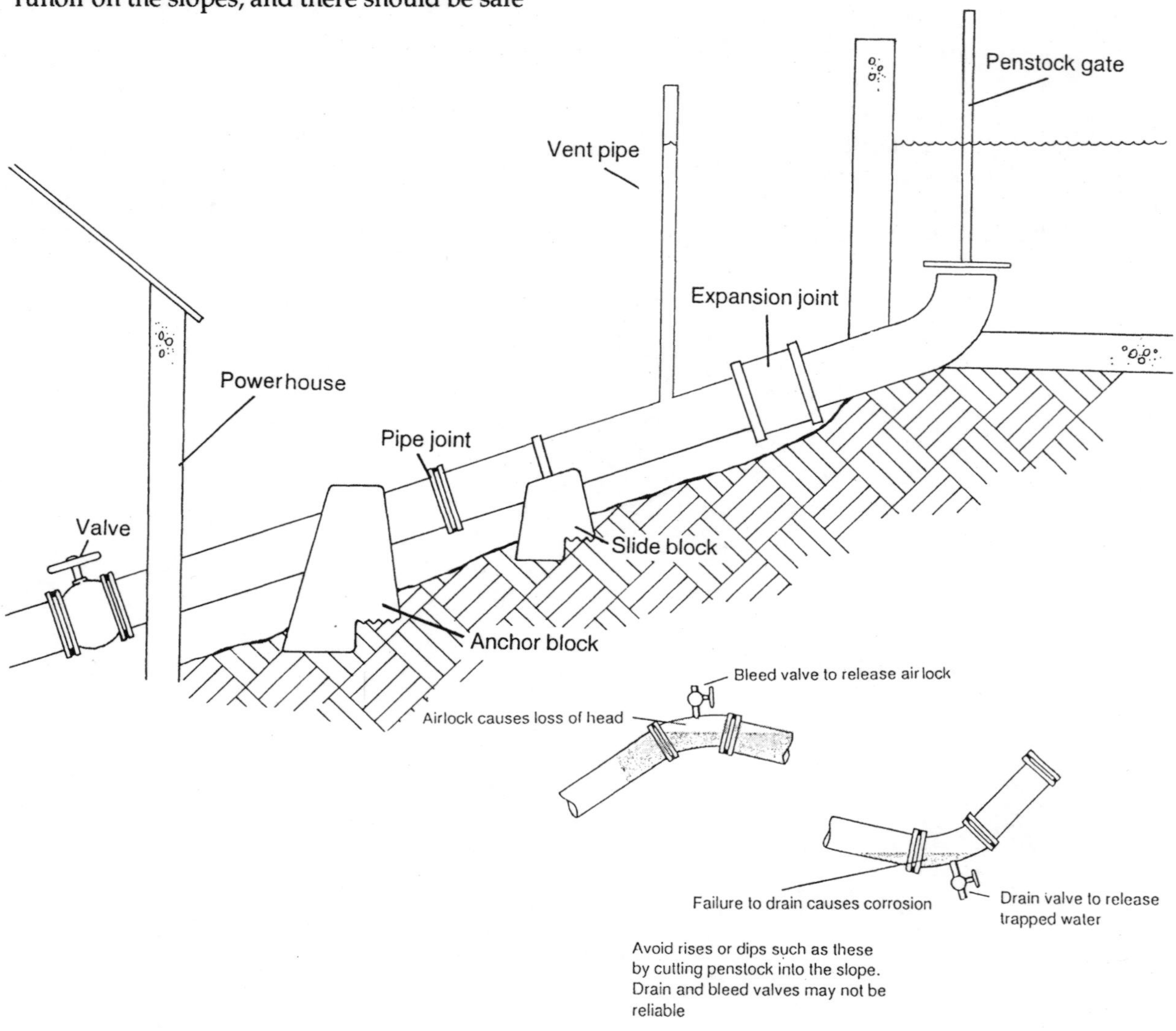

Fig 3.7.1 Components of penstock assembly. Penstocks must be laid in such a way as to prevent airlocks forming inside them. These airlocks act as obstructions in the penstock and cause a pressure drop across them. If a danger of airlocks exists because the ground rises and the penstock cannot be cut in, an air bleed valve must be fitted as shown. Similarly water drain valves may be needed. Always avoid the use of valves since after some years they can become unreliable.

3.8

Materials

The following factors have to be considered when deciding which material to use for a particular project:

- Design pressure
- Method of jointing
- Diameter and friction loss
- Weight and ease of installation
- Accessibility of the site
- Terrain
- Soil type
- Design life and maintenance
- Weather conditions
- Availability
- Relative cost
- Transport to site

The following materials are likely to be used for the penstocks of micro-hydro schemes.

- Mild steel
- Unplasticized polyvinyl chloride (uPVC)
- High-density polyethylene (HDPE)
- Medium-density polyethylene (MDPE)
- Spun ductile iron
- Asbestos cement
- Prestressed concrete
- Wood stave
- Glass reinforced plastic (GRP)

These materials are briefly discussed here. The first four on the list, mild steel, uPVC, HDPE and MDPE, are the most common.

Table 3.8.1 shows the relative merits of each type, and Table 3.11.4 shows the physical characteristics of the more common materials.

Mild steel

Mild steel is perhaps the most widely used material for penstocks in micro-hydro schemes. It is relatively cheap, and may be fabricated locally, requiring machinery common to most medium-sized steel fabrication workshops. It is made by rolling steel plate into a cylinder, and welding the seam. It can be made in a variety of diameters and thicknesses as required. It has medium friction loss characteristics, and provided it is well protected by paint or other surface coating can have a life of up to 20 years. It is resistant to mechanical damage, but buried steel penstocks are at greater risk from corrosion. Mild steel pipes are relatively heavy, but can be manufactured in lengths convenient for transport and installation if needed. Mild steel pipes may be joined by flanges, on-site welding or mechanical joints.

Table 3.8.1 **Comparison of common materials**

• = Poor ••••• = Excellent

Material	Friction loss	Weight	Corrosion	Cost	Jointing	Pressure
Ductile iron	••••	•	••••	••	•••••	••••
Asbestos cement	•••	••••	••••	•••	•••	•
Concrete	•	•	•••••	•••	•••	•
Wood stave	•••	•••	••••	••	••••	•••
GRP	•••••	•••••	•••	•••	••••	•••••
uPVC	•••••	•••••	••••	••••	••••	••••
Mild steel	•••	•••	•••	••••	••••	•••••
HDPE	•••••	•••••	•••••	•••	••	••••
MDPE	•••••	•••••	•••••	•••	••	•••••

Unplasticized polyvinyl chloride (uPVC)

This pipe is one of the most widely used in micro-hydro schemes throughout the world. It is relatively cheap, widely available in a range of sizes from 25 mm to over 500 mm, and is suitable for high-pressure use. Different pressure ratings are obtained by varying the wall thickness of the pipe, but generally the outside diameter remains constant for a range of pressure ratings in a given diameter. It is light, and easy to transport and lay. It has very good friction loss characteristics, and is resistant to corrosion. It is, however, relatively fragile, and prone to mechanical damage from falling rocks, etc, particularly at low temperatures.

The main disadvantage is that uPVC deteriorates when subjected to ultraviolet light, causing surface cracking which can seriously affect the pressure rating. It must, therefore, always be protected from direct sunlight, by either burying, covering with foliage, wrapping, or painting.

It is recommended that in all cases uPVC pipe is supported all along its length either by laying it on the ground or burying it. It is in fact best always to bury uPVC. Continuous support is needed because the pipe is vulnerable to stress fatigue, due to vibrations caused by water flow, if it is under a bending stress such as caused by carrying a weight of water when suspended between support blocks. Fatigue failure of this sort may occur after 5 or 10 years of life. When it is buried it must be laid on a specially prepared soft bed of sand or fine gravel to avoid stress concentrations caused by sharp rocks or bricks – any such stresses will also give rise to fatigue failure.

It is also necessary to avoid the temptation to impose bends or arcs in the line followed by the pipe when laying it, as these impose bending stresses, in addition to inducing hydrostatic forces. uPVC pipe sections are jointed by spigot and socket and a special glue, called uPVC pipe cement, or spigot and socket with a flexible sealing ring. In both cases it is not possible to disconnect sections once they are connected.

uPVC has some immediate advantages over most other materials. It is smoother than most, so inducing less friction for a given diameter, with the result that smaller diameter uPVC pipe can be used for a given application; this can reduce transport and installation costs.

Because it is a very elastic material it does not induce high surge pressures, although this is offset to an extent by smaller diameters and higher water velocities. Because it is a light material it can be handled, transported, and installed more easily. This difference may be dramatic: a steel penstock might cost as much as 50% more because installation costs involve for instance the cutting in of a new road. Transport by truck is often costed by volume, so that one technique for reducing cost is to design a penstock with half its length of a smaller diameter so that the smaller sections can be packed inside the bigger sections.

Concrete

Concrete pipes are unsuitable for use at even moderate pressure; they are heavy, and hence difficult to transport and lay. Friction characteristic can vary from good to very poor. Jointing is generally by rubber ring joints. Steel reinforced concrete pipes, especially when prestressed, can be a cost-effective solution for low and medium head sites.

Spun ductile iron

Ductile iron has largely replaced cast iron, although the latter may still be found on older schemes. Ductile iron pipes are sometimes coated with cement on the inside, which gives good protection from corrosion and a low friction loss. It is a heavy material, hence difficult to install, and tends to be costly. Ductile iron pipes are usually joined by mechanical joints (bolted gland), push-in spigot and socket with a flexible seal, or occasionally flanged.

Photo 3.8.1 Steel penstocks are rolled from sheet and seam-welded in a small workshop. (Nepal)

Photo 3.8.2 A mild steel penstock. (Sri Lanka)

Asbestos cement

Asbestos cement pipes are made from cement, reinforced with asbestos fibre. They are unsuitable for use at even moderate pressure, and are fragile, and hence transporting and laying them requires care. They are light, have good friction loss characteristics, and do not corrode. The dust caused by cutting asbestos pipes is a health hazard, so adequate protective clothing and masks should be provided for people working with asbestos pipes.

Wood stave

Wood stave pipe is made from strips of wood (generally pine or cedar), tongued and grooved, and held together by steel hoops, in much the same way as a barrel. Provided it is kept wet, wood stave has a long life, with a good friction characteristic that may improve with age. It is light to transport and lay, but is not suitable for high pressure applications. Its main disadvantage is poor availability, as it is largely only manufactured in the USA and Canada.

Glass reinforced plastic (GRP)

GRP pipes are made from resin, reinforced with glass fibre wound spirally and an inert filler, such as sand. They can be used at high pressures. They are light, and have good friction characteristics.

GRP pipes are fragile, and care is needed during installation. They are best protected by burying and backfilling with selected fine material. There is evidence that GRP may be weakened by water absorption by osmosis over a long time span. Joints are usually spigot and socket with a flexible seal. Depending on availability and relative cost, GRP may be a good option, but has yet to gain wider acceptance.

High and medium density polyethylene (HDPE and MDPE)

M/HDPE pipes offer a good alternative to uPVC, although somewhat more costly. They are available in diameters from 25 millimetres up to over one metre. Small diameters (up to 100 mm) are usually supplied in rolls of 50 or 100 metres, and are particularly useful for small schemes due to the ease of installation. M/HDPE pipes have excellent friction loss and corrosion characteristics, and do not deteriorate when subjected to sunlight. Pipes are generally jointed by heating the ends and fusing them under pressure using special equipment, which is a disadvantage, but for smaller diameter pipes mechanical compression fitting joints are an economical option.

Photo 3.8.3 Steel penstocks can be welded on site if skilled operators are available. (Peru)

3.9

Penstock jointing

Pipes are generally supplied in standard lengths, and have to be joined together on site. There are many ways of doing this, and the following factors should be considered when choosing the best jointing system for a particular scheme.

- Suitability for chosen pipe material;
- Skill level of personnel installing the pipes;
- Whether any degree of joint flexibility is required;
- Relative costs;
- Ease of installation.

Methods of pipe jointing fall roughly into four categories, all of which are discussed below:

- Flanged;
- Spigot and socket;
- Mechanical;
- Welded.

Flanged joints

Flanges are fitted to each end of individual pipes during manufacture, and each flange is then bolted to the next during installation. A gasket or other packing material, usually rubber, is necessary between each flange of a pair. Flange jointed pipes are easy to install, but flanges can add to the cost of the pipe. Flange joints are generally used on mild steel pipes, and occasionally to join ductile iron. Flanges should conform to some recognized standard; for instance, British Standard (BS) or International Organization for Standardization (ISO). Flange joints do not allow any 'flexibility' or deflection.

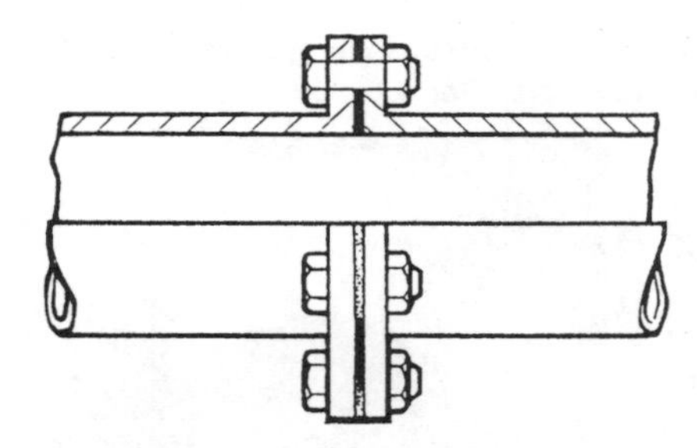

Fig 3.9.1 Flanged joints

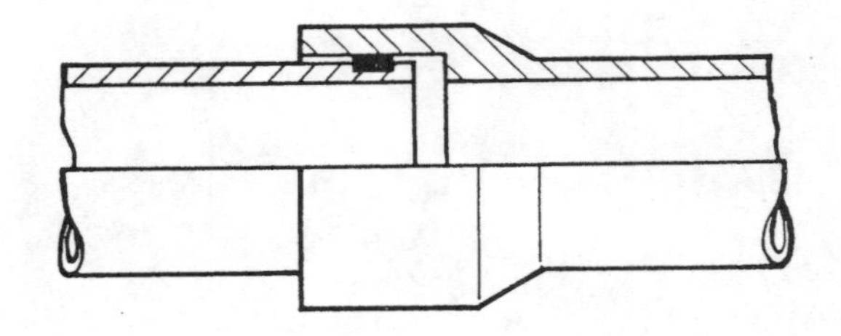

Fig 3.9.2 Spigot and socket joints

Spigot and socket joints

Spigot and socket joints are made by either fitting a collar to, or increasing the diameter during manufacture of, one end of each pipe, such that the internal diameter of the collar or increased internal diameter of the pipe is the same as the external diameter of the pipe. The plain end of each pipe can thus be pushed into the collar or 'socket' in the next. A good seal is required between each pipe section, and this is achieved by either providing a rubber seal or using a special glue called solvent cement, depending on the material of which the pipes are made. Previously a seal was achieved by packing the joint with lead chippings, but this method is now only used to repair existing old joints in cast iron penstocks.

Rubber seal joints fall generally into two types: 'O' ring seals and single or multiple 'V' lip seals. They are generally used to join ductile iron, PVC, or GRP pipes. Rubber seal joints generally permit a few degrees of deflection.

A few precautions are necessary when installing this type of joint:

- The seal must be clean before assembly.
- A special lubricant must be used. Never use an oil-based grease as this will cause the seal to rot. If the special lubricant is not available, soap may be used.
- For pipes over 50 mm, always use clamps and a ratchet pulley to 'pull' the two halves of the joint together.
- Ensure that the joint is properly aligned before final coupling. 'V' ring seal joints are extremely difficult and often impossible to take apart.

Solvent cement welded spigot and socket joints are used on uPVC pipes. A special chemical paste, which dissolves the plastic material, is applied to the spigot half of the joint, which is then inserted into the socket. The chemical action 'welds' the two halves together.

The following precautions should be observed:

- The joint must be dry. Unless unavoidable do not attempt to install a pipe using this method of jointing when it is raining.
- The joint must be clean. Special cleaning fluid and clean rag should be used.
- Pipe clamps and a ratchet puller should be used on pipe diameters greater than 200 mm.
- Take great care to align the joint correctly. It is impossible to take apart solvent-welded joints.
- The fumes from solvent welding cement are highly toxic. Avoid prolonged exposure to them, and ensure adequate ventilation.

Mechanical joints

Mechanical joints are rarely used on penstocks because of their cost. In some cases their extra cost is justified by savings made in installation cost (eg on-site welding) and because they allow slight misalignments in the penstock. This facilitates installation and allows later movement of anchors/foundations to occur (due to subsidence, landslip, earthquake, etc) without causing undue stress in the penstock.

One important application is for joining pipes of different material (eg mild steel and uPVC) or where a slight deflection in the penstock is required that does not warrant installing a bend. Some types of mechanical joint cannot take any strain in the direction of the pipe, and have to be restrained by anchor blocks.

Welded joints

Welded joints are used on mild steel penstocks and, using special techniques, on M/HDPE. One advantage of welding on site is that changes in the direction of the pipe can be accommodated without preparation of a special bend section. Mild steel pipes are brought to the site in standard lengths, and then welded together, generally using an arc welder, on site. It is a relatively cheap method, but has the disadvantage of needing skilled site personnel, and the problems of getting an arc welder and power supply (which may mean a generator) into remote and difficult terrain.

It is essential to have a competent person doing the welding to ensure sound joints. The major disadvantage is that a site-welded steel pipe usually ends up with a low quality finish with respect to corrosion-protection.

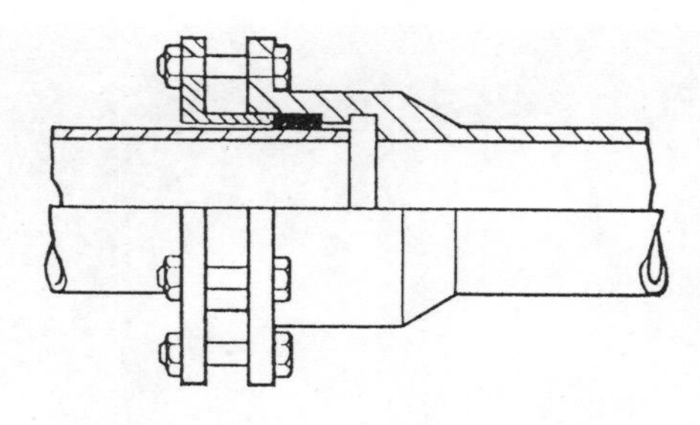

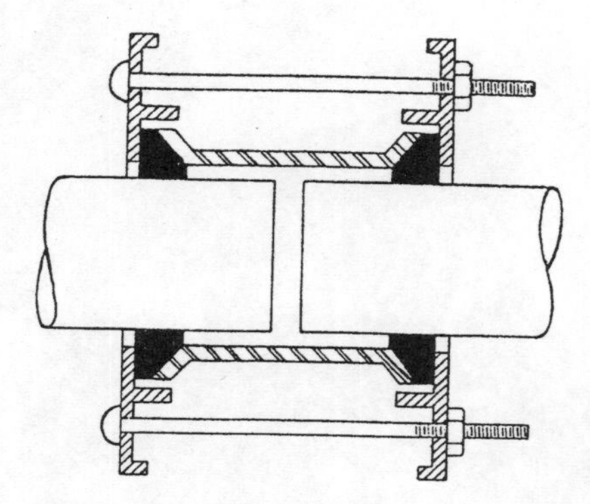

Fig 3.9.3 Mechanical joints

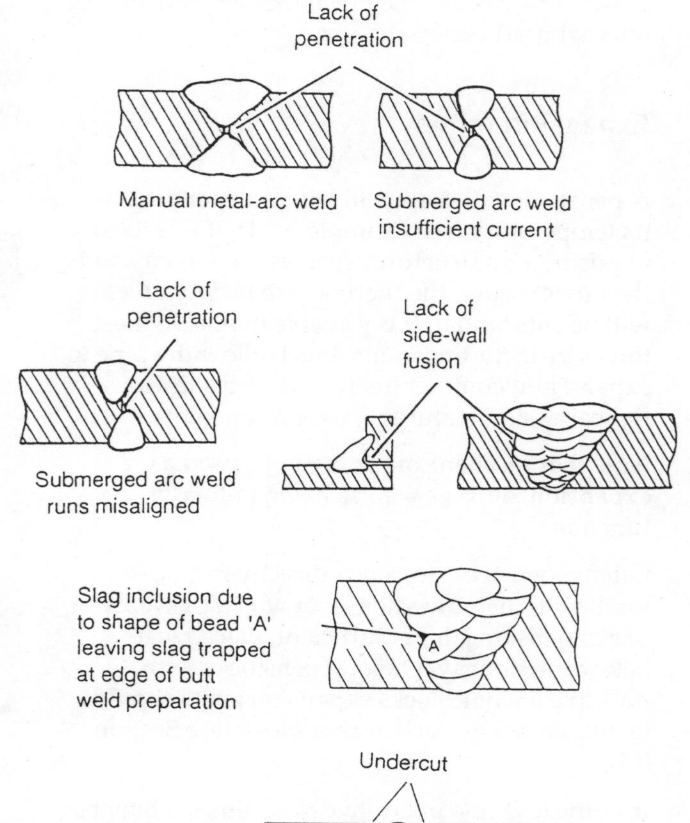

Fig 3.9.4 Weld defects

The welding process removes paint near the joints. Any subsequent attempt to shot-blast and repaint the pipe inside and out is difficult to get to an adequate standard, with the result that corrosion occurs early in the life of the pipe. A further disadvantage of welded joins compared to bolted flanges is that sections of pipe cannot be removed later for repair or maintenance, except by cutting, although it should be remembered that glued and push-fit uPVC joins usually are permanent as well.

The welding of M/HDPE also requires some skill, and special equipment that is expensive to buy but can often be rented from the pipe manufacturer. The two ends of the pipe to be joined are fixed in a special jig, and heating pads are applied to each extremity. The temperature of the heating pad and the length of time it is applied are quite critical in obtaining a good joint. When the material at the end of the pipe is semi liquid, the two ends are brought together with some force, which causes them to 'fuse' together. The process is called 'fusion welding', and with practice can be done very rapidly.

The disadvantage of fusion welding is that beads form inside the pipe at the join and cause friction losses. These beads may be removed with an internal bead cutter.

Expansion joints

A penstock will change in length depending on its temperature (see Example 3.9.1). If it is fixed in position by structures such as the forebay and the powerhouse, the thermal expansion forces will be substantial. It is possible to relieve these forces by including joints which allow the pipe to expand and contract freely. Fig 3.9.5 shows typical design features of expansion joints.

Note that mechanical joints can be used as expansion joints as well as having other functions.

Often a penstock descends directly from the forebay to the powerhouse, in which case only one expansion joint is sufficient, placed just below the forebay. On long penstocks with multiple anchor blocks expansion joints should be placed below each anchor block (see Section 3.12).

In some low cost micro-hydro schemes it may be possible to allow the pipe to slide to and fro inside the forebay wall, so eliminating the cost of an expansion joint at the forebay. A small leak may occur but this would be an insignificant water loss and would act to lubricate the sliding expansion socket. In the case of a straight penstock and a sliding forebay socket like this, no expansion joints will be needed for the penstock. Care must be taken to provide a paved drain below the leaking forebay socket joint to avoid damage to the forebay tank foundations. Although this idea is mentioned here speculatively, it is possible that some existing schemes already operate successfully in this manner by default if not by design.

It is also possible to eliminate expansion joints by allowing pipe movement and stresses to be taken up by bends in the pipe which are not fully anchored – this can only be done if the effects of the full range of forces listed in Section 3.12 are considered carefully. It is also possible to consider elimination of expansion joints by allowing the rubber gaskets of ordinary pipe section joints to take up thermal movement. This will also require some careful engineering design. It is important to consider the time of installation. Pipes installed in hot weather will contract later in winter or night conditions; pipes installed in cold weather will expand later. Bolts can only be tightened in cold weather, if the intention is to allow the pipe to squeeze the gasket during hot weather. The forces on the structures holding the pipe will be in different directions during hot or cold weather, and it may be worthwhile to install pipes at one or the other extreme of temperature.

Photo 3.9.1 An expansion joint is best situated just below an anchor block.

Sizing and installing an expansion joint

Example 3.9.1

The following equation tells you how many expansion joints you will need and how much movement each one will have to accommodate:

$$x = a\,(T_{hot} - T_{cold})\,L \quad \text{metres}$$

where x = Expansion of pipe (m)

a = Coefficient of expansion given in Table 3.11.4 (m/m°C)

L = Length of pipe (m)

T_{hot} = Highest temperature the pipe will experience (°C)

T_{cold} = Lowest temperature the pipe will experience (°C)

$x = a(T_{hot} - T_{cold})L$

For instance, suppose an 87 metre long steel penstock will experience a 5° temperature at night in the coldest part of the year, and 40° in the day in the warmer season.

$$x = a\,(T_{hot} - T_{cold})\,L \quad \text{metres}$$

$$x = 12 \times 10^{-6} \times (40 - 5) \times 87 = 37 \text{ mm}$$

It is normal to provide a safety factor for over-expansion of between 1.4 and 2.0. On this basis it is recommended that all the expansion joints in this penstock (assuming it is anchored only at the top and bottom) be capable of accommodating a total change of length of at least 50 mm. In other words, either use one expansion joint which can move 50 mm, or for instance two which can move 25 mm each.

If the penstock is anchored at more points than at the top and bottom, then each anchored section must be taken separately for calculation of expansion. For instance, if there is an anchor on this penstock at a bend 50 metres from the forebay and no other anchor until the powerhouse, then do the above calculation for the 50 metre length first (this would be 'L' above) and then do it again for the remaining 37 metres (again 'L' above).

Think carefully when you will be installing. If installing in hot weather, adjust the pipe lengths so that the pipes can contract without overstretching the expansion joint when the weather turns cold; vice-versa for cold weather.

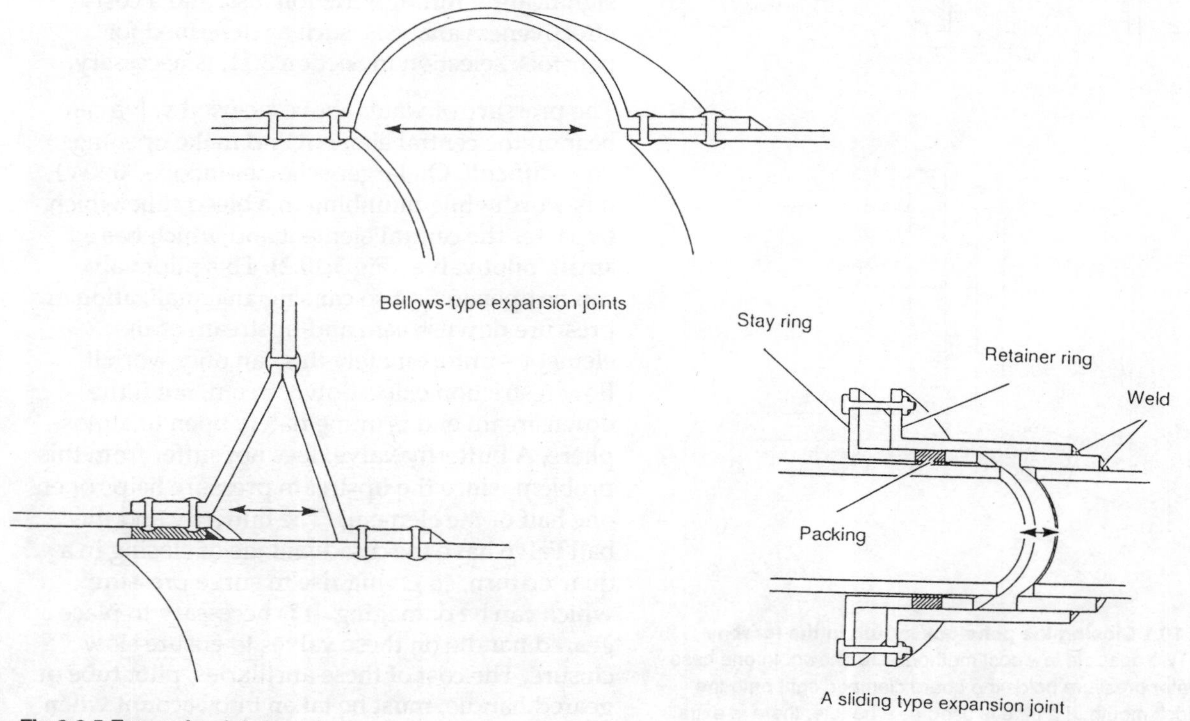

Fig 3.9.5 Expansion joints

3.10 Valves

Valves control the water flow through the penstock. It is sensible to place a valve at the entry to the turbine, to permit uncoupling of the turbine with the penstock full. It is possible to save the expense of the valve, but work associated with the turbine or shutting off power can be awkward if the penstock has to be emptied. A valve is not usually necessary at the top of the penstock, because water can be diverted from the penstock in other ways, eg by diverting from the channel, by closing the mouth of the penstock with a flat board faced with rubber (see Fig 3.10.1) or by opening a sluice in the forebay tank. In all cases, whether or not a valve is placed in the upper section of the penstock, an air vent must be placed downstream of the penstock inlet as shown in Figs 3.7.1 and 3.10.1, to avoid vacuum collapse of the penstock while it is emptying, or when the penstock mouth is accidentally blocked. If a valve is fitted, the vent must be just downstream of it.

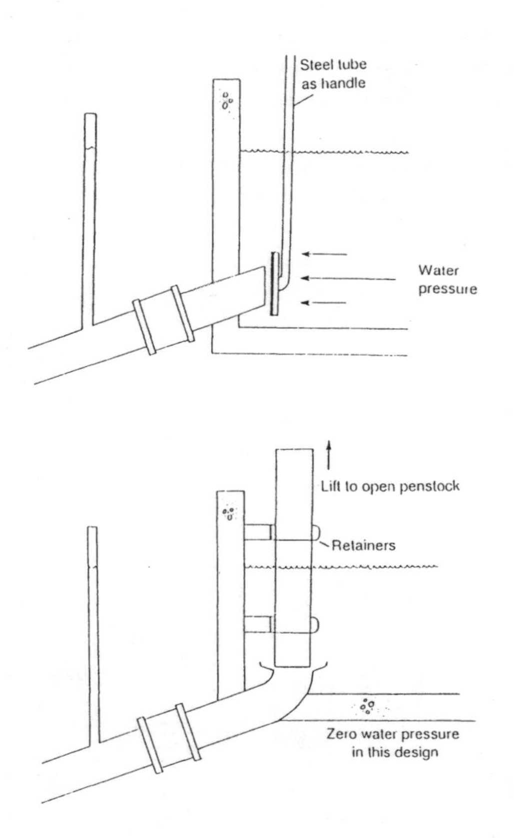

Fig 3.10.1 Closing the penstock mouth in the forebay tank Two possible low cost methods are shown. In one case the water pressure holds the board clamped tight onto the penstock mouth. If a pipe is used as a handle, there is extra venting for safety.

Some types of valves are shown in Fig 3.10.2. For micro-hydro applications the most usual types are gate and butterfly valves. The choice of the best valve is made on a basis of cost, availability and pressure handling capacity.

In Section 3.11 advice is given on the calculation of surge pressure. The valve will be subjected to the same total pressure as the penstock, the sum of surge pressure and normal pressure. When buying a valve, total pressure must be specified. To ensure trouble-free operation the most durable material should be chosen, preferably bronze or brass, second choice cast iron, finally steel depending on budget constraints. The 'k' factor governing friction loss should be ascertained for the valves under consideration.

Valves become enormously expensive as their size increases. One of the advantages of the multi-jet Pelton is that there is no need to buy a large valve since a smaller valve can be placed on each jet passage. In this case a number of smaller valves can be cheaper than one large valve. Expense can be saved by under-sizing the valve so that its internal diameter is less than the penstock internal diameter. This reduces power significantly through friction loss, and a cost-effectiveness analysis, such as described for penstock selection in Section 3.11, is necessary.

The pressure of water above a closed valve can bear on the central element and make opening very difficult. On larger schemes (above 30 kW) it is worthwhile plumbing in a bleed tube which bypasses the central element and which has a small 'pilot valve' (Fig 3.10.2). This pilot valve can be opened first so causing an equalization of pressure downstream and upstream of the element – unfortunately this can only work if flow restriction exists downstream, not if the downstream end is immediately open to atmosphere. A butterfly valve does not suffer from this problem (since the upstream pressure helps open one half of the element). The butterfly and the ball valve have the disadvantage of closing in a quarter-turn, so giving rise to surge pressure which can be damaging. It is necessary to place a geared handle on these valves to ensure slow closure. The cost of these ancillaries, pilot tube or geared handle, must be taken into account when comparing options.

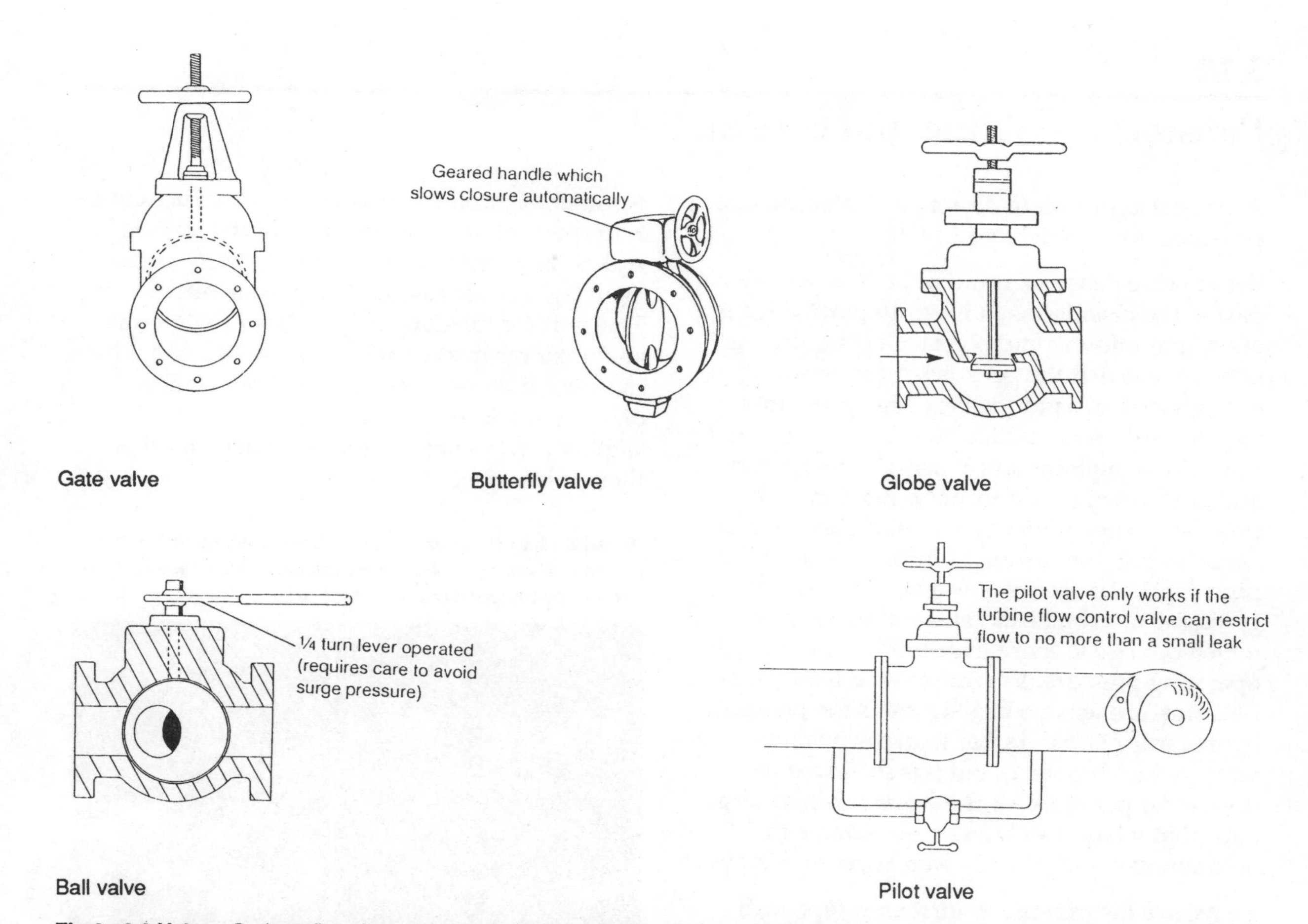

Fig 3.10.2 Valves On buttefly valves always use a geared mechanism to slow the rate of closure. The final moments of closure give rise to dangerous surge pressures.

Photo 3.10.1 A gate valve (Sri Lanka)

Photo 3.10.2 A gate valve with pilot valve and gearing for slow closure. (Sri Lanka)

Photo 3.10.3 A cylinder valve design which can be used for the penstock entry or for sluicing of silt basins. (Nepal)

3.11

Penstocks – Sizing and costing

A general approach to design and selection of a penstock is given in Note 3.11.1.

Because the penstock is often the most expensive part of the installation, it is worth putting some effort into minimizing its cost. One reason it is expensive is that the penstock must be strong enough to withstand the very high pressures which result from sudden blockage of the water flow. These high pressures are only temporary and are known as 'surge' or 'water hammer' pressures. They travel in waves of positive and negative pressure throughout the length of the pipe. Fig 3.11.1 shows how a sudden blockage gives rise to a total head of water in the pipe which can rise to more than 150% of the normal operating pressure. In some cases it is the negative pressure wave which destroys the penstock by inward collapse, rather than the positive wave (which bursts it), but it is sufficient to design the penstock for the worst positive surge, such that it has a wall thickness suitable to accommodate negative as well as positive surge.

To reduce the expense of increased pipe wall thickness, it is possible to design a hydro scheme to minimize the extent of surge pressure. For instance, choosing to use a multi-jet Pelton as the turbine (with separate valves on each jet) can have a beneficial effect on penstock cost. This is because a sudden blockage will most likely stop the flow through only one of the jets, such that the surge pressure resulting will be much less than if the full flow was obstructed. The low surge pressure can lead to the choice of a thinner walled and cheaper penstock.

Another example of good design to reduce surge is to specify that the valve used at the turbine end of the penstock has a slow-closing mechanism (eg it is not closed by a quarter-turn lever). This reduces the danger of high surge pressure and a thinner-walled penstock can be used with more confidence. In terms of the calculation described in Note 3.11.3, this means that a slightly lower safety factor (SF) can be used as there is less risk.

Photo 3.11.1 Plastic pipe (HDPE) being transported by foot in Nepal. Micro-hydro economic decisions are affected strongly by transport costs in some areas.

Note 3.11.1 — **Selection of a penstock**

1. Prepare a table similar to Table 3.11.1. Various different pipe materials and sizes can be compared on the table. Each option will have different implications in terms of both overall capital cost and in terms of performance achieved.
2. For each pipe considered, calculate the loss of energy due to friction in the pipe. The calculation is described in Note 3.11.2 and Example 3.11.2.
3. For each option, calculate the wall thickness necessary to withstand pressures in addition to normal operating pressure, as described in Note 3.11.3, and Example 3.11.3.
4. Calculate the total cost of each option, and assess the performance associated with each option, to allow a choice to be made of the most cost-effective or economic option (Example 3.11.4). Note that a penstock with high friction loss implies smaller and less costly electro-mechanical equipment as well as less pipe cost, but will deliver less power. The delivery of less power may be a disbenefit (if all power is sold, for instance) or it may be unimportant (if peak power is hardly ever used). The final comparison of costs and benefits is therefore dependent on local conditions and the specific application.

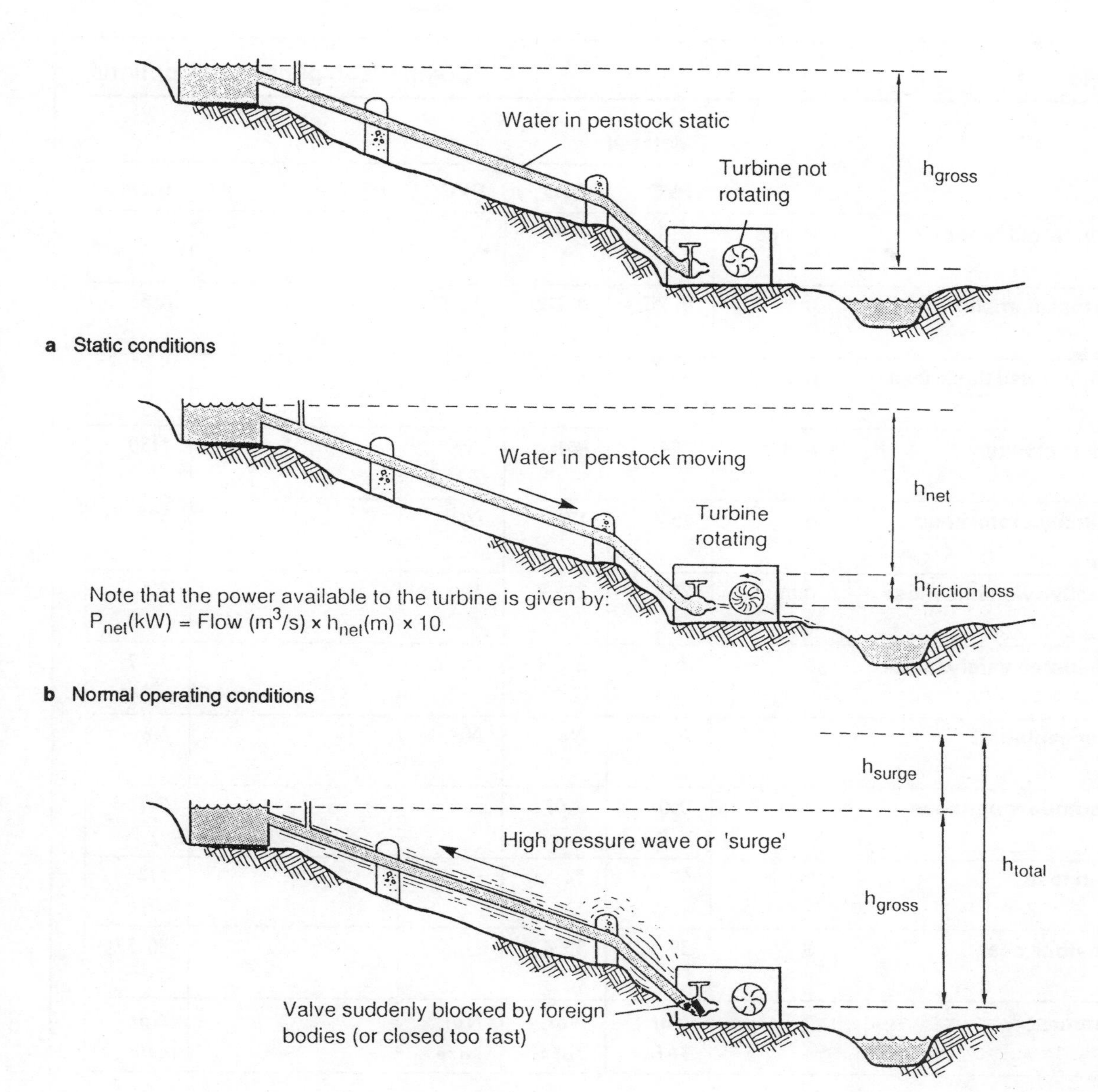

Fig 3.11.1 Surge pressures and friction loss.

A burst penstock may cause expensive damage to the turbine, and generator, and civil works of an installation. To avoid this the penstock should be pressure-tested under controlled conditions up to the worst expected total head. This can be done as soon as it is installed on site. The worst head it will experience is the sum of static head (h_{gross}) and surge head (h_{surge}), as shown on Fig 3.11.1. A simple 'stirrup pump' and gauge can be used to produce this total head. (A stirrup pump is a hand operated positive displacement pump using the plunger and one-way valve principle, usually arranged to suck water from a bucket. At times it has been used for firefighting.)

Flanges must be built onto the entry mouth of the penstock so that it can be closed off for pressure testing (the vent pipe is usually built into the penstock after pressure testing; alternatively it can be closed off or used as the pump inlet).

When pressure-testing, build up the pressure in stages and check for leaks at each stage. A leak indicates a point at which a full burst is likely unless corrective action is taken. For penstocks longer than 100m, it is often convenient to test in sections by blanking off each test section using plates (spades) at suitable flanges.

Table 3.11.1 **Comparison of penstock options**

		Material					
		PVC	PVC	PVC			Steel
Nominal diameter	inches	16	14	12			12
Internal diameter	m	0.352	0.312	0.277			0.3
Nominal wall thickness t	mm	23	21	18			6.4
Wave velocity a	m/s	396	401	395			1150
Estimated total head h_{total}	m	156	176	202			392
Effective wall thickness $t_{effective}$	mm	23	21	18			29
Calculated safety factor SF		2.3	2.1	1.8			1.7
SF acceptable?		No	No	No			No
Assumed roughness k	mm	0.01	0.01	-			0.1
Head loss	%	4%	7%	-			11%
Penstock cost	$	39.520	33.670	-			30.576
Comment		Not safe	Not safe	Not safe			Not safe

Assuming a $70,000 scheme using 14" PVC as a reference:

Additional cost	$	5.850	0	-			-3.094
Additional benefit	kW	2.1	0	-			-2.8
Total cost	$	75.850	70.000	-			66.906
Total benefit	kW	72.1	70	-			67.2
Comparison	$/kW	1.052	1.000	-			996

The above items are defined in Notes 3.11.2, 3, 4. For instance, $h_{total} = h_{gross} + h_{surge}$ as defined in Note 3.11.3.

All the pipes considered here have wall thicknesses which give unacceptable safety margins. Pipes with thicker walls should be chosen.

The method suggested here for calculating wall thickness (Note 3.11.3) makes use of a safety factor (SF).

The designer of the penstock has a certain amount of freedom in ascertaining the value to be adopted for the safety factor, SF. In general it is recommended that a value of 3.5 is used. (Manufacturers of PVC and HDPE pipes often recommend pressure ratings which are based on a safety factor of 3.) In some cases a significant cost saving can be made by selection of slightly thinner-walled pipe which implies a reduced safety factor. For instance, the safety factor can be relaxed to less than 3.5 (but never less than 2.5) if the designer is sure that *all* of the following conditions will be met:

1 The effects of a burst are not likely to be dangerous or very costly to repair (eg low head, gentle slopes).

2 Design to remove risk of instant stoppage of flow (eg multi jets and slow closing valves).

3 Assembly and welding operations are being undertaken by experienced staff with a track record of success on installation of penstocks handling similar pressures.

4 The penstock is pressure tested before use to the full value of total head and action is taken in response to any signs of weakness.

The 'effective wall thickness' factors listed in Note 3.11.3 refer to certain conditions in manufacturing of steel pipe. For instance, the manufacturer may quote the thickness of the original steel sheet from which the pipe is formed. Lower quality sheet may in practice vary in thickness, so an allowance is made for this. Secondly, welding of steel can have the effect of weakening the steel wall adjacent to the weld. This is especially where the pipe sections are welded in difficult conditions on site, so that the quality of the welding is reduced.

The method described in Note 3.11.2 for calculating friction loss makes use of a factor 'k' which describes the roughness of the inside wall of the pipe. It is possible to inspect a pipe, and estimate the value of k yourself, if an aged or very rough pipe is considered. k is measured in mm and describes the diameter of grains of sand which if glued to the surface of a smooth pipe interior would produce the same friction loss as the inspected penstock. To estimate k, measure the depth of protrusions on the pipe wall. Experience has shown that roughness can become considerably worse as time goes on. In the case of steel and cast iron this is due to corrosion. In the case of concrete or asbestos cement vegetation growths can have this effect (as well as increasing friction by reducing internal diameter). In the case of PVC algal growths have this effect. Table 3.11.2 summarizes these ageing effects.

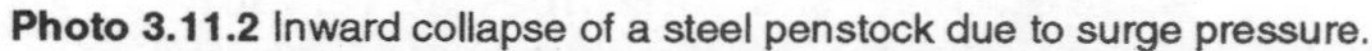

Photo 3.11.2 Inward collapse of a steel penstock due to surge pressure.

Photo 3.11.3 and 3.11.4 PVC pipe for small schemes can be handled and installed with relative ease. (Sri Lanka)

There are two results of roughness increasing over time. The first is that the gross hydraulic power is reduced. The second is that the optimum operating speed of the turbine is reduced. This can spoil the speed matching between the penstock and the turbine, creating inefficiency and so further reducing power.

For these reasons it is recommended that the 'normal' figures shown in Table 3.11.2 are used for design purposes, referring to typical roughness after 5 or 10 years of use. The penstock manufacturer may be quoting values of k for new pipe which are not appropriate for design purposes.

The need to match the turbine speed to the penstock friction loss often has an important influence on the final selection of penstock diameter and material. This is true in schemes where reduced water flow during the dry months of the year will be used to operate the turbine at reduced power, and where electricity is generated, so requiring that the turbine rotates at a given speed. If the friction loss is very high, the seasonal change in flow will cause a change in net head and so change the ideal speed of the turbine. As a result the turbine will run inefficiently. On the other hand, it is sensible in most cases to design the turbine drive to allow high efficiency performance during low flow, and power wastage during high flow, because usually power is most valuable when it is scarce. A compromise is therefore needed.

On larger schemes this compromise often results in losses in the order of 5-15%. In cases where the turbine is run at full flow all year, there is less reason not to use an inefficient penstock. Sometimes in very small schemes the most economic penstocks can have losses as large as 20 or 30% because local costs of handling and purchase of small diameter standard pipes are very low compared to larger pipes. In general where speed regulation for an alternator is not in question, ie mechanical drive of a grain mill, there is no limit to friction loss, since variations can be taken up by adjustment of the mill and the loading rate.

To choose the best diameter and type of penstock a comparison must be made which depends on the following factors:

1 **The purchase cost** of the penstock and ancillary costs (jointing, supports, delivery, installation, design etc).
2 **Maintenance cost** of the penstock such as repainting, drain clearing etc.
3 **The power** delivered by the penstock, after its friction loss is taken into account.
4 **The 'knock-on' costs/savings,** that is, the financial implications of the power delivered by the penstock, for instance where a penstock with significant loss allows use of smaller drive systems and alternator or other driven device. These are the electro-mechanical cost implications.

5 **Part flow** The relative importance of demand on the hydro installation at part flow or full flow. If stream flow is particularly low during the dry season, but more than sufficient in the wet season, the best penstock may be the one which is most cost-effective during part-flow, and its efficiency at full flow may be allowed to fall off.

6 **Matching** Identification of the limits to penstock efficiency imposed by the need to match penstock net head and turbine speed at both part flow and full flow, as explained in the text above.

7 **Peak power.** Peak power is the final 10% or 20% of power produced. In some installations it is particularly important that this power is available because no other power source is present to run all appliances at certain crucial stages of a manufacturing process. For instance an agricultural process may require 7kW of power at certain times and not be able to function if only 6kW is provided. In this case an expensive low-friction penstock may be a necessary choice. In contrast, this same process may be powered partly by a diesel engine or grid electricity and use of hydro power only as an auxiliary energy source. In this case the extra 1kW can be provided by the other energy source and a cheaper hydro can be installed without any disbenefits resulting. This factor can be thought of alternatively as the minimum power limit.

8 **Additional benefit.** This is what you get in return for paying out extra costs for a penstock with less head loss. Additional benefit is usually a mixture of qualitative and quantitative factors:

8.1 **Additional benefit – qualitative.** The idea of additional benefit covers the benefits mentioned already, for instance whether or not the improved penstock provides good matching (factor 6) and satisfies the peak power requirement (factor 7). It is possible to have zero benefit or even disbenefit.

8.2 **Additional benefit – quantitative.** In certain types of installation, additional benefit can be calculated in actual sums of money. For instance, your installation serves as a power source to a factory which is already powered by grid electricity. Every kW it produced earns revenue in the form of savings made by not using grid electricity. If you install a bigger penstock which has a 3% head loss instead of one with a 7% head loss, you might produce 5kW more power. Over ten years this will earn perhaps $7000 in extra income.

9 **Additional cost.** This is the extra amount you pay for a bigger penstock which has less head loss. It is also the extra money you save or pay in 'knock-on' costs (factor 3 above). Additional cost is always measured in money terms, unlike additional benefit. Continuing our example immediately above, we might find that the 3% penstock (the bigger one) costs an extra $5000. It earns $7000, so it is a better choice. But it may happen that the smaller penstock producing 5kW less power allows use of a smaller turbine/alternator set which costs $4000 less. The smaller penstock will then be better, so long as factors 6 and 7 – qualitative benefits – are also satisfied.

10 **The total costs** of the complete hydro scheme, as projected in approximate terms. This can also be expressed as the target figure for cost per kW delivered by the scheme. If previous schemes in the same region have for instance cost $1,400 per kW and have been successful in operation and economic return, you might adopt this figure as a target.

There are a number of different ways of using these factors to choose the most cost-effective penstock. Usually it is recommended that benefit is calculated as revenue, but it is possible to use a shorthand method instead which saves time. This is described in Note 3.11.4. Any method like this must be used only if the qualitative factors such as peak power and matching limits are taken into account.

Photo 3.11.5 Corrosion inside a steel penstock. Here the roughness index 'k' can be estimated by observation.

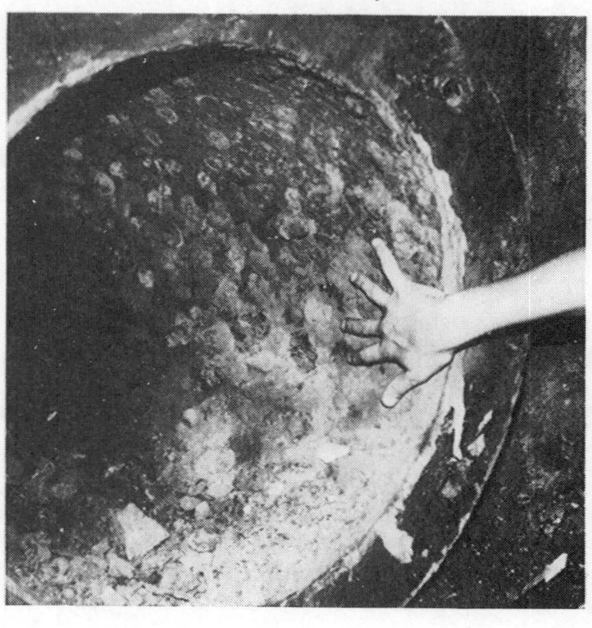

Note 3.11.2 **Friction loss and diameter**

1 Establish the gross head (see Fig 3.11.1) and turbine design flow value, Q_{net}.

$$Q_{net} = Q_{gross} - Q_{irrigation} - Q_{seepage}$$

2 Measure or estimate the length of penstock needed. If you are doing a detailed calculation, use a value measured on the actual proposed site of the penstock. If you are doing a rapid costing/feasibility study, and an accurate map is available, the penstock length can be estimated very roughly from the map. The penstock length (L_{pipe}) is given by simple trigonometry:

$$L_{pipe} = \sqrt{L_{horizontal}^2 + h_{gross}^2}$$

where $L_{horizontal}$ is the horizontal distance from forebay to powerhouse measured on the map. The contours give h_{gross}.

3 Choose a material and establish which internal diameters are available for the penstock, picking one for a first try. This is 'd'.

4 Estimate the extent of pipe inside wall roughness (k) or use the 'normal' values given on Table 3.11.2. Calculate k/d, using the same units top and bottom. Also calculate 1.2 Q/d, using SI units. Read the friction factor value 'f' from the Moody chart (Fig 3.11.2). See footnote*

5 Calculate the head lost due to pipe wall friction ($h_{wall\ loss}$):

$$h_{wall\ loss} = \frac{f\ L_{pipe}\ 0.08\ Q^2}{d^5}$$

6 Calculate the velocity of water in the penstock (v):

$$v = \frac{4Q}{\pi\ d^2}$$

7 Usually turbulence losses are minor compared to the effect of wall friction. (In the case of multi-jet Peltons this is not always true, since manifold bend losses can be high and should be calculated.)
If you are doing a quick initial sizing calculation, and you think turbulence effects are minor, skip step 7. Referring to Table 3.11.3, calculate turbulence losses ($h_{turb\ loss}$) in entrance sections, bends, valves, and other obstructions, such as manifolds:

$$h_{turb\ loss} = \frac{v^2}{2g}\left(K_{entrance} + K_{bend\ 1} + K_{bend\ 2} + K_{contraction\ 1} + \ldots + K_{valve}\right)$$

where K is a factor associated with the bend or obstruction, and g is 9.8. Sum the friction losses to obtain total friction loss ($h_{friction\ loss}$):

$$h_{friction\ loss} = h_{wall\ loss} + h_{turb\ loss}$$

8 Calculate the percentage loss of head due to friction and the net head (h_{net}):

$$\%\ loss = \frac{h_{friction}}{h_{gross}} \times 100 \qquad h_{net} = h_{gross} - h_{friction}$$

9 Enter details on a table such as Table 3.11.1.

10 Calculate wall thickness as described in Note 3.11.3.

11 Choose another material and/or diameter and repeat steps 3 to 11.

---

* The Moody chart appears in some text books on fluid mechanics in a slightly different form, with a friction factor which is 4 times smaller than the values given in Fig 3.11.2. If such a chart is used, the friction factor must be multiplied by 4 before use in the equation for wall loss given above. In all text books the Moody chart is given in terms of 'Reynolds number' on the horizontal axis. In fact 1.2 Q/d is simply the Reynolds number for water at a nominal temperature.

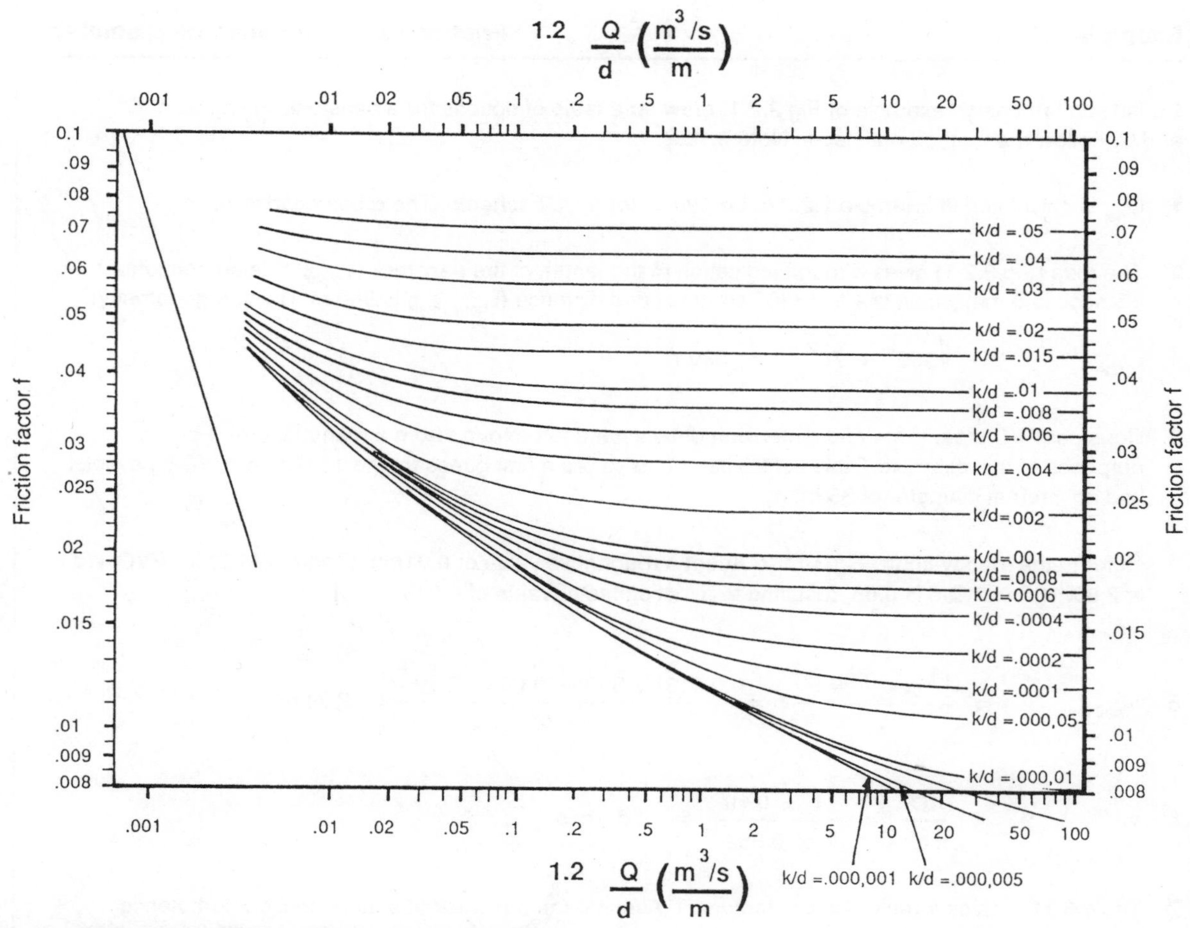

Fig 3.11.2 The Moody Chart for wall friction losses in pipes

Roughness values, k mm — **Table 3.11.2**

Use **'normal condition'** for design purposes

Material		Age/condition: Good (< 5 years)	Normal (5-15 years)	Poor (>15 years)
Smooth pipes PVC, HDPE, MDPE, Glass fibre		0.003	0.01	0.05
Concrete		0.06	0.15	1.5
Mild steel	- Uncoated	0.01	0.1	0.5
	- Galvanized	0.06	0.15	0.3
Cast iron				
New		0.15	0.3	0.6
Old	- Slight corrosion	0.6	1.5	3.0
	- Moderate corrosion	1.5	3.0	6.0
	- Severe corrosion	6.0	10.0	20.0

Example 3.11.1 **Friction loss and penstock diameter**

Continuing the design example of Fig 2.2.1, draw up a table of options for a penstock joining point B1 and C. Follow the method outlined in Note 3.11.2.

1 Q_{net} is calculated in Example 2.2.2 to be 190 l/s for a Q57 scheme. The gross head is 75 m.

2 The map (Fig 2.2.1) gives a rough indication of the length of the penstock (L_{pipe}). Scaled horizontal distance is 6 mm, scale is 1 to 63360 so horizontal distance ($L_{horizontal}$) is 380 m. Using trigonometry:

$$L_{pipe} = \sqrt{380^2 + 75^2} = 390 \text{ m}$$

3 Consider PVC pipe. Price and dimension details need to be found from a manufacturer. For the purposes of this example Table 3.11.5 can be used. As a first guess try the 10 bar 16" PVC pipe which has an internal diameter of 352mm.

4 Refer to the Moody chart, Fig 3.11.2. Adopt a roughness value of 0.01mm (Table 3.11.2) for PVC; k/d is 2.8×10^{-5}. 1.2 Q/d is 0.65, resulting in a friction factor value of 0.013.

5 $$h_{wall\ loss} = \frac{f\ L_{pipe}\ 0.08\ Q^2}{d^5} = \frac{0.013 \times 390 \times 0.08 \times 0.19^2}{0.352^5} = 2.71 \text{ m}$$

6 $$v = \frac{4Q}{\pi d^2} = \frac{4 \times 0.19}{\pi \times 0.352^2} = 2.0 \text{ m/s}$$

7 Table 3.11.3 gives turbulence loss factors, k. Assume the penstock has no bends, but turbulence occurs at a fully open gate valve ($K_{valve} = 0.1$) and at a sharp cornered flush entrance ($K_{entrance} = 0.5$). Assume the turbine is not a multi-jet Pelton, so manifold losses will not be significant. Head loss due to turbulence ($h_{turb\ loss}$) is:

$$h_{turb\ loss} = \frac{v^2}{2g}(0.5 + 0.1) = 0.12 \text{ m}$$

$$h_{friction\ loss} = h_{wall} + h_{turb\ loss} = 2.71 + 0.12 = 2.8 \text{ m}$$

8 $$\%\ loss = \frac{2.8}{75} = 4\% \qquad h_{net} = h_{gross} - h_{friction\ loss} = 75 - 2.8 = 72.2 \text{ m}$$

9 The details are entered onto Table 3.11.1

10 The wall thickness required of this pipe is calculated in Example 3.11.2 in order to make sure that the thickness provided by the manufacturer (shown on Table 3.11.5) is adequate to accommodate the pressures that may be imposed.

11 Next, try the 10 bar 14" pipe. Calculation results are in Table 3.11.1. A third PVC pipe diameter and a steel pipe are also calculated.

Note: Appendix 2 contains a simple pocket computer program suitable for automatic calculation of friction factor and pipe wall friction loss.

Turbulence losses in penstocks **Table 3.11.3**

The total turbulence head loss ($h_{turb\ loss}$) is given by

$$h_{turb\ loss} = \frac{v^2}{2g}\left(K_{entrance} + K_{bend} + K_{contraction} + K_{valve}\right)$$

Note that there may be more than one bend or contraction. K in all cases represents a head loss coefficient.

Head loss coefficients for intakes ($K_{entrance}$)

Entrance profile

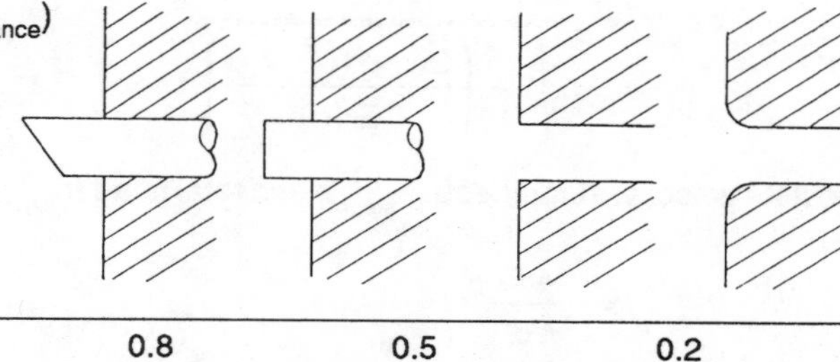

$K_{entrance}$	1.0	0.8	0.5	0.2

Head loss coefficients for bends (K_{bend})

Bend profile

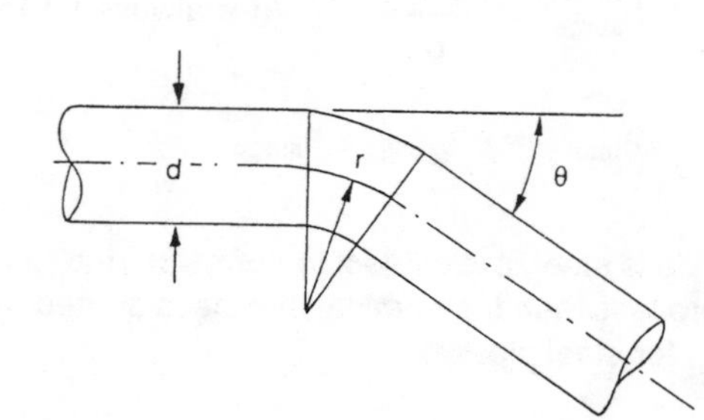

r/d		1	2	3	5
K_{bend}	(Θ = 20°)	0.36	0.25	0.20	0.15
K_{bend}	(Θ = 45°)	0.45	0.38	0.30	0.23
K_{bend}	(Θ = 90°)	0.60	0.50	0.40	0.30

Head loss coefficents for sudden contractions ($K_{contraction}$)

Contraction profile

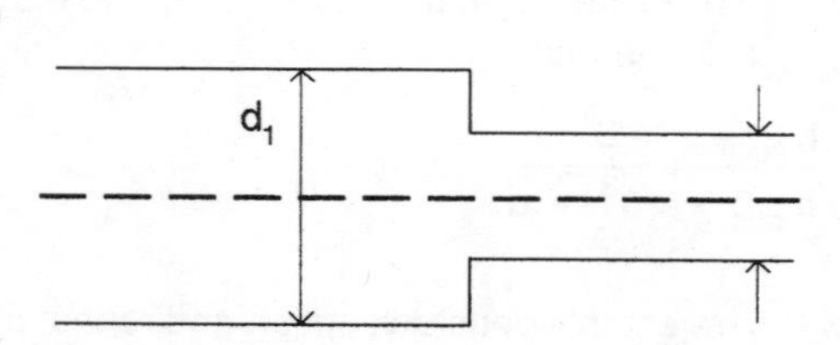

d_1/d_2	1.0	1.5	2.0	2.5	5.0
$K_{contraction}$	0	0.25	0.35	0.40	0.50

Head loss coefficients for valves (K_{valve})

Type of valve	Spherical	Gate	Butterfly
K_{valve}	0	0.1	0.3

Note 3.11.3 **Wall thickness of penstock**

1 Consult a pipe manufacturer's table, and choose a pipe with internal diameter 'd' and associated wall thickness 't'.

2 Calculate the pressure wave velocity 'a' from the following equation, using Table 3.11.4 to find 'E'. Remember all units are SI, so the wall thickness t is expressed in metres.

$$a = \frac{1400}{\sqrt{1 + \left(\frac{2.1 \times 10^9 \times d}{E \times t}\right)}}$$

3 Calculate velocity, surge head (h_{surge}), and total head (h_{total}):

$$v = \frac{4\,Q}{\pi\,d^2}$$

$$h_{surge} = \frac{a\,v}{g} \qquad \text{(g is always 9.8)}$$

$$h_{total} = h_{gross} + h_{surge}$$

4 If the pipe is steel, it is subject to corrosion and to welding or rolling defects. Its effective thickness is therefore less than the nominal thickness quoted by the manufacturer. To find effective thickness ($t_{effective}$) for steel pipes:

Welding	divide by 1.1
Flat-rolled	divide by 1.2
Corrosion	subtract for 10 years life: 1 mm subtract for 20 years life: 2 mm

If the pipe is PVC refer to the pipe manufacturer for low temperature thickness correction factors. For instance, $t_{effective}$ may be $0.5 \times t$ if temperatures are sub zero. Avoid bending and point stresses. In tropical climates, if the PVC pipes are carefully laid, $t_{effective} = t$.

5 Calculate the safety factor (SF) from the equation below, using Table 3.11.4 to find S and remembering that all units are SI, so $t_{effective}$ is in metres:

$$SF = \frac{t_{effective} \times S}{5 \times h_{total} \times 10^3 \times d}$$

6 If the safety factor is below 3.5, reject this penstock option and repeat the above calculation for stronger walled options. In certain circumstances it is legitimate to accept a safety factor of 2.5 but only if the surge head is calculated as above and all of the following conditions are met:

- Staff are experienced with similar pressures and materials
- Slow closing valves and design to avoid "emergency slam" stoppage of flow
- Low damage costs and safety risks (gentle slopes, low heads)
- Careful pressure testing to total head before commissioning

7 Enter the calculation results in a table such as Table 3.11.1 together with head loss and cost-benefit calculations.

Wall thickness of penstock — Example 3.11.2

1 Consider the 10 bar 16" PVC pipe (Table 3.11.5). It has a wall thickness of 23 mm, and an internal diameter of 0.352 m.

2 From Table 3.11.4:

$$a = \frac{1400}{\sqrt{1 + \left(\frac{2.1 \times 10^9 \times d}{E \times t}\right)}} = \frac{1400}{\sqrt{\left(1 + \frac{2.1 \times 10^9 \times 0.352}{2.8 \times 0.023}\right)}} = 396 \text{ m/s}$$

3 Calculate surge and total head:

$$v = \frac{4Q}{\pi d^2} = \frac{4 \times 0.19}{3.14 \times 0.352^2} = 2.0 \text{ m/s}$$

$$h_{surge} = \frac{a v}{g} = \frac{396 \times 2}{9.8} = 81 \text{ m}$$

$$h_{total} = h_{gross} + h_{surge} = 75 + 81 = 156 \text{ m}$$

4 If we imagine that this PVC pipe is used in a tropical climate where temperatures do not fall below zero, and the pipe is laid carefully without bending or point stresses, then $t_{effective} = t = 23$ mm.

5 From Table 3.11.4:

$$SF = \frac{t \times S}{5 \times h_{total} \times 10^3 \times d} = \frac{0.023 \times 28 \times 10^6}{5 \times 156 \times 10^3 \times 0.352} = 2.3$$

6 This safety factor falls below the recommended value of 3.5. The pipe may fail under pressure surge conditions. A pipe with greater wall thickness should be selected.

Physical characteristics of common materials — Table 3.11.4

Material	Young's Modulus (E) N/m²	Coefficient of linear expansion (a) m/m °C	Ultimate tensile strength (S) N/m²	Density (ρ) kg/m³
Steel	200×10^9	12×10^{-6}	350×10^6 *	7.8×10^3
uPVC	2.8×10^9	54×10^{-6}	28×10^6	1.4×10^3
HDPE/MDPE	$0.2 - 0.8 \times 10^9$	140×10^{-6}	$6 - 9 \times 10^6$	0.9×10^3
Ductile iron	170×10^9	11×10^{-6}	350×10^6	0.7×10^3
Cast iron	100×10^9	10×10^{-6}	140×10^6	7.2×10^3
Asbestos cement	variable	8×10^{-6}	variable	$1.6 - 2.1 \times 10^3$
Concrete	20×10^9	10×10^{-6}	variable	$1.8 - 2.5 \times 10^3$

* In some countries steel of a lower strength may be supplied. If you are uncertain it is possible ask for samples to be tested by university laboratories. Other materials such as PVC can also be independently checked for strength (S) and elasticity (E).

Table 3.11.5 **Pipe dimensions and prices**

PVC pipe prices ($) per metre length

Nominal diameter		**Maximum working pressure 6 bar**			**Maximum working pressure 10 bar**		
		Internal diameter	**Minimum wall thickness**	**Price per metre**	**Internal diameter**	**Minimum wall thickness**	**Price per metre**
inches	mm	mm	mm	$	mm	mm	$
6	160	148	6	14	140	9	22
9	225	208	8	27	198	13	43
11	280	259	10	42	246	16	67
11	315	292	11	46	277	18	74
14	355	329	13	53	312	21	85
16	400	370	14	64	352	23	100

Section lengths: 6m
Price per joint: $4.
Handling/delivery: allow $2 per section.
Installation (including burying): allow $2 per section.

Mild steel pipe prices ($) per metre length unprepared/unpainted

Nominal diameter	inches	**6**	**8**	**10**	**12**	**14**	**16**	**18**
	mm	**150**	**200**	**250**	**300**	**350**	**400**	**450**
Thickness	**1/8**	26	27	31	33	34	44	48
inches	**3/16**	29	32	38	40	43	54	61
	1/4	31	36	42	46	50	61	69
	5/16	39	46	55	60	64	79	91

Shot blasting/painting/flange welding: Increase the prices above by 15%.
Handling/delivery: allow $5 per metre.
Installation: allow $5 per metre.

Flanges for mild steel pipe prices ($), pipes are bought in 2 m long sections

Nominal pipe diameter	**Inner diameter**	**Outer diameter**	**Thickness**	**$ per pair of flanges with bolts and seals**
inches	mm	mm	inches	
6	150	250	3/8	19
8	200	300	3/8	20
10	250	350	1/2	28
12	300	400	1/2	31
14	350	450	3/4	50
16	400	500	3/4	52
18	450	550	3/4	56

Note: These prices are based approximately on quotations in Sri Lanka in 1990.

Economic optimization of penstock — Note 3.11.4

1. Assuming that you have already entered all details into Table 3.11.1 for a range of penstock options for the site specified, now calculate the penstock cost (factors 1 and 2) for each of the options. Do not omit delivery, installation, or jointing costs.
2. Estimate an acceptable economic target for the entire installation (factor 10). This is done by projecting similar costs per kW as found economic in other schemes in the region (or by actually calculating budget from knowledge of revenues and acceptable payback times).
3. Choose any option to be a reference option and cost this under the 'total cost' heading at the nominal cost figure.
4. For each of the other options calculate any 'knock-on' cost implications they may have (factor 4) and enter the difference in prices between them and the reference under 'additional cost'. Then enter the difference in total scheme cost in the column 'total cost'.
5. Calculate the difference in kW delivered between each option and the reference, enter it under 'additional benefit', and enter the implied total kW delivered under 'total benefit'.
6. Consider carefully whether any of the qualitative factors listed, such as peak power (factor 7) or matching (factor 6), should be entered under 'comment', possibly having the effect of ruling out some of the options or emphasizing their desirability.
7. Divide total cost by total benefit and enter the ratio under 'comparison'. If appropriate enter a comment instead based on qualitative factors. Choose the option giving the lowest overall cost/kW ratio, or the best cost-benefit performance based on qualitative factors.

Economic optimization of penstock — Example 3.11.3

See Note 3.11.4.

1. Prices entered into Table 3.11.1 are calculated from the prices given in Table 3.11.5.
2. Assume in this case that $1000 per kW is an economic target, so that the whole budget for a nominally 70kW scheme will be $70,000. The nominal 70kW is calculated from the equation

$$P_{net} = e_o \times H \times Q \times 10 \text{ kW}$$

where e_o is 0.5 representing a nominal overall efficiency of 50%.

3. Choose the14" PVC pipe as a reference. Any option could be chosen. Enter the nominal total cost figure, in this case $70,000. Under the 'total benefit' column write the nominal kW produced, in this case 70kW, and then write the target ratio of 70,000/70 =1,000 $/kW in the 'comparison' column.
4. The 12" steel pipe produces less power because of the greater head loss. This may make a difference to the cost of turbine, generator and transmission. Assume here that it doesn't.
5. See Table 3.11.1. Note that the 12" steel pipe produces less power because it suffers a greater head loss.
6. It may be that the power produced by the inefficient penstock is not enough for the needs of the community, in which case enter a comment 'insufficient power' and reject this option. There may be matching difficulties, as described in the text. Assume that in this case all power produced is being sold at a fixed price, so that all options are considered equally.
7. The lowest overall cost/kW ratio indicates the most economic penstock. The entries shown in Table 3.11.1 illustrate this (12" steel pipe gives the lowest ratio), but because none of the entered pipes fulfill the wall thickness safety factor recommendation the entries are invalid. Pipes with thicker walls or stronger material should be specified to complete this exercise.

3.12

Penstocks – Supports and anchors

Anchors, slide blocks, and thrust blocks are used to constrain movement of the penstock. They should be placed on original soil and not fill. The bearing area must be calculated to support the pipeline without exceeding the safe bearing load of the soil. Drainage should be provided to prevent erosion of support foundations.

Distance between blocks

Both anchors and slide blocks act to support the penstock. It is possible to calculate the maximum spacing of blocks so that the pipeline does not collapse between supports when filled with water. This depends on predicting crumpling of the top of the pipe which is difficult to do precisely. It is preferable to use Table 3.12.1 for spacing of supports. Generally, if in doubt use one sliding support for each pipe length.

The important criterion in the case of steel pipes is the jointing system. For any flexible coupling method, one support per length is needed. For flanges to British Standard (min thickness 16mm) the pipes can be considered as one piece and Table 3.12.1 is relevant. For more cost effective flange thicknesses (1 or 2 times pipe thickness) one support per length is needed.

In the case of PVC the first rule is to bury it wherever possible. If you are forced to use the PVC pipe above ground, pipe should never be allowed to bend as it is susceptible to fatigue weakening. Follow makers' recommendations, almost always 1 support per length. Cast iron, ductile iron and concrete pipes commonly use spigot and sockets joints. These can take minimal bending so use one support per length.

Photo 3.12.1 An anchor block is placed at a bend and a slide block above it. (Nepal)

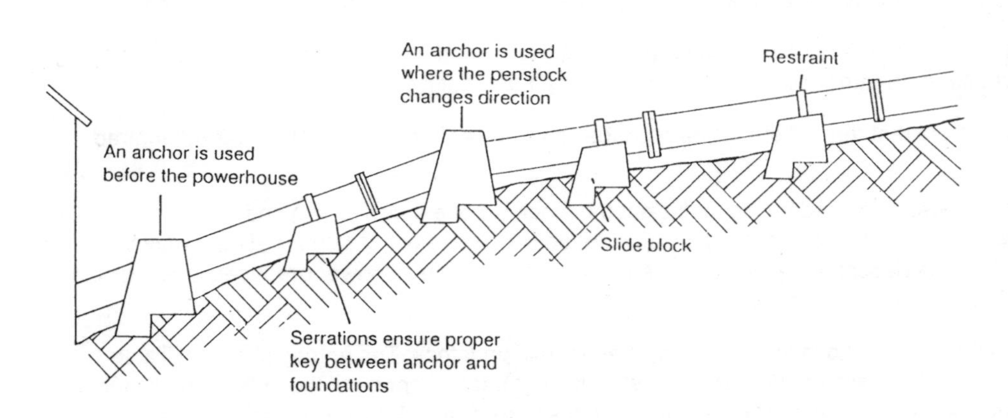

Fig 3.12.1 Slide blocks and anchors

Slide blocks

A slide block carries the weight of pipe and water and restrains the pipe from upward and sideways movement, but allows it to move longitudinally. A slide block is used to remove bending stresses caused by the weight without incurring the cost of a full anchor. It also stops the pipe from pulling apart due to upwards or sideways movement caused by pressurization inside. It can be a light construction so long as it is designed for very low longitudinal friction (see Fig 3.12.3) in contact with the pipe, so that it does not have to withstand significant longitudinal forces. If the block is not designed for low longitudinal friction, or does not function correctly, it is effectively an anchor block and should be sized as described for an anchor block.

If the penstock is buried slide blocks are unnecessary, but make sure instead that the pipe is laid into a trench on a bed of sand or gravel of consistent quality, with no big stones which could cut into the pipe wall or cause stress concentrations on the pipe wall. Stress concentrations will lead eventually to failure of PVC pipe due to vibrations caused by the water movement. Another danger posed by stress concentrations is that a vehicle passing over the buried pipe may then burst it.

Photo 3.12.2 Angle iron rocking frames.

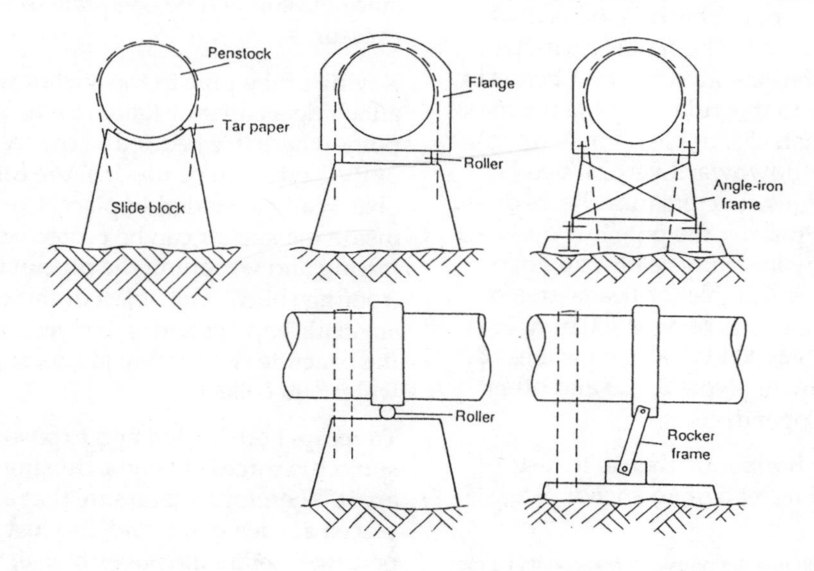

Fig 3.12.2 Slide blocks Use of rollers or tar paper to reduce friction and abrasion on a slide block. Restraint hoops are used to prevent upwards or sideways movement; these are shown by dashed lines.

Support spacing in metres — **Table 3.12.1**

This applies only to penstocks fabricated from mild steel, welded or flanged to British Standard. In other cases use one support per pipe length.

Diameter mm	**100**	**200**	**300**	**400**	**500**
Thickness mm					
2	2	2	2.5	3	3
4	3	3	3	4	4
6	4	4.5	5	6	6

Anchor blocks

An anchor block is usually a mass of reinforced concrete keyed to the penstock so that the penstock cannot move in any way relative to the block.

An anchor should always be placed at a vertical or horizontal bend in the penstock. An anchor block should always be placed at the point at which the penstock enters the powerhouse, to protect the turbine. Anchors are also placed along straight sections of pipe, each one next to an expansion joint.

On small low/medium head installations (below 20 kW and 60 m head) with straight penstocks (whether buried or not buried) anchors can be designed according to this simple rule: place an anchor every 30 metres, by keying the pipe to a cubic metre of concrete for every 300 mm diameter of pipe – in other words, a 200 mm pipe only needs 0.67m^3 of concrete. This method is illustrated in Example 3.12.1. The "cubic metre per 300 mm" rule can be extended to cover bends in the pipe according to this rule: "double the mass for bends of less than 45°, treble it for sharper bends". In cases of downward vertical bends make sure that all this concrete mass bears down on the pipe, either placing the pipe low down in a larger block, or by inserting reinforcing bars. This approach is not suitable for heads greater than 60 m, pipe diameters of over 300 mm, or schemes of more than 20 kW rating; for these do a full analysis following Note 3.12.3 and other texts as listed in Appendix 3.

If a buried bend is horizontal then a 'thrust block' can be used instead of an anchor. See Note 3.12.2 for an illustration method for sizing of thrust blocks. The thrust block transmits forces to the surrounding soil rather than relying on its own weight. Remember that a buried *downward* bend will still need an anchor block since the soil will be adequate to take the bearing pressures.

Photo 3.12.3 In this scheme the penstock rests on its flanges and forebay spill is directed along the penstock bed. (Indonesia)

Photo 3.12.4 An anchor block in the foreground provides mass above a bend. Slide blocks are used on the steep slope. Note the expansion joint. (Nepal)

Keying of the pipe to the anchor is best done by attaching a collar or flange to the pipe at the point where it is bedded in concrete. In the case of PVC glue can be used; in the case of steel any object can be welded in place. Corrosion of steel inside the anchor can be protected against by tarring and wrapping the pipe in tar sheet (roofing sheet). The object of this is to avoid a small air gap appearing between the steel and the concrete (which will shrink slightly) in which water will collect.

To recap, both buried and exposed pipe are subject to forces at bends, causing joints to pull apart. Therefore, make sure that anchors are placed at each bend, and also just before the penstock enters the powerhouse, to reduce stress on the turbine casing.

Force calculations

On larger schemes (above 20kW) the quick method is not reliable, and a full calculation of forces is necessary. To do this, first distinguish between the portion of pipe upstream of the anchor (which has length termed L_u), and the portion downstream (length L_d). Calculate forces for each portion separately, then add the results.

Fig 3.12.4 shows an anchor and above it a flexible coupling or expansion joint, allowing longitudinal movement of the pipe to take place without affecting the pipe further upstream. The presence of this coupling defines the end of the upstream

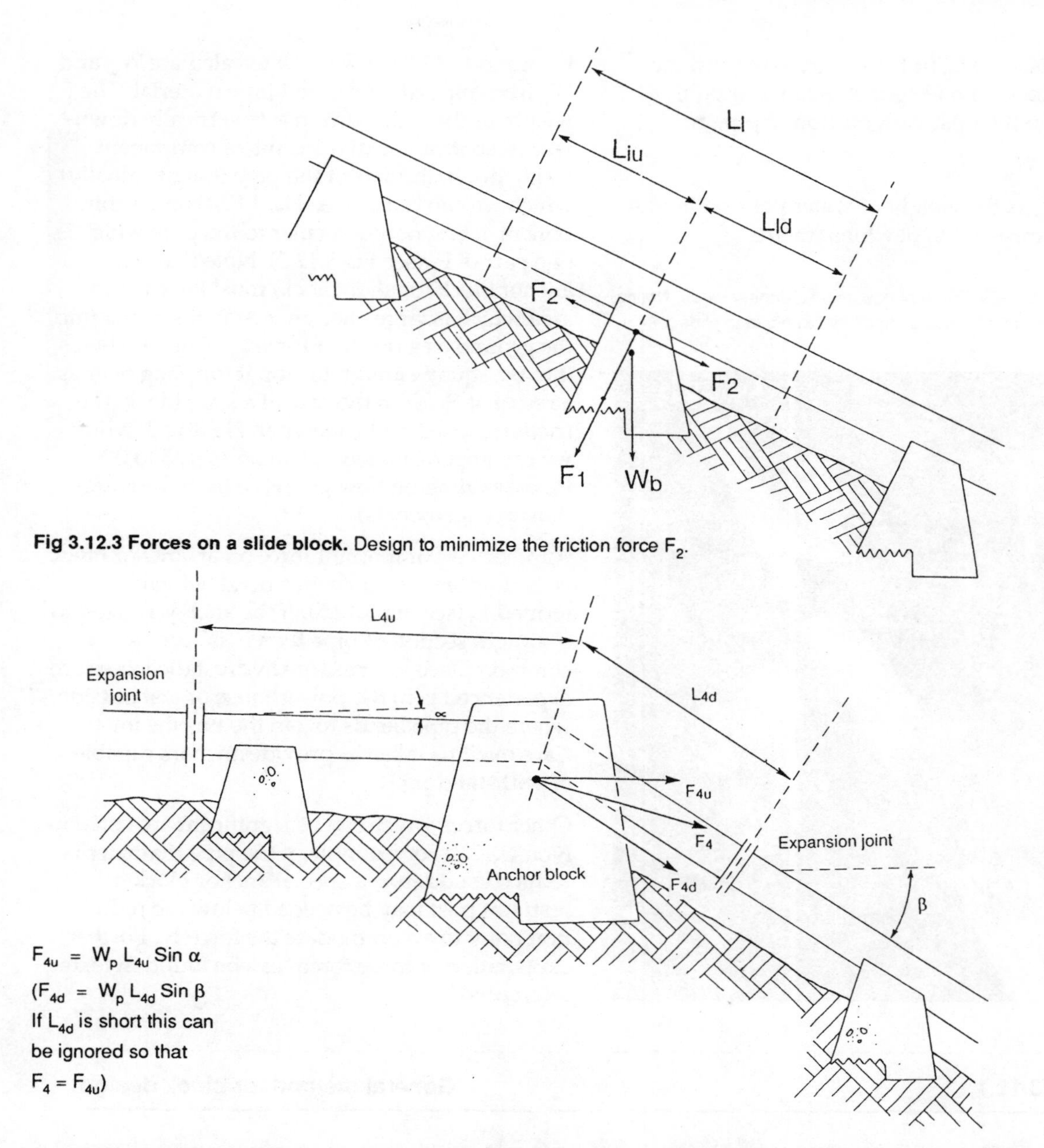

Fig 3.12.3 Forces on a slide block. Design to minimize the friction force F_2.

$F_{4u} = W_p L_{4u} \sin \alpha$

($F_{4d} = W_p L_{4d} \sin \beta$

If L_{4d} is short this can be ignored so that

$F_4 = F_{4u}$)

Fig 3.12.4 Force F_4 due to the weight of the pipe. Note that the length of pipe is different on either side. Note also that the weight of water does not create an axial force at this point but instead at the turbine where the force is used to rotate the turbine.

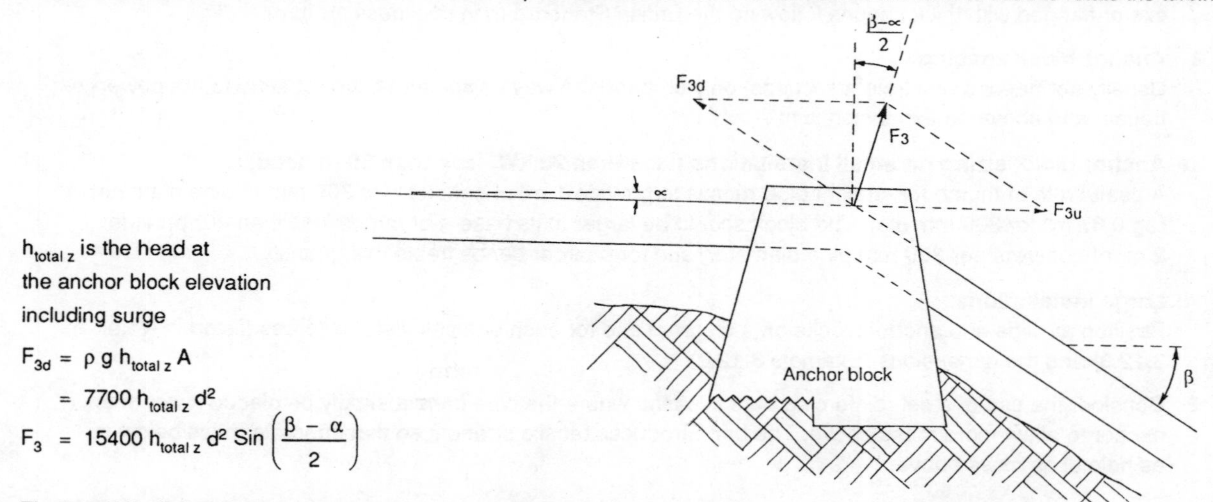

$h_{total\,z}$ is the head at the anchor block elevation including surge

$F_{3d} = \rho\, g\, h_{total\,z} A$

$= 7700\, h_{total\,z}\, d^2$

$F_3 = 15400\, h_{total\,z}\, d^2 \sin\left(\frac{\beta - \alpha}{2}\right)$

Fig 3.12.5 Hydrostatic force F_3 Note this is a vertical bend but the equations are the same for a horizontal bend.

pipe at distance L_u (u for 'upstream') from the anchor block. The weight of water and pipe material in the upstream portion of pipe is:

$$(W_p + W_w) \times L_u$$

where W_w is the weight of water per unit length of pipe expressed in newtons/metre.

Photo 3.12.5 A PVC pipe bridge using catenary wires. Notice the supports have apertures to allow the pipes to slide. (Colombia)

Example 13.12.2 shows how to calculate W_w and W_p from pipe diameter and pipe material. The weight of the pipe portion acts vertically downwards, so it can be divided into a component acting down the axis of the pipe (longitudinally) which we can call F_4 (see Fig 3.12.4) and a component acting perpendicular to the pipe which is one part of F_1 (see Fig 3.12.3). Note that our anchor block (or slide block) must have a base area which is large enough to stop it sinking into the ground as a result of force F_1. This also helps it to be 'squat' enough to stop it toppling over as a result of F_4 (or in the case of a slide block, the friction caused by F_2 shown in Fig 3.12.3, which we can approximately calculate as 0.25 to 0.5 × F_1, depending on how good the friction-reduction arrangement is).

Another very important force on an anchor block is the hydrostatic force on a bend, which is termed F_3 (see Fig 3.12.5). If the anchor is fixed to a straight section of pipe then F_3 is zero. A very common place where large hydrostatic forces are experienced is in the powerhouse or just outside, where the pipe bends to join the turbine inlets. Care must be taken to provide anchors capable of withstanding F_3.

Other forces which can be significant are listed in Note 3.12.3. For example, if the pipe diameter is reduced suddenly, a special anchor block or restraint post may be needed below the reduction point to accommodate the force F_9. Further explanation of these forces can be found in texts referenced.

Note 3.12.1 — **General method for block design**

1 **Slide block spacing.**
In most cases place one slide block for each length of pipe. Only use Table 3.12.1 if pipe is welded on site or flanged with thick flanges following the British Standard (min thickness 16 mm).

2 **Anchor block spacing.**
Usually not needed on straight sections, only on bends. Always place an anchor at entry to the powerhouse, and above an expansion joint.

3a **Anchor block sizing on small installations (less than 20 kW, less than 60 m head).**
A useful rule of thumb for straight pipe runs is to provide 1 m^3 of concrete for 300 mm of pipe diameter (eg 0.67 m^3 for 200 mm dia). The block should be larger at its base. For bends less than 45° provide 2 m^3 of concrete per 300 mm pipe diameter, and for sharper bends treble this quantity.

b **Large installations.**
Position all slide and anchor blocks on a diagram and for each one calculate all forces (listed in Note 3.12.3) and the dimensions (Example 3.12.2).

4 Consider the use of steel reinforcing bars in cases where the pipe cannot simply be placed lower in an anchor to allow more mass above. The bar introduces tensile strength so that concrete mass below is as helpful as mass above.

Thrust block sizing

Note 3.12.2

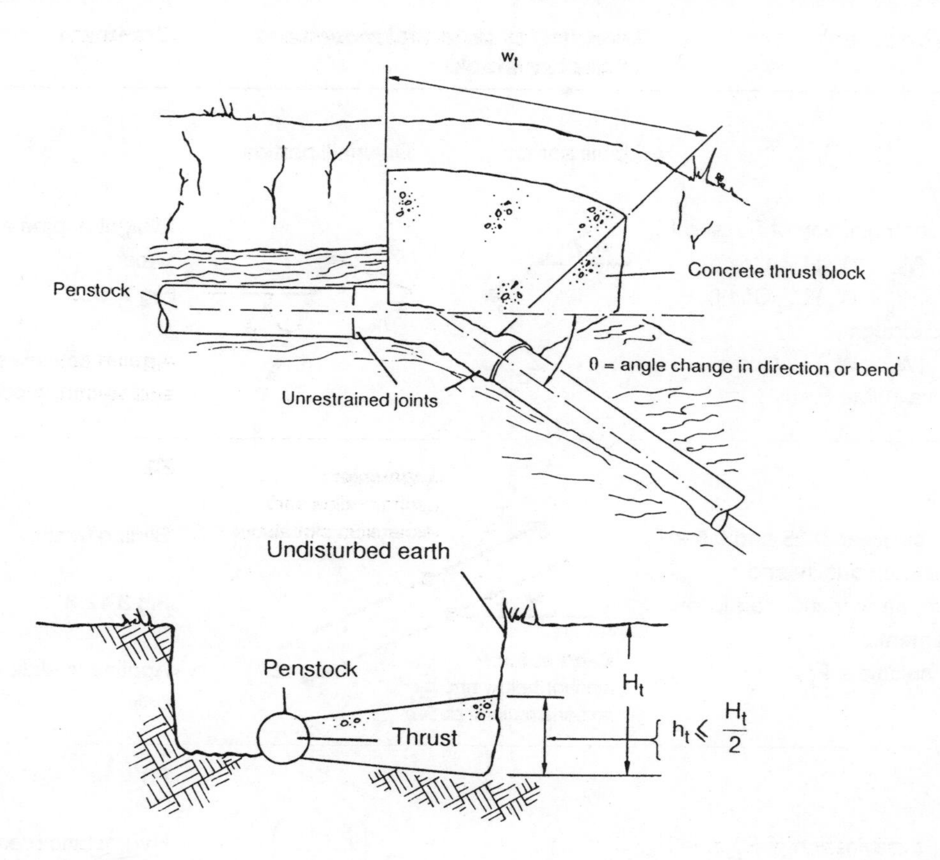

Pipe size mm	Nominal thrust bearing area (A_n) in m²	
	45° Bend	90° Bend
100	0.2	0.2
150	0.3	0.5
200	0.3	0.8
250	0.5	1.2
300	0.7	1.7
350	0.9	2.3
400	1.6	3.0

Multipliers	
Soil type	m
Clay	4
Sand	2
Sand/gravel	1.3
Gravel/shale	0.4

To design a thrust block on small penstocks (below 20 kW and 60 m head) you must first find from the table the nominal area (A_n) which bears against the soil. The type of soil also has an influence, so use the multiplier (m) to find the thrust bearing area (A_t):

$$A_t = m \times A_n$$

The height (h_t) must be less than half the full depth of the trench (H_t) as shown above. Having determined h_t from this rule, the width W_t can be found:

$$W_t = \frac{A_t}{h_t}$$

On larger schemes or schemes with higher heads the value of hydrostatic force must be found (force F_3 in Note 3.12.3) and the thrust area found following the method for sinking stability (Note 3.12.4 and Example 3.12.2).

Note 3.12.3 **Forces on anchor and slide blocks**

Force (Newtons)	Direction of potential movement of anchor block	Comment
F1 F_1 = combination of F_{1u} and F_{1d} $F_{1u} = (W_p + W_w)\, L_{1u} \operatorname{Cos} \alpha$ $F_{1d} = (W_p + W_w)\, L_{1d} \operatorname{Cos} \beta$ If pipe is straight, $F_1 = (W_p + W_w)\, L_1 \operatorname{Cos} \alpha$ Always calculate F_1	Uphill portion Downhill portion 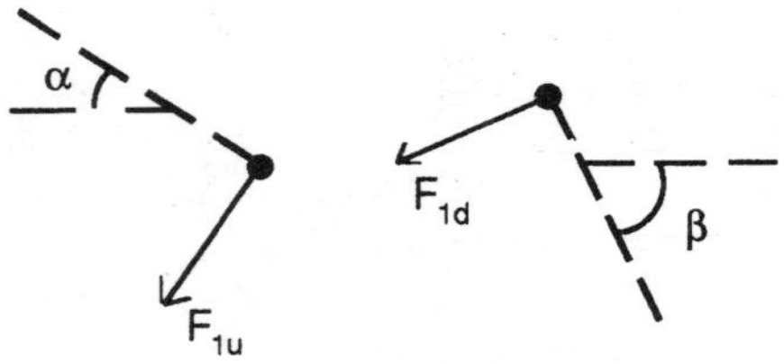α F1u F1d β	**F1** Weight of pipe and water Fig 3.12.3 Applies to slide blocks and anchor blocks
F2 F_2 = between 0.25 and 0.5 × F_1 depending on quality and deterioration of friction reduction arrangement. Always calculate F_2	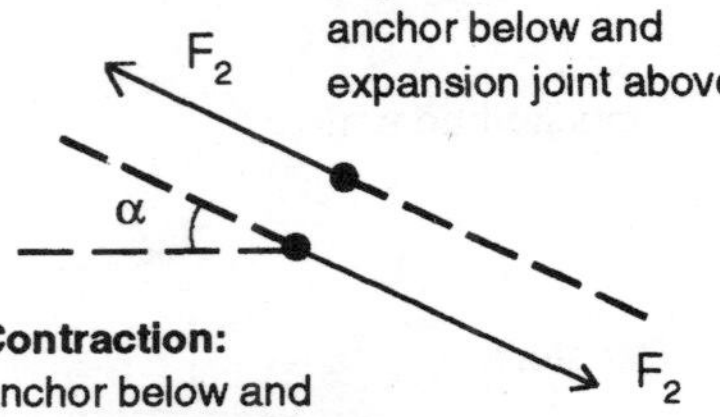**Expansion:** anchor below and expansion joint above F2 α **Contraction:** anchor below and expansion joint above F2	**F2** Sliding friction Fig 3.12.3 Applies to slide blocks only
F3 F_3 = combination of F_{3u} and F_{3d} $F_3 = 15400\, h_{total\,z}\, d^2 \operatorname{Sin}\left(\frac{\beta - \alpha}{2}\right)$ For right angle bends, $F_3 = 1.414 \times F_{3u}$ (diff directions) For 45° bends, $F_3 = 0.76 \times F_{3u}$ (diff directions) Always calculate F_3	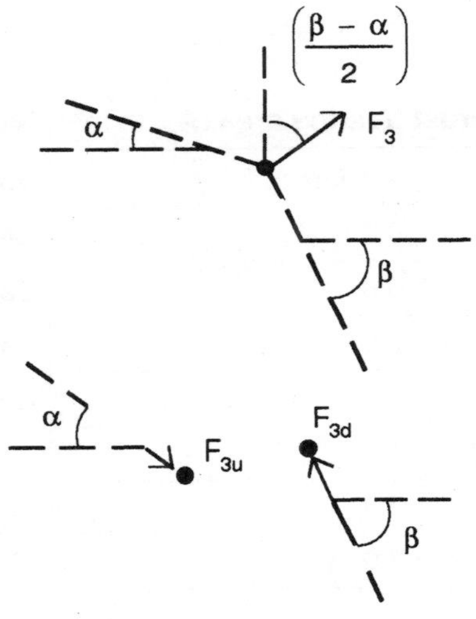(β − α / 2) α F3 β α F3u F3d β	**F3** Hydrostatic pressure. Only applies to horizontal and vertical bends. Do not use F_3 if already using F_{3d} and F_{3u}. Fig 3.12.5 Note that '$h_{total\,z}$' is the head or pressure at the anchor block (elevation Z) including surge head Applies to anchor blocks only
F4 F_4 = combination of F_{4u} and F_{4d} $F_{4u} = W_p\, L_{4u} \operatorname{Sin} \alpha$ $F_{4u} = W_p\, L_{4d} \operatorname{Sin} \beta$	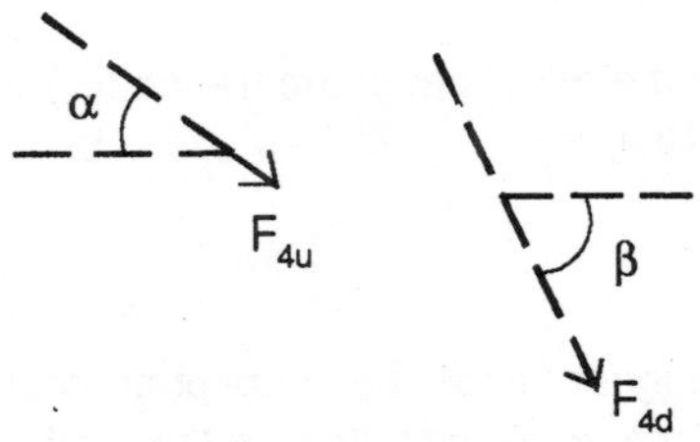α F4u β F4d	**F4** Component of pipe weight acting down length of pipe. Ignore unless the pipe is heavy and both angles more than 20°. Fig 3.12.4 Applies to anchor blocks only

continued opposite

Forces on anchor and slide blocks — **Note 3.12.3** *continued*

Force (Newtons)	Direction of potential movement of anchor block	Comment
F5 $F_5 = E\,a\,(T_{hot} - T_{cold})\,\pi\,d\,t_{wall}$ (As usual all units SI so t_{wall} is in metres). See Table 3.8.2 for values of E and a	Uphill portion / Downhill portion 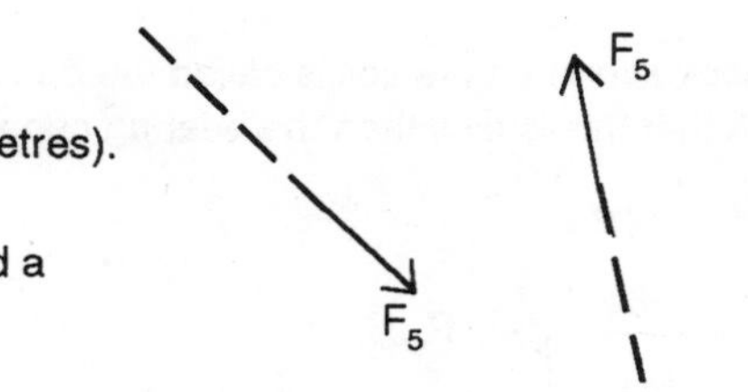	**F5** When no expansion joints included F_5 is the thermally induced stress restrained by the anchor block. Applies to anchor blocks only
F6 $F_6 = d \times 10^4$ Usually insignificant	F_6 directions as F_5	**F6** Friction in expansion joint. The F6 force is felt because the joint will resist sliding. Applies to slide blocks only
F7 $F_7 = 3 \times 10^4\,h_{total}\,dt_{wall}$ Usually insignificant	F_7 directions as F_9	**F7** Hydrostatic pressure on exposed ends of pipe within expansion joint Applies to anchor blocks only
F8 $F_8 = 2.5 \times 10^3 \left(\frac{Q^2}{d^2}\right) \text{Sin}\left(\frac{\beta - \alpha}{2}\right)$ Usually insignificant	F_8 directions as F_3	**F8** Dynamic pressure at a bend due to change in direction of moving water. Velocities are usually low in penstocks so this force is small. Applies to anchor blocks only
F9 $F_9 = h_{total} \times \left(d_{big}^2 - d_{small}^2\right) \times 10^4$ Usually insignificant	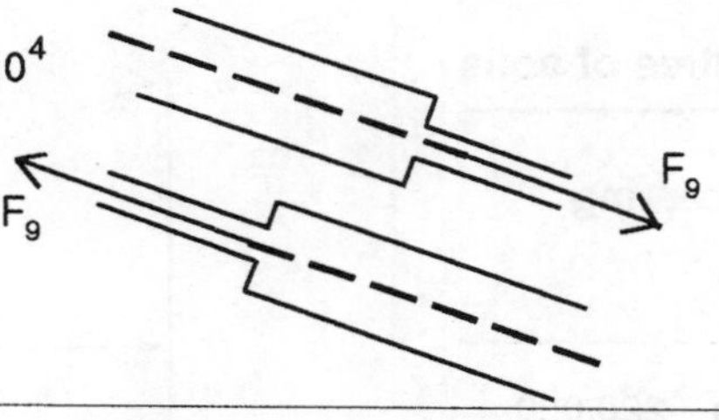	**F9** Reduction in pipe diameter from d_{big} to d_{small} Applies to anchor blocks only
W_b $W_b = Vol_{block} \times \rho_{block} \times g$ Note the density of concrete ($\rho_{concrete}$) = 2300 kg/m³ Always calculate W_b	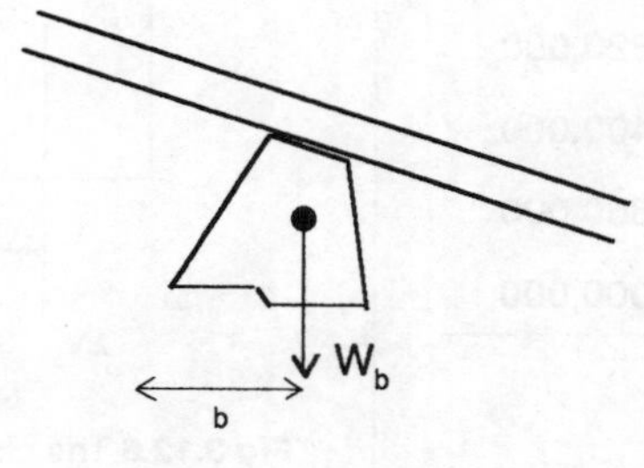	**W_b** Weight of block. Note that the weight does not necessarily act in the middle Applies to anchor and slide blocks

Note 3.12.4 — Conditions for stability

A support or anchor is shaped and sized in such a way that it cannot move subject to the forces imposed by the penstock. It must not sink into the ground, slide along the ground, or turn in the ground (allowing the penstock to sag above it). To test that these three conditions will be met the following equations must be satisfied:

1 Soil sinkage

The pressure the base of the block exerts on the soil is called the base pressure (P_{base}). The base area (A_{base}) must be big enough such that this is less than the bearing capacity of the soil (P_{soil}).

The condition for stability is $P_{soil} > P_{base}$

$$P_{base} = \frac{\Sigma V}{A_{base}}\left(1 + \frac{6e}{L_{base}}\right) < P_{soil}$$

Values for P_{soil} are given in Table 3.12.2. ΣV is the sum of the vertical forces (Fig 3.12.6). e and L_{base} are also illustrated in Fig 3.12.6.

2 Sliding

If the sum of the horizontal forces on the block are greater than the frictional resistance of the soil, the block will slide. Therefore for stability:

$$\Sigma H < \mu \Sigma V$$

Where μ is the coefficient of friction between soil and concrete, normally taken as 0.5. Design of the block base with serrations or a step in it will also help avoid sliding.

3 Toppling

All the vertical forces on the block can be resolved into one single force (ΣV) as shown in Fig 3.12.6. If this is too far toward the edge of the block it will tend to topple the block or turn it in the ground causing it to sink slightly (a slight sinkage can cause expensive damage to the penstock). The rule of thumb usually used to protect against this is to design the block so that this force acts within the middle third of the length of the base.

The rule is

$$e < \frac{L_{base}}{6}$$

Table 3.12.2 Bearing capacities of soils

Soil type	Maximum bearing pressure P_{soil} (N/m^2)
Clay	180,000 to 220,000
Sand	200,000 to 320,000
Sand/gravel	300,000 to 400,000
Sand/gravel/clay	350,000 to 650,000
Rock	600,000 to 1,000,000

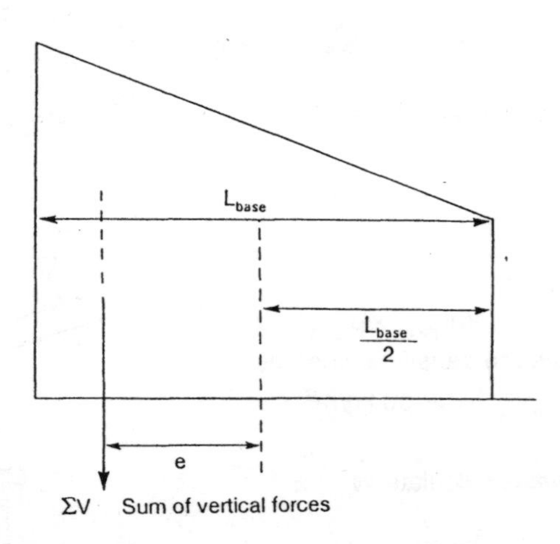

Fig 3.12.6 The dimension L_{base} and the eccentricity 'e' are defined by the diagram.

Sizing of slide and anchor blocks – quick method **Example 3.12.1**

A 100 mm internal diameter PVC penstock is chosen for a 2 kW (output power) site. The pipe is to be buried in a 1 metre deep trench in sandy/gravelly soil most of its length on a hillside, as shown in the diagram, until the final 5 metres before it enters the powerhouse. It then reduces to 50 mm flexible pipe which takes a sharp right-angle bend in order to form a join with a pump-turbine installation. The head is 50 metres and the pipes are delivered in 6 metre lengths. The pipe wall thickness is 5 mm. There are one horizontal and three vertical bends. Position and size all blocks.
Refer to the method outlined in Note 3.12.1.

1 **Slide block.** Not needed for support, instead a carefully prepared gravel or sand bed must be laid in the trench on which to rest the pipe safely. It should not contain any sharp stones or large items such as bricks. Restraints (to stop upward and sideways movement) will be needed while testing the penstock for leaks (before burying).

2 **Anchor block spacing.** Since this installation is small (less than 20 kW) the rule of thumb given in Note 3.12.1 for securing the penstock can be used. The penstock is to be buried, but the soil is loose and therefore may allow some movement of the pipe even underground. A sensible precaution would be to include anchor blocks every 30 metres. Note that these will be useful when pressure testing the pipe before filling the trench. An anchor must be placed at entry to the powerhouse; in this case it could be at the final 45° vertical bend 5 m prior to the powerhouse. Three more anchors are needed at bends, two more buried vertical bends and one more buried horizontal bend (referred to as the thrust block).

3 **Anchor block sizing.** For straight sections, following the '1 m³ per 300 mm dia' rule, each block would be 0.3 m³, dimensions 0.7 m × 0.7 m × 0.7 m. For vertical bends, simply double the mass of concrete calculated for straight runs if the bend is less than 45°, and treble it in cases of 45° or sharper bends. In all cases of downward vertical bends make sure the concrete mass is above the pipe, to restrain upward movement. This can be done either by insertion of steel bars or by increasing the height of the block and placing the pipe low in the block. The soil cannot be trusted to restrain movement, so even in the case of the final vertical bend, the anchor block must be sized as for other vertical bends or a second thrust block used here.

Thrust block size. For design of a thrust block on the horizontal bend, see Note 3.12.2. The bend angle is 45° so the bearing area will be $1.3 \times 0.2 = 0.26$ m². The trench is 0.5 m deep. Therefore top of the thrust block should be at least 0.25 metres below the surface and its height could be 0.25 metres. Its width is therefore 1 metre.

Exposed vertical bend. This is the bend after the valve just as the penstock enters the turbine. Note the diameter is reduced. Calculate forces F_{3u} and F_{3d} separately following Note 3.12.3 and Fig 3.12.5. Horizontal and vertical restraint can then be supplied, for instance, in the form of angle irons bolted to the powerhouse floor, or by ensuring the pipe is connected at either end by joints capable of withholding the tension force. Alos calculate F_9 and add to F_{3u}.

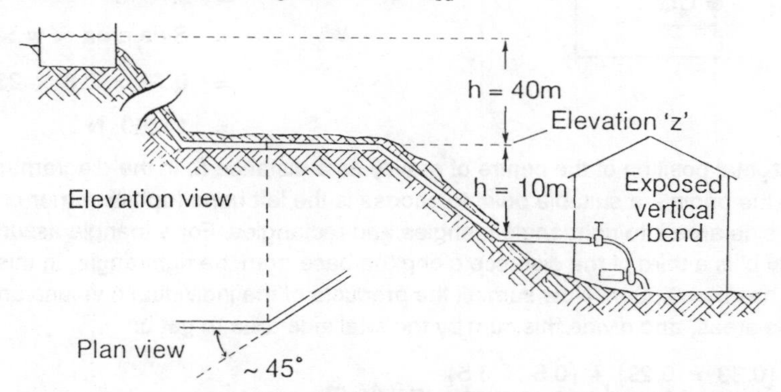

4 **Reinforcing bars.** Concrete blocks are strong in compression but weak in tension. To improve their performance in tension, either oversize the block (so using gravity to reduce tension forces) or use steel reinforcing bars. In all cases the concrete must be prepared correctly to ensure some strength in tension.

Example 3.12.2 **Sizing of slide blocks – full method**

For a particular penstock, a slide block is proposed with the shape shown here.

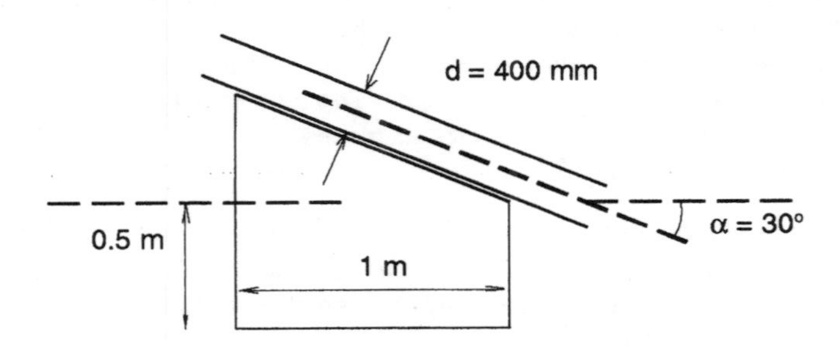

It has a width (w) of 0.8 m. It supports a mild steel penstock of diameter 400 mm and with wall thickness of 6 mm. Distance between slide blocks is 6 m. Slope angle (α) is 30°. Expansion joint position: assume an expansion joint is fitted above the slide block before an anchor but not below.

Decide whether the block can safely support the penstock without sinking, slipping, or overturning. Consider conditions of thermal expansion and contraction. Redesign the block if necessary.

1 Calculate sizes and directions of all forces listed in Note 3.12.3. Decide that only F_1, F_2 and W_b are of interest here. The pipe is straight (α = β) so $F_1 = (W_p + W_w)L_1 \cos \alpha$.

The density of steel (ρ_{steel}) is found from Table 3.11.4. See Fig 3.12.3.

$$W_p = \pi d t_{wall} \rho_{steel} g = 3.14 \times 0.4 \times 0.006 \times 7800 \times 9.81 = 577 \text{ N}$$

$$W_w = \pi \left(\frac{d^2}{4}\right) \rho_{water} g = 3.14 \left(\frac{0.4^2}{4}\right) \times 1000 \times 9.81 = 1233 \text{ N}$$

$$F_1 = (1233 + 577)6 \cos 30 = 9405 \text{ N}$$

Assume deteriorated friction reduction :

$$F_2 = 0.5 \times F_1 = 4703 \text{ N}$$

The weight of the block is found by calculating its side area first :

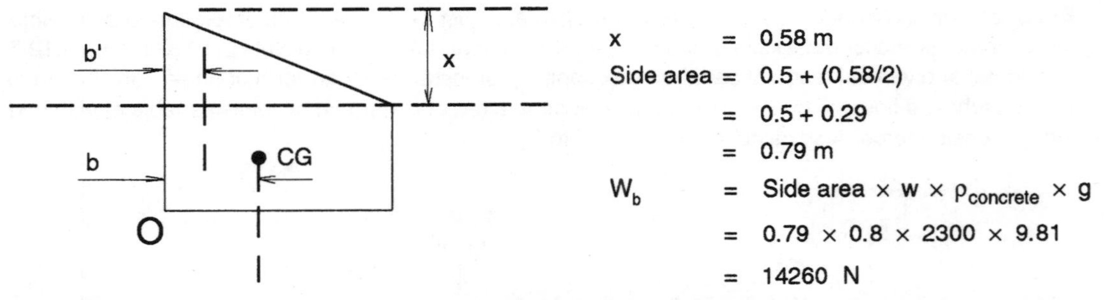

$x = 0.58$ m

Side area $= 0.5 + (0.58/2)$

$= 0.5 + 0.29$

$= 0.79$ m

$W_b = \text{Side area} \times w \times \rho_{concrete} \times g$

$= 0.79 \times 0.8 \times 2300 \times 9.81$

$= 14260$ N

Calculate the horizontal position of the centre of gravity as a distance (b in the diagram) measured from a point defined as the origin – a suitable point to choose is the left hand (uphill) corner of the block. To do this, divide the side area into right angle triangles and rectangles. For a triangle assume the horizontal centre of gravity b' is a third of the distance along the base from the right angle. In this case the triangle at the top has b' = 0.33 m. Now sum all the products of the individual b values and their corresponding side areas, and divide this sum by the total side area to get b:

$$b = \frac{(0.33 \times 0.29) + (0.5 \times 0.5)}{0.79} = 0.44 \text{ m}$$

In this case the block has a simple shape and we could have assumed a centre of gravity in the centre (b = 0.5). For more complex shapes which are used to save expensive concrete while increasing stability, the above procedure is necessary.

continued opposite

Sizing of slide blocks – full method

Example 3.12.2 *continued*

When the penstock is expanding it moves uphill over the slide block and the force F_2 is experienced by the slide block as an uphill pull. Resolve the forces F_1, F_2, W_b into x and z components. Negative signs indicate leftward x and upward z directions.

Forces in Newtons (N)	x	z
$F_1 = 9405$	F_1 Sin 30 = – 4702	F_1 Cos 30 = + 8145
$F_2 = 4703$	F_2 Cos 30 = – 4073	F_2 Sin 30 = – 2352
$W_b = 14260$	0	+ 14260
Total	ΣH **= – 8775**	ΣV **= + 20053**

Draw a scaled diagram to measure moment arms to the origin 'O' and resolve all moments around the origin. Clockwise moments are positive, anticlockwise moments are negative. Note that F_1 acts through a point halfway along the sloped surface, its position is not dependent on the centre of gravity.

$$\Sigma M = (F_2 \times F_2 \text{ arm}) + (F_1 \times F_1 \text{ arm}) + (W_b \times W_b \text{ arm})$$
$$= -(4703 \times 0.93) + (9405 \times 0.04) + (14260 \times 0.44)$$
$$= 2277 \text{ Nm}$$

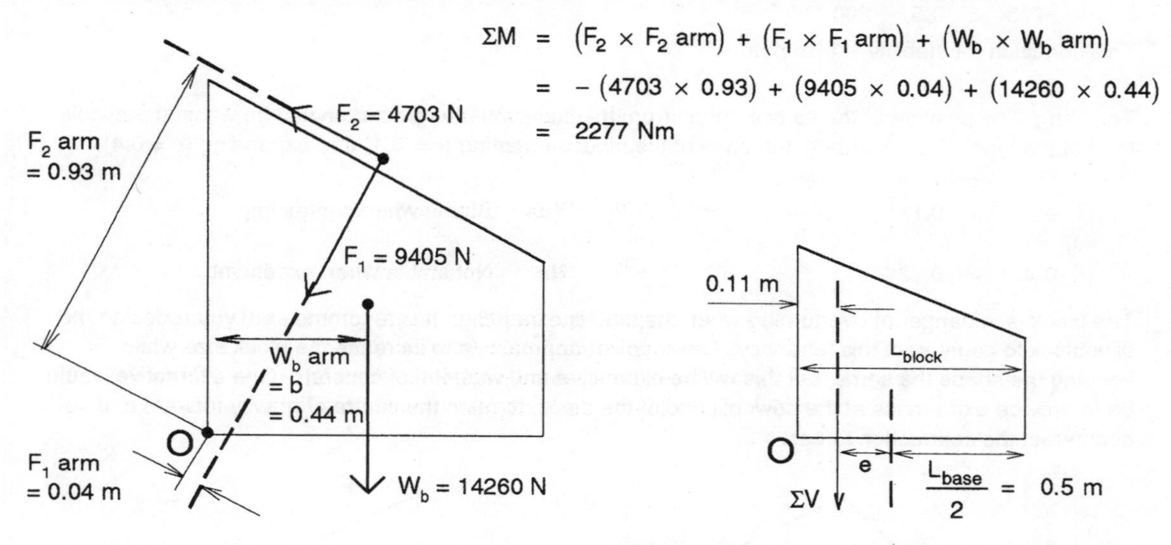

Replace the moment by moving the vertical force sideways. This gives a vertical force at a distance of 2277/20053 = 0.11 m from the origin.

Calculate the distance between the line of the new vertical force and the centre line of the block. This distance is called the eccentricity (e) of the vertical force. In this case e is 0.39 m.

In the case of a contracting penstock, everything is the same except the direction of F_2 which is now positive. Therefore repeating the x-z table :

Forces in Newtons (N)	x	z
$F_1 = 9405$	– 4702	+ 8145
$F_2 = 4703$	+ 4073	+ 2352
$W_b = 14260$	0	+ 14260
Total	ΣH **= – 629**	ΣV **= + 24757**

$$\Sigma M = (F_2 \times F_2 \text{ arm}) + (F_1 \times F_1 \text{ arm}) + (W_b \times W_b \text{ arm}) = 11024 \text{ Nm}$$

Replace this moment with the vertical force acting at 11024/20053 = 0.55 m from the origin, such that e is 0.05 m.

continued overleaf

Example 3.12.2 *continued* **Sizing of slide blocks – full method**

The conditions for block stability can now be tested:

1 Soil movement

$$P = \frac{\Sigma V}{A_{base}}\left(1 + \frac{6e}{L_{base}}\right)$$

$$\therefore \; P_{base} = \frac{20053}{0.8} \times 3.4 = 85425\ \text{N/m}^2$$

$\Sigma V = 20053$ N

$A_{base} = 0.8\text{m}^2$

$e = 0.4$ m

$L_{base} = 1$ m

Refer to Table 3.12.2. P_{soil} is 200000 to 320000 N/m² for sand. Therefore this value of P_{base} is not dangerous on sandy soil since the condition $P_{base} < P_{soil}$ is met.

2 Sliding ΣH is greatest when the pipe is expanding

$\Sigma H < \mu \Sigma V$ where $\mu = 0.5$

$8775 < 0.5 \times 20053$?

This condition for stability is also met.

3 Toppling. The position of the vertical thrust from the block into the ground should be within the middle third of the block. Consider both the case of the pipe contracting (e = 0.1) and expanding (e = 0.4).

$e = 0.11 < \frac{L_{base}}{6} = 0.17$? **Yes** Stable when contracting

$0.4 < 0.17$? **No** Not stable when expanding

The block is in danger of overturning when the pipe is expanding. It is recommended you redesign the slideblock to counteract this tendency. The simplest approach is to increase the block size while keeping the shape the same, but this will be expensive and wasteful of concrete. One alternative would be to provide extra mass at the downhill end of the block, to push the centre of gravity forward and so counteract the contraction forces.

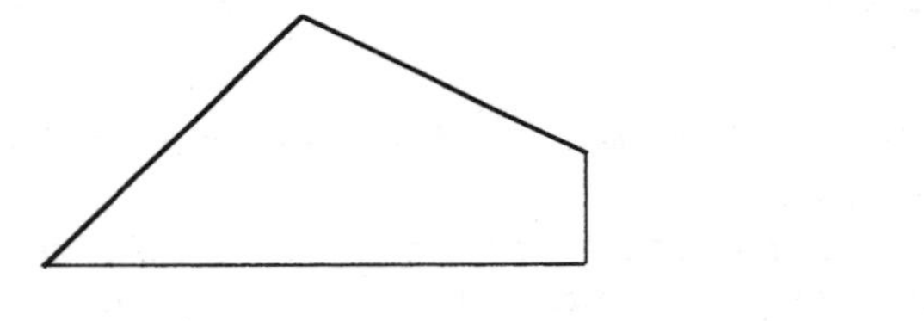

Another alternative is to provide buried 'footings'. These increase the base area so lessening the tendency to topple (as well as increasing base area for sinking stability). Care must be taken to reinforce with steel if the concrete columns and footings are slender, otherwise cracks will separate the footings from the column. A useful refinement also is to provide steps or serration in the base, to reduce sliding effects.

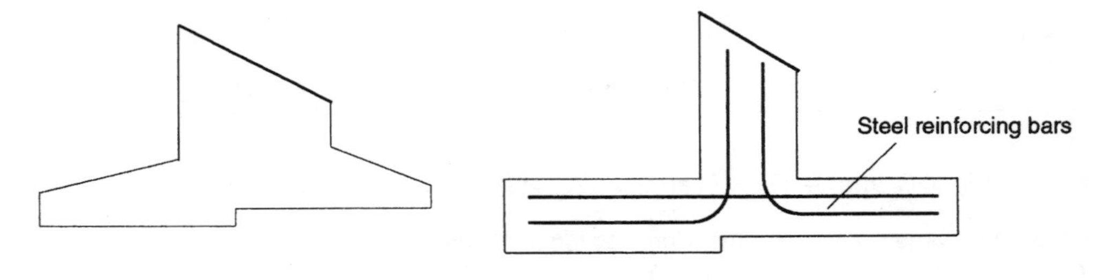

Commercial engineering

4

4.1 Introduction

The purchase of equipment and services involves large sums of money and therefore considerable financial risk. The risks are reduced and managed by the 'commercial engineering' aspects of the project, which are just as important as its technical and economic co-ordination.

4.2 The initial enquiry

The purchasing engineer first prepares the initial specification, as shown in Note 4.2.1. This is sent to various potential suppliers, with a note attached to say that a more detailed specification will be negotiated later. Each supplier replies by proposing a system and quoting its cost. Following this, it is essential that all technical matters are clarified. A checklist is given in Note 4.2.2; some of the items on the list are specified by the supplier, and some by the purchaser. If you are the purchasing engineer, you can use this list as a basis for a letter sent to the supplier in response to the supplier's initial quote, inserting the items as questions where appropriate.

When both parties are satisfied with the specification, it can be considered alongside other finalized quotations, perhaps in a formal tendering procedure as described in Section 4.5.

How tight should this detailed specification be? The supplier will prefer it to be quite general and flexible. This approach usually leads to a greater follow-on cost, because sorting out difficulties with performance and operation becomes very time consuming. If the specification is detailed and the correct paperwork is in order, it is relatively cheap and quick to sort out responsibilities. It is also much easier to monitor the performance of the equipment and evaluate the scheme's viability.

It is not advisable for the purchasing engineer to attempt to define exactly the equipment supplied by the turbine manufacturer. But it is necessary for the turbine manufacturer to define the equipment in detail on the request of the purchaser, so that the purchaser has a clear and comprehensive specification to refer to.

Note 4.2.1 **Initial request for budget prices for turbine/generator set**

The initial request is made by providing the following information

1. Gross head available.
2. Maximum or design flow, and the net head available at this flow.
3. Minimum power at the shaft or at the alternator terminals under full flow conditions.
4. Part flow efficiency – this can be initially indicated by a single design point. Specify a part flow with corresponding net head and, if necessary, specify a minimum power at this flow. If necessary, supply penstock flow/head curves and request turbine performance curves for the complete flow range.
5. Preferred governor – specify if a particular method (load or flow control) is preferred or required.
6. Power demand characteristic – describe the loads to be driven and their power factors.

Note 4.2.2 **Request for a detailed quotation**

1 - 6 Items 1-6 in the initial request (Note 4.2.1) must be included in expanded form.

7 **Water quality**
Specify whether chemical characteristics of the water are normal or whether it is corrosive.

8 **Silt particle size**
Ask the supplier to specify the particle size which can be accommodated by the turbine. Inform the supplier what particle size is achievable by the silt basins.

9 **Connection, water end**
Specify type and dimensions of joints on penstock end, where turbine is to be connected.

10 **Isolating valve**
Clarify whether turbine is fitted with isolating valve shutting off water supply; type of valve, closing time.

11 **Drawings**
Both schematic sketches and detailed drawings of equipment must be supplied, with dimensions and weights.

12 **Drive system**
Ask supplier:
- whether drive direct or geared – if geared, provide information on all components, working life, and maintenance procedures;
- the characteristics of the bearings supplied with the turbine: whether they are capable of taking side loads (radial loads) and their working life for expected loads and similarly for the alternator if supplied;
- which spares are advisable.

13 **Instrumentation**
Specify a pressure gauge on turbine inlet manifold. If the turbine is the Francis type, request a pressure gauge also on the draught tube. List other instruments required.

14 **Spare parts**
Request supply of spare parts on short-life components. Examples given here include alternator parts, drive system parts.

15 **Tools and manuals**
Maintenance procedures must be clearly described in written manuals. Suitable special tools must be provided. One important special tool is a lifting hoist for turbine installation and maintenance.

16 **Governor**
- If a governor is supplied, full specification is given in Chapter 6. Take special care in the case of mechanical governors to procure service support.
- If no governor is supplied, it is likely that you are planning to buy an ELC independently. If so send the same alternator specification (see Note 8.10.1) to both the ELC manufacturer and alternator supplier (or the turbine supplier if this is the source of the alternator as well). Both sources should approve the specification; ask the alternator supplier to check suitability of the thyristor load of the ELC, the type of AVR, and so on.

17 **Alternator**
- If the turbine manufacturer is supplying the alternator, provide the alternator specification (Note 8.10.1) and protection requirements.
- If the turbine manufacturer is not supplying the alternator, specify the radial load on turbine shaft. Request drawings showing positions of turbine bearings, position of pulley mount.

18 **Draught tube**
If a reaction turbine is to be supplied, specify suction head for 'draught' required. Request range of suction heads allowed for efficient turbine performance, and power output curves for this range. This allows planning of civil works in installation of turbine.

Note that when specifying an alternator 'kVA' is listed as well as 'kW'. This is because current is a critical design parameter with respect to alternator operation.

In setting specifications, do not underestimate the ability of the seller to identify attractive alternative ways of implementing the scheme. For example, the seller may be able to supply and utilize a reconditioned alternator, with slight modifications to the rest of the scheme. The use of second-hand or available cheaper equipment can significantly improve the return on capital investment. Modifications to a specification in order to accommodate such alternatives can be difficult to assess but may prove worthwhile.

Detailed specification for a generator

Example 4.2.1

Our reference: MH003

GENERAL SUPPLIES LTD

For the attention of the Sales Department

We would be pleased if you would tender us for the generator detailed below, by Friday 30th February 1999.

Type of turbine, drive system	Pelton wheel, belt drive
Number off	One
Turbine input power	35 kW
Generator rating	50kVA
Duty	Maximum continuous rated
Speed	1500rev/min
Overspeed	180% continuously
Voltage, phases	415V, 3-phase
Frequency	50Hz
Mounting	Vertical shaft, skirt mounted
Type	Salient pole, synchronous
Bearings	Grease lubricated, drive end suitable for radial thrust of 400kg continuously
Excitation	Shunt with forcing current transformers (CTs)
AVR	Mounted internally, voltage regulation = ± 2.5% with under/overvoltage and under/overfrequency protection
Ambient temperature	40°C
Altitude	500 metres asl
Humidity	90%
Loading	Electronic load control (thyristor load) Power factor 0.7

The generator should be designed and rated in accordance with BS4999/BS5000, or equivalent, and should be provided with a class F insulation system. Please advise the weight of your proposed generator, and provide an outline drawing with your tender. The following accessories should be included in the price of the generator: slide rails, holding down bolts. Spares should include rectifier sets, diodes, bearings and AVR.

Please show additional prices for: anti-condensation heaters; winding temperature detectors; tropical paint finish; shaft-mounted speed switch.

Your tender should be submitted subject to the following commercial conditions.

Conditions of sale	Your tender should be based on I Mech E/IEE Model form B2 conditions of contract.
Prices	Your prices should be firm and fixed for the validity period stated & include works testing, packing for shipment, & delivery CIF (Port of delivery). Your prices should be quoted in US$ or UK£.
Validity	120 days from the date of tender.
Delivery	Your delivery time CIF (Port) should be quoted from receipt of order. Our estimated delivery time is 15 weeks.
Terms of payment	100% of contract value on delivery CIF (Port).
Penalty clause	A penalty would apply at a rate of 0.5% of the contract value per week of delay up to a maximum of 5.0% of the contract value.
Customs & import duties	All duties and taxes to be paid by the supplier.
Guarantee	All items to be covered by a two-year guarantee to include parts and labour; the guarantee to operate from the date of delivery.
Documentation	Supply to include full installation, operation and maintenance manual in triplicate and in English language

Please don't hesitate to contact us if you need any further help.

Yours faithfully

Purchasing manager

4.3

Terms and conditions for specifications, orders and contracts

Standard conditions of sale

Standard conditions of sale, correctly defined and applied, are an effective safeguard against the commercial problems which may occur in a project involving the purchase of engineering equipment. Both the purchaser and the seller should understand, accept and agree to work to the same set of 'rules' controlling the contract. In the event that the purchaser and the seller disagree on a commercial matter, or technical matter which has an effect on the commercial conditions, they must initially follow the instructions in the conditions of contract, and may finally have to accept the decision of an external arbitrator nominated in the conditions of contract.

Typical UK conditions of contract that are applied and accepted throughout the world are listed below.

BEAMA 'AE' Conditions of sale – for the sale of *individual* items of electrical and in some cases mechanical equipment, exclusive of full delivery to site, or erection. These are published by The British Electrical and Allied Manufacturers Association, of 8 Leicester Street, Leicester Square, London.

In addition, Conditions of Contract are available from the British Institution of Mechanical Engineers and Institution of Electrical Engineers.

These are widely accepted and govern the supply and delivery f.o.b. (free on board) of equipment, but not delivery to site. They are available in various forms depending on the extent of the seller's involvement at site.

Many major sellers seek to apply their own conditions of contract, which puts the responsibility for accepting these conditions on the purchaser. The disadvantage to the purchaser is that the conditions will almost certainly favour the seller, but the advantage is that the prices should be slightly lower as the seller should accept his own conditions.

It is best for the purchaser to apply the same commercial conditions to all sellers, as this reduces the variety of contract management required. This may have the effect of increasing the prices slightly as sellers may have to accept commercial conditions they might consider unfavourable to them. There is a balance to be made between commercial rigidity and obtaining minimal prices.

There are many other acceptable conditions of contract, operated by EEC, North American, and Australasian sellers, and a good standard against which they might be compared is the equivalent UK set of conditions.

Limits to the scope of supply

To avoid any errors and, more importantly, the omission of any item of equipment, it is important to state the limits to the scope of supply of equipment at the start. Enquiry documents should state in words (and on drawings if possible) a precise list and brief description of every major item of plant that is to be included in the contract. Where auxiliary plant is required to make an item of plant work, it should be stated that the auxiliary plant is also to be provided. A good way of defining the limits of supply is to describe all of the interface points for each major item of plant.

- **Example Generator**
  Coupling to water turbine
  Main terminal box
  Auxiliary terminal box
  Mounting feet and bolts

Having detailed the full scope of supply the job of tendering is easier as the seller can decide easily which equipment he is to include. Also, the purchaser can more easily decide whether or not a tender is the same as the enquiry document.

Prices

Where estimates of prices are required for budgeting purposes, a purchaser may request budget prices from sellers. This type of price is not contractual and is usually accurate to within 10% of the actual price of the equipment.

Where a purchaser asks for prices against which to place an order, these are said to be firm prices. These prices are contractually binding if the offer is accepted and would not change unless there are contract variations which are dealt with in the conditions of contract.

Firm prices can be given in two different ways, Fixed or CPA (contract price adjustment).
A fixed price is one which does not vary at any time, regardless of the effects of inflation during the period of the contract.

- **Example Fixed price**
  Tender price $1000 on 1st January
  Order date 1st April
  Despatch date 1st July
  Inflation 5% from 1st Jan to 1st July
  Payment $1000

CPA takes account of any increases (or decreases) in the costs to the seller of materials and labour. A CPA price is stated at the time of tender and given a reference date. The amount paid for the plant is changed by a formula which has a Materials element, a Labour element and a Fixed element. It is important to agree both the formula and the indices to be used for materials and labour at the enquiry stage.

- **Example CPA price**
  Tender price $1000 on 1st January
  Order date 1st April
  Despatch date 1st July
  Calculated rise 5% from 1st Jan to 1st July
  Payment $1050

Validity

The validity of a tender is the period of time in which the purchaser can place the order on the seller. The validity may be stated as a number of days/months from the date of the tender or from a specific date.

Once a tender has passed its validity expiry date it is no longer contractually binding on the seller and any increase in the validity period may increase the prices.

When specifying a validity period, the purchaser should indicate as long a time as is required for checking the tenders received and for the documentation involved in placing the order.

Despatch, delivery and completion times

If you are a purchaser with a specific programme of construction for the installation, you may wish to indicate in the enquiry the despatch, delivery and completion dates which are to be followed. Such aspects as the weather or availability of transport or labour may dictate the programme dates. Where possible, plant despatch periods should be estimated in advance through discussion with the sellers.

Important dates, such as despatch or completion, may be subject to financial penalty if they are not met by the seller. Seeking to impose large penalties for failure to meet these dates may result in fewer tenders being submitted and higher prices from those who do tender.

Estimated despatch and completion dates should always be stated in tenders.

The tender document can encourage better planning by requesting detailed bar charts to show activities and time scales. The tender document should also ask for supervision to be budgeted for in detail.

Penalties

A penalty clause in a contract is a specific attempt to persuade the seller to complete the contract satisfactorily. Failure to do so would result in the seller paying a specific sum to the purchaser. This sum may not be related to the contract value and may not be in proportion to the actual loss suffered by the purchaser. Note that the seller would receive full payment for the contract.

It is vital to include realistic penalty clauses. Penalty clauses are typically based on potential lost earnings. A final date should be set for when penalty payments are no longer acceptable and the seller loses the contract. In practice it is very inconvenient to terminate a contract at an intermediate stage as the purchaser will have difficulty finding someone prepared to finish the job within the required time.

An exception clause should be included for events outside the control of seller and purchaser such as natural disasters, civil unrest or changes in the political climate.

Penalty clauses are sometimes imposed by the seller on the purchaser. These protect the seller from delayed payments or from the purchase being unable to accept delivery at the agreed date.

Contingency

Where the purchaser is bound by complex rules on payments (eg a government body), it is often convenient to include a contingency (say 10%) to cover unforeseen costs. These save expensive delays by allowing the seller to file small claims without lengthy procedures.

Terms of payment

The terms of payment decide the dates on which the seller will receive payment. The dates chosen for progress payments usually relate to specific events during or on completion of manufacture. Purchasers should remember that sellers have to buy materials before assembly and, in the case of a small firm, may require part payment either on order or shortly afterwards to finance their purchases.

Care should be taken not to overpay since the contractor loses the incentive to complete. It is necessary to hold back about 10% of the contract value for at least three months after the completion of the work. This gives the purchaser some hold on the seller for unfinished work.

Payments for stock items or low-cost items are usually made 100% on despatch or delivery.

The conditions of contract will make allowances for variations to the payment dates due to delays in the running of the contract.

Letters of Credit

The purchaser may often find that overseas sellers require that payments are made against Confirmed Irrevocable Letters of Credit. Compliance with this would require the purchaser to instruct his or her bank to issue a credit in favour of the seller; this in turn is confirmed by a bank in the country of the seller. The effect of this is to provide security to the seller that payment will be made, and failure to provide a Letter of Credit may result in the seller withdrawing from the contract. Provision for the costs associated with this should be made by the purchaser and, when implementing a Letter of Credit, the expiry date should make allowance for the issue and transfer of documentation, otherwise the credit may expire before the goods are shipped.

Currency and exchange rates

Sellers, particularly small firms, will prefer to tender in, and receive payment in, their own currency. Wherever possible, the purchaser should attempt to make payments in that currency. If this proves difficult, some sellers may accept payment in a currency of a third country, provided that it is a 'strong' currency, such as US dollars.

Taxation

Purchasers should investigate and make allowance for all local taxes which will arise from the installation. Any such taxes which will apply to the seller (and personnel during erection/commissioning) should be made known at the enquiry stage.

Sellers should take account of any taxes for which they are liable in their own country. Any taxes or duties arising through the export of the plant should be identified and the purchaser should be made aware of the resultant additional costs.

Customs and import duties

Where these are applicable, the method of payment and whether they are to be paid by the purchaser or the seller should be stated at the enquiry stage. A seller who is to be responsible for these may over-estimate them at the tender stage thus increasing the cost of the plant. Either way, the purchaser pays these and may as well pay the precise sum known at the time of shipping rather than an estimate which will possibly be high.

The seller must make an effort to establish in advance the cost of customs and import duties.

Guarantees

As a minimum, the purchaser should obtain from the seller a 12-month guarantee period covering the satisfactory operation of the plant supplied.

Of particular importance is the information given by the purchaser to the seller at the enquiry stage. Equipment can only be designed and manufactured on the basis of the data given at this stage, and it will be on this information that the guarantee will be based. In the event that the equipment is operated under different conditions to that for which it was designed, the guarantee will not be valid. Similar constraints will be placed on the maintenance of the plant to keep the guarantee valid.

Another important point to cover in the pre-contract correspondence is the date from which the guarantee operates; for example, from date of manufacture, date of despatch, or date of completion of erection. Most sellers will limit the final date of expiry in relation to the manufacture date in the event that there are delays in shipping and installation which are beyond their control.

Equipment is normally guaranteed for a period of 12 months from commissioning but no longer than a period of 18 months from the completion of manufacture.

Documentation

Most sellers will issue a limited number of copies of arrangement drawings and Operation and Maintenance Manuals. Any particular requirements which the purchaser has in respect to documentation should be made clear in the enquiry document. Requests for large amounts of documentation will probably lead to higher tendered prices and may result in a delay in the completion of the contract.

Spares

Very often, the raising of money to purchase equipment is a difficult task. It is thus important to make provision for spares to cover commissioning and the first few years of operational life of the plant in the original order. The purchaser may ask the seller to provide the recommended spares for, say, two years' operation. Where possible, the purchaser should investigate the experiences of similar installations in the region and draw up a detailed list of the components which are likely to require replacement.

Training

The provision of spares cannot guarantee to cover every possible fault which may occur. In order to run the equipment satisfactorily the plant operators should be trained to operate and maintain the equipment in accordance with the seller's instructions. The combination of correct operation within the design rating of the plant along with regular routine maintenance should increase the lifetime of the equipment.

The most suitable time for training is usually during or just after commissioning of the plant by the seller's skilled engineers. If training is required as a part of the contract then this should be indicated in the enquiry document.

4.4
Limits of responsibility

The completion of a contract is said to be the point where the seller is no longer responsible for the equipment which has been supplied, other than under the terms of the guarantee.
At this time the purchaser thus directly assumes full responsibility for the equipment. The most common limits of responsibility are as listed here.

- **Ex-works** means that the sellers' only responsibility is to make the goods available at their premises (ie works or factory). The purchaser bears the full cost and risk involved in bringing the goods from there to the desired destination. So this term represents the minimum obligation for the seller.
- **f.o.b.** means 'free on board'. The goods are placed by the seller, free of cost to the purchaser, on board a ship at a port of shipment named in sales contract.
- **f.o.r.** and **f.o.t.** mean 'free on rail' and 'free on truck'. These terms are synonymous since the word 'truck' relates to the railway wagons. They should only be used when the goods are to be carried by rail.
- **c. & f.** means 'cost and freight'. The seller must pay the costs and freight necessary to bring the goods to the named destination.
- **c.i.f.** means 'cost, insurance and freight'. This term is basically the same as C&F but with the addition that the seller has to produce insurance against the risk of loss or damage to the goods during carriage. The seller contracts with the insurer and pays the insurance premiums.
- **c. & f. to site.** While C&F is used for goods which are to be carried by sea, the term 'freight or carriage paid to...' is used for land transport only, including national and international transport by road, rail and inland waterways. The above limits are indicated in Fig 4.4.1.

The purchaser must take responsibility for the equipment beyond the limit provided for by the seller and must make provision for the completion of the transport of the equipment to site. In general, the purchaser will know the transport conditions, route and haulage contractor better than an overseas seller; thus the costs of transport within the purchaser's country will probably be lower if arranged by the purchaser rather than the seller.

The purchaser must also decide how the equipment is to be erected and commissioned once it arrives at site. This will depend on the level of skill of the purchaser's own staff and of the skill and availability of local skilled personnel.

If there is a requirement for skilled erection and commissioning engineers from the seller then this should be decided early in the contract period. Normally, sellers can provide erection and commissioning engineers who will supervise labour provided by the purchaser. Alternatively, there may be a local firm with the necessary skills that would be able to perform the installation in accordance with the seller's drawings and manuals.

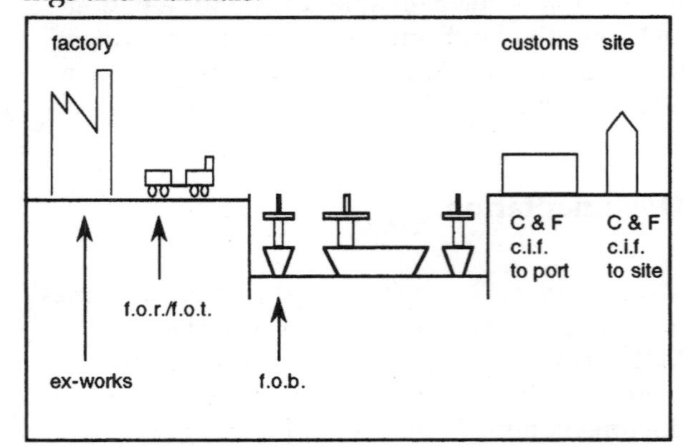

Fig 4.4.1 Limits of responsibility

4.5 Tender mechanism

The purchaser must incorporate those items that are considered important from Sections 4.3 and 4.4 into an enquiry document, which supplements the schedule of equipment, and specifications of equipment required for the project. The letter sent with the enquiry, and the enquiry document, should state the correct receiving address for tenders, and the closing date by which tenders are to be received (beyond the closing date the purchaser has the right to return tenders to the sellers unopened, and without consideration). The enquiry should then be sent to as many sellers of equipment as possible. It is a good idea to ask sellers to confirm receipt of the enquiry, and their intention to submit their offer by the specified closing date.

During the tendering period it is possible that there will be sellers who will require clarification of the specification or commercial conditions, and who will contact the purchaser to obtain this information. Such questions should be treated professionally, and in confidence. There will always be sellers who wish to tender standard equipment which may not exactly meet the specification. Provided that they are prepared to state in their tender where they do not comply with the specification, and that there is commercial or technical advantage in offering other equipment, there is no reason why a seller should be prevented from doing this.

Such deviations should be treated as confidential, as they may represent a commercial advantage to both the seller and the purchaser. It is good practice to state in the enquiry and afterwards to assume that 'unless clearly stated by the tenderer, the offer will be deemed to be technically and commercially compliant with the specifications and commercial conditions in the enquiry document'.

On receipt of all tenders for equipment, the offers must be assessed for technical and commercial compliance. A quick check will establish an order of merit from most to least expensive. Those which are clearly too expensive may be set aside at an early stage, provided that this leaves enough cheaper bids from otherwise respectable companies.

It is crucial that all of these exchanges of information should be kept confidential, in telex or written form, and that different sellers should not be able to gain access to detailed information from other tenders. Even if by doing this purchasers are able to lower the prices on offer, they will in the longer term simply discourage companies from tendering. There is no harm in going back to *all* sellers, or those whose bids are otherwise attractive and seeking reductions in price, if it assists the project to proceed. Subject to money being available, the purchaser is then in a position to place an order for the equipment.

Turbines

5

5.1 Introduction

A turbine converts energy in the form of falling water into rotating shaft power. The selection of the best turbine for any particular hydro site depends on the site characteristics, the dominant factors being the head available and the power required. Selection also depends on the speed at which it is desired to run the generator or other device loading the turbine. Other considerations, such as whether or not the turbine will be expected to produce power under part-flow conditions, also play an important role in the selection. All turbines have a power-speed characteristic, and an efficiency-speed characteristic. For a particular head they will tend to run most efficiently at a particular speed, and require a particular flow rate.

Often the device which is driven by the turbine, say an electrical generator, needs to be rotated at a speed greater than the optimum speed of a typical turbine. This leads to a need for speed-increasing gears or pulleys and belts, linking the turbine to the generator. It is preferable to minimize the speed-up ratio in order to reduce transmission costs and difficulties. (Transmission of power at low rpm involves heavier shafts, bearings and belts.) As a rule of thumb, ratios of more than 3 : 1 should be avoided, and the designers should aim for less than 2.5 : 1. If, for example, a 1500 rpm generator is to be used, the selected turbine should run at design speeds greater than 500 rpm. In some cases it is possible to select a turbine giving the exact speed

Photo 5.1.1 A Pelton turbine. (IT/Lindel Caine – Nepal)

required, so that no speed change is required and the generator can be coupled directly to the turbine. Manufacturers will advise on this, and commonly supply complete sets of this sort. Usually in micro-hydro installations it is cheaper to purchase separate units which must then be linked with a transmission drive.

A turbine's correct design speed depends upon:

- the power rating (size);
- the site head;
- the type (shape of turbine).

Within the micro-hydro power range, then, we can crudely classify turbines as high, medium or low-head machines as shown in Table 5.1.1. The operating principle also divides the turbines into two groups.

Table 5.1.1 Groups of impulse and reaction turbines

Turbine runner	Head pressure		
	High	Medium	Low
Impulse	Pelton Turgo Multi-jet Pelton	Crossflow (Mitchell/Banki) Turgo Multi-jet Pelton	Crossflow (Mitchell/Banki)
Reaction		Francis Pump-as-turbine (PAT)	Propeller Kaplan

In Fig 5.1.1, you will notice that in the case of a reaction turbine the rotating element (known as the 'runner') is fully immersed in water and is enclosed in a pressure casing. The runner and the casing are carefully engineered so that the clearance between them is minimized.

The runner blades are profiled so that pressure differences across them impose lift forces which cause the runner to rotate. In contrast, the flow through an impulse runner experiences no pressure change, so the runner can be operated 'open' without close running tolerances.

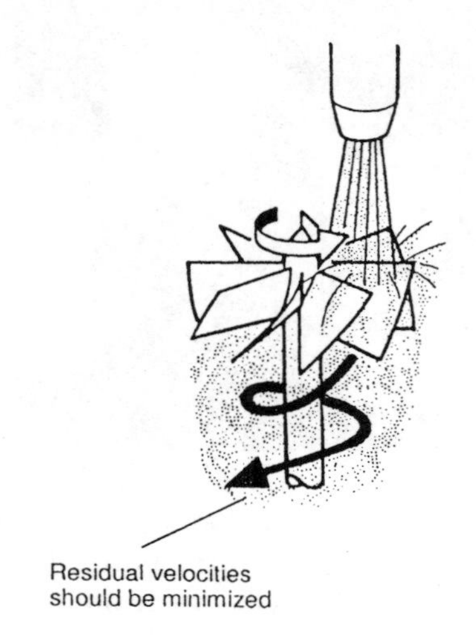

a Impulse

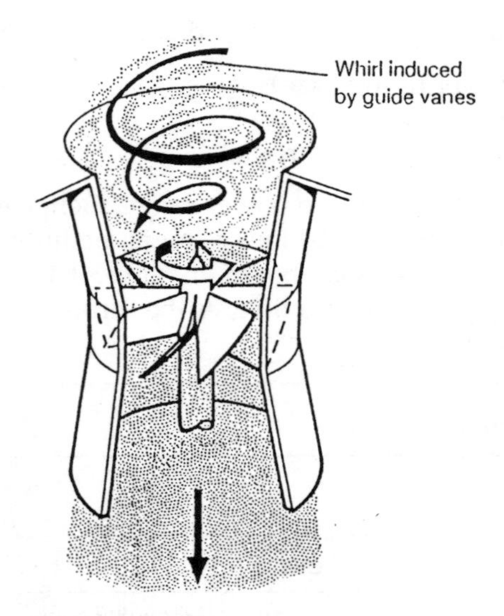

b Reaction

Fig 5.1.1 Turbine types Turbines operate on the principle of either 'impulse' (equal pressure on each side of the runner) or 'reaction' (pressure drops across the runner). In the case of reaction turbines, whirl is introduced by guide vanes.

Often, as in the case of a Pelton wheel, the runner can operate in air and the water remains at atmospheric pressure before and after making contact with the runner blades. In this case pressure energy is converted first in a nozzle into the kinetic energy of a high-speed jet of water, which is then converted to rotation in contact with the runner blades by deflection of the water and change of momentum. An impulse turbine needs a casing only to control splashing and to protect against accidents. Usually impulse turbines are cheaper than reaction turbines because no specialist pressure casing and no carefully engineered clearances are needed.

A convenient method for selecting a suitable turbine for a particular site is given in Fig 5.1.2. One example of use of the nomogram is given. Notice that use of a Pelton turbine is not always restricted to high head – if the power transmitted is low, then the Pelton will also run on low heads, although at slow rotational speed.

Notice also that the nomogram allows you to choose the shaft speed produced by the turbine, and to calculate when a speed-increasing gear (or pulley and belt arrangement) is needed, together with the speed increase gear ratio.

Fig 5.1.3 is useful in estimating the size of the runner required of a turbine for any application. In general the smaller the runner, the less material used, and the faster it rotates – requiring less gearing, leading to lower cost. Casing costs will also tend to be less for a small diameter runner, whichever type of turbine is considered. In practice there are limits to how small the runner can be, to handle the flow needing to pass through or over it, and a well-designed turbine always has the smallest runner permissible within these limits. Manufacturing costs can increase as runners become very small. Impulse turbines are usually cheaper than reaction types even in cases where their runners are larger.

The different turbine types are described in more detail in the sections following. An important aspect is their performance at part-flow.

Small turbines designed for micro-hydro applications often will have no method of altering the flow rate of water. On larger machines, some method of altering flow is normal. For instance, a multi-jet Pelton can be run with some jets shut off. Crossflow and Francis turbines have guide vanes which alter the water flow rate. A single-jet Pelton may have a spear valve which alters the nozzle orifice size. Alternatively, a single-jet Pelton could be run in the wet season, when flows are high, with one nozzle fitted, and then

run in the dry season at lower flows with a second, smaller nozzle replacing the wet season nozzle.

If flow control devices are fitted to the turbine, then the same head of water can be maintained above the turbine while flow reduces. Different turbine types respond differently to changed flow at constant head. Typical efficiency characteristics are given in Fig 5.1.4.

Photo 5.1.2 Building a crossflow. (IT/Jeremy Hartley – Nepal)

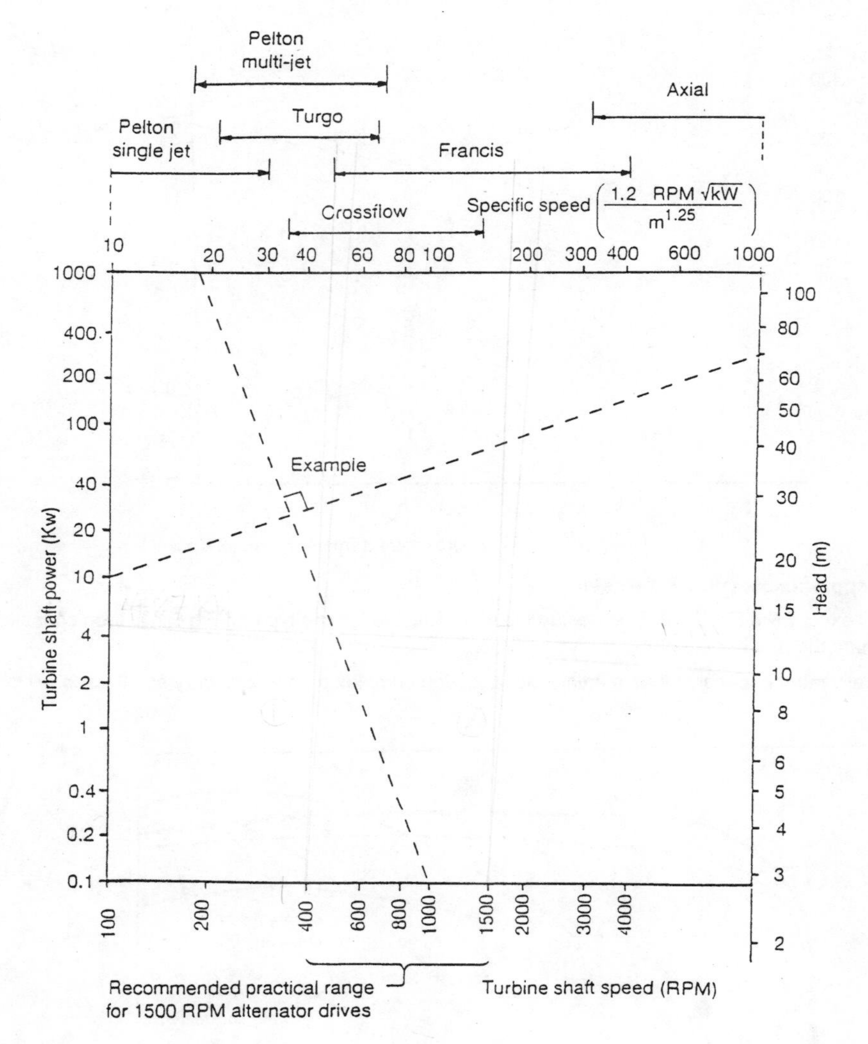

Fig 5.1.2 Selection of a turbine for a hydro site

The nomogram shown is used as follows:

- From knowledge of the proposed site, mark head available, calculate turbine output power from the following equation and rule line across.
- Turbine output power = $e_{turbine} \times 10 \times Q \times H$ kW (Q in m^3/s; H in m)
  Assume $e_{turbine}$ = 65 to 80 per cent; 65 per cent for locally made crossflow, 75 per cent for Pelton, and Turgo, 80 per cent for reaction types.
- Choose desired turbine running speed (400 is slowest speed advised if an alternator is to be run at 1500 rpm) and draw a line from this point which is at right angles to the head-power line.
- Choose a turbine shaft speed of 1500 rpm for direct drive between turbine and generator (50Hz 4 pole systems)

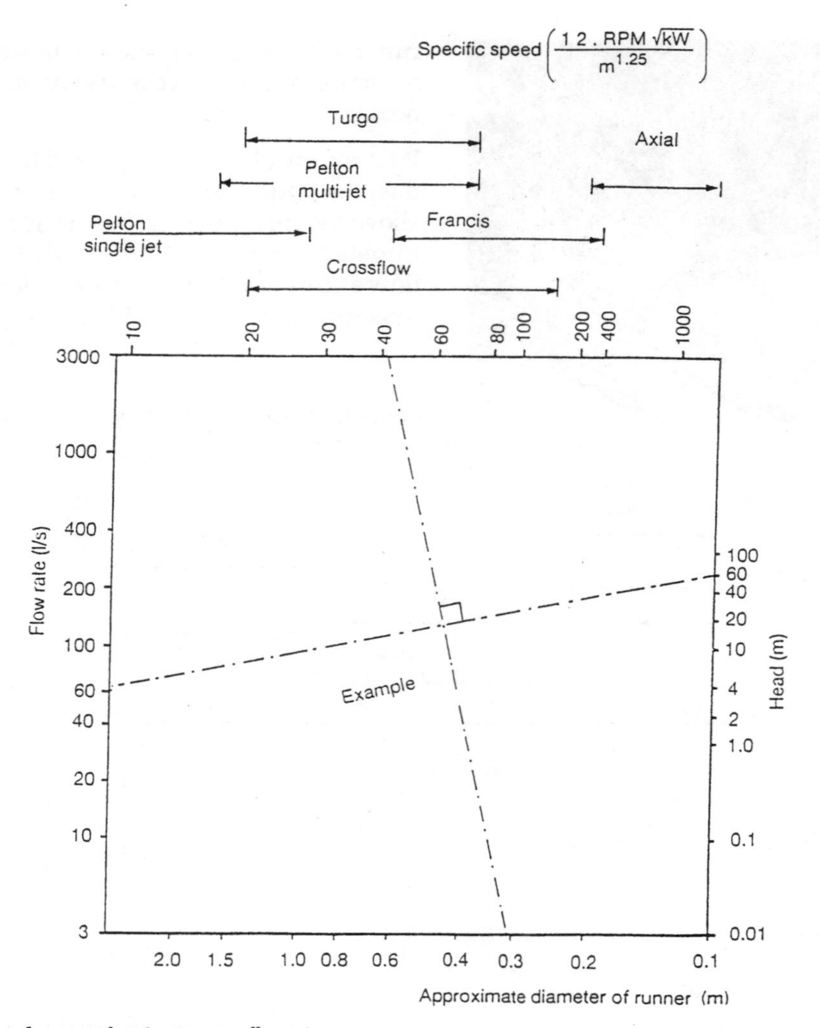

Fig 5.1.3 Estimate of approximate runner diameter

This nomogram is used exactly like Fig 5.1.2. First decide the most suitable turbine type on Fig 5.1.2, then consult this nomogram to approximate the runner diameter.

A large runner is generally more expensive than a small one, although complex profiles and very small sizes will be expensive.

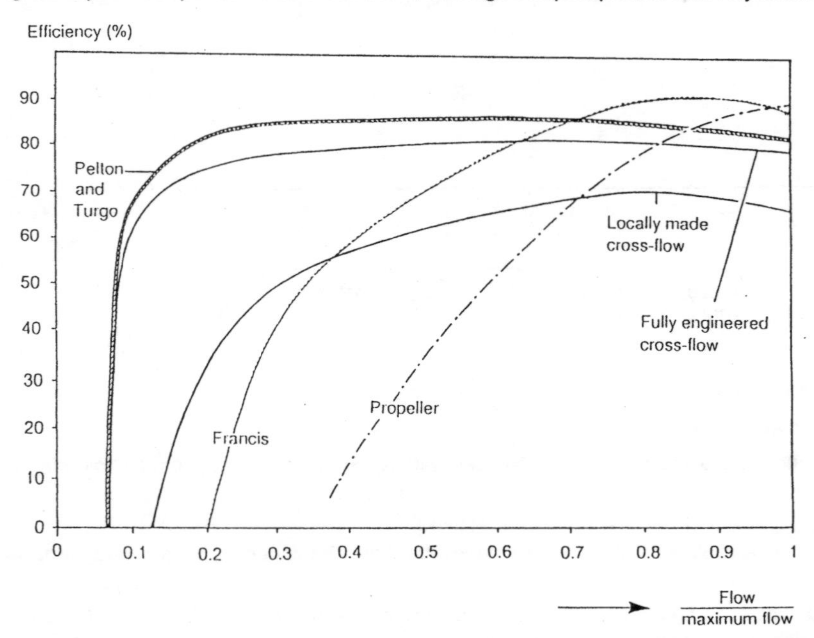

Fig 5.1.4 Part-flow efficiency of various turbines These curves assume turbines which have facilities for varying water flow rate at constant head (for example, spear valves or multiple jets on Peltons, partition devices or flow control vanes on crossflows).

Example 5.1.2 shows the importance of part-flow performance in selecting equipment for a site.

Assuming that flow-control devices are fitted, an important point to notice is that the Pelton and crossflow turbines retain high efficiency when running below design flow; in contrast the Francis drops in efficiency, producing very poor power output if run at below half the normal flow, and fixed pitch propeller turbines are very poor except at 80 to 100 per cent of full flow.

The Francis is a popular turbine in larger hydro schemes, although it is generally a more complex and expensive machine, and has poor part-flow efficiency. Fig 5.1.2 indicates that the Francis is one of the few turbines which turns at a reasonable speed at certain power and head combinations. An impulse turbine operated under these conditions of head and flow would be much larger, expensive as a result of its size, cumbersomely slow-turning and would need a greater speed increasing transmission.

In addition to giving high speed at low head-to-power ratios, reaction turbines are particularly suited to low head applications for a second reason. Since power conversion is caused partly by pressure differences across the blades, the drop in head below the blades (known as the 'suction head') is as effective in producing power as is the head above the turbine. It is generally difficult or expensive to situate a micro-hydro turbine lower than about 2 metres above the surface level of water downstream of the turbine. On a low head site of, say, 10 metres the suction head then represents 20 per cent of the power available at the site, which is likely to be very significant in terms of the overall economy of the scheme. The tubing or casing holding the suction pressure downstream of the runner is known as the draught tube.

In contrast, low cost impulse turbines do not usually make use of any suction head as their casing runs at atmospheric pressure. However, sophisticated crossflows on low heads often do use suction heads.

Having noted the advantage of using a suction head, it should also be observed that the magnitude of the usable suction is limited. This is because very low water pressures are induced on the blades of a reaction turbine running under high suction. These can be low enough to vaporize the water into pockets (or 'cavities') of vapour attached to the internal surfaces of the turbine. The cavities form and collapse at a very high rate which after a period of time can cause serious pitting and cracking of the blades. The phenomenon is known as 'cavitation'. In practical terms great care must be taken to situate the runner at a position which prevents the possibility of cavitation damage. Often the need to avoid cavitation leads to setting the runner lower than desired; because of the constraints of draught tube design, this can lead to additional civil works costs.

An indication of the costs of the various turbine types is given on Table 5.1.2. Cost variation with turbine size is also indicated. Costs are given in terms of the range from lowest expected to highest expected cost.

A further check on costings may be obtained by using the information in Chapter 1, which provides a guide to the expected proportion of total scheme cost taken by electro-mechanical equipment. Total scheme costs will depend on power and head; the following formula giving a useful guide:

$$\frac{\$}{\text{kW}} = \frac{3500 - 4500}{(\text{power kW})^{0.3} \times (\text{head m})^{0.15}}$$

Table 5.1.2 **Costs of turbines in units of US $1000** excluding alternator and drive

Shaft power kW	**Crossflow**	**Francis**	**Single-jet Pelton**	**Multi-jet Pelton**	**Turgo**	**Propeller**
2	1 - 2	4 - 6	1 - 4	1 - 3	2 - 4	4 - 6
5	2 - 6	8 - 10	2 - 8	2 - 6	5 - 8	8 - 10
10	2 - 10	15 - 20	2 - 15	2 - 10	8 - 14	15 - 20
20	3 - 14	20 - 30	3 - 20	3 - 15	12 - 20	20 - 30
50	5 - 30	25 - 70	5 - 50	5 - 30	35 - 50	25 - 70
100	30 - 50	40 - 100	40 - 80	15 - 60	55 - 80	40 - 100
150	50 - 80	60 - 120	60 - 100	30 - 80	80 - 100	60 - 120

Note 5.1.1 **Part-flow system efficiency**

In Fig 1.3.2, repeated as Fig 8.5.6, the caption suggests that the overall efficiency of a hydro scheme is usually 0.5. This is a useful rule of thumb when considering turbines used at the full flows they are designed for, *but it does not apply when there is less water available* and the turbine is run at part-flow (for instance in the dry season). First be clear what is 'efficiency':

$$\text{Efficiency of a component} = \frac{\text{output power}}{\text{input power}}$$

When running at part-flow there can be a reduction in the efficiency of each component, the turbine, the drive belt and the generator; these reduced inefficiencies combine together to give a very poor overall system efficiency – sometimes so poor that you cannot expect any significant power supply to consumers.

This situation is most common in small electrical schemes since generators of less than 5 kW capacity can become inefficient when operated at low power. Bigger schemes are also affected depending on the design of the turbine.

The important thing is to know, or estimate, the part-flow performance of the turbines you are considering using. If the manufacturer cannot tell you, or test results cannot be found from university laboratories, then assume the turbine is about 20% less efficient at part-flow than similar machines made by other manufacturers.

In the case of electrical generators, either induction or synchronous, it is necessary to ask the manufacturer for test data at part power. Sophisticated machines of all sizes will maintain 70 or 80% efficiency at half power, and reduce to 60 to 75% at quarter power; less expensive machines can produce worse efficiency at quarter power. In general generator efficiency drops very low at less than one fifth power.

Drive systems often lose a fixed amount of power. That is to say, a 95% efficient drive on a 10 kW scheme loses 0.5 kW. If the same drive is used to transmit 2.5 kW (quarter power) it will still lose 0.5 kW, and its efficiency is therefore $(2.5 - 0.5)/2.5 = 80\%$.

A quick way of calculating output power at part-flow is given in Example 5.1.1.

One component of the system actually improves in efficiency as the flow reduces. This is the penstock. On schemes above 5 kW, it is usually economic to choose penstock diameters giving 90% efficiency (10% friction loss in the penstock) when runing at full flow. But on smaller schemes it can be more cost-effective to use standard pipes at efficiencies as low as 80 or 70%. When flow is reduced the efficiency improves; for instance an 80% efficient penstock (at full flow) becomes 95% efficient at half flow. If the penstock is 90% efficient at full flow it becomes 97% efficient at half flow. The effect of this is to provide the turbine with more input power in the form of *increased head* at half flow. This will tend to speed the turbine up which may cause it to operate further away from its best efficiency point and so make matters worse rather than better. This is one reason why it is best to design variable power systems with penstocks of fairly high efficiency at full flow, and possibly to run at slightly below best efficiency speed (which is given by the 'rpm equation' in Note 5.3.1) at full flow. In electrical systems where frequency is dependent on turbine speed this can be important.

Photo 5.1.3 A spear valve allows flow reduction in a Pelton. (Peredeniya University, Sri Lanka)

Photo 5.1.4 The equivalent for a crossflow is the guide vane. (BYS/SKAT T8 model)

Part-flow system efficiency — Example 5.1.1

You wish to design a hydro scheme to provide 3 kW to a village. The turbine you are considering has a full flow efficiency of 70%, and you know the generator is 80% efficient when used between half power and full power. You will use a drive system which you estimate is 95% efficient.

You know that for some of the year there is only half the water available because of lack of rainfall, and to get some power to the village at this time would be very useful. You would also like to get some power in the driest month when only a quarter flow is available.

1 What is the power input required for the turbine during full flow conditions?
2 What is the output power of the generator at half flow?
3 And at quarter flow?

Note: Ignore the effect of changing flow on penstock efficiency in this example.

1 Full flow, power input to turbine

$$= \frac{\text{output power}}{\text{efficiency}} = \frac{\text{output power}}{e_{gen} \times e_{drive} \times e_{turb}} = \frac{3}{0.8 \times 0.95 \times 0.7} = \frac{3\text{ kW}}{0.53} = 5.6\text{ kW}$$

2 To answer this question you need part-flow performance data for the turbine. The manufacturer, or an independent test, may give you this, or you might guess by assuming it is a bit worse than other published data for that type of turbine. Suppose you find out it is 60% efficient at half flow and 25% efficient at quarter flow.

You also need to know the part-power efficiency of the generator; you already know this is 80% at half power. Assume that you learn it is 60% at quarter power.

Input power to turbine at half flow $= 0.5 \times 5.6 = 2.8$ kW

Efficiency of turbine at half power $= e_{turb\ (0.5)} = 0.6$

The power lost by the drive at full power = loss

$= \text{turbine output} \times (1 - e_{drive}) = (0.7 \times 5.6) \times (1 - 0.95) = 0.2$ kW

Efficiency of generator at half flow $= e_{gen\ (0.5)} = 0.8$

Input power to generator at half flow $= \text{gen input}_{(0.5)}$

$= \text{turbine input}_{(0.5)} \times e_{turb\ (0.5)} - \text{loss}_{drive} = 2.8\text{ kW} \times 0.6 - 0.2\text{ kW} = 1.5$ kW

Output power from generator at half flow

$= \text{gen input}_{(0.5)} \times e_{gen\ (0.5)} = 1.5\text{ kW} \times 0.8 = 1.2$ kW

Note that the combined efficiency has reduced from 53% at full flow to 43% (1.2/2.8) at half flow. Overall system efficiency (e_o) will be this multiplied by other component efficiencies.

3 For quarter flow do the same calculation:

Turbine input at quarter flow $= \text{turbine input}_{(0.25)} = 0.25 \times 5.6 = 1.4$ kW

Output power $= (\text{turbine input}_{(0.25)} \times e_{turb\ (0.25)} - \text{loss}_{drive}) \times e_{gen\ (0.25)}$

$= (1.4 \times 0.25 - 0.2) \times 0.6 = 0.1$ kW

Combined efficiency has now reduced to 7% (0.1/1.4) and it would be unwise to promise the village any useful power at all during the driest part of the year. Overall efficiency (e_o) will be even lower because of penstock losses and transmision losses.

This example illustrates an important rule: do not use a simple estimate of overall efficiency (for instance 50%) for part-flow conditions. When using the power equation ($P = e_o \times Q \times H \times 10$ kW) always consider the value of e_o following this example.

Note 5.1.2 **Specific speed**

The concept of 'specific speed' is mentioned often in hydro power texts and discussed amongst hydro engineers. It is useful to understand it for this reason. In actual fact it is quite reasonable never to use the concept in the field of micro-hydro. A chart like the one given in Fig 5.1.2 can be used to select a turbine without reference to the specific speeds which are marked on it.

The relationship between head (H), turbine output power (P_o), and actual turbine speed (N_{turb}) is expressed by the following equation which applies to all turbines. N_s is a constant which describes the particular machine as fabricated, running at its design/peak efficiency, and has a value which can only be supplied by the turbine manufacturer. Larger or smaller turbines of the same shape would have identical N_s. N_s is known as the turbine's 'specific speed'.

$$N_s = N_{turb} \times 1.2 \times \frac{P_o^{0.5}}{H^{1.25}}$$

Care must be taken to establish which units are being used. N_{turb} is in rpm, H in metres, and P_o in kW. The equation is easily understood if the particular turbine considered is imagined as scaled down to produce 1 kW of power at a head of 1 metre. In this case its runner will rotate at the specified speed ($N_{turb} = N_s$).

In principle the equation allows selection of a turbine which runs exactly at the required speed. From known head and power output (calculated from input power by assuming a turbine efficiency of, say, 0.7), the desired specific speed N_s can be calculated, and a turbine of this characteristic can be ordered from a manufacturer. A standard-size turbine will be cheaper but will not have exactly the required specific speed. If a standard or 'off-the-shelf' turbine is considered, a speed-increasing transmission G will probably be needed. For instance, G is 3 if the turbine runs at 500 and the alternator speed (N_{alt}) is 1500 rpm.

$$\text{Since } N_{turb} = \frac{N_{alt}}{G} \quad \text{therefore } N_s = \frac{N_{alt}}{G} \times 1.2 \times \frac{P_o^{0.5}}{H^{1.25}}$$

Since G can be varied, there is now allowance for adjustment or 'fine tuning', in possible values of specific speed. As an example, a hydro site offering a head of 25 metres and a flow of 200 l/s implies a turbine output power (at assumed 70 per cent turbine efficiency) of $0.7 \times 50 = 35$ kW. The alternator speed is to be 1500 rpm. With direct drive (no drive system involving speed change) N_s is 190; with gearing stretched to a maximum of 3:1, N_s can reduce to 63. Looking at Fig 5.1.2, this implies that the only suitable turbine for direct drive would be a Francis, whereas with a 3:1 gear ratio a crossflow or multi-jet Pelton would be the least costly options.

If the specific speed of a Pelton runner with a single jet is known, the specific speed of multi-jet versions of this turbine can be calculated:

$$N_s \text{ (Multi-jet having } n_j \text{ jets)} = N_s \text{ (single-jet)} \times \sqrt{n_j}$$

A belt drive can be troublesome if G is larger than 3. Three ways to limit the value of G are (1) choose a turbine with a higher specific speed, (2) in the case of Pelton turbines, increase the number of jets or (3) divide the power between two identical rotors (sometimes mounted on the same shaft) which halves the value of P_o.

Photo 5.1.5 A 3-jet Pelton made in Sri Lanka.

Photo 5.1.6 A 3-jet Pelton with deflectors viewed from below. (Peru)

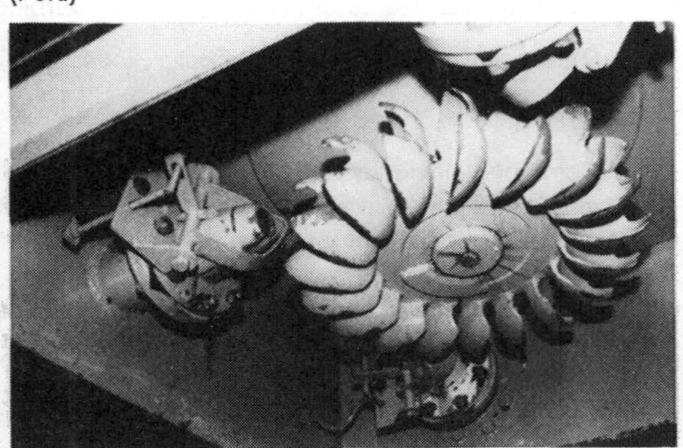

Selection of a variable flow turbine and generator — Example 5.1.2

In Example 5.1.1 we saw that the overall efficiency (e_o) of a scheme running at part-flow may be lower than expected. This example illustrates the same point but it also shows how to use a Flow Duration Curve (FDC) to plan your hydro scheme.

You wish to select a turbine and generator for a site. Your hydrology study concludes with the FDC shown in Fig 5.1.7 below. The users of the hydro installation wish to run an electric crop-dryer for 6 months of the year. They would like to know whether to buy the smallest dryer (3 kW) or if they can use a bigger one, a 6 kW unit. They wish to have at least 1 kW available all 12 months of the year to provide lighting to several households. There is no irrigation or other water use requirement to consider. The gross head is 30 metres. You can expect the penstock will lose 10% of head when running at the design flow rate.

A person living at the site has already suggested purchase of an 8 kW crossflow and 7 kW generator from local manufacturers. He has used the simple power equation (Example 1.3.1) to calculate that at an estimated overall efficiency of 50%, 30 metres head and 40 l/s flow the generator output will be 6 kW, while at 10 l/s flow (which is available all year round) the generator output will be 1.5 kW. He has suggested therefore that the 1 kW lighting demand can be met and the bigger dryer purchased but is asking your advice to confirm this.

To answer this enquiry, first obtain reliable part-flow performance data for the turbines and generators you might be using. If the manufacturer does not have the data, arrange to collaborate on tests to produce the data. In this case you may decide to collect data for a 4 kW generator and 5 kW crossflow as well as for the two larger sizes proposed. Illustrative data is shown below by way of example. (Remember you cannot use these graphs in another situation; each manufacturer's turbines will perform differently under part-flow, so you must collect your own data.)

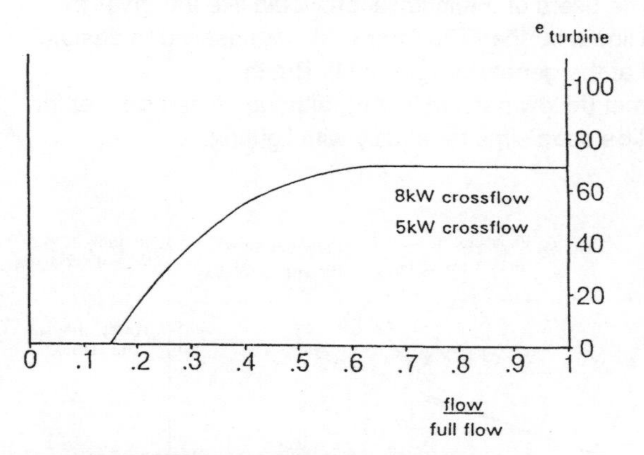

Fig 5.1.5 **Example of manufacturer's data for a crossflow's part-flow performance**

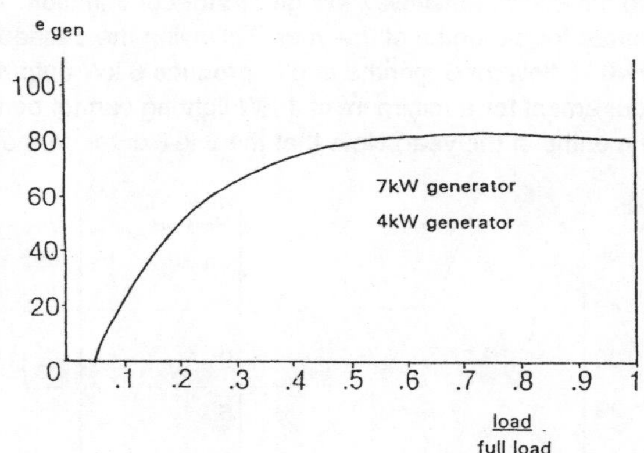

Fig 5.1.6 **Example of manufacturer's data for a generator's part-load performance**

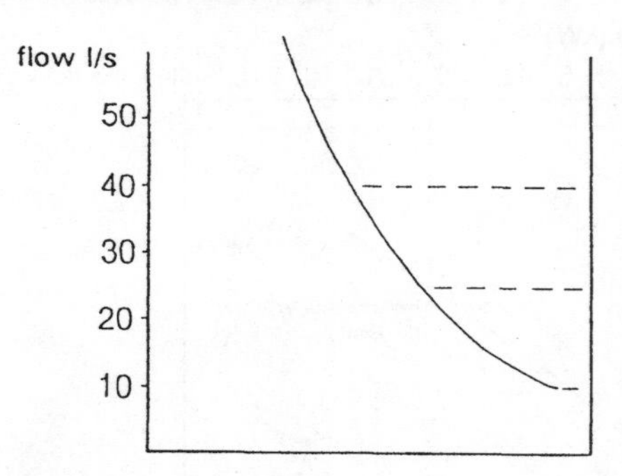

Fig 5.1.7 The FDC for the site in this example Note that an annual hydrograph can also be used for this purpose; both can be modified to show water diversion to other uses such as irrigation.

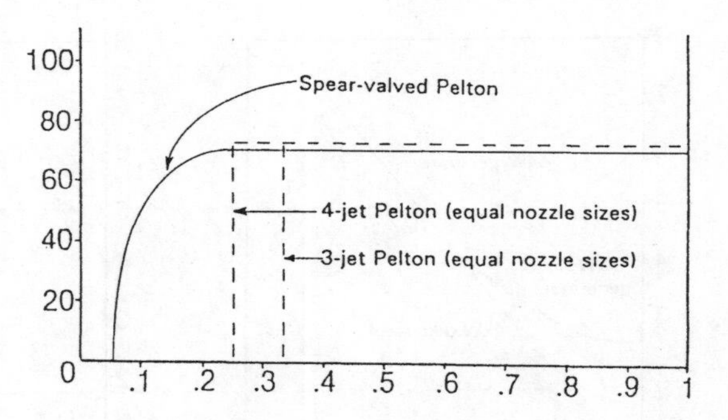

Fig 5.1.8 Example of manufacturer's data for spear-valved and multi-jet Peltons

continued overleaf

Example 5.1.2 *continued* **Selection of a variable flow turbine and generator**

In Example 1.5.1 the flows shown on the FDC and hydrograph are converted to power outputs simply by assuming an overall system efficiency of 0.5. When considering part-flow systems this is misleading; instead develop a multiple-graph as shown below, representing the efficiencies of each component of the system. Firstly the penstock will become more efficient as flow reduces, since the head lost due to friction is dependent on flow^2, and will therefore be less at part flow by a factor of (part flow/full flow)^2. If the loss factor ($h_{wall\ friction}/h_{gross}$) at full flow is termed 'L', then the power at the entry to the turbine is:

$$\text{Power at entry to turbine} = Q \times 10 \times h_{gross} \times \left(l - L \left(\frac{\text{part flow}}{\text{full flow}} \right)^2 \right)$$

Plot the FDC as shown below (graph A) and then plot the power at entry to the turbine next to it (graph B). In this example L is 0.1 (the head loss in the penstock is assumed to be 10% at full flow). Note that the variation in penstock efficiency at varying flow is quite small in this case. In fact losses of 20% or more at full design flow are often economic solutions in very small schemes and therefore penstock efficiency at part-flow will become significant in such schemes. See Note 5.1.1 on this topic.

Next plot the turbine efficiency (Fig 5.1.5) in terms of power entering and power output as in graph C. Finally plot generator efficiency (Fig 5.1.6) in terms of turbine output power and generator output power (graph D). Do this for each of the turbines and generators you are considering.

System 1

Use the 8 kW crossflow/7 kW generator combination. The users of the installation would like the dryer to operate for 6 months of the year. Following the dashed line **a** on the FDC (graph A) it is possible to design for 40 l/s flow for 6 months and to produce 6 kW output at the generator (graph D). But the requirement for a minimum of 1 kW lighting cannot be met (follow dashed line **c**), although it can be met for 10 months of the year. Note that the 6 kW dryer cannot be used simultaneously with lighting.

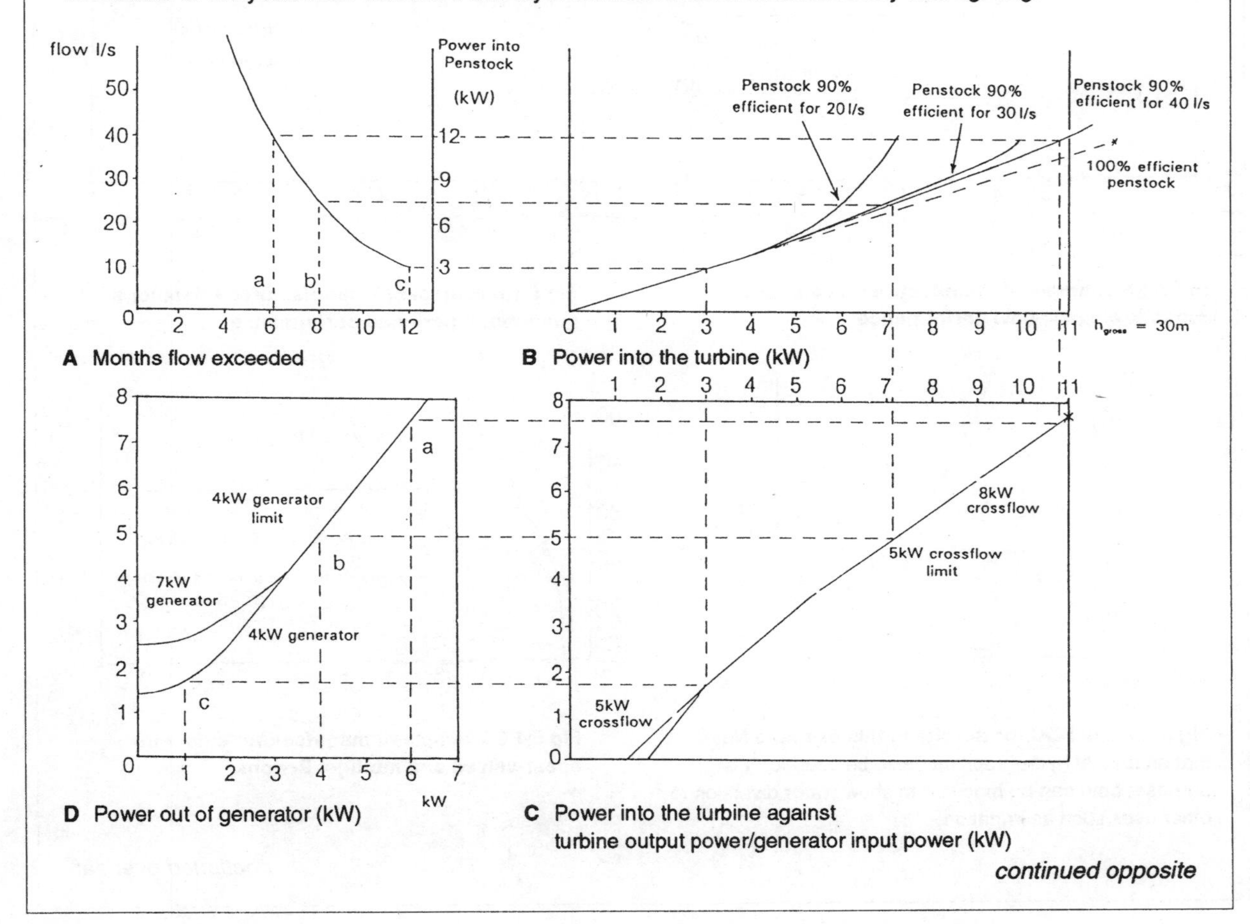

A Months flow exceeded

B Power into the turbine (kW)

D Power out of generator (kW)

C Power into the turbine against turbine output power/generator input power (kW)

continued opposite

Selection of a variable flow turbine and generator **Example 5.1.2** ***continued***

System 2

Use the 5 kW crossflow and 4 kW generator. Follow dashed line **b**. This represents a flow of 26.7 l/s which drives the crossflow to its full 5 kW capacity and drives the generator to its full 4 kW capacity. This 4 kW output could provide (for more than 6 months of the year) the minimum 3 kW required for the smallest dryer, while still providing 1 kW for lighting. For 12 months of the year, the minimum 1 kW is available, as shown by dashed line **c**.

What is your conclusion? Clearly if the 1 kW all-year-round supply is really a high priority then the smaller turbine/generator combination is preferable. These small machines and the smaller penstock will also have the advantage of costing less, so finance will be available to solve the requirement for drying in other ways. Notice that the assumption of 0.5 overall efficiency used in the simple power equation (Example 1.3.1) does not hold good in this example. Dashed line **c** represents an overall efficiency ratio of

$$\frac{\text{power out}}{\text{power in}} = \frac{1}{3} = 0.33$$

One consideration still remains. The 4 kW generator will be used at full capacity if it powers a 3 kW dryer and 1 kW lighting load at the same time. In Chapter 8 certain practical reasons are given for not using a generator at full capacity. Consequently, a larger generator will be needed here and, in order to produce 1 kW at part flow, it will be necessary to find one with a better part-load performance than shown on Fig 5.1.6. Alternatively the 4 kW generator could be used to provide only 3 kW at any one time, in which case it would be possible to purchase an even smaller crossflow and fit an even smaller penstock.
See Chapter 8 for advice on use of over-rated generators.

A multi-jet Pelton turbine could be used on this site. If it has a spear-valve, the efficiency curve may be as represented as Fig 5.1.4. If it has fixed nozzle sizes with no spear valve, it will have a stepped efficiency characteristic. In general a 4-jet Pelton, whether a low cost or an expensive one, will maintain an efficiency of 70% when running on any number of jets. It is a worthwhile exercise to plot these two types of characteristics on the graph above, and try the calculations with different numbers of jets, and with variations in the nozzle sizes and combinations of nozzle sizes used.

5.2 Impulse turbines

The three impulse turbines considered here are the crossflow, Turgo and Pelton.

These are on the whole more suitable for micro-hydro applications than are reaction turbines, as they have the advantages listed below. Some of these features have already been discussed, while others will become clear in due course. Compared to reaction turbines, impulse turbines:

- are more tolerant of sand and other particles in the water;
- allow better access to working parts;
- are easier to fabricate and maintain;
- are less subject to cavitation (although at high heads, high velocities can cause cavitation in nozzles or on the blades or buckets);
- have flatter efficiency curves if a flow control device is built in (eg nozzle area change, spear valve, change of number of jets, guide vanes, partitioning of flow).

The major disadvantage of the impulse turbines is that they are mostly unsuitable for low head-to-power ratios. Fortunately, the crossflow, Turgo and multi-jet Pelton are impulse turbines which are suitable for medium head-power ratios as characterized in Fig 5.1.2. An impulse turbine can operate on a low head, if the power transmitted is also low and a slow speed is acceptable.

Photo 5.2.1
A crossflow. (Nepal)

5.3
Pelton turbines

A Pelton turbine has one or more nozzles discharging jets of water which strike a series of buckets mounted on the periphery of a circular disc (Fig 5.3.1). In large hydro-power installations Pelton turbines are normally only considered for gross heads above 150 metres. For micro-hydro applications, however, Pelton turbines can be used at much lower heads. For instance (Figs 5.1.2 and 5.1.3) a small diameter Pelton rotating at high speed can be used to produce 1 kW on a head of less than 20 metres. At higher powers and low heads the rotational speed becomes very low and the runner is large and unwieldy in relation to the power generated. From Fig 5.1.2 it can be seen that a Pelton running at 200 rpm on a 3 metre head produces 200 watts output. The runner of such a turbine is large for the power produced (Fig 5.1.3) and turns slowly. If runner size and low speed do not pose a problem for a particular installation, then a Pelton turbine can be used with low heads.

Photo 5.3.1 A single jet Pelton with spear valve.

If a higher running speed and smaller runner are required, then there are two main design options open:

- **Increasing the number of jets.**The use of two or more jets will allow a smaller runner for a given flow of water and hence an increased rotational speed. The required power can still be attained. The part-flow efficiency of a multi-jet Pelton wheel is especially good as it can be run on a reduced number of jets.
- **Twin runners.** Two runners can be used side by side on the same shaft or can be placed on either side of the generator on the same shaft. This design option is unusual in micro-hydro schemes, but occurs often with single jet Peltons in larger hydro schemes. In micro-hydro schemes this option is usually considered only when the number of jets per runner has already been maximized, but the rotational speed is still considered too low and the runner too bulky. Care must be taken to allow plenty of space on either side of a Pelton runner to allow deflected water to exit without splash interference (Fig 5.3.3).

The principle of the Pelton turbine is to convert the kinetic energy of a jet of water into angular rotation of the buckets it strikes. In order to do this at the best efficiency, the water leaving the buckets after striking them should have little remaining kinetic energy. It should transfer most of its energy. This implies that the water has just sufficient speed to move out from between the buckets and fall away under gravity from the wheel.

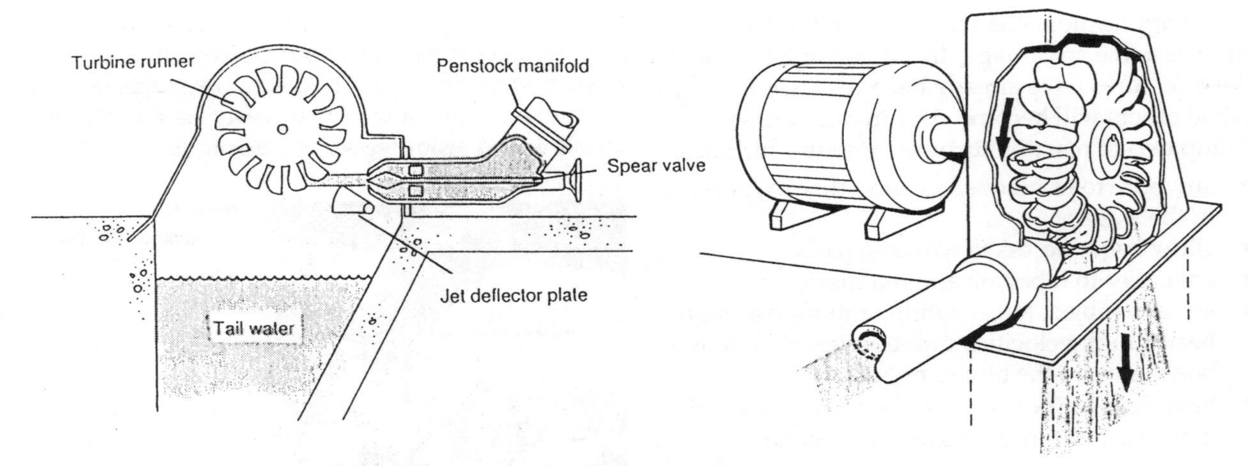

Fig 5.3.1 The single-jet Pelton wheel Spear valves are not essential components. They can be replaced by variable nozzle sizes, or variable numbers of jets.

The Pelton bucket is designed to deflect the jet back through 165 which is the maximum angle possible without the return jet interfering with the next bucket. This water leaving the rotor represents a loss of efficiency in energy conversion since a small part of the original kinetic energy is retained.

The momentum of the water jet is best used if the water is deflected back by the bucket at the same speed at which it arrives. The bucket travels forward, and the jet is therefore deflected back at incoming speed in order for it to have almost no forward speed. For optimum efficiency the jet velocity therefore needs to be about twice the speed of the buckets. Peak efficiency is found in practice when bucket speed is approximately equal to 0.46 × jet speed. Note 5.3.2 shows how this relationship gives rise to the fundamental sizing equations of the Pelton, which are used for practical calculations in Note 5.3.1 and Example 5.3.1.

The profile and shape of the Pelton bucket often receives comment. Essentially it has been evolved for maximum efficiency through experience and theoretical modelling over many years. The bucket is split into two halves so that a central area does not act as a dead spot incapable of deflecting water away from the incoming jet. The cutaway notch on the lower lip allows the following bucket to move further into place before interfering with the jet which is still propelling the earlier bucket (see fig 5.3.4).

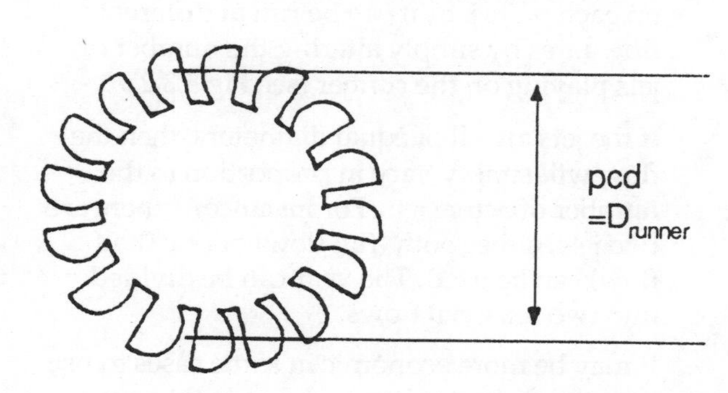

Photo 5.3.2 A Pelton runner. (Nepal)

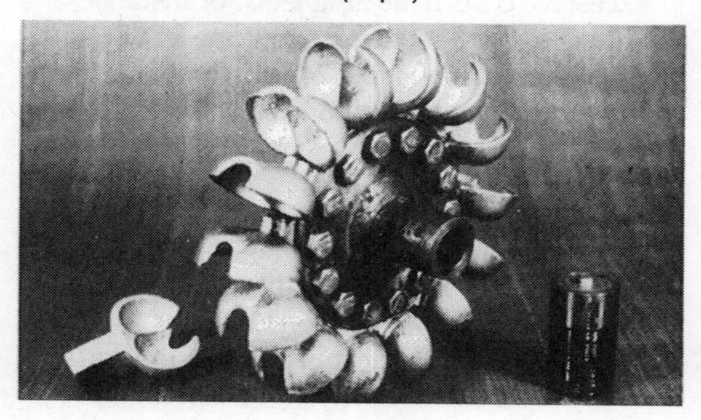

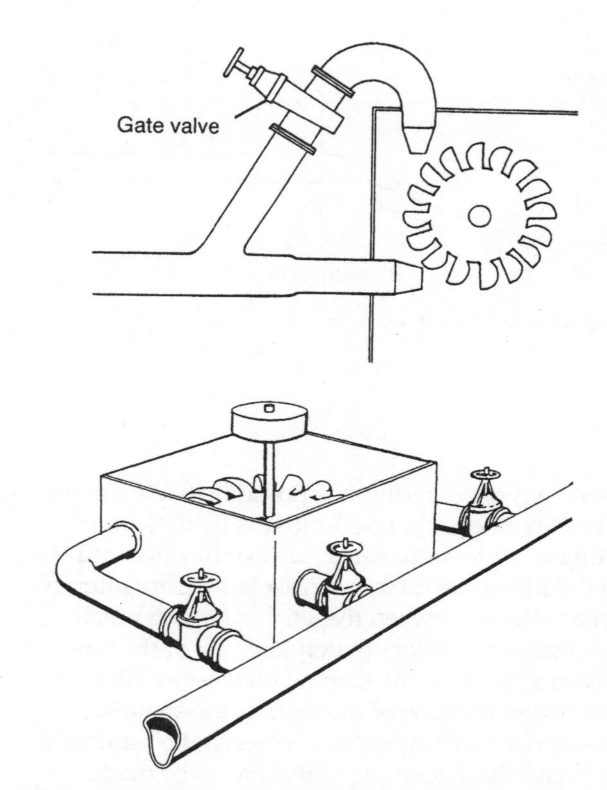

Fig 5.3.2 The multi-jet Pelton wheel The spear valve is not necessary as flow changes are made by varying the number of jets in play and possibly also varying nozzle diameters. If the Pelton has more than two jets it is convenient to mount the shaft vertically.

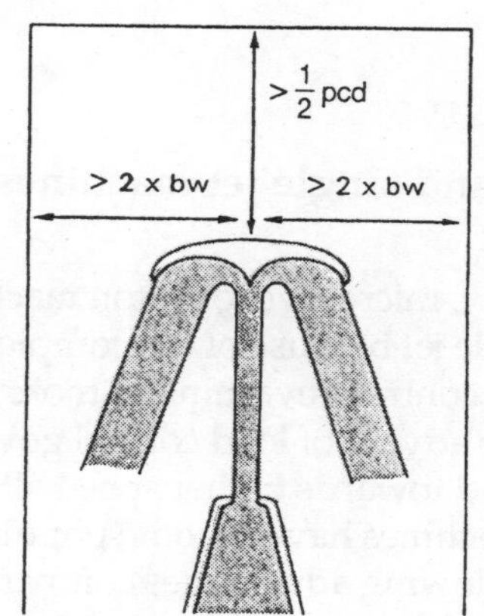

'pcd' = pitch circle diameter of Pelton

'bw' = bucket width

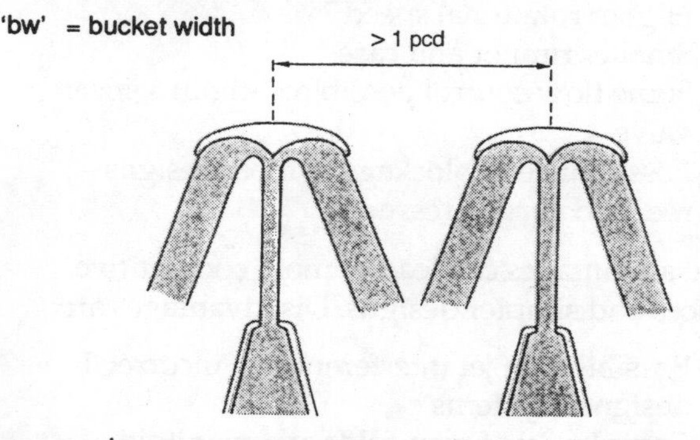

Fig 5.3.3 Clearance between bucket and case is essential to avoid splash-back in both axial and radial directions. In the case of twin runners care must be taken to avoid splash interference.

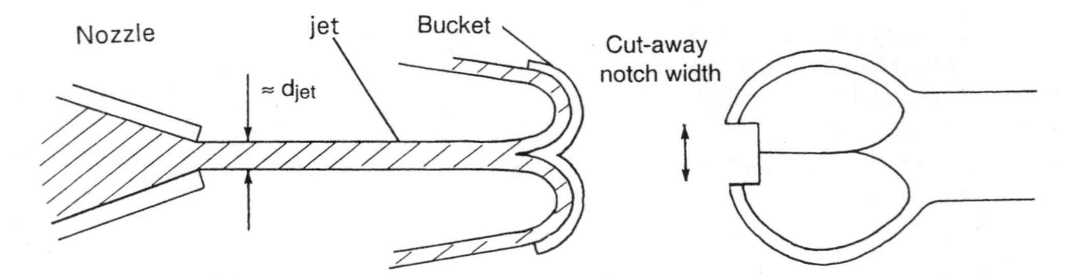

Fig 5.3.4 The Pelton buckets

As mentioned earlier it is possible to use smaller (and therefore cheaper) runners by dividing the flow of water between a number of jets – 'multi-jet' Peltons can have up to six jets before interference effects between the jets lead to too much inefficiency. Unfortunately the cost of the non-moving parts of the turbine increases with each jet added because of the extra components needed, so a compromise between the number of jets and the size of the wheel has to be made.

Note 5.3.1 and the examples following give initial guidance on the selection of single and multi-jet Peltons for micro-hydro schemes.

Multi-jet and single jet machines

Traditionally, micro-hydro Pelton machines were always single jet because of the complexity and cost of flow control governing of more than one jet. With the advent of load control governing, and the trend towards higher speed alternators, multi-jet machines have become popular. They have the following advantages compared to single jet machines:

- Higher rotational speed
- Smaller runner and case
- Some flow control possible without a spear valve
- Less chance of blockage in most designs – reduced surge pressures

The advantages can lead to more competitive prices and simpler designs. Disadvantages are:

- Possibility of jet interference on incorrectly designed systems
- Complexity of manifolds and manifold friction losses
- If flow control governing is required, becomes prohibitively complex.

There are many control options of which the designer should be aware:

1 **Replacement of nozzles**

A wet season flow will require a larger nozzle size than a dry season flow. It is possible to divide the yearly flow variation with two, three or more parts and make a nozzle for each flow. The turbine operator can then remove one nozzle and replace it with a more appropriate one to suit a seasonal change. This is a very low-cost method of controlling flow, especially if simple orifice plates are used rather than expensively profiled nozzles. It is important to be certain that sufficient operator skill and input will be available – and it must be remembered that the plant has to be stopped (by diversion of flow) when nozzles are changed.

2 **Varying the number of jets**

If a multi-jet turbine has shut-off valves fitted on each of its jets, it can be run at different flow rates by simply altering the number of jets playing on the runner (see Fig 5.3.2).

If the jets are all of equal diameters, then the flow will simply vary in proportion to the number of active jets. For instance, if there are two jets, either both (full flow) or one (half flow) can be used. The year can be divided into two seasonal flows.

It may be more economic in some cases to use unequal diameter jets, and divide the year into more seasonal flows while not incurring excessive costs in adding jets. For instance, suppose a two-jet turbine has one jet of 15mm diameter and one jet of 20 mm diameter. Full flow would be provided by the two jets acting together. Medium flow would be provided by the 20 mm jet acting alone, and low flow by the 15 mm jet acting alone. This would be a 1/3, 2/3, full power arrangement, strictly 36%, 64%, 100% of power. The reader can use the 'nozzle equation' given in Note 5.3.1 (showing that Q is proportional to the square of jet diameter) to clarify this relationship between power (also proportional to jet diameter squared) and combinations of different jet diameters.

It is possible also to utilize the nozzle replacement approach in a multi-jet turbine, to provide greater variation in flow. This will reduce the cost of adding further jets, but will require increased input from an operator.

3 Shut-off valves

It is usual to place a valve, either a gate valve or a butterfly valve, in the turbine manifold. Some comments on these valves are given in Chapter 3. Pelton wheels are often driven by long penstocks in which surge pressure effects, due to valve closure, can be very dangerous and lead to damage caused by bursting of the penstock. If valves are fitted, they must always be closed slowly, particularly during the last phase just before shut off. Count to ten slowly while completing the closure.

Gate valves are sometimes used mistakenly to regulate flow, by partially closing them. This causes damage on the valve plate due to cavitation effects.

4 Deflector plate

The water jet can be deflected away from the buckets of the runner if a plate is rotated into its path, as shown in Fig 5.3.1. This is a very useful device for stopping the turbine since it is very quick and it does not require the shutting off of the flow in the penstock, with consequent pressure surge dangers. The process of stopping the turbine is then completed by diverting the flow at the forebay. (An alternative is to divert immediately at the forebay especially if convenient access is available). Deflector plates (or jet deflectors) are used in many flow control governors to give rapid power changes.

Photo 5.3.3 3-jet Pelton. Notice the load controller, ballast tank, and belt guard. (Peru)

5 Spear valves

Fig 5.3.1 illustrates the needle valve or spear valve, which is so called because a streamlined spear head is arranged to move within the nozzle, allowing variation in effective orifice cross-sectional area without introducing energy losses. The spear is moved either by turning a thread manually by a handle, or automatically by a mechanical speed governor (described in Chapter 6).

Traditionally the spear valve has been a very important component of Pelton turbines because it has been linked to a mechanical speed governor. Modern micro-hydro plant can be controlled by electronic load controllers which eliminate the need for the mechanical speed governor and remove one historical reason for use of spear valves.

The mechanical governor is an expensive device requiring a high level of skilled maintenance. The spear valve is also expensive, and presents the danger of instantaneously blocking the nozzle should it become detached while the turbine is operating. Although these are reasons for avoiding spear valves, they are nevertheless very effective where continuous flow regulation is considered necessary, and their use in such cases can be cost effective. It is worth commenting that the spear valve will stop a foreign body which has fallen into the penstock from blocking the nozzle. When a spear valve is fitted, instantaneous closure should be protected against, by fitting a spacer onto the spear shaft.

In many cases of turbine selection, there will be a choice presented between a single or twin-jet Pelton fitted with one spear valve, and a multi-jet Pelton. A decision can be made on the basis of initial cost projections and maintenance cost projections.

In general, the multi-jet Pelton is increasingly popular because it lends itself more easily to local manufacture, avoids the danger of accidental nozzle blockage, and provides adequate flow regulation for most micro-hydro sites.

Note 5.3.1 **Sizing of Pelton turbines: two approaches**

There are two situations to consider:

A A Pelton is required for a new site and you wish to give the required dimensions to a local manufacturer who already has some standard bucket sizes.

B You already have in hand a range of Pelton runners of specific diameters, complete with buckets. You wish to choose the best runner for a particular site.

For each situation a particular approach is suitable. In situation **A** you will need to specify the dimensions of the Pelton. Example 5.3.1 illustrates this using the approach given here:

Approach A

1 Optimize the penstock diameter to calculate net head (see Section 3.11).

2 Use the turbine selection chart (Fig 5.1.2) to see if a Pelton is suitable and to estimate drive gearing requirements. (A line shaft with pulleys allows the turbine to drive mechanical loads during the day as well as generators in the evening, so it is often as wise to allow gearing as it is to design for a direct drive unit.)

Find the ideal runner diameter from the rpm equation:

$$D_{ideal} = \frac{38 \times \sqrt{H}}{\text{Pelton rpm}} \quad \text{or} \quad D_{ideal} = \frac{38 \times \sqrt{H_{net}} \times G}{\text{load rpm}}$$ *The rpm equation*

3 Draw up a table as shown in Example 5.3.1, calculating the nozzle diameter (d_{jet}) from the following equation. (Nozzle and jet diameter can be assumed the same, although orifice plates and tapered nozzles produce contraction; see Note 5.3.2.)

$$d_{jet} = 0.54 \times \frac{\sqrt{Q}}{H_{net}^{0.25}} \times \frac{1}{\sqrt{n_{jet}}}$$ *The nozzle equation*

Consider which gear ratios are actually available and calculate D_{ideal} as in 2 above. Buckets are always made to suit a particular runner diameter, and should always be marked with this in mm. (If not calculate the intended diameter from the shape of the bucket root). Try to find out what D/d_{jet} ratio was originally intended for the bucket. A common ratio is 10:1. (The width of the cut-away notch can be used to estimate maximum jet size -subtract 5mm to allow an alignment tolerance- but this will be misleading if the bucket was designed for a wheel with large clearances between the buckets. Only use this approach if you can check that the return jet will not interfere with the next bucket and so reduce efficiency). Enter the correct D_{runner} for the bucket onto the table and then calculate the actual D/d_{jet}. Adjust either D_{runner} or d_{jet} or your choice of G to maintain the ratio within acceptable limits; for instance, for a 10:1 ratio keep within the range 9:1 to 14:1, and check that the notch width is adequate to accomodate the jet.

4 Consider also part-flow requirements and speed regulation requirements in order to choose the number of jets. Consider jets of varied diameter, and interchangeable jet sizes.

Approach B (see Example 5.3.2)

1 Find the turbine speed:

$$\text{Pelton rpm} = \frac{38 \times \sqrt{H_{net}}}{D_{runner}}$$ *The rpm equation*

2 Calculate the gear ratio G:

$$G = \frac{\text{load rpm}}{\text{turbine rpm}}$$

3 Consider whether the net head can be adjusted to achieve a desirable turbine speed.

4 Consider whether the load can operate at a revised speed.

5 Calculate the flow drawn by the turbine and choose the number of jets which best gives the desired flow:

$$Q = 3.43 \times d_{jet}^2 \times \sqrt{H_{net}} \times n_{jet}$$ *The nozzle equation*

6 Consider practical modifications of nozzle size and number of jets in order to achieve the desired flow.

Specifying dimensions of a Pelton — Example 5.3.1

The site you are considering for a Pelton (Fig 2.2.1) has the following characteristics:

Gross head	75 m
Turbine flow	from 70 to 190 l/s depending on the time of year
Alternator speed	1500 rpm

1 Assume in this case penstock friction loss is 11% and therefore net head is 67 m.

2 Approximate turbine output power is: $e_{turbine} \times 10 \times Q \times H = 95$ kW for a 75% efficient Pelton

The selection chart (Fig 5.1.2) indicates a multi-jet Pelton is suitable and, in any case, gearing may be easy to provide if a line shaft is used for alternative loads on the turbine.

$$D_{ideal} = \frac{38 \times \sqrt{H_{net}} \times G}{\text{alternator rpm}} = \frac{38 \times \sqrt{67} \times G}{1500} = 207 \times G \text{ mm}$$

The gear ratio G is chosen from standard pulley sizes available and will often involve a fractional number, for instance you might choose between 2.31, 2.65, 2.86, etc.

3 The flow varies from 70 to 190 l/s, implying that two or more jets will be useful to obtain part-flow operation of the turbine. Draw up a table first using the nozzle equation given in Note 5.3.1 to find the ideal jet size. Refer to section 3 of note 5.3.1; in this example it is assumed the buckets used are designed for a D/d_{jet} ratio of 10:1 such that actual wheel diameter and jet size can be adjusted only as long as the ratio stays in the range 9:1 to 14:1. Remember to recalculate flow and therefore also power if you adjust the jet size; expand the table to allow several iterations. Check that the notch width on the bucket allows the jet size with some tolerance for alignment; assuming this checks, the example shows that various options are possible. The final choice depends on step 4.

n_{jet}	2	2	3	3	4	4
d_{jet} mm (ideal)	58	58	47	47	41	41
Actual G	3.0	2.5	2.5	2.0	2.5	2.0
D_{ideal}	621	518	518	414	518	414
Actual D_{runner}	600	500	500	450	450	400
Actual D/d_{jet}	10.3	<9	10.6	9.6	11.0	9.8
Revised d_{jet}						

4 Note that with a 3-jet turbine the lowest flow obtainable is 190/3 = 63 l/s. This satisfies the flow range specified, 70 to 190 l/s. A 4-jet turbine could be selected, so allowing flow to reduce to 190/4 = 48 l/s. Considering the FDC for the site (eg Fig 2.2.11 or Table 2.2.2) this indicates reduced power output available for a further three months each year. Remember 4-jet turbines are generally harder to build and will incurr greater expense.

At this stage you should find from the manufacturer the generator efficiency at part-load and combine it with turbine efficiency data to estimate the actual part-power performance (Example 5.1.1). It may turn out cheaper to use three jets since the fourth jet will not give an improvement in performance due to this decline in overall system efficiency. (Note that penstock losses will reduce at reduced flow so offsetting the decrease in overall efficiency, but also causing the best operating point of the turbine to rise; it may be necessary to somehow correct the speed.)

Regulation of flow: a spear valve could be used on one jet. To save the expense of a spear valve, the three jets could be used to regulate flow to 1/3, 2/3, or full flow. Greater variation is possible by using varied nozzle sizes. For instance, one jet could be sized for 30%, and one jet for 50% of full flow. By varying their combinations, six flow settings are possible. A further possibility is to make seasonal changes of the nozzle diameters on a given jet.

Example 5.3.2 **Matching a Pelton to a site**

This example illustrates approach **B** described in Note 5.3.1:

You have available a Pelton with two 13 mm jets and a 15 cm runner. You have a site in mind where the net head is 17 m and the flow range is 5 to 14 l/s. You would like to drive an induction generator. Pulleys which give ratios of 1.4 or 1.6 are at hand.

1 Find the turbine speed:

$$\text{Pelton rpm} = \frac{38 \times \sqrt{H_{net}}}{D_{runner}} = \frac{38 \times \sqrt{17}}{0.15} = 1044 \text{ rpm}$$

2 Calculate the gear ratio: A 4-pole induction generator runs at around 1575 rpm for 50 Hz (see Fig 8.6.1) so a gear ratio of 1575/1044 = 1.5 is needed. (A direct drive is possible if a 6-pole generator is used.)

3 To achieve a desirable turbine speed in step **1** above, consider altering H_{net} by raising or lowering the forebay or by use of a different penstock diameter, which creates a different head loss.

$$H_{net} = \left(\frac{D_{runner} \times \text{generator rpm}}{38 \times G}\right)^2$$

In this case a gear ratio of 1.4 demands a net head of 20 m and, when G = 1.6, net head reduces to 15 m. (At this stage you may decide it is easiest not to use the gear pulleys you have at hand, but instead make up or purchase a special pulley for the job, for instance giving the 1.5 ratio calculated in step **2**; but first consider step **4**...)

4 The easiest solution may be to generate electricity at a non-standard frequency. It is unwise to go below 49 Hz since transformers and generator coils will overheat. But it is usually possible to run most loads at up to 65 Hz. Some loads will tolerate an even higher frequency. See Section 6.3. The available gear pulleys perhaps could then be used, following the rpm equation written for the generator (or 'load'):

$$\text{load rpm} = \frac{38 \times \sqrt{H_{net}} \times G}{D_{runner}}$$

Output frequency must be within the range 50 to 65 Hz

In this case the 1.4 pulley gives 1462 rpm and so from N = 120 f/p (see Fig 8.6.1) frequency is 48.7 Hz and not quite acceptable. The 1.6 pulley gives 1670 rpm and a frequency of 55 Hz which is preferable.

5 Assume the 6-pole direct drive option is chosen. The flow drawn by the 2-jet turbine is given by the nozzle equation:

$$Q = 3.43 \times d_{jet}^2 \times \sqrt{H_{net}} \times n_{jet} = 3.43 \times 0.013^2 \times \sqrt{17} \times 2 = 4.8 \text{ l/s}$$

Some modification will be necessary if the turbine is to accommodate the full 14 l/s flow. If there are 16 to 20 buckets the D/d_{jet} ratio is probably 10:1 (more buckets indicate a higher ratio). Assuming 10:1 consider increasing the nozzle diameters to 15 mm:

$$Q = 3.43 \times d_{jet}^2 \times \sqrt{H_{net}} \times n_{jet} = 3.43 \times 0.015^2 \times \sqrt{17} \times 2 = 6.4 \text{ l/s}$$

This does not make sufficient difference. Consider adding two more jets and increasing each nozzle to 16 mm diameter: the full flow is now just over 14 l/s. (Do not increase jet size below a D/d_{jet} ratio of 9:1 if the buckets were designed for 10:1). Care must be taken to align the nozzles to the buckets since the 4 mm clearance is less than recommended. The lowest flow setting will be 3.2 l/s (one jet operating). The turbine has been modified to produce around 1.8 kW whereas in its original form it would have produced only around 0.6 kW on a 17 metre head (assuming 75% effficiency).

The same turbine could also be used for a higher head and flow, for instance 40 metres to produce up to 6.6 kW, with a runner rotating at 1600 rpm. High heads increase the runner speed, giving rise to mechanical strains and shortened bearing life. The transmission of 6.6 kW (or more power on even higher heads) may exceed the mechanical strength limits of the turbine shaft and components.

Note 5.3.2 — Fundamentals of the Pelton

In this note the two fundamental equations which govern Pelton design are explained. They both derive from Bernoulli's energy law and the 'orifice discharge equation' seen already in Chapter 3. In this chapter it is assumed that the load on the turbine is adjusted to ensure that the turbine runs at the speed given by the rpm equation given in Note 5.3.1. Any deviation in this speed, caused by excess or insufficient load, will cause the turbine to produce less than its optimum power output. The rpm equation is therefore the 'optimum rpm equation'. Maintenance of turbine efficiency is the subject of Chapter 6 on governing.

From the rpm equation, it is clear that the size and the number of the water jets driving the Pelton make little difference to the speed of the jets and the optimum rpm of the runner. Given a particular runner diameter and a particular pressure of water (the net head), the Pelton will need to be rotated at approximately the same shaft speed whatever nozzle diameter is used, however many nozzles there are or whatever position the spear valve is set to. There will be some variation in optimum runner speed as flow varies, because friction losses in the penstock change, and therefore net head changes, although gross head remains the same.

For instance, as the jets of a multi-jet Pelton are opened successively, flow increases, but manifold losses and pipe wall losses decrease net head, and optimum runner speed decreases. The manifold must therefore be designed carefully to avoid this effect causing major speed changes (as well as power loss). The pipe bend friction loss factors given in Chapter 3 can be used to do this.The law of conservation of energy (Bernoulli) dictates that:

$$\text{jet speed} = \sqrt{2\,gH_{net}} = 4.43\sqrt{H_{net}}$$

As described, bucket speed is equal to 0.46 times jet speed for peak (or best) efficiency. Since the rpm of the Pelton is equal to the bucket speed multiplied by $60/\pi D$ the 'rpm equation' is easily derived. The diameter of the jet can be expressed in terms of the flow (Q):

$$Q = \text{jet speed multiplied by jet area} = \sqrt{2\,gH_{net}} \times \frac{\pi d_{jet}^2}{4}$$

The 'nozzle equation' follows from this expression.

The diameter of the orifice or nozzle producing the jet is not necessarily exactly the same as the diameter of the jet. For instance a flat plate orifice (a hole in a plate) produces a jet of smaller diameter than the diameter of the orifice itself.

Losses occur because of friction in the nozzle and turbulence in the contraction region just outside the nozzle. A simple orifice will create edge turbulence effects. These effects are expressed by a coefficient of kinetic energy loss or velocity loss, (c_v):

$$v_{jet} = c_v \times v_{nozzle}$$

The compression of the water in the contraction region is expressed as a contraction coefficient (c_c) reducing the cross-sectional area of the jet:

$$A_{jet} = c_c \times A_{nozzle}$$

The two effects together (c_c is generally 1 in the case of simple orifices) cause a reduction in flow expressed by a coefficient of discharge (c_d):

$$Q_{jet} = c_d \times Q_{orifice} = c_v \times c_c \times Q_{orifice}$$

Both c_v and c_d are directly measurable in laboratory tests. Typical figures found in practice are:

sharp-edged plate orifice:	$c_c = 0.61$, $c_v = 0.98$	damaged, worn nozzle	$c_v = 0.7 - 0.8$
tapered nozzle	$c_c = 1.00$, $c_v = 0.98$	profiled nozzle	$c_c = 0.98$, $c_v = 0.97$
spear valved nozzles: almost closed	$c_v = 0.92$	mid-position	$c_v = 0.97$

The friction loss associated with turbulence in the jet can be found from an estimate of c_v:

$$\text{Head loss} = H_{net}\left(1 - c_v^2\right)$$

To be strictly accurate, the nozzle and rpm equations should include the relevant coefficients:

$$D = 38c_v \frac{\sqrt{H}}{\text{rpm of Pelton}} \qquad d_{jet} = \sqrt{c_d} \times 0.54 \times \frac{\sqrt{Q}}{H^{0.25}} \times \frac{1}{\sqrt{n_{jet}}}$$

When sizing the nozzle, the contraction effect should be allowed for: $d_{nozzle} = d_{jet}/c_c$

The number of buckets around the periphery of the wheel is related to the D/d_{jet} ratio. Typically:

8:1 ratio, 12 to 14 buckets; 10:1 ratio, 16 to 20 buckets; 12:1 ratio, 22 to 26 buckets

5.4

Turgo turbines

The Turgo turbine is an impulse machine similar to a Pelton turbine (Fig 5.4.1). However, the jet is designed to strike the plane of the runner at an angle (typically 20°). In this turbine, water enters the runner through one side and exits through the other. As a consequence of this, the flow that a Turgo runner can accept is not limited by spent fluid interfering with the incoming jet (as is the case with Pelton turbines). Hence, a Turgo turbine can have a smaller diameter runner than a Pelton for an equivalent power. It therefore runs at a higher rpm. Like the Pelton, the Turgo is efficient over a wide range of speeds and needs no seals with glands around the shaft. It also shares the general characteristics of impulse turbines listed for the Pelton.

The Turgo does have certain disadvantages. Firstly, it is more difficult to fabricate than a Pelton, since the buckets (or vanes) are complex in shape, overlapping, and more fragile than Pelton buckets. Secondly, the Turgo experiences a substantial axial load on its runner which must be met by providing a suitable bearing on the end of the shaft. This can be a deep groove bearing or spherical roller bearing (an example is given in Chapter 7), both of which are more convenient than a conventional thrust bearing.

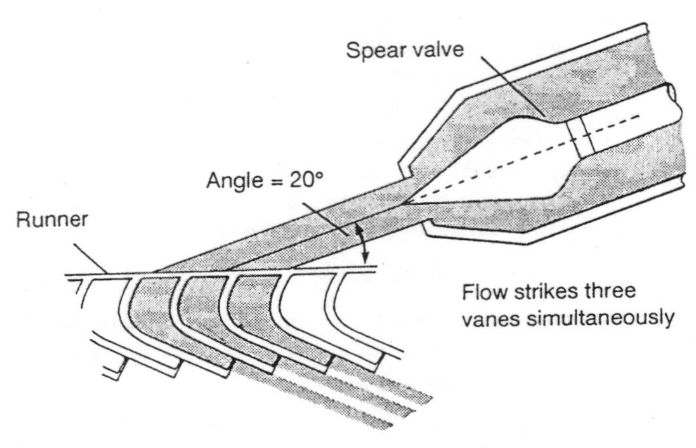

Fig 5.4.1 The Turgo turbine Although this is a compact turbine, the blade shape is complex and less suitable than the Pelton for local manufacture.

Photo 5.4.1 A turgo runner

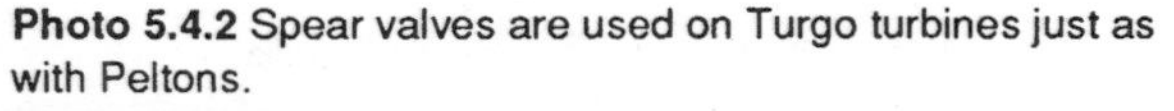

Photo 5.4.2 Spear valves are used on Turgo turbines just as with Peltons.

5.5

Crossflow turbines

Crossflows are also called Banki, Mitchell or Ossberger turbines. A crossflow turbine (Fig 5.5.1) comprises a drum-shaped runner consisting of two parallel discs connected together near their rims by a series of curved blades. A crossflow turbine has its runner shaft horizontal to the ground in all cases (unlike Pelton and Turgo turbines which can have horizontal or vertical orientation). In operation a rectangular nozzle directs the jet to the full length of the runner. The water strikes the blades and imparts most of its kinetic energy. It then passes through the runner and strikes the blades on exit, imparting a smaller amount of energy before leaving the turbine.

The effective head driving the crossflow runner can be increased by the inducement of a partial vacuum inside the casing. This is done by fitting a draught tube below the runner which remains full of tail water at all times. Any decrease in the level induces a greater vacuum which is limited by the use of an air bleed valve in the casing.

Careful design of the valve and casing is necessary to avoid conditions where water might back up and submerge the runner. This principle is in fact applicable to other impulse-type turbines but is not often used on any other than the crossflow. It has the additional advantage of reducing spray damage on the bearings since the internal vacuum causes air to be sucked in through the bearing seals, so impeding movement of water out of the seals.

Because of the symmetry of a crossflow turbine the runner length can theoretically be increased to any value without changing the hydraulic characteristics of the turbine. Hence, doubling runner length merely doubles the power output at the same speed. The lower the head, the longer the runner becomes, and conversely on high heads the crossflow runner tends to be compact.

There are practical limits to length in both cases. If the blades are too long they will flex leading quickly to fatigue failure at the junction of blade

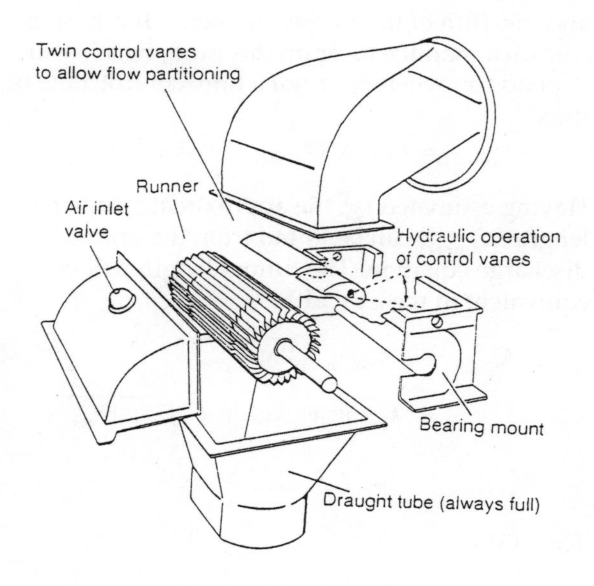

a A sophisticated crossflow which achieves efficiencies up to 80 per cent through use of flow partitioning and suction pressure maintained by an air inlet valve.

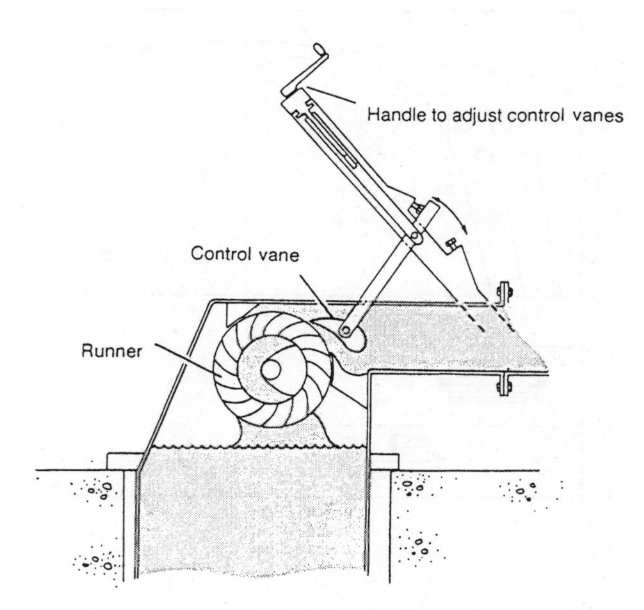

b A simpler but highly successful design. Flow can be controlled by change of orifice size, but with less improvement of low flow efficiency (see Fig 5.1.4).

Fig 5.5.1 The crossflow turbine

and disc. Intermediate bracing discs placed along the length of the runner prevent this but reduce turbine efficiency by interfering with the water flow. In the case of a short runner operating on high head, efficiency losses at the edges become considerable.

The efficiency of a crossflow depends on the sophistication of its design. A feature such as vacuum enhancement is necessarily expensive as it requires the use of air seals around the runner shaft as it passes through the casing, and an airtight casing. Sophisticated machines attain efficiencies as high as 85 per cent. Simpler ones range typically between 60 and 75 per cent. Part-flow efficiency down to less than a quarter of full flow can be maintained at high values by the arrangement for flow partitioning illustrated in Fig 5.5.2. Some designs (Example 5.5.1) maintain high efficiency down to 50% of full flow without the inclusion of a partitioning device, but with a standard guide vane.

Two major attractions of the crossflow have led to a considerable interest in this turbine. Firstly, it is a design suitable for a wide range of heads and power ratings (see Fig 5.1.2). Secondly, it lends itself easily to simple fabrication techniques, a feature which is of interest in developing countries. The runner blades, for instance, can be fabricated by cutting pipe lengthwise in strips.

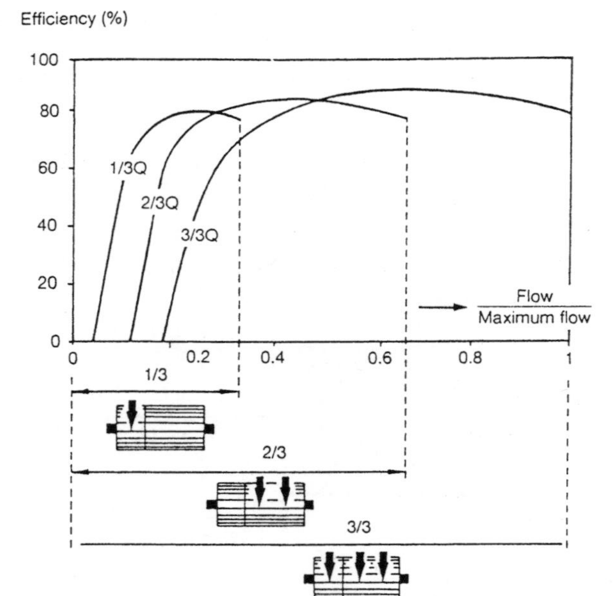

Fig 5.5.2 Crossflow efficiency is maintained over a wider flow range by a partition device which allows a third or two-thirds of the runner to be closed off. Overall efficiency at low flow will be worse than indicated here, because the part-load efficiency of the generator may be poor. Always find out the generator part-load efficiency (see Example 5.1.2).

Photo 5.5.1 A very short runner length is suitable for relatively high heads. (IT/Jeremy Hartley – Nepal)

For sizing calculations on the crossflow see Example 5.5.1 and refer to the bibliography in Appendix 3. The dimensions of interest are the runner length (L_{runner}) and diameter (D_{runner}), and jet thickness (t_{jet}). The width of the rectangular jet orifice is always equal to runner length, while the second cross-sectional dimension, the jet thickness, is designed for optimum performance. If Q is flow in m^3/s, H_{net} is net head in metres, then an approximate feel for the physical dimensions (all in metres) can be obtained using the method described here:

First, decide on the preferred running speed and calculate the approximate runner diameter:

$$D_{runner} = 40 \frac{\sqrt{H_{net}}}{\text{rpm crossflow}}$$

The jet thickness is usually between one tenth and one fifth of the runner diameter. It is best to consult a manufacturer on this ratio, which also depends on whether or not a flow control vane is fitted.

$$t_{jet} = 0.1 \times D_{runner} \rightarrow 0.2 \times D_{runner}$$

Having estimated t_{jet}, the approximate runner length (L_{runner}) can be found from the orifice discharge equation. The runner length will be equivalent to the jet width:

$$Q = A_{nozzle} \sqrt{2\,g\,H_{net}}$$
$$= t_{jet} \times \text{jet width} \times \sqrt{2\,g\,H_{net}}$$
$$= t_{jet} \times L_{runner} \times \sqrt{2\,g\,H_{net}}$$

Therefore:

$$L_{runner} = \frac{Q}{t_{jet}\sqrt{2\,g\,H_{net}}}$$
$$= \frac{0.23\,Q}{t_{jet}\sqrt{H_{net}}}$$

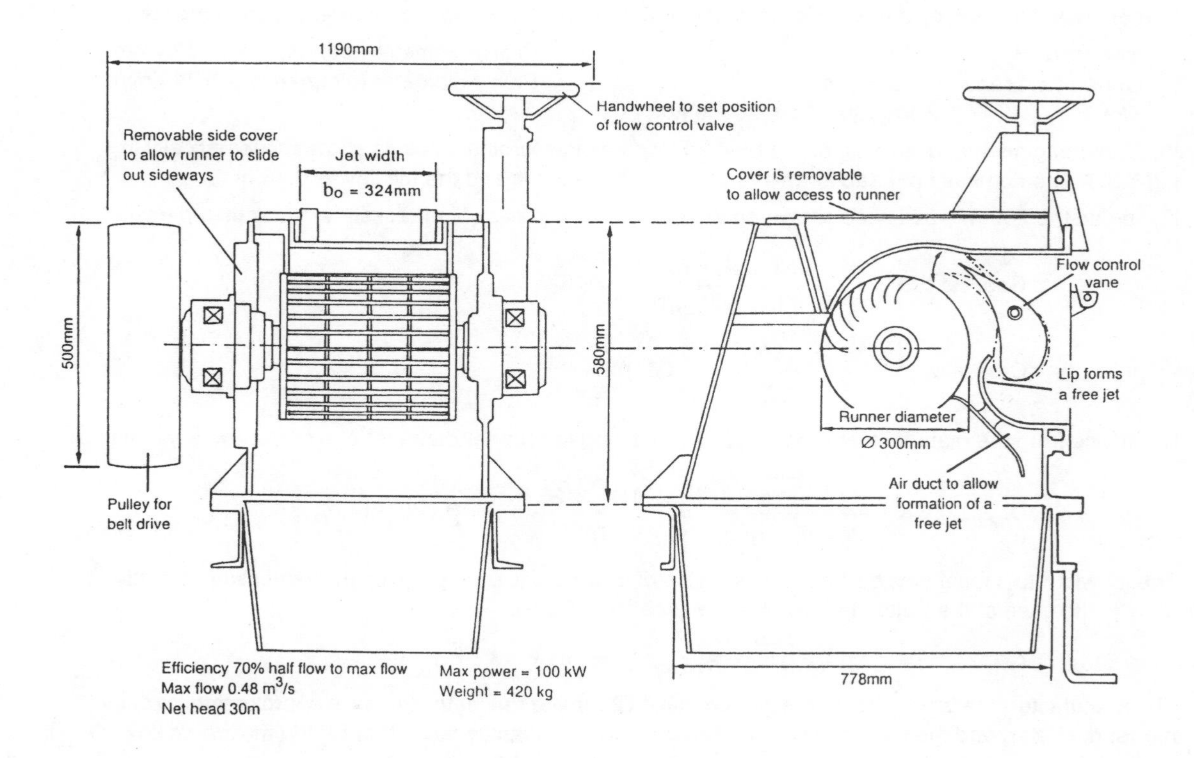

Fig 5.5.3 The SKAT T12 crossflow turbine (see Appendix 3, bibliography) made by BYS in Kathmandu, Nepal. The turbine is designed for minimum vibration. All moving parts can be disassembled from a permanently installed frame.

Photo 5.5.2 A crossflow design by Promihdec with an overhung rotor (both bearings are on one side). (Peru)

Example 5.5.1 **Sizing of a crossflow turbine**

You are asked to specify a turbine for local manufacture for a site with the following characteristics:

Net head (H_{net})	26 m	Pulley diameter (turbine)	500 mm
Generator speed	1500 rpm	Pulley diameter (alternator)	250 mm
Flow	Variation from 200 to 400 l/s		

You have engineering drawings for the SKAT T12 crossflow turbine sized as shown schematically in Fig 5.5.3. The nominal net head for this 100 kW turbine is 30 m and the full flow is 0.48 m^3/s.

If you use the 500 mm pulley as shown on the drawing the speed of the crossflow is calculated easily:

$$\text{Gear ratio } G = \frac{\text{alternator rpm}}{\text{turbine rpm}} = 2$$

$$\text{Crossflow rpm} = \frac{1500}{2} = 750 \text{ rpm}$$

On advice from the manufacturer, the equation for sizing the runner diameter of a crossflow is taken to be:

$$D_{runner} = \frac{41\sqrt{H_{net}}}{\text{crossflow rpm}} = \frac{41 \times \sqrt{26}}{750} = 0.279 \text{ m}$$

The jet width (b_o) must now be found. The orifice discharge equation governs this dimension, exactly as it does in the case of the Pelton impulse turbine (see Note 5.3.2).

$$Q = c_d \sqrt{2\,g\,H} \times A_{nozzle} = c_d\, 4.43 \sqrt{H}\, b_o\, t_{jet}$$

The manufacturer tells you that in the case of the T12 turbine the jet thickness is approximately 1/5 the runner diameter, and the coefficient of discharge is a given quantity, such that for the maximum flow (Q_{max}):

$$Q_{max} = 0.9 \times \sqrt{H} \times b_o \times D_{runner} \qquad \text{(for T12 crossflow)}$$

$$b_o = \frac{Q_{max}}{0.9 \times \sqrt{H} \times D_{runner}} = \frac{0.4}{0.9 \times \sqrt{26} \times 0.279} = 0.312 \text{ m}$$

It is now possible to modify the drawings yourself to allow for a runner diameter of 279 mm and a jet width of 312 mm. It is possibly easier and safer to maintain at least one of the original dimensions on the drawing, or even better, both of them.

For instance, to keep the runner diameter to 300 mm change the gear ratio:

$$\text{Crossflow rpm} = \frac{41\sqrt{H_{net}}}{D_{runner}} = \frac{41 \times \sqrt{26}}{0.3} = 697 \text{ rpm}$$

$$\text{Gear ratio } G = \frac{\text{alternator rpm}}{\text{turbine rpm}} = \frac{1500}{697} = 2.15$$

Note that if D_{runner} is maintained at 300 mm then the jet width must be recalculated:

$$b_o = \frac{Q_{max}}{0.9 \times \sqrt{H} \times D_{runner}} = \frac{0.4}{0.9 \times \sqrt{26} \times 0.3} = 0.29 \text{ m}$$

If no modifications are made, and both the runner diameter and the jet width are maintained at 300 mm and 324 mm respectively, then

$$Q_{max} = 0.9 \times \sqrt{H} \times b_o \times D_{runner} = 0.9 \times \sqrt{26} \times 0.3 \times 0.324 = 0.45 \text{ m}^3/\text{s}$$

The maximum flow actually available is 0.4 m^3/s so in this case the turbine would never run at its full design flow. The manufacturer specifies that efficiency is maintained at 70% at partial flow down to 50% of full flow, which is 0.24 m^3/s. In your application performance will therefore be acceptable in the range 240 to 400 l/s, but will deteriorate at below 240 l/s. If your low flow performance (in the 200 to 240 l/s range) is important, then modifications as described above are necessary. There will be a loss of efficiency therefore when the turbine is run at 0.2 m^3/s, since this is less than 50% of the full design flow.

5.6

Reaction turbines

The reaction turbines considered here are the Francis, and the propeller. A special case of the propeller turbine is the Kaplan. In general, reaction turbines will rotate faster than impulse types given the same head and flow conditions. The propeller will rotate even faster than the Francis (its specific speed is higher). These high speeds have the very important implication that reaction turbines can often be directly coupled to an alternator without any speed-increasing drive system. Some manufacturers make turbine-generator sets of this sort. Significant cost savings are made in eliminating the drive so that maintenance of the hydro unit is very much simpler.

Following Fig 5.1.2 it is clear that the Francis type is suitable for medium heads, while the propeller is more suitable for low heads.

On the whole, reaction turbines need more sophisticated fabrication than impulse types, because they involve the use of larger more intricately profiled blades. The extra expense involved is offset by high efficiencies and the advantage of high running speeds at low heads from relatively compact machines.

Fabrication constraints make these turbines less attractive for use in micro-hydro in developing countries. Because of the importance of low-head micro-hydro, work is nevertheless being undertaken to develop propeller machines which are simple to construct, having for instance non-profiled runner blades.

All reaction turbines are subject to the danger of cavitation, and tend to have poor part flow efficiency characteristics.

Photo 5.6.1 Low cost reactive turbines of many kinds are used in China.

Photo 5.6.2 A 5 kW Francis turbine driving an induction generator with a load controller.

5.7

The Francis turbine

Fig 5.7.1 illustrates the Francis turbine. The runner blades are profiled in a complex manner and the casing is scrolled to distribute water around the entire perimeter of the runner.

In operation, water enters around the periphery of the runner, passes through the guide vanes and runner blades before exiting axially from the centre of the runner. The water imparts most of its 'pressure' energy to the runner and leaves the turbine via a draught tube.

The Francis turbine is always fitted with guide vanes as shown in Fig 5.7.1. These regulate the water flow as it enters the runner, and usually are linked to a governor system which matches flow to turbine loading, in the same way as a Pelton spear valve. As the flow is reduced, the efficiency of the turbine drops off, as is evident in Fig 5.1.4.

Photo 5.7.1 A Francis turbine in a tea estate. (Sri Lanka)

Photo 5.7.2 A linkage mechanism to control flow. (Peredeniya University, Sri Lanka)

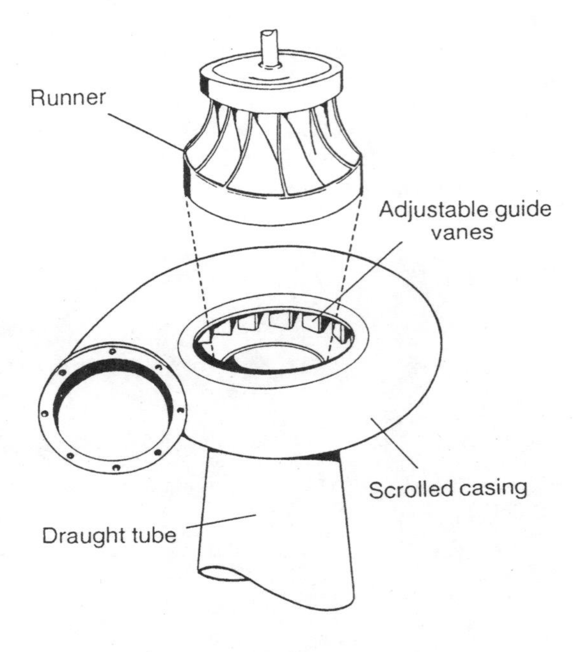

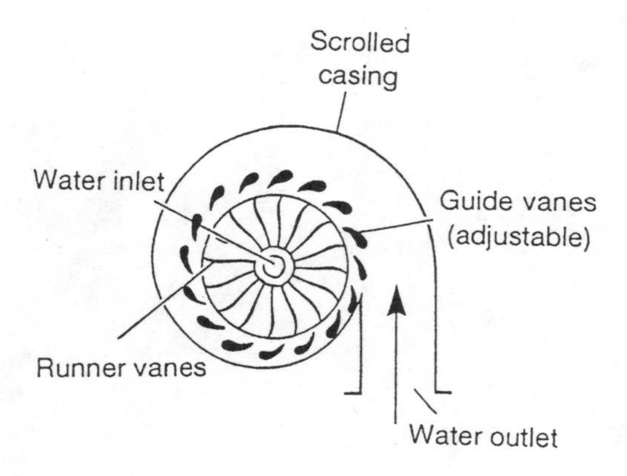

Fig 5.7.1 The Francis turbine. The complex blades and scroll casing shape and the need for close tolerances cause this to be an expensive turbine. Guide vanes are rotated to change flow by a linkage mechanism.

5.8
The propeller turbine and Kaplan

The basic propeller turbine consists of a propeller, similar to a ship's propeller, fitted inside a continuation of the penstock tube, its shaft taken out where the tube changes direction. Usually three to six blades are used, three in the cases of very low head units. Water flow is regulated by the use of swivelling gates ('wicket gates') just upstream of the propeller. The part-flow efficiency characteristic tends to be poor. This kind of propeller turbine is known as a 'fixed-blade axial flow' turbine, since the geometry of the blade does not change. Although traditionally the propeller is profiled to optimize the effect of pressure lift forces acting on it, designs have been produced with flat section blades which offer less efficiency but are more easily fabricated. This kind of design can be considered seriously for micro-hydro applications where low cost and ease of fabrication are priorities. It is also possible to consider casting the propeller casing in concrete (Fig 5.8.1).

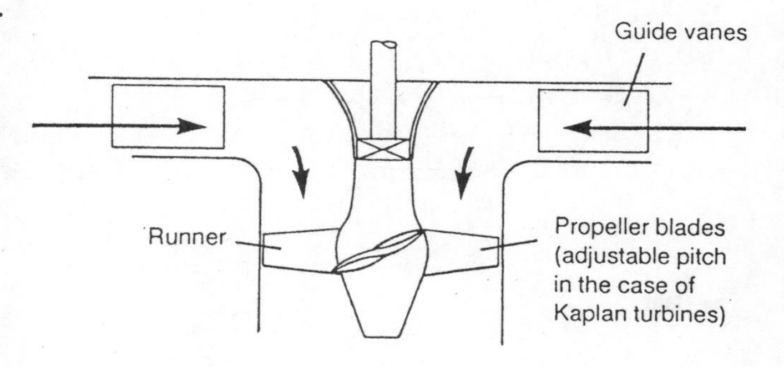

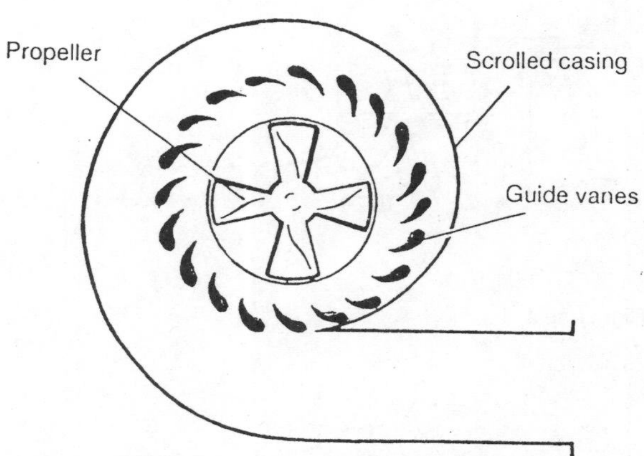

Fig 5.8.2 Propeller and Kaplan turbines. In either case the water can enter radially or axially through guide vanes. The Kaplan is distinguished by having adjustable pitch runner blades.

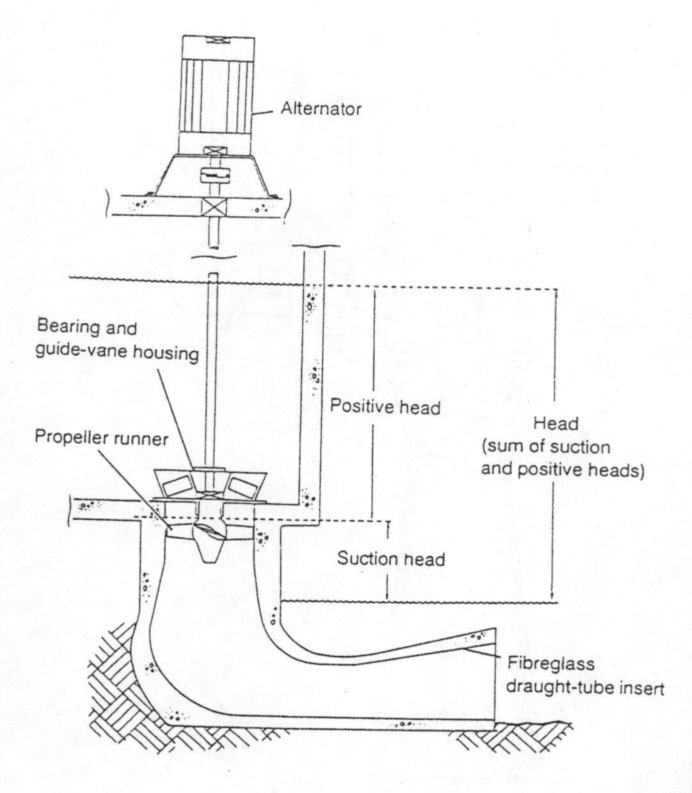

Fig 5.8.1 Low cost 'open flume' propeller turbine. Note the extra head achieved by the draught tube and its profile.

Large-scale hydro sites make use of more sophisticated versions of the propeller turbine. Swivelling (or varying the pitch) of the propeller blades simultaneously with wicket gate adjustment has the effect of maintaining high efficiency under part-flow conditions. Such turbines are known as 'variable pitch' propeller types or Kaplan turbines. Wicket gates are carefully profiled to induce tangential velocity or 'whirl' in the water. Common designs include the use of a scroll casing feeding water flow radially from the periphery inwards, as illustrated in Fig 5.8.2. Whether radial or axial flow, variable pitch designs involve complex linkages and are usually not cost-effective in any except the largest of micro-hydro applications.

Fig 5.8.3 shows some methods of installing propeller turbines, which are in the mini-hydro range but which can be reduced in scale by the manufacturers or adapted for local manufacture.

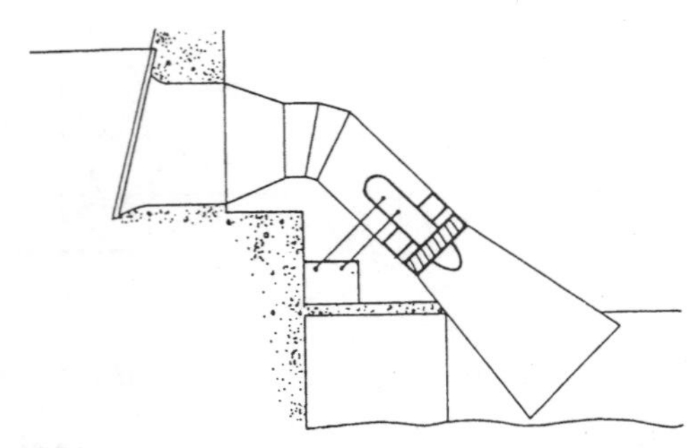

a Bulb

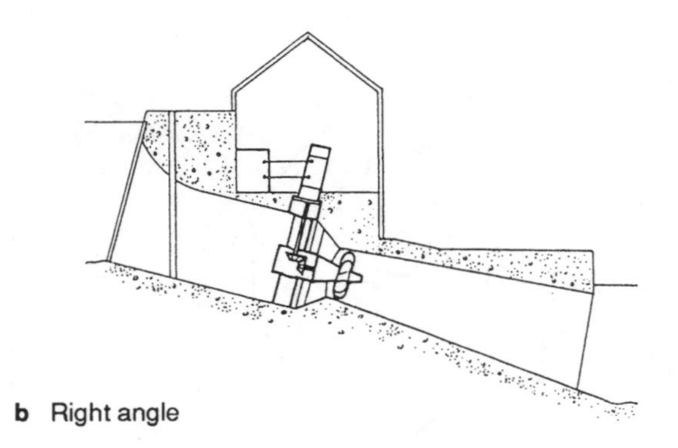

b Right angle

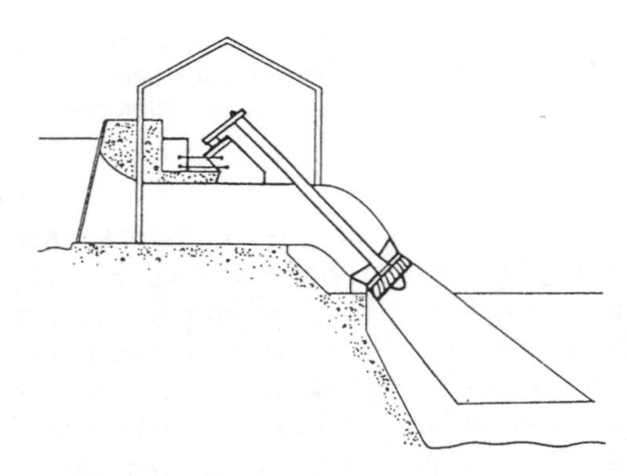

Fig 5.8.3 Conventional propeller turbine arrangements. Notice that the 'bulb' design eliminates the need for a bend in the water passage or a complex right angle drive. This arrangement can also be miniaturized for micro-hydro, as in the case of the Chinese 'Pocket Generator' which is available in capacities as small as 1 kW. It is also reasonable to adapt standard pipework sections to make a low capacity version of a tubular turbine.

Photo 5.8.1 A small Chinese propeller turbine is used to draw power from a drop in an irrigation channel.

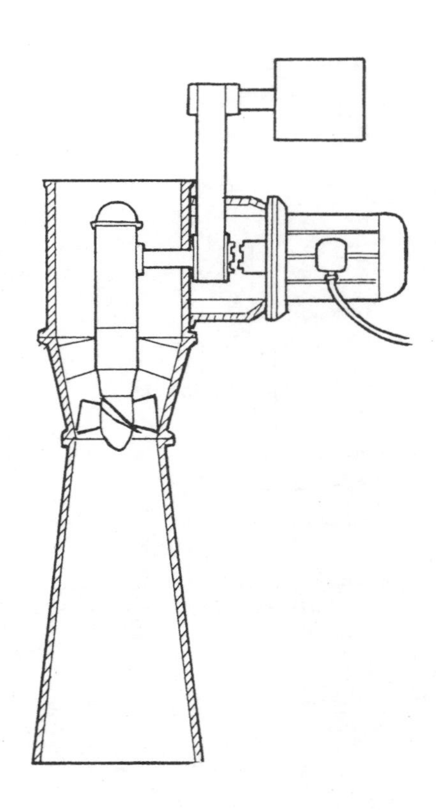

Fig 5.8.4 A small propeller turbine design with right angle drive

5.9 Draught tubes

Reaction turbines and the partially-evacuated crossflow can benefit, as has been mentioned, from an enclosure below the runner known as the draught tube. One purpose of the enclosure is to maintain a column of water between the turbine outlet and the downstream water level. The second purpose is the recovery of velocity energy or kinetic energy in the water leaving the runner. Since water has to leave the turbine runner at a relatively high velocity in order to exit from the turbine, it still possesses a substantial proportion of the available kinetic energy. To recover this energy efficiently, the water velocity must be reduced gradually while friction losses are minimized. If the velocity were not reduced, water would jet out of the turbine outlet into the tailrace and energy would be lost as turbulence in the tailrace.

This recovery of velocity head can be obtained by a gradual increase in the cross-sectional area of the draught tube, resulting in a gradual reduction of the velocity. In general the diameter of tube outlet is about twice that of its inlet, and the angle between opposite walls of an expanding draught tube should be between 7° and 20° to give optimum energy recovery. Most manufacturers either supply a correctly designed draught tube or give detailed specifications for the construction.

The draught tube usually contains pressure below atmosphere (a partial vacuum) so the outlet must remain below the water surface at all times to prevent air from being sucked into the tube and displacing the water column. The difference in level between runner exit and tailwater level is called the static suction head.

The use of draught tubes with reaction turbines requires knowledge of cavitation limits. The turbine manufacturer will advise on the correct position of the runner above (or below) tailwater level to avoid cavitation. Suitable texts on the subject are referenced. In summary, the maximum allowable distance in metres between the runner and the tailwater level (Z_{max}) is given by

$$Z_{max} = H_a - H_v - \sigma_{TC} H$$

where H_a is atmospheric pressure expressed in metres. H_v is the vapour pressure in metres of water at the site water temperature, H is the net head in metres, and σ_{TC} is a characteristic constant of the runner, known as its 'critical Thoma sigma' value. Like specific speed, σ_{TC} is a value which the manufacturer can provide. Z_{max} will be positive in cases where the runner is positioned above the tailwater level. The runner must be positioned below this value of Z_{max} if cavitation is to be avoided.

Photo 5.9.1 and 5.9.2 Francis turbine draught tube. (Sri Lanka)

5.10
Reverse pumps

Centrifugal pumps can be used as turbines. Research is being undertaken to allow the performance of 'pumps and turbines' to be predicted. Potential advantages are low cost owing to mass production, local production, availability of spare parts and wider dealer/ support networks. Disadvantages are as yet poorly understood characteristics, lower typical efficiencies, unknown wear characteristics, and poor part-flow efficiency.

In many countries pumps are manufactured in large quantities for water supply and irrigation purposes, whereas there may be no local manufacture of water turbines. In these countries 'pumps-as-turbines' (PATs) may be economical for a wide range of heads and flows.

In general, PATs are most appropriate for medium head sites, which normally require a slow-running, multi-jet Pelton or small crossflow turbine. For outputs below 10kW, the cost of a PAT is likely to be significantly lower than a crossflow or Pelton for heads of less than 30 metres.

The use of a PAT has the disadvantage that the flow rate is fixed for a particular head. This can be overcome at some cost by using two units of different size, and switching between them depending on the flow rate.

Photo 5.10.1 A standard centrifugal pump/motor can be used in reverse as a turbine/generator. (Pakistan)

An important disadvantage of using a PAT is that it is difficult to characterize the turbine performance. Some simple formulae are given below for converting pump data to turbine performance data, but these are not always very accurate. A number of pump manufacturers have run tests on their machines working as turbines and so are able to offer PATS complete with data.

An advantage of the PAT is that it is easy to purchase and install, particularly if a close-coupled unit is used and the motor runs as an induction generator. Usually spare parts are easily available, and knowledge of pump maintenance is widespread.

Types of pump suitable for micro-hydro application

Investigations on various types of PAT suggest that there are several which are suitable for micro-hydro, running in reverse with an induction generator. Of the main types, end-suction centrifugal pumps are nearly always suitable for this application. Axial flow pumps are suitable for low-head applications, but small sizes (below 30kW) are not commonly available.

End-suction centrifugal pumps are the most commonly available, used extensively for water supply and irrigation. They are of relatively simple design, and are easily maintained. In-line and double-suction pumps may be suitable but sometimes they have considerably lower efficiencies in turbine mode. Self-priming pumps are not suitable for PATs since they contain a non-return valve which prevents reverse flow.

A centrifugal pump may have its own bearings, or may be close-coupled to an induction motor which contains the necessary bearings. Close-coupled units are sometimes referred to as 'monobloc' pumps. On smaller sizes of end-suction pump, the impeller is screwed on to the shaft. However, the impeller will not unscrew in turbine mode, unless a solid object causes the impeller to stick, because the torque is applied in the same direction as in pump mode. Care must be taken with certain types of seal which incorporate a spring to ensure that the spring will operate effectively with reverse rotation.

Dry-motor submersible pumps are widely used for construction site drainage and for pumping from open wells. The motor is integral with the pump, and is cooled by the pumped water flowing either through a jacket or across fins on the motor housing. Those with fin-cooling are not suitable for use as turbines, as they will overheat unless submerged below water level. Also, some pumps of this type have rubber linings on the diffuser parts, which can prevent the impeller from running in reverse. Wet-motor submersible borehole pumps are of a more specialized design and are not normally suitable for use as turbines. They usually contain a non-return valve, and may also incorporate a thrust bearing designed only for pump operation.

For many head-flow combinations, different PAT options are available, depending on the pump speed chosen. Fig 5.10.1 shows the approximate range of application of monobloc pumps from one UK manufacturer. This chart can only be used as a rough guide for selecting a PAT.

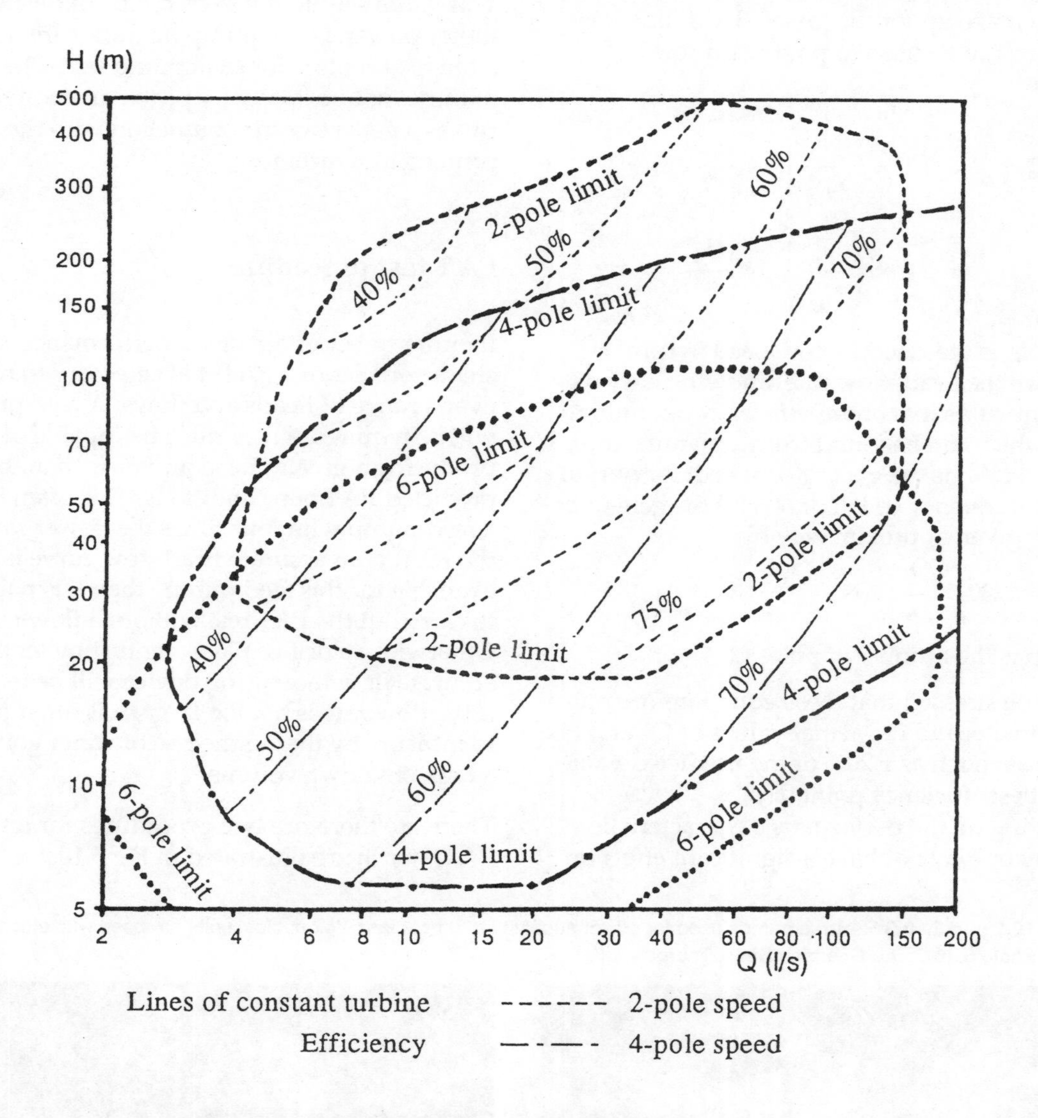

Approximate direct-coupled reverse motor and pump turbine speeds

	2-pole	4-pole	6-pole
50 Hz	3100	1550	1050
60 Hz	3700	1850	1250

Fig 5.10.1 PAT selection Application of standard 'monobloc' centrifugal pumps as turbines, based on information from one pump manufacturer in the UK.

Selection of a PAT to match site conditions

The running conditions, in terms of head and flow, for best efficiency as a turbine are very different from the rated pump output, although the efficiency in each case will be approximately the same. Friction and leakage losses within a centrifugal pump result in a reduction of head and flow from the theoretical maximum. When running as a turbine, the head and flow required will be greater than the theoretical values in order to make up for the losses. The following equations can be used to predict turbine head and flow:

$$Q_T = 1.1 \times \left(\frac{N_g}{N_m}\right) \times \frac{Q_P}{e_p^{0.8}}$$

$$H_T = 1.1 \times \left(\frac{N_g}{N_m}\right)^2 \times \frac{H_P}{e_p^{1.2}}$$

where N_m is the rated motor speed in rpm. H_p, Q_p, e_p are the head, flow rate, and efficiency of the pump at its maximum efficiency operating point, which can be found from a manufacturer's data sheet. N_g is the generator speed: for output at rated frequency (f) (50 or 60 Hz) the generator speed is given approximately by:

$$N_g = 240 \times \frac{f}{p} - N_m$$

where p is the number of poles (2, 4, 6...).

It must be stressed that these equations are only approximate, and the actual values of Q_T and H_T may be as much as ± 20% of the predicted value for the best efficiency point (*bep*).
Depending on the performance characteristics, this may or may not have a significant effect on the PAT output. It is therefore recommended that, after initial selection, the pump is tested under turbine conditions to find what output power will be produced at the available head. The method for testing is described below.

Care must be taken to ensure that the induction machine is not overloaded when running as a generator. The electrical output of the generator is approximately:

$$P_{gen} = e_{motor}\, e_p\, 9.81\, Q_T\, H_T \text{ kW}$$

This value should not exceed 80% of the rated motor power. Often pump manufacturers supply a range of motors for each pump size. The largest of these motors is usually powerful enough to run as a generator in conjunction with the pump running as a turbine.

PAT test procedure

In order to be certain of the performance characteristics of a PAT it is necessary to test it over a range of heads and flows. A feed pump will be required which must be capable of producing a greater head and flow than the predicted PAT bep conditions. This often implies a feed pump with four times the power rating of the PAT. If an accurate head-flow curve is available for this feed pump , then it is possible to carry out the PAT test without a flow meter. Otherwise an orifice plate, venturi-meter or other accurate flow measuring device will be required. If the flow rate is not too large, it is often possible to measure by timing the rise of water going into a tank of known volume.

There are therefore two procedures for testing a PAT, which are illustrated in Fig 5.10.2.

Photo 5.10.2 and 5.10.3 Reverse pumps used for power supply to a remote farm. 'Monobloc' units are used with electronic load control. (Photo 5.10.2 The Guardian/Denis Thorpe – UK)

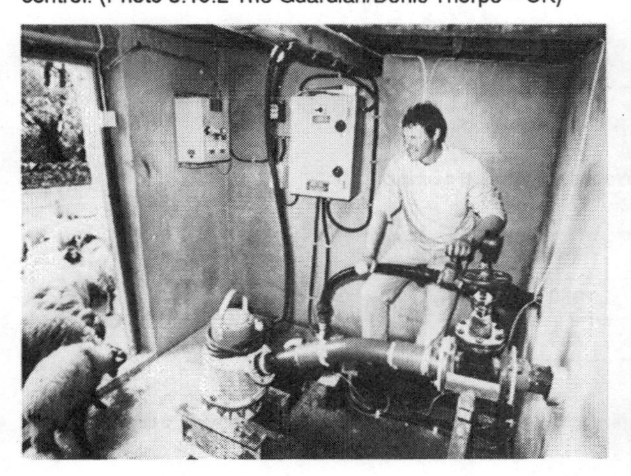

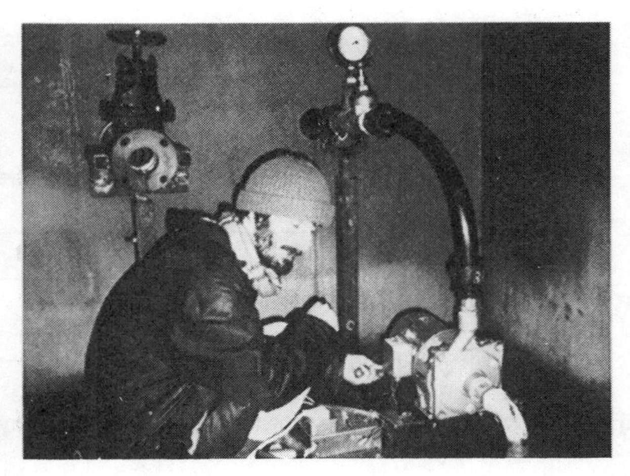

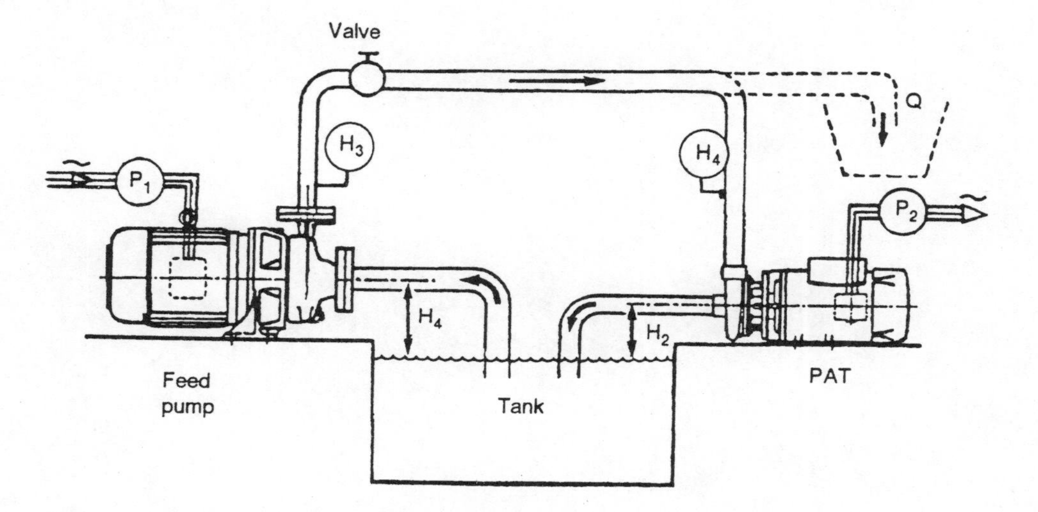

i Test without flow meter, using feed pump H-Q characteristic curve.

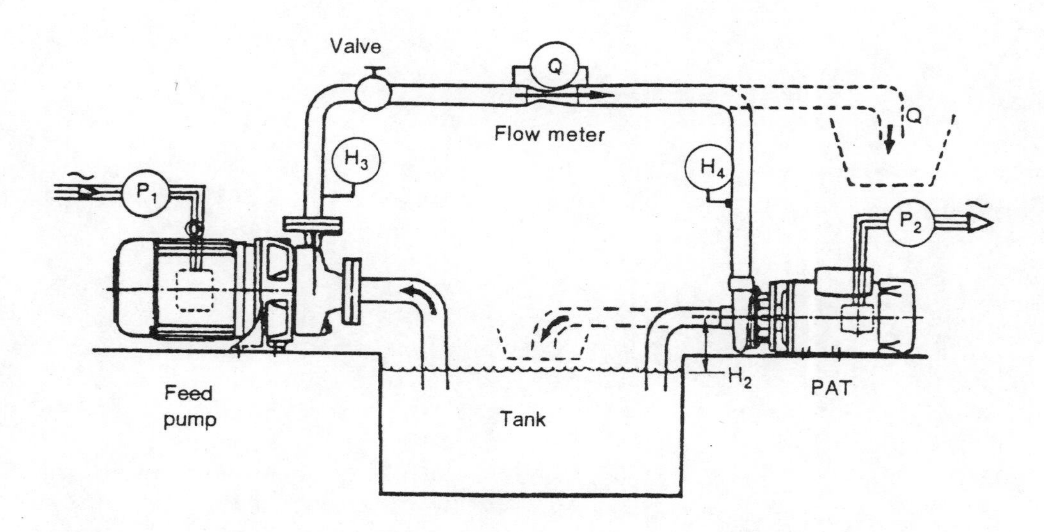

ii Test using a flow meter

Fig 5.10.2 Laboratory testing for final selection of a pump-as-turbine

i Using feed pump H-Q data

For this test it is not necessary to measure the flow rate. A pressure gauge is mounted at the outlet from the feed pump. Using the pump H-Q curve, the value of Q can be determined from using the feed pump head (H_3+H_4 in Fig 5.10.2). A separate pressure gauge will be required to measure the PAT head (H_1+H_2), as there will be head loss in the pipe and valve between feed pump and PAT, particularly when a valve is used to constrain the flow.

ii Using measured flow

For this test, the value of Q_T is obtained from a flow meter, and the value of head (H_1+H_2) is obtained using a pressure gauge at the inlet to the PAT.

In the case of either test, the power output from the PAT can be measured using a voltmeter and ammeter to measure the current and voltage to a resistive load. Alternatively, the induction generator can be connected to the mains supply, but in this case a wattmeter will be required to measure the output power.

The overall efficiency of the PAT will be given by the following formula:

$$e_T = \frac{P_{out}}{P_{in}} = \frac{P_{out}}{9.8\ H_T\ Q_T}$$

If these values are plotted against Q_T or H_T it is possible to find the best efficiency operating point for the PAT and generator.

Governing

6

6.1 Introduction

Governors are used to control the speed of turbines. Some governing or control of the turbine speed is often required to ensure proper operation of the mechanical or electrical end use machinery.

Until recently, all governing devices for hydro power plants adjusted the flow of water through the turbine to vary the water power input on the plant. In this case, as more electrical power is required of the plant, the load on the generator increases, reducing speed.

Through either mechanical means (eg using flyball mechanisms) or electrical means, the governor senses this speed reduction. It then opens the appropriate valves to admit more water, and thus increases the power through the turbine to satisfy the increased demand. Similarly, if less power is required, the governor senses an increase in speed resulting from the excess power available and causes the valve to reduce the water flow through the turbine.

6.2 Purpose of governing

When electricity is generated at an isolated site by a synchronous generator, its frequency is determined by the speed of the generator and the number of poles. A four-pole generator, for example, generates two cycles per revolution (rev) of its shaft. To generate 50 cycles/s (or 50 Hz), it must run at 25 rev/s, or 1500 rev/min. If this speed increases or decreases, the frequency generated also increases or decreases. Although most generators have some form of voltage regulation, the voltage output is also affected somewhat by speed changes.

Electrical equipment is designed to operate at a specific voltage and frequency. Operation at frequencies and voltages other than the design values can cause serious damage. For example, an electric motor will run hot if the frequency is too low or may burn out rather than start if the voltage is too low.

Some control of generator speed is therefore needed. As the speed of the generator is determined by the turbine, the speed of the turbine must be regulated. The governing device controls the speed of the turbine/generator equipment in response to changing external electrical loads placed on the generator.

Where water power drives mechanical equipment directly, a governor is also needed if the speed at which this equipment is driven is critical. For many mechanical uses, this is not the case, and governing is not necessary.

Photo 6.2.1 This generator provides lighting in a village in Nepal. (IT/Jeremy Hartley)

6.3

Specifying the governor

In order to specify a governor, it is necessary to find out the tolerance of the end use machinery to variations in frequency and voltage. Although most electrical equipment can tolerate up to ± 10% voltage fluctuation, frequency is usually held much closer to the nominal figure. The effects of such variations and the allowable limits are described in this section.

Voltage and frequency tolerance

Electrical loads

a **Heating**
Heating loads are most tolerant of variations in supply. Frequency variations do not affect these loads at all. Under-voltage increases the life of heating elements but reduces heat output. Over-voltage can generate excess heat and cause the elements to burn out. For example, an over-voltage of 10% can cause an increase of 21% in heating output.

b **Lighting**
Incandescent bulbs are not affected by frequency. Under-voltage decreases light output very sharply but significantly increases bulb life, unless there are significant fluctuations in supply voltage. On the other hand, over-voltage greatly reduces bulb life. For example, an over-voltage of only 5% reduces the life of bulbs by up to 50%.

Fluorescent lamps are affected by both voltage and frequency variations. If voltage is more than 15% down, the lamp will not light. If the lamp is already operating, it will flicker more as voltage decreases. If voltage drops more than 25%, the lamp may go out.

Over-voltage may lead to the choke overheating, but this is likely to be less limiting than the effect of over-voltage on incandescent bulbs. Fluorescent lamps should operate correctly in the frequency range -5% to +10% of nominal frequency.

c **Transformers**
Transformer losses appear as heat. As losses increase, so does the heat generated within the transformer and, consequently, its temperature.

At fixed frequency, all these losses are very approximately related to the square of the voltage. Over-voltage therefore can pose a problem, and voltage is usually allowed to increase about 5% at rated load. However, under-voltage does not pose a problem.

At fixed voltage, decreases in frequency lead to increased losses and heat generation and operation at below rated frequency at rated voltage should be avoided. Operation at up to 20% over the rated frequency presents no problem as losses decrease.

Motor loads

Induction motors and transformers are affected in similar ways. Under-frequency with steady voltage causes high currents and overheating. Under-voltage at steady frequency has a similar effect on both these and other types of motor.

Overvoltage at rated frequency should be limited for induction motors in the same way as for transformers. However, such motors may be operated at frequencies up to 10% above the rated value unless they drive loads whose torque increases rapidly with speed, such as fans. With such loads the frequency should not rise to more than 5% of the rated value.

Inductions motors take up to six times their rated current in order to produce their rated torque when starting. Thus manufacturers often specify that motors will start and operate satisfactorily if the voltage is within 10% of their rated value. However, motors in rural workshops have operated with completely ungoverned turbines; voltage momentarily drops significantly when a motor is switched on but as, during start-up, minimal loads are imposed on these motors, they come up to speed rapidly.

If a motor is started under a large load, such as a motor driving a compressor in a refrigerator, the longer period of low speed and high current may cause the windings of the motor to overheat and fail.

A large voltage drop may result in insufficient torque to start the motor. When considering voltage regulation, the effect of transmission line resistance must be included – where possible large motors should be placed close to the generator.

Summary

It is recommended that for an all-purpose electrical supply system the voltage and frequency tolerances should be:

Voltage: ± 7% of rated value;

Frequency: up to 5% above, but not below the rated value.

In a small system the voltage and frequency may go outside these limits when loads are switched on or off; but this does not matter so long as the fluctuations are not excessive and do not last for more than two or three seconds.

Where the electrical supply system supplies only limited types of load, such as lighting only, wider voltage and frequency limits may be used as described earlier in this section.

The frequency limits chosen will normally determine what governing is required for the turbine/ generator set if a modern automatic voltage regulator is used. This will maintain a steady voltage over a range of frequencies. If no automatic voltage regulator is used, the voltage as well as the frequency will be dependent on speed; voltage increasing with increasing speed.

Photo 6.3.1 Hydraulic governor, belt-driven from the turbine. (Sri Lanka)

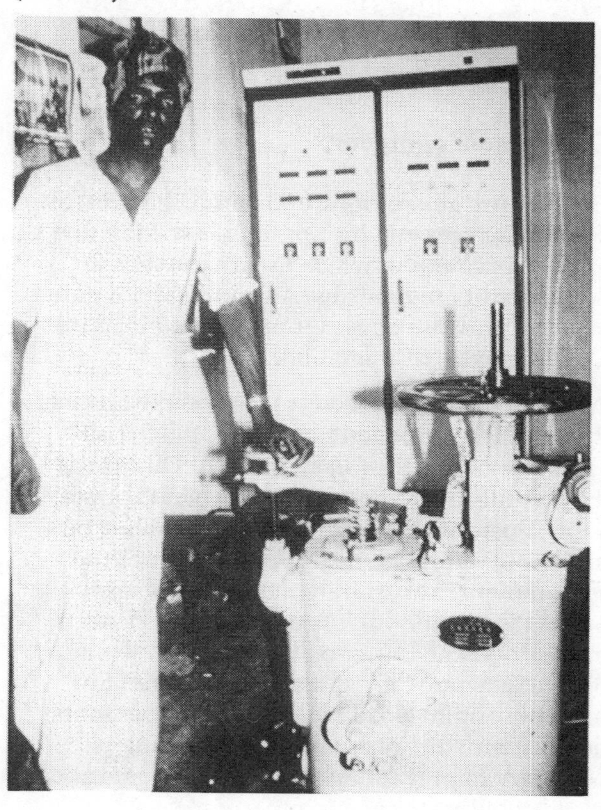

6.4

Approaches to governing

The approaches to governing may be classed in two categories :

- conventional
- non-conventional

The conventional approaches can be used with plants of all sizes, perform to high standards, and may be complex and costly. Such systems are most convenient where the power generated will be put to a variety of uses, where capital exists to cover the cost of a suitable governing system and where the skill to maintain and, if necessary, operate it can be found. Conventional approaches originated from industrialized countries and are designed to meet the standards imposed by the needs of these countries. However, the more recent load controller governor systems have been designed bearing in mind the possibility of their use in developing countries and are much cheaper and easier to maintain than the more traditional turbine control systems. All the conventional systems require sophisticated components from industrialized countries and the hardware may be difficult to manufacture in developing countries without importing some of these components.

Non-conventional approaches are usually used only with plants in the micro-hydro range where costs must be kept down. Non-conventional governing relies on proper manipulation of the turbine or the load. It therefore usually requires more labour to operate than automated, conventional approaches. Output is usually of lower quality, with larger speed, frequency and voltage variations, but it can still be adequate to satisfy most common end uses.

Conventional governing

The older, conventional governor systems originated in the eighteenth century and use mainly mechanical and hydraulic, but latterly electrical and electronic, systems to control turbine output power. The more recent load controller governors use electronic control to impose a constant electrical load on the generator.

Oil pressure governor

Oil pressure governors are so named because oil kept under pressure by a pump is used to drive a piston/servomotor, which in turn moves the flow-control mechanisms. A simplified diagram of an oil pressure governor presented in Figure 6.4.1 illustrates its operation.

When an increased load is placed on the turbine, the resulting reduction in speed of the flyball assembly causes the flyball to drop. This forces the floating lever to raise the sleeve in the pilot valve, opening the upper part and allowing oil under pressure into the upper chamber of the servomotor. The resulting motion of the piston opens the turbine valve, permitting an increase in the flow. The increased flow allows the turbine to generate the extra power required to meet the original load increase. A compensating device consisting of a dashpot and spring is necessary to prevent over-travel of the gate and the instability that would otherwise occur because of the system's inertia. The manual control is used to set the operating speed to start or shut down the system.

Fig 6.4.1 shows how the turbine speed is mechanically sensed by means of a flyball mechanism, but other electro hydraulic governors are in use where turbine speed is sensed as an electronically measured frequency or voltage. The speed measurement is compared continuously with a reference value and the difference results in electronic commands that are then transmitted to the hydraulic part of the governor by electrical actuators.

Although oil pressure governors are used extensively on large hydro projects, they are sophisticated devices whose costs do not decrease in proportion to the size of the turbine they govern. They also require regular maintenance:

- leaking seals and worn pins and bushings must be replaced;
- all bolts and connections should be checked regularly for looseness;
- all levers must be checked for friction or binding;
- all pivots , cam surfaces and linkage rod ends should be oiled frequently and any accumulated grit, rust or dust removed;
- dust, which serves as an insulation on electrical parts, must be guarded against;
- governor characteristics, such as servomotor timing, should be checked periodically and appropriate adjustment made;
- pistons should be checked for wear;
- the hydraulic oil must be kept very clean and filters changed regularly.

The cost and sophistication of oil pressure governors detract from their appropriateness in rural areas. They should be considered only where minimizing water flow is important such as when water storage reservoirs are available. Note that most micro-hydro plants do not have water storage facilities due to the high construction costs these entail.

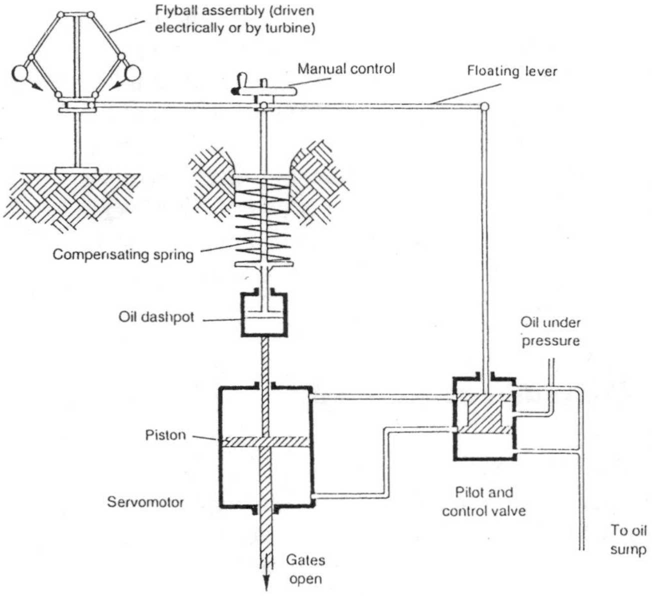

Fig 6.4.1 Hydraulic mechanical governor.

The governor usually consists of a vertical shaft that has counter opposed flyweights suspended from the top of the vertical shaft. As the speed of the shaft increases or decreases, the flyweights move up or down the shaft, respectively. The flow control gates are then opened or closed in response. For instance, if the turbine speeds up due to decreased load, the hydraulic governor will close the gates just enough to maintain constant speed.

To follow the operation of the control linkages, imagine first that the flyweights are dropping as the turbine speed decreases due to a load increase. This causes oil under pressure to flow to the upper half of the servomotor so forcing the piston down and the flow gates to open.

The oil dashpot acts to damp sudden flow changes so reducing pressure surge in the penstock, and further, it eliminates the tendency for the governor to 'hunt'.

Mechanical governor

With micro-hydro power turbines the forces needed to operate guide vanes, inlet valves or jet deflectors can be quite small in comparison with larger turbines. In these cases, direct acting mechanical governors can be used without the need for additional hydraulic actuators. Such governors are used mainly to move jet deflectors on Turgo and Pelton wheels. They are not normally used on crossflows or Francis machines because of the higher forces required and the need for damped response to reduce surge. Thus, they cannot usually control total flow.

Typically, mechanical governors incorporate a flyball arrangement driven by the turbine shaft. The output from this assembly is used directly to drive a jet deflector. The flyballs associated with a mechanical governor are much more massive. This is due to the significantly larger force required to drive the deflector compared to that required to drive a pilot valve of an oil pressure governor.

Although cheaper than the oil pressure governor, the mechanical governor cannot usually control flow and has been largely superseded by electronic load control governors.

Load controller

The load controller is an electronic device that maintains a constant electrical load on a generator in spite of changing user loads. This permits the use of a turbine with no flow-regulating devices and their governor control system. The flow through the turbine is set at a constant value and the load controller ensures that a constant electrical load is supplied by the generator. The turbine power output is then constant and thus its speed is constant.

The load controller maintains a constant generator output by supplying a secondary 'ballast' load with the power not required by the main load. Figure 6.4.2 shows its operation. (In this case the controller type is the phase-angle ELC which is discussed further in Section 8.5.) Take the case of a plant running at near-full load and governed by a load controller. If the consumer switches off some appliances or other outputs the following events occur:

1 Less electrical power is used while the same water power is maintained to the plant.

2 The turbine and generator speed and frequency generated begin to increase.

3 The change is sensed by the load controller.

4 The load controller adds a ballast load of sufficient resistance at the generator output to dissipate power equivalent to that which was switched off.

Therefore, in spite of a change in consumer load, the total load on the generator remains constant.

A load controller can maintain the system speed (and thus frequency) at its design value regard-

Example 6.4.1 **Rehabilitating an old turbine**

An old turbine set has to be rehabilitated and it is necessary to carry out the following work on the set. Overhaul the turbine and alternator; patch up the mild steel penstock which is leaking; do extension work on the switchboard and carry out expensive repairs to the flow control governor with readily available local components. The people living nearby would like to know whether or not hot water could be provided by the hydro system.

Your thoughts may be along the following lines:

- old turbines can generally be repaired.
- old alternators can also be reconditioned.
- the mild steel penstock has to be patched up where it is leaking
(however, most of the penstock will be rusty and will not be able to withstand a high surge pressure induced by the flow control governor).
- the flow control governor can be repaired with locally available parts but experiences with such repairs show poor results – the chances of it working well are small.
- any faults on the repaired flow control governor could cause surges which may damage the weak penstock.

In this case the best solution may be to use an electronic load controller (ELC) which is likely to give a higher reliability without causing surge pressures. An ELC will also provide some hot water if used with a ballast tank for the dump load.

less of fluctuating mechanical and electrical loads.

Load controllers normally sense turbine speed by measuring frequency of the voltage waveform. The principal advantage of a load controller is that the overall system becomes simpler and less costly. It not only eliminates the need for an intricate governor and an actuating mechanism, but it allows the design of the turbine to be simplified. A less sophisticated system increases the chances for long-term viability and reduces equipment cost considerably for plants in the micro-hydro power range. For example, rather than a cost of US$10,000 for an oil pressure governor for an 8 kW plant, materials for a single phase controller (excluding ballast loads) for the same plant could cost less than US$1000.

Load controllers require no maintenance and have no moving parts. If electronic components fail in the field, they probably cannot be repaired there and then; however, a well-designed unit is composed of separate printed circuit boards that can be replaced easily.

As mentioned earlier, when a load controller is used, flow through the turbine is set at some constant level (but not necessarily fully open) and all the water is used for power generation. The water used to generate power that is not used productively by the consumer is wasted. For 'run-of-river' schemes, this is irrelevant as there is no provision for the storage of unused water for later productive use. However, maximum use can be made of the energy generated in a load controller system as all the 'ballast load' is available in the form of heat. A well-planned scheme will use this for purposes which can generate income and, by increasing the load factor on the plant, this can significantly reduce the cost of the energy generated.

Load controllers alone cannot be used with schemes that have to use a limited quantity of water efficiently. Where a reservoir is included to store excess water during periods of low power demand, a governor or other device that regulates the low flow of water must be used.

Photo 6.4.1 An electronic load controller (ELC). (Peru)

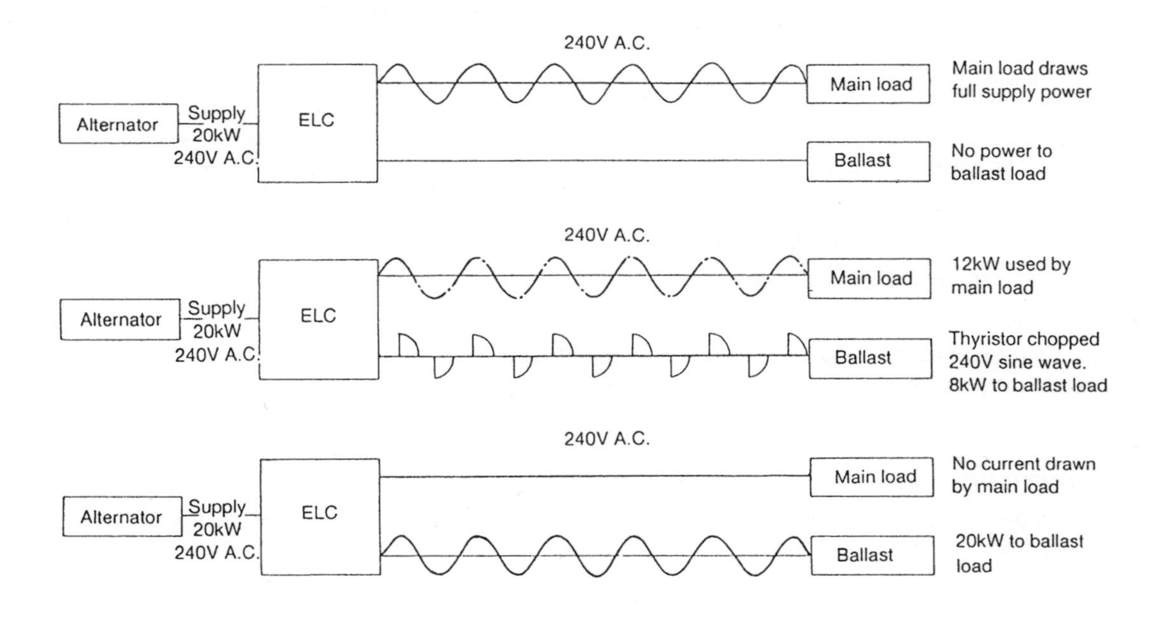

Fig 6.4.2 Electronic load controller (ELC)

Induction generator controller (IGC)

When a load controller is used to govern the speed of a turbine/generator set, an automatic voltage regulator will also normally be required to ensure a steady voltage level when the main load power factor changes. However, recent work on the use of induction generators in place of the normally used synchronous generators has led to the development of an induction generator controller. This is designed specifically for use with induction generators and combines the functions of both a load controller and an automatic voltage regulator. It thus controls the speed of a turbine/induction generator set in a similar way to a load controller and the notes above as to suitable schemes for a load controller also apply in this case.

The induction generator controller is described in more detail in Chapter 8.

Non-conventional governing systems

If a high-quality electrical supply is required for sensitive equipment, then one of the automatic methods listed above must be used. However, where quality of supply is not so critical, some compromise in performance may be justified, particularly with very small plants (less than 10 kW). Lower cost, technically less sophisticated approaches to controlling speed are possible.

It must be kept in mind that the use of non-conventional approaches usually implies less control over speed. However, this is generally not a valid argument against such approaches, especially in the initial years of a project.

In rural areas, electricity is a new commodity and loads build up gradually. Non-conventional approaches permit access to electricity with reduced cost and with reduced sophistication. If demand grows and the sophistication of consumers increases, a load controller can later be included with virtually no modification to the existing system. By that time, both users and plant operators will have gained some familiarity with the hydro power plant, and more sophisticated devices will have a better chance of being operated and maintained successfully.

Constant load

This approach permits only the use of fixed loads, such as village lighting. However, it allows for a bit more flexibility than may at first

Using an electric motor **Example 6.4.2**

A mains-powered factory has an auxiliary non-mains supply in the form of an old 40 kVA alternator driven by a Pelton connected to a mechanical governor. The governor operates electrically depending on supply from the alternator. The alternator is connected to a large induction motor 30 kW through a manual star/delta starter. When the motor is started the turbine and alternator begin to speed up without starting the motor. However, on the mains supply the motor starts without any problem.

Consider the following circumstances:

- 40 kVA alternator;
- 30 kW motor which is started by manual starter – star/delta;
- the motor starts and runs on the main supply;
- the turbine is governed by an electric governor;
- the turbine speeds up when the motor is started.

From the last consideration you know that the governor does not function when the motor is started. As the governor is electrically operated, it can be concluded that during start-up something cuts off the electric supply to the governor. However it is also noted that the motor works perfectly on the mains supply.

Since the motor is a large inductive load the starting current is very high and during start-up the motor connected across the generator terminals acts as a short circuit across these terminals. If the generator terminals are short circuited there is no voltage across the terminals which also means that the electric governor has no voltage across it. Therefore it does not govern the turbine which then speeds up.

One solution to the problem is to get the motor started on the mains supply and use a contactor type change-over to change over the supply from mains to hydro when the motor is running.

simplest method of ensuring constant frequency and voltage. In this form, no switches are included in the circuit. Power is switched on by opening the valve to the turbine sufficiently wide to attain the nominal frequency and/or voltage. Closing the valve when power is no longer required switches off the electricity. In this approach, neither switches nor power outlets should be included in the system. Frequency may be accurate to only ± 10% but modern automatic voltage regulators permit good voltage stability.

One variation of this approach permits more than one constant load to be used. A two- way switch can be used so that switching off one load (eg lighting) when it is no longer needed switches on a second load of similar magnitude (eg a water heater). In this manner a constant load is maintained on the generator despite the use of several loads. Each switch must only control a few per cent of the load. There are numerous variations of this approach.

Manual control

With this approach, an operator is required to maintain a relatively constant frequency manually. This can be done by adjusting:

- the flow of water to the turbine; or
- the total load imposed on the generator.

It is important to remember that it is not necessary for the operator to make adjustments continually during the operation of the plant. Small deviations from nominal frequencies have no adverse effect on the plant or the load.

- **Flow modification** requires a turbine that includes a flow regulating valve. Load variation can be noted by observing voltage or frequency readings on meters. (Where lighting loads are present, light intensity is a good voltage indicator. This warns the operator that the flow to the turbine should be increased or decreased. Reference to an electric meter may be necessary when flow is read-

Example 6.4.3 **An ELC problem**

At a certain site where a 50 kW turbine drives a single-phase 240 volts 50 Hz alternator, the governing is done by an electronic load controller. When no load is connected to the alternator and when the flow through the turbine is fully opened, the meters on the ELC indicate the following:

phase volts: 240
frequency: 55 Hz
current: 0 amps
ballast meters show equal readings.

If you open the ELC box and adjust the frequency knob, the frequency does not respond. However, if a load is connected to the generator, the speed drops to 50 Hz and when the frequency knob is adjusted, the frequency of the machine responds. What could be the problem in the equipment?

Consider the following circumstances:

- ELC type governor;
- turbine flow fully open;
- speed 55 Hz should be 50 Hz;
- when a load is connected the speed drops to 50 Hz;
- when the speed is 50 Hz the adjustment of the frequency knob makes the machine respond.

From the above you can conclude that the hydraulic power input is not matched (equal) to the power absorbed by the ballast.

By connecting an extra load it is possible to bring the total of the power dumped into the ballast plus the power taken by the load to be equal to the input power. This brings the ELC back into control.

The likely cause is excess hydraulic power at the turbine or a partially burnt out ballast load.

justed in order to return to nominal operating frequency.)

- **Load modification**: The manual modification of load is similar in approach to that of using an automatic load controller. This method can be used with a turbine that has no flow-regulating valves. If the flow into the turbine remains constant, the load imposed on the generator must also remain fairly constant to prevent the frequency from varying significantly. This requires a ballast load, also connected across the generator, to be increased or decreased manually as the user load varies. The ballast loads might be a series of light bulbs or perhaps heating elements in either air or water. A volt or frequency meter can be used to monitor the generator speed so that ballast loads can be added or removed when necessary.

Any manually controlled governor system is vulnerable as a turbine is a very responsive machine often capable of accelerating from an underspeed condition to full overspeed in around one second. As a result, the machine may run away to a high speed and damage could be caused if, for example, the electrical load is accidentally removed. Despite this risk, many small hydro plants are operated for a few years using manual control until funds are available for a governor. The effects of 'runaway' may be minimized in two ways as follows.

- **Operation on the back side of the power curve.** This method avoids the need for frequent manual adjustment as well as reducing the runaway problem.

A turbine with parabolic power output against speed curve shown in Figure 6.4.3(a) is assumed in the following description. The description is valid, however, for any curve. Whatever the curve is, there are numbers of possible operating points along it.

Conventionally, a generator is coupled to the turbine so that it generates a standard frequency of, for example, 50 Hz at the turbine speed corresponding to peak efficiency and power (Figure 6.4.3(b)). If the electrical load is removed from a turbine which has no governor, then the generator is driven at approximately twice normal speed. The generator and voltage regulator would no longer function properly at these high speeds.

The gearing ratio between the generator and turbine may be changed so that the required operating frequency is generated, for example when turbine speed is halfway between nominal and runaway speed as shown in Figure 6.4.3(c). From zero to full load, speed variation amounts to about ± 30% of the nominal speed. This variation of speed and frequency might still be too large, but it permits full use of plant capacity without even manual control and without risk of damage. An automatic voltage regulator is required to avoid excessive voltage variation.

This approach is only suitable for impulse machines. Reaction turbines may suffer from cavitation when used in this way. The turbine must be capable of operating continuously near its runaway speed.

- **Turbine flow modification:** Some form of hydraulic overspeed braking can be incorporated into turbines. Shrouds are sometimes fitted to Pelton wheels, for example, which spoil the performance at overspeed, reducing the runaway speed to around 130%. There are performance and cost penalties, however.

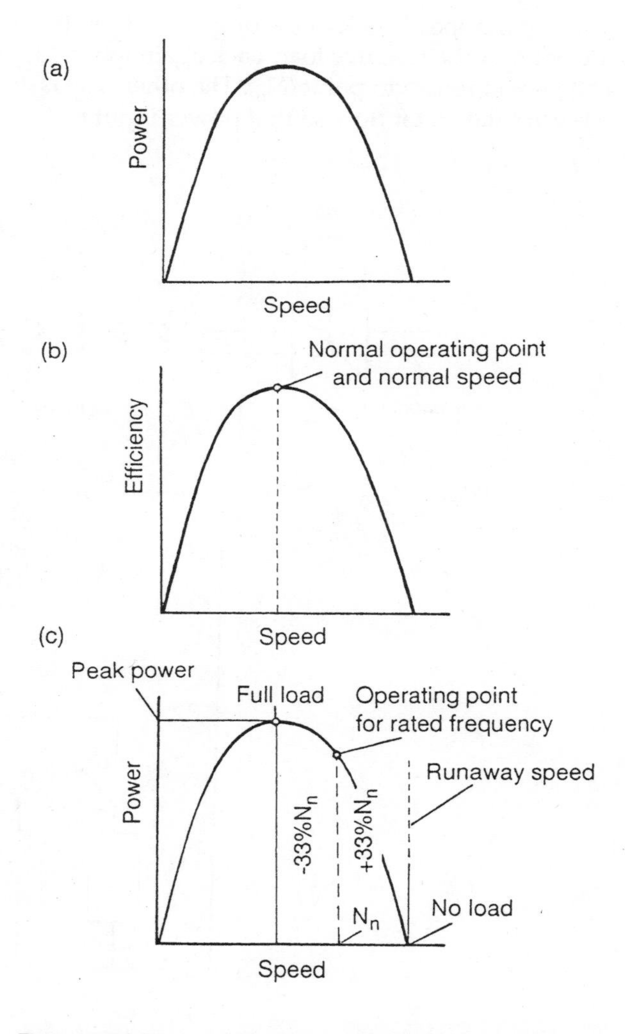

Fig 6.4.3 Turbine output characteristics at constant flow rate.

Effect of varying flow and load on speed

Consider a turbine coupled to a generator which in turn is connected to a resistive load as shown in Fig 6.4.4. The water flow to the turbine can be controlled manually and also individual loads can be switched on and off manually.

When the valve is open to allow some flow, the turbine operates at power P_o and speed N_o. The amount of power generated is dissipated by the resistive load. The speed remains stable at 'N_o' as long as there is no change to the input power or output power.

If you introduce a change to the system, for example disconnecting one resistive load from the bank, it will now be observed that the input power is greater than the output power, so the operating point of the turbine must change again to a stable point. This is achieved where the power input is the sum of the power output and power loss due to inefficiencies.

To bring the speed back to the original values, if you connect the resistive load back again the turbine will return to point 'N_o'. The other way is to reduce the water flow so that power input is again equal to the new power output. Similarly, if an extra load is added the speed will drop, and it is possible to bring the speed of the turbine back to 'N_o' by increasing the flow.

From the above we can observe that the speed of the turbine can be controlled either by adjusting the flow or by adjusting the load.

Adjusting the flow: in the run-of-the-river schemes, if the flow is adjusted or decreased the excess water will flow over the weir and is not used.

Adjusting the load: if the alternator is connected to a resistive load bank and only a portion of the heat power produced is used, the balance can be lost as waste heat.

The only difference is that in the first case the water flows over the weir without generating any power and in the second case the water flows through the turbine generating a fixed amount of power and only part of it is used. In both cases the turbine is kept at constant speed; that is, governing is achieved.

If the scheme has a water storage reservoir then a flow-adjustment approach has the benefit of saving the stored water.

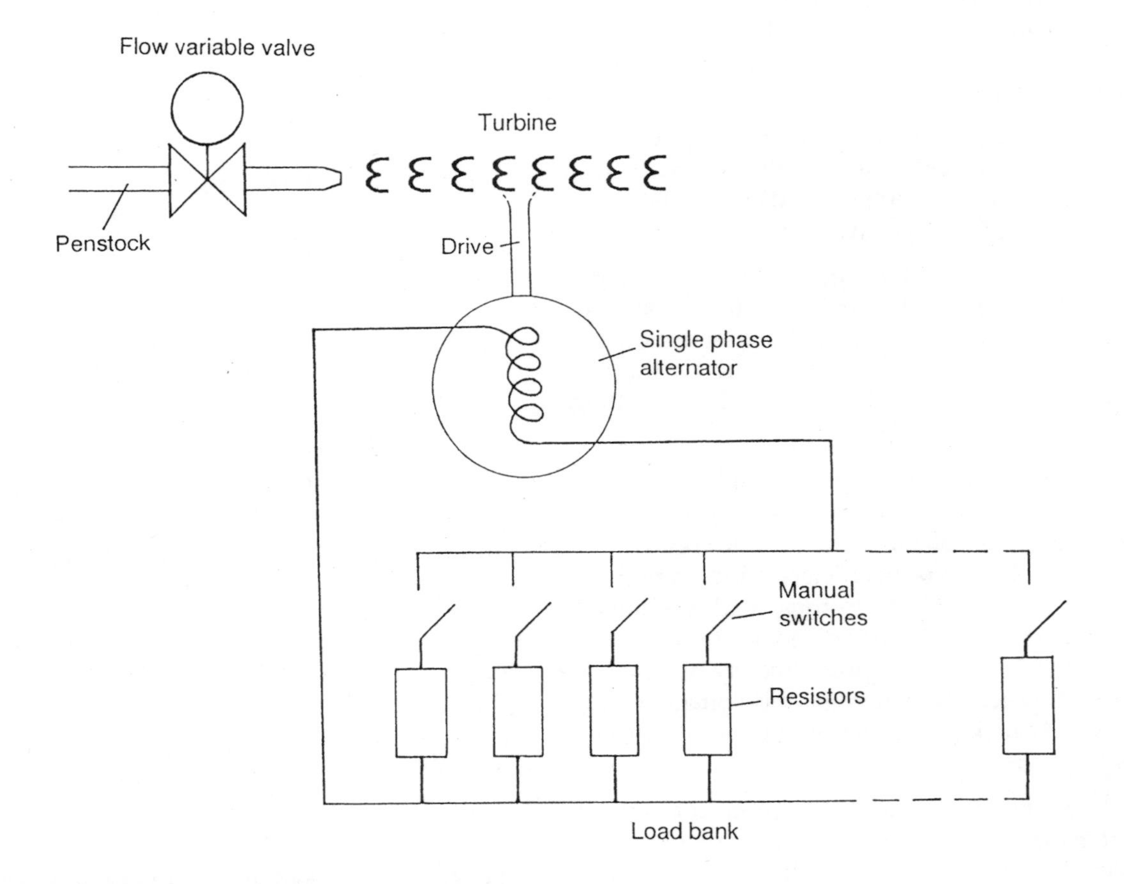

Fig 6.4.4 Alternator with resistive load.

Drive systems

7

7.1 Introduction

Drive systems comprise the generator shaft, turbine shaft, the bearings which support those shafts, couplings to connect shafts, any extra shafts, bearings, pulleys, belts, gearboxes or other components used to change speed or orientation of the shafts. The function of the drive system is to transmit power from the turbine to the generator at the correct speed for the generator and in a suitable direction.

Some of the many possible arrangements for a drive system are shown in Figs 7.1.1 to 7.1.6 with comments on each. Irrespective of the arrangement chosen, the drive system must carry the required loads. Additionally, the turbine, generator and foundations must be able to carry the loads imposed by the drive.

Common purchased components in drive systems are bearings and belts. These have properties unique to particular manufacturers. It is essential to obtain manufacturers' catalogues and use them in conjunction with the remainder of this chapter. The same will apply to any seals, gearboxes or couplings which are needed.

The components in any drive system are insignificant in relation to the reservoir, penstock, turbine, generator, controller and switchgear. It is nevertheless absolutely vital that all drive components are strong enough to give a long, trouble-free life. They must be assembled correctly, and they must receive suitable maintenance during their service life, as detailed in Chapter 10. Great care must be taken over small details.

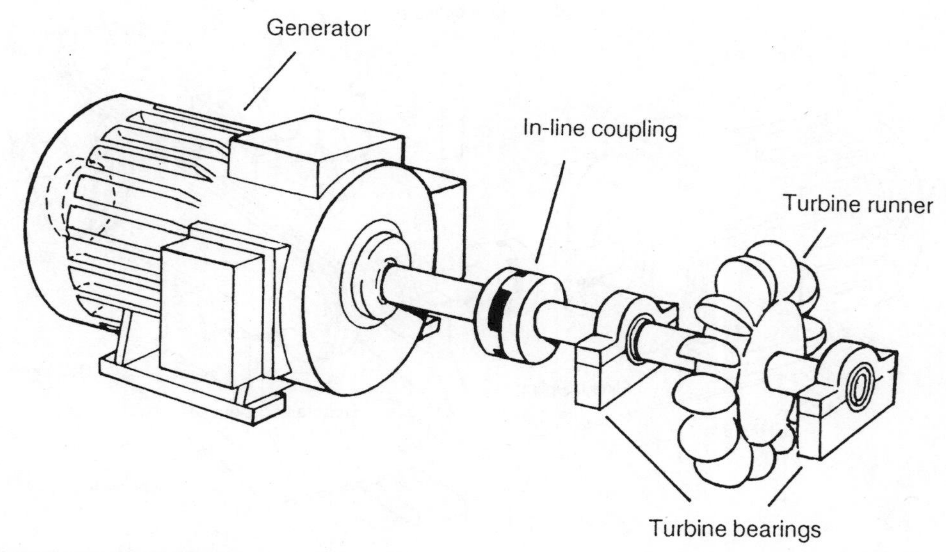

Fig 7.1.1 Direct coupled drive system

- Compact, simple; efficiency approaches 100%
- Drive imposes no additional loads on bearings.
- Both turbine and generator may be bolted to concrete foundation but metal packing is required to bring both shafts to the same height.
- Alignment of shafts must be correct to prevent failure of coupling or bearings.
- Flexible in-line coupling is needed because shaft alignment is never perfect. The coupling may be made locally (see Fig 7.2.2).
- No speed change between turbine and generator is possible.

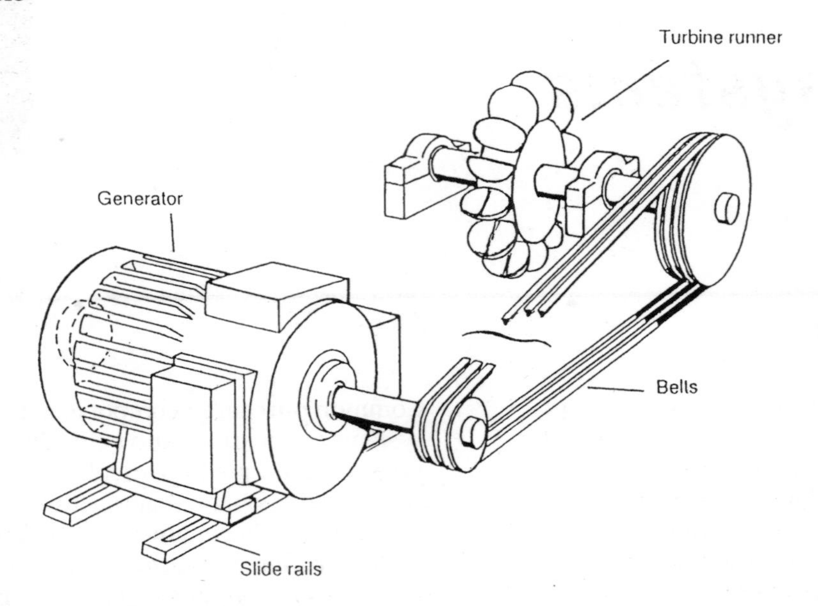

Fig 7.1.2 Wedge belt drive system (overhung shafts)

- An alternative arrangement could have flat belt instead of wedge belts. Axes of both turbine and generator could be vertical for both types of belts.
- Generator is mounted on slide rails to obtain belt tension.
- This drive applies loads to generator and turbine bearings. Check that the bearings are adequate.
- Turbine and generator may run at different speeds.
- Turbine and generator may be at different heights. Shafts must be parallel and pulleys in line but alignment is not as critical as for drive shown in 7.1.1.
- Direction of rotation should pull on the lower part of the belts.

Note Slide rails are available from most stockists of electric motors and are well worth purchasing.

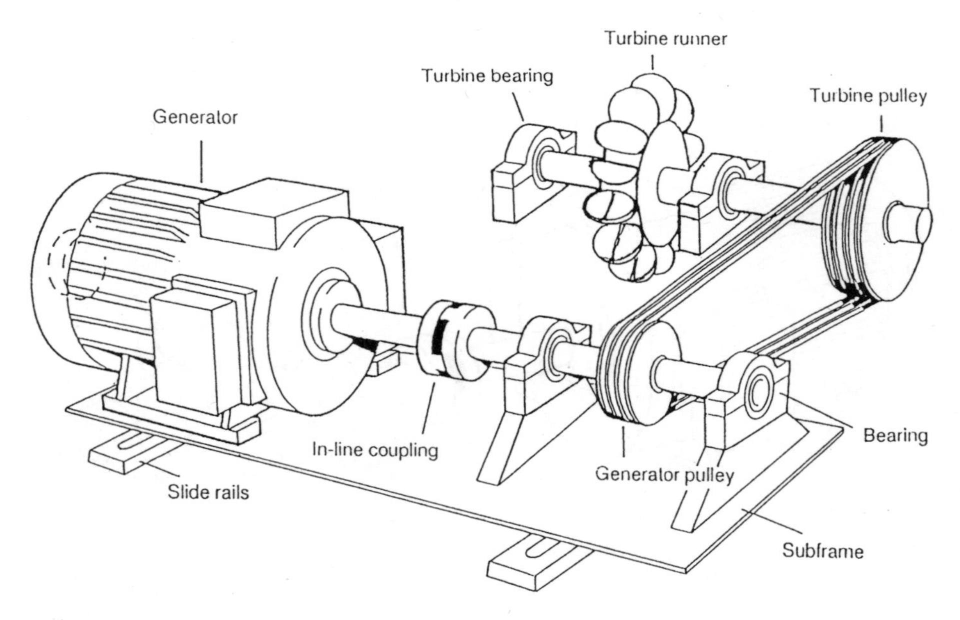

Fig 7.1.3 Wedge belt drive system with extra bearings

- Alternative arrangement could have flat belt instead of Wedge belts.
- Turbine may have an extra shaft and bearings as shown for the generator.
- The generator extension shaft must be removable from the bearings to enable belts to be changed.
- Turbine and generator may run at different speeds.
- Flexible in-line coupling required.
- Direction of rotation should pull on the lower part of the belts.
- Belt tensioning must be achieved with a jockey pulley (not shown) or the generator with its extra shaft and bearings must be mounted on a subframe which can slide to obtain belt tension as shown.

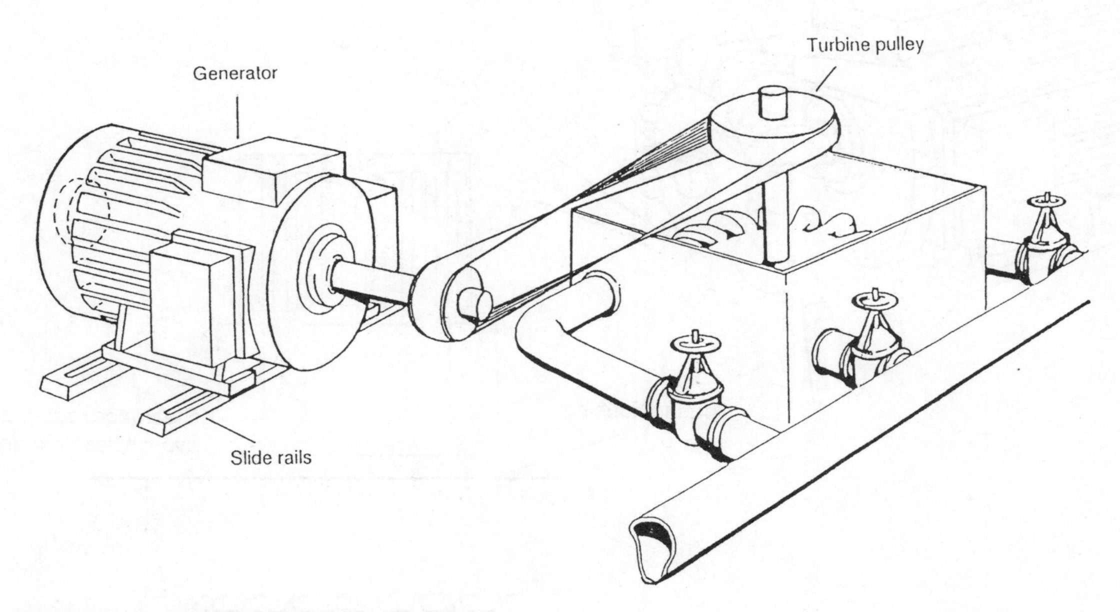

Fig 7.1.4 Quarter turn belt drive (overhung shafts)

- Alternative arrangement could have wedge belts in place of flat belt.
- Generator mounted on slide rails to obtain belt tension. This drive applies loads to the generator and turbine bearings. Check that the bearings are adequate.
- Extra bearings, shafts and couplings as shown in 7.1.3 may be used.
- Turbine and generator pulleys must be aligned with care.
- Not always permissible – consult manufacturer.

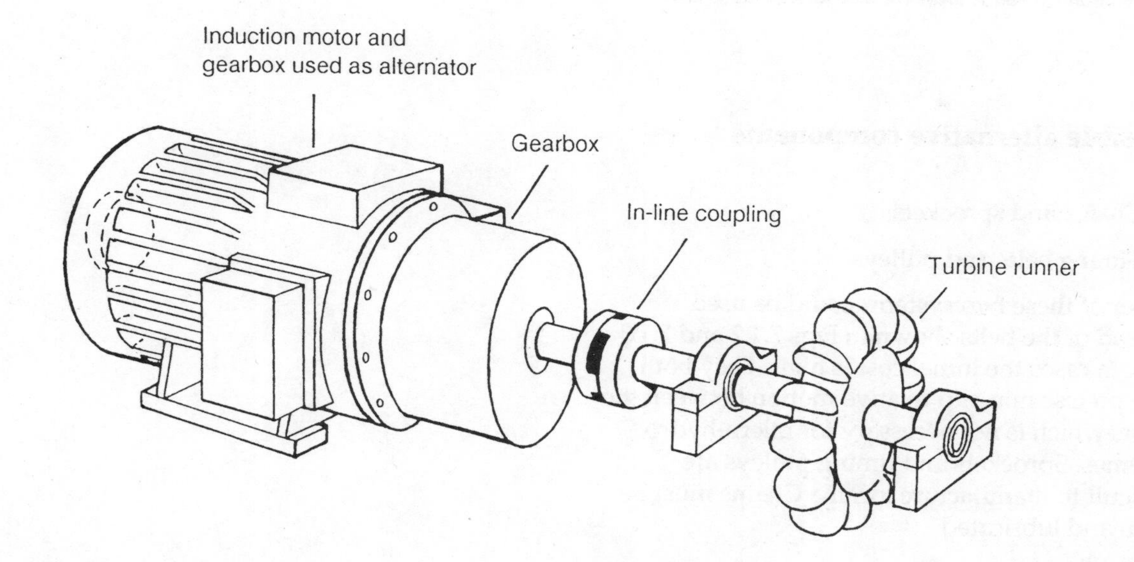

Fig 7.1.5 Direct coupled turbine and geared motor used as alternator

- The use of induction motors as alternators is considered in Chapter 8. Geared motor units are available from many manufacturers in wide ranges of powers and speed ratios. Various types of gears are used. Spur gears, helical gears and bevel gears are suited to speed increasing drives. Worms and wormwheels are not suitable because they will not run backwards.
- The cost of a geared motor unit is significantly less than the cost of separate motor and gearbox.
- The gearbox provides the speed change between turbine and generator. Otherwise the comments for Fig 7.1.1 apply.

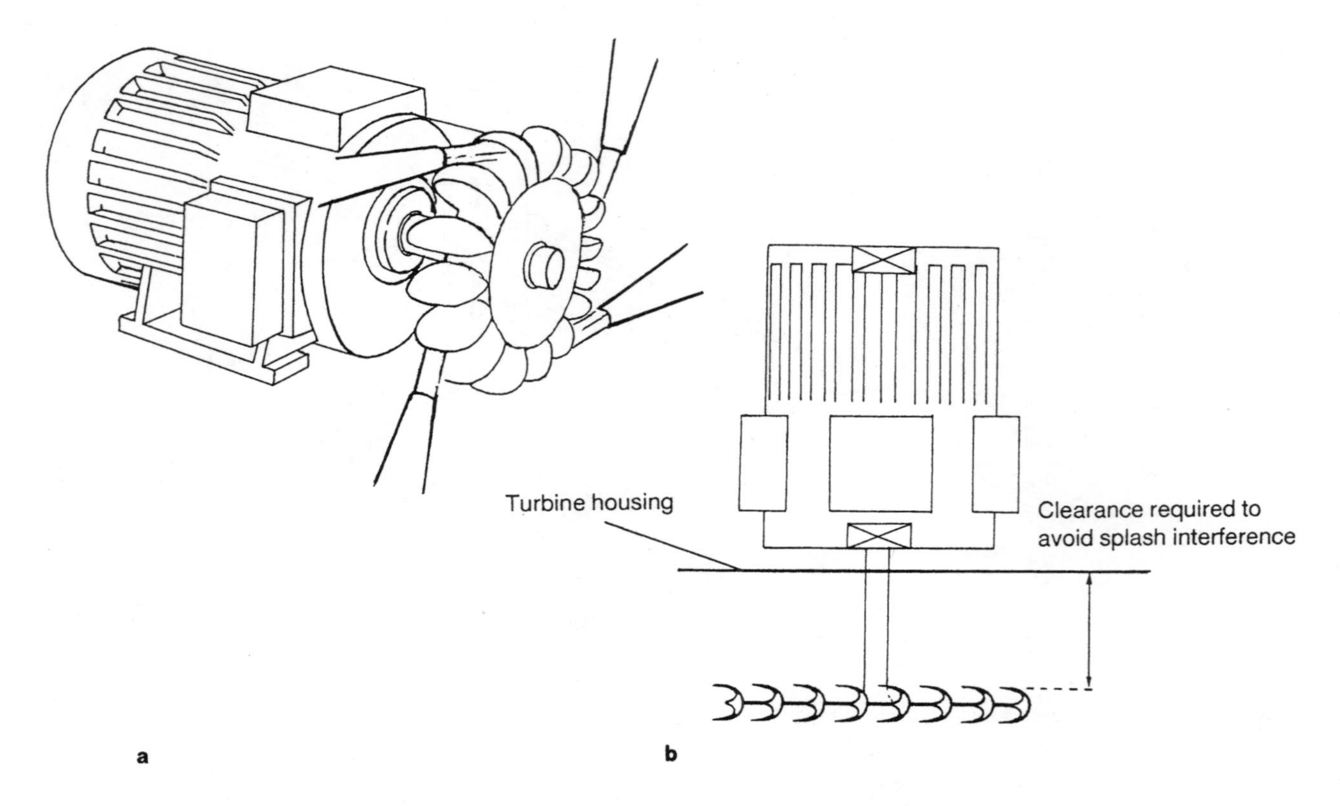

Fig 7.1.6 Turbine rotor mounted on generator shaft. (Alternatively a geared motor unit may be used as an alternator, as in Fig 7.1.5. The turbine rotor would then be mounted on the gearbox shaft)

A very neat, trouble-free installation results, providing:

a the speeds match;

b the shaft is long enough to give clearance (see **b**);

c the bearings can tolerate the side load of the turbine.

Possible alternative components

- Chains and sprockets
- Timing belts and pulleys

Either of these two systems could be used instead of the belts shown in Figs 7.1.2 and 7.1.3. In both cases the initial cost is high. They both give precise non-slip relative motion between the shafts which is not necessary for micro-hydro schemes. Sprockets and timing pulleys are difficult to manufacture locally. Chains must be clean and lubricated.

Neither system is recommended for micro-hydro schemes.

Sample procedure for design of a drive system — Note 7.1.1

1. Calculate power transmitted by drive system (see Example 7.1.1).
2. Calculate speed ratio between turbine shaft and driven shaft (eg alternator shaft).
3. Choose most economic and reliable type of drive system: consider Figs 7.1.1 to 7.1.6.
4. If a belt drive is chosen, decide whether to use a wedge belt or a flat belt. This will depend on local conditions – local availability, reliability of guarantees, relative cost, and so on.
5. Calculate belt width, or number of belts.
6. Calculate the loads put onto the shafts by the belts when running under load.
7. Calculate whether the generator and turbine shaft bearings are sufficiently large to have an adequate life.
8. If answer to 7 is no, decide whether the bearings can be modified or whether extra shaft, bearings and couplings must be used.
9. Calculate initial tension to be applied to the belts.
10. Decide how that tension should be applied and monitored throughout the belt life.
11. Decide how to keep the belts clean and dry (from both oil and moisture) while running.
12. Decide what regular maintenance is required to ensure trouble-free operation, what spare parts should be kept in stock, and how and when bearings should be lubricated.

Drive problem — Example 7.1.1

This example gives the information necessary to start designing a drive system.

An alternator rated at 50 kW output, 1500 rpm and 70% efficiency is to be driven by a turbine running at 450 rpm.

Power transmitted to alternator

$$= \frac{\text{Output power}}{\text{Efficiency}} = \frac{50}{0.7} = 71 \text{ kW}$$

This will be the power required of the drive.

Speed ratio

$$= \frac{1500}{450} = 3.33$$

Note that the driving shaft rotates slower than the driven shaft. It is therefore a speed-increasing drive.

The following list indicates which drive arrangements should be considered:

Fig 7.1.1 unsuitable
7.1.2 suitable in both flat and wedge belt alternatives
7.1.3 suitable in both flat and wedge belt alternatives but use only if arrangement 7.1.2 proves unsuitable after calculations (see later sections of this chapter).
7.1.4 suitable but calculations and other considerations are similar to 7.1.2. and 7.1.3
7.1.5 suitable but very expensive to purchase and maintain
7.1.6 unsuitable

Assume the decision to be:

Fig 7.1.2 preferred
7.1.3 if necessary
7.1.4 if necessary

7.2

Direct coupled drives

Direct coupled drives are a neat solution, as is clear from Figs 7.1.1 and 7.1.5. A further significant advantage is that the drive imposes no extra loads on bearings. The main disadvantage is that of speed. A gearbox with the correct speed ratio (7.1.5) overcomes that problem but otherwise special generators are required which will run at the same speed as the turbine. The cost of any specially bought plant is always high but it could well be cost effective because of savings arising from there being no other drive components except a coupling. Direct drives are very easy to maintain and once set correctly require no adjustment. This eliminates the difficulty of arranging for regular maintenance by trained operators. It is therefore worth considering the purchase of a specially designed and built generator to run at the same speed as the turbine.

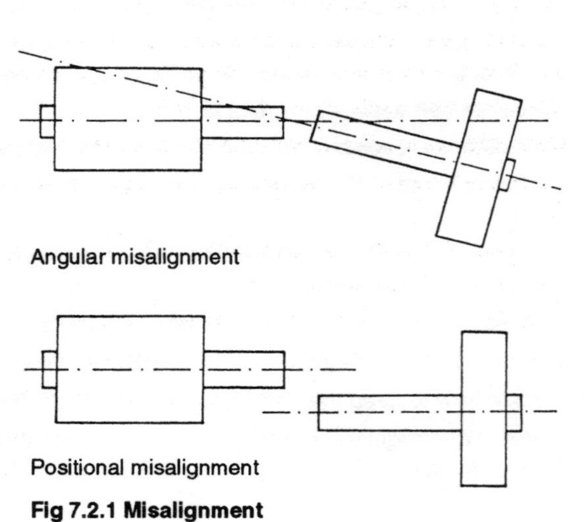

Angular misalignment

Positional misalignment

Fig 7.2.1 Misalignment

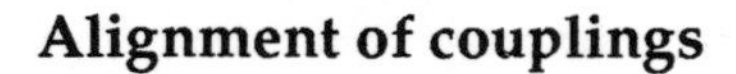

Alignment of couplings

There may be angular or positional misalignment between two shafts (see Fig 7.2.1). Frequently they are combined.

Any attempt to connect those shafts with a strong rigid coupling will give rise to bending of the shafts, causing fatigue breakage as well as imposing very high loads on bearings.

Since accurate alignment is difficult to obtain, the solution is to use flexible couplings. Examples are shown in Fig 7.2.2. Type **c** will accept angular misalignments up to 0.5° and positional misalignments up to 0.25 mm.

Errors of greater magnitude will cause loss of efficiency and wear. Considerable effort should be made to reduce errors to a minimum.

Any shaft is usually mounted on two bearings. The rigid connection of two shafts with a solid coupling results in the equivalent of a single shaft having four bearings. The four bearings can never be aligned with sufficient accuracy and should *never* be used.

The same applies to three bearings on a single shaft. Belt drives to be considered later put loads onto bearings and it is very tempting to use a third bearing to take that load. This arrangement must likewise *never* be used.

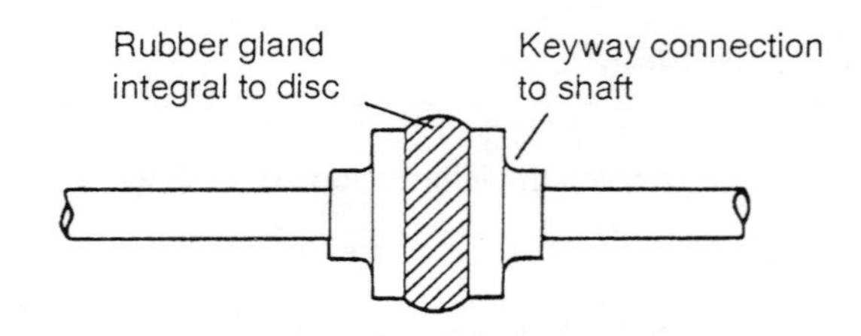

a One-piece couplings are not recommended. They cannot be dismantled, maintained or constructed locally

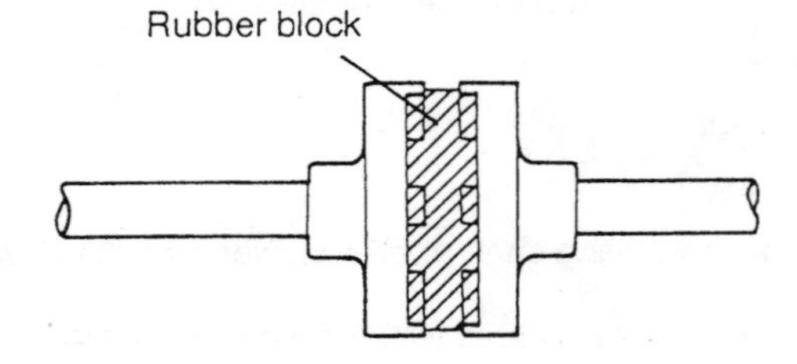

b Two-piece commercial coupling

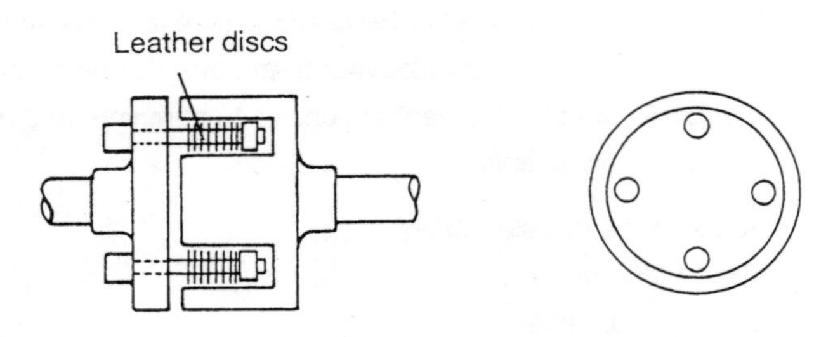

c Two-piece coupling design that can be made locally

Fig 7.2.2 Flexible couplings

7.3 Belt drives

Belts of various types have much to recommend them for micro-hydro schemes. The turbine and generator may run at different speeds so that each runs at maximum efficiency. The speed ratio is obtained by using pulleys of different diameters on the turbine and generator shafts.

It is usually convenient to locate the generator and turbine parallel to each other which is ideal for a belt drive (Figs 7.1.2 and 7.1.3). Other arrangements may be used; see, for example, Fig 7.1.4.

Belts work by using friction. As the driving pulley turns, it grips the belt by friction and pulls it. In turn the moving belt grips the driven pulley causing it to turn. The grip between a belt and pulleys depends upon the tension in the belt and the coefficient of friction between the belt and pulleys.

It is important that belts are kept clean and dry because oil or water on the belt will affect the grip and may cause the belt to slip.

Well-designed modern belt drives can have an efficiency of 95% to 97%. Incorrect tensioning of the belt, poor alignments or use of poor-grade components can reduce the drive efficiency to 70%.

To ensure interchangeability and uniformity, belt manufacturers use standards. Only industrial standard drives are relevant to micro-hydro schemes. Agricultural or automotive standards are not suitable.

Belts are of two main types – flat and Vee – but there are variations within each group.

Flat belts

The traditional flat belt was made from 'balata', leather, cotton or canvas. These materials are relatively weak, are dimensionally unstable with different weather conditions, and have a low coefficient of friction with pulleys.

Weakness of the materials means that belts must be thick or wide. Thick belts require to be run only on large diameter pulleys. Wide belts impose high loads on the bearings of overhung shafts or require a complete arrangement with extra shafts, couplings and bearings. The low coefficient of friction is overcome by use of 'dressing' (application of a sticky substance between belt and pulley) which requires skill to estimate how much and how frequently it should be applied.

Flat belts using traditional materials, and the use of dressings, are not recommended for micro-hydro schemes.

Modern flat belts made by specialist companies are much stronger, are dimensionally stable, and have high coefficients of friction with pulleys. Nylon or terylene is used for the core which is coated with a rubber compound. Belts are made up by taking an appropriate length of material and bonding the ends together. This will usually be done by the supplier although in some cases special fixtures and adhesives are available for the customer to make the bonds required.

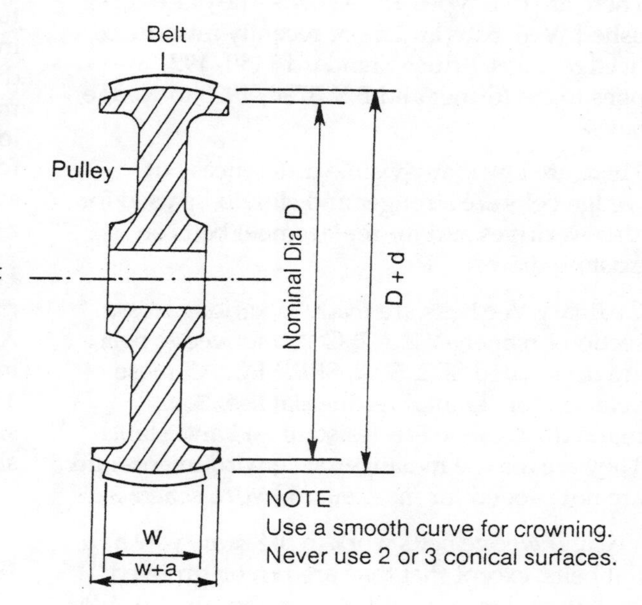

Belt width mm	a mm
up to 100	10
100 - 200	20
200 - 400	30

Pulley diameter D mm	d mm
100	0.6
200	1.2
300	1.6
400	2.0
600	2.6
1000	4.0

Fig 7.3.1 Pulley crowning and width

The strength and high coefficient of friction can result in high efficiency drives of small dimensions imposing relatively low loads on bearings.

Any type of flat belt must have some guidance to prevent it from falling off the pulleys. Crowning of pulleys is the usual way, as shown in Fig 7.3.1. It is preferred that both pulleys are crowned but if necessary the larger one only need be crowned. Both pulleys must be crowned for quarter turn drives (Fig 7.1.4) in which case a higher crown of 1.5 times the value given in Fig 7.3.1 is needed. Belt manufacturers' catalogues should be consulted.

Flat belts should not be run on pulleys with side flanges. The belt invariably creeps to one side, rubbing against the flange which in turn causes excessive wear on the edge of the belt. In extreme cases the belt will 'climb' onto the flange and actually run itself off the pulley.

Vee belts

There are two types of Vee belts, the old-established 'Vee' belts and more recently introduced 'wedge' belts. British Standard 1440/1971 applies to the former and B.S. 3790/1973 to wedge belts.

There are few if any visible differences but wedge belts are stronger and should be used for all new drives and for replacement belts on existing drives.

Ordinary Vee belts are made in various cross-sections, namely Y,Z,A,B,C,D; and wedge belts are designated SPZ, SPA, SPB, SPC. All these belts are for standard industrial use. Belts marked FRAS are fire-resistant and anti-static. They are for use in dangerous environments and are not needed for most micro-hydro schemes.

Vee and wedge belts work in the same way as flat belts, except that they are run on grooved pulleys and the grip is between the tapered sides of the belts and the tapered sides of the grooves. The actual coefficient of friction for Vee and wedge belts is similar to that for flat belts but the grip is about three times higher.

Numbers of Vee or wedge belts are run together in parallel on multiple grooved pulleys. In such cases belts should be ordered in complete sets and used only as a complete set. Should one belt fail in a set it is recommended to replace all the belts in the set.

Pulleys in a Vee or wedge belt drive must be properly aligned or the belts will rub on their sides, resulting in wear and premature failure of the drive.

It is essential that the belts are used with pulleys which have the correct groove shape and size. Belts must never rest on the bottom of pulley grooves. Suitable pulleys may be made locally but belt suppliers stock a wide range of standard sizes. These stock pulleys frequently have special tapered arrangements for easy fitting to shafts and are relatively cheap.

Vee or wedge belts must not be forced over the outside diameters of pulleys. That can seriously damage the belt and failure will follow. It is usual to provide either centre distance adjustment or to install a jockey pulley. In either case it is necessary to provide about 60 mm adjustment.

Positions of pulleys on shafts

Belt tension determines the sideways loads applied to shafts. That is, the belt tension applies a radial load to the shaft. The radial load must in turn be carried by the bearings holding the shaft in place. Moments (forces multiplied by their distances from a point of rotation or resistance) must be applied in order to calculate the radial load to be carried by bearings. It therefore follows that moment arms (the distances of applied loads from the bearings) must be kept to a minimum.

Fig 7.3.2 shows a typical arrangement with the pulleys on overhung shafts. The distances X_T and X_G are moment arms, which must be kept short in order to minimize the loads on the bearings. There must be only two bearings on any one shaft. *Never* attempt to use extra bearings on the shaft ends at A or B.

Belt tensioning methods

The easiest way to tension belts and provide adjustment for installation is to use slide rails under the generator feet. The turbine is connected to the penstock so that it is fixed. The generator may be moved because electrical connections are flexible. Slide rails have already been illustrated in Figures 7.1.2 and 7.1.4. Testing for belt tension is shown in Fig 7.3.3.

The alternative to slide rails is to fix the position of the generator and install a jockey pulley and supports into the drive. The principle of a jockey pulley is shown in Fig 7.3.4.

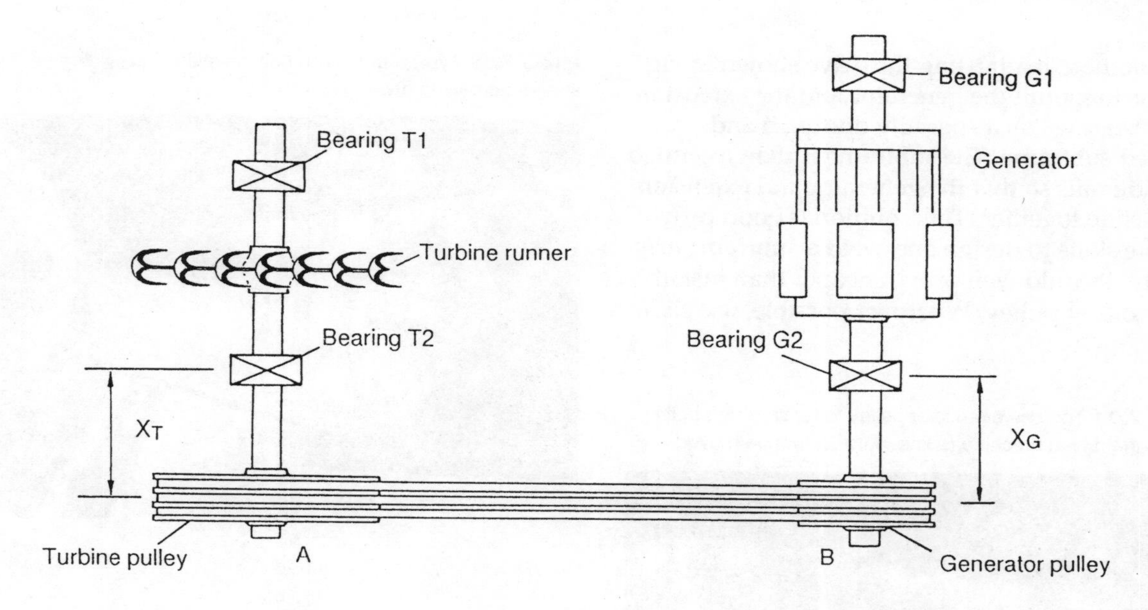

Fig 7.3.2 Pulley on shafts

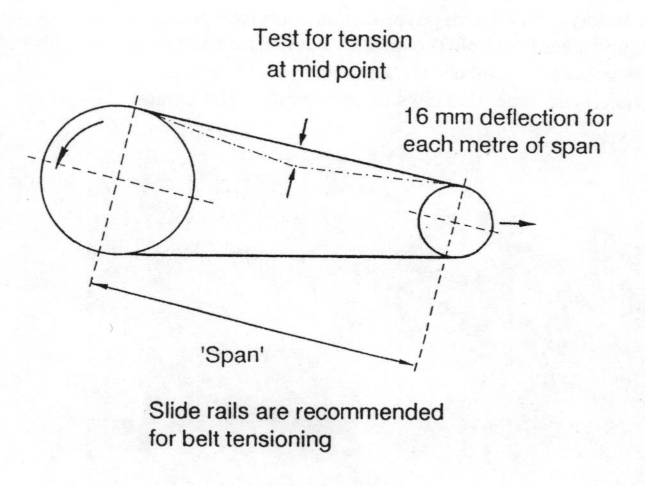

Fig 7.3.3 Testing belt tension

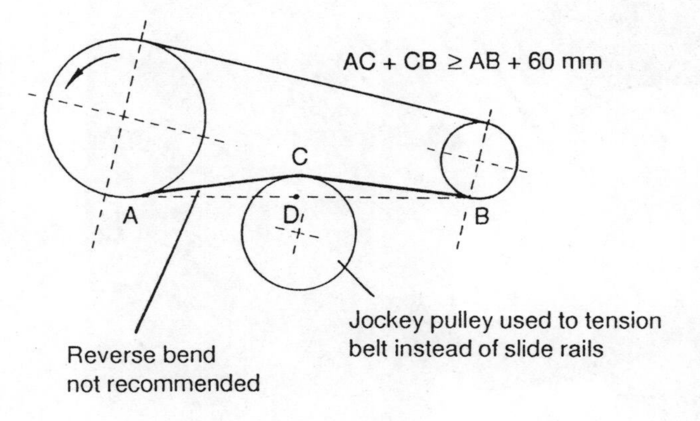

Fig 7.3.4 Position of jockey pulley

A jockey pulley can be used if slide rails are not possible. Jockey pulleys must always be installed on the slack side of a belt. They must never be on the tight side. Fig 7.3.4 shows a situation where the jockey pulley is causing the belt to bend in the reverse direction to the bends around the pulleys. This arrangement is not approved for wedge belts although it is acceptable for flat belts and Vee belts.

The reverse bending of wedge belts causes fatigue and the belts will have a shorter life than normal. It could be a sensible sacrifice to make in order to provide belt adjustment and tension.

The diameter of jockey pulleys should be as large as possible and never smaller than the diameter of the small pulley of the drive. Jockey pulleys running on the outside of a drive should be plain (ie no grooves or crowning).

Placing the jockey pulley inside the drive eliminates the problem of reverse bends in the belt but it decreases the arc of contact of the belt on the small driven pulley. Use the smallest arc of contact when sizing the drive.

Jockey pulley mountings must be designed to carry forces of about 100 Newtons per belt in the case of wedge belts. (65 N is the catalogue figure.) For flat belts use about 3 N per mm belt width. This force is applied at an angle so the frame holding the bearings must carry horizontal as well as vertical forces.

Use of a jockey pulley is not cheap. The cost of pulley, shaft, bearings, bearing housings and mounting frame must be taken into account. Even so, it could be a way to provide adjustment and belt tension for drives needing extension shafts as in Fig 7.1.3.

The method of adjusting the drive shown in Fig 7.1.3 is to mount the generator and the extension shaft bearings on a specially designed and welded subframe. The subframe is then mounted on slide rails so that the generator and extension shaft slide together. This solution is good providing the skills to design and weld a subframe exist locally. It could well prove cheaper than installing a jockey pulley. Wherever possible, use slide rails.

Photo 7.3.1 The use of an idler pulley to tension Vee belts where the driven machine (a lineshaft) cannot be moved.

Photo 7.3.2 Typical multi-Vee belt installation providing speed-increasing drive. (Nepal)

Photo 7.3.3 Rehabilitated 60 kW 1955 Francis machine in a tea factory. The original governor has been replaced by electronic load control. The ballast tank can be seen in the background. Original direct mechanical drive has been replaced with multi-Vee belts to an alternator. (Sri Lanka)

7.4 Belt drive calculations

The basic data from Example 7.1.1 will be used for the calculation of a wedge belt drive (Example 7.4.1) and a modern flat belt drive (Example 7.4.2). The traditional leather flat belt drive is also calculated to prove how unsuitable it would be.

Basic data from Example 7.1.1 used in 7.4.1 to 7.4.4:

Turbine shaft speed 450 rpm
Generator shaft speed 1500 rpm

Therefore,

speed ratio = 3.33 : 1 increasing
Generator output 50 kW
Generator efficiency 70%

Therefore,

power transmitted = 71 kW

$$\text{Torque at generator} = \frac{9550 \times 71}{1500} = 452 \text{ Nm}$$

(Note 9550 converts kW to W and rpm to radians per second).

Manufacturers catalogues must be used when designing a belt drive. Full catalogues and local supply information can be obtained from:

Fenner International Ltd
Marfleet, Hull HA9 5RA, UK.

Habasit (GB) Ltd
Phoenix Parkway,
Corby, Northants NN17 1DT, UK.

Extracts from both the Fenner and the Habasit belt catalogues are included here to provide information for the worked examples.

Fig 7.4.1. shows extracts from Fenner catalogues.

Fig 7.4.2 shows extracts from Habasit catalogues.

Comment on calculated results

See examples 7.4.1, 7.4.2, 7.4.3, 7.4.4. The traditional leather flat belt (7.4.4) is far too wide to be practical, so the choice is between the wedge belt drive (7.4.1) and the modern flat belt drive (7.4.2).

The recommended and standard pulley diameters for the wedge belt drive are smaller than those recommended for flat belts. Furthermore, the pulley width is less for the wedge belt drive. The narrower pulley width has a distinct advantage when considering shaft overhangs (Fig 7.3.2) and moments used to calculate bearing loads which are considered later.

The difference in shaft loads calculated to belt manufacturers' methods must be considered further. There is a contradiction of both theory and accepted opinion.

The use of an inaccurate shaft load at the design stage could result in an unsatisfactory micro-hydro scheme.

Consider the theory of belt drives:

Torque at pulley shaft
= effective pull × pulley radius

Tensioning forces **Table 7.4.1**

Extracted from Fenner catalogue (see Fig 7.3.3). You also need to calculate shaft loading.

Belt section	Force required to deflect belt 16 mm per metre of span: Small pulley diameter (mm)	Force (N)	Belt section	Force required to deflect belt 16 mm per metre of span: Small pulley diameter (mm)	Force (N)
SPZ	56 to 95	13 to 20	SPB	112 to 224	45 to 65
	100 to 140	20 to 25		236 to 315	65 to 85
SPA	80 to 132	25 to 35	SPC	224 to 355	85 to 115
	140 to 200	35 to 45		375 to 560	115 to 150

Effective pull
= tension on tight side - tension on slack side

Shaft load
= tension on tight side + tension on slack side

An 'ideal' belt drive would have no tension on the slack side, so the shaft load becomes equal to the tension on the tight side. This in turn can be calculated from the torque and the diameter of the pulley on that shaft:

For a 236 mm diameter pulley,

$$\text{Shaft load} = \frac{\text{torque}}{\text{pulley radius}}$$

$$= \frac{452 \text{ Nm}}{0.118 \text{ m}} = 3830 \text{ N}$$

For a 280 mm pulley, the load would be 3230 N. Real belt drives will always give shaft loads far greater than this.

The Fenner figure of 12480 N is conservative, designed for trouble-free use with a long life.

The Habasit figure of 6510 N is considerably lower in this case but does require a larger diameter, wider drive pulley. This flat belt uses grooves on the belt surface to prevent air being trapped, and very flexible materials. Not all modern flat belts will give such low shaft load figures.

The belt drive is only a minor part of the cost of a micro-hydro scheme but the whole scheme is useless if any components fail. *Never* use cheap substitutes.

TABLE 1

Speed of faster shaft rev/min	*Minimum Pulley Diameter (mm) Design Power (kW)																			
	up to 1	3,0	4,0	5,0	7,5	10,0	15,0	20,0	25	30	40	50	60	75	90	110	130	150	200	250
500	56	90	100	112	125	140	180	200	212	236	250	280	280	315	375	400	450	475	500	560
600	56	85	90	100	112	125	140	180	200	212	224	250	265	280	300	335	375	400	475	500
720	56	80	85	90	100	106	132	150	160	170	200	236	250	265	280	300	335	375	450	500
960	56	75	80	85	95	100	112	132	150	180	180	200	224	250	280	280	300	335	400	450
1200	56	71	80	80	95	95	106	118	132	150	160	180	200	236	236	250	265	300	335	355
1440	56	63	75	80	85	85	100	112	125	140	160	170	190	212	236	236	250	280	315	335
1800	56	63	71	75	80	85	95	106	112	125	150	160	170	190	212	224	236	265	300	335
2880	56	60	67	67	80	80	85	90	100	112	125	140	160	170	180	212	224	236	–	–

* This table is intended as a guide to selection only. Bearing loads should be carefully considered when using small motor pulleys. This is particularly important when using the small pulleys associated with C.R.E. belts.

TABLE 2

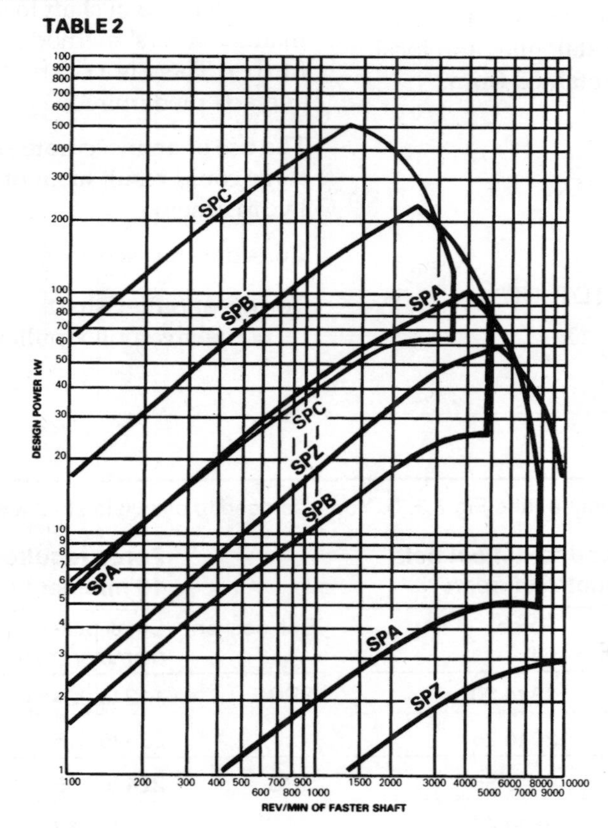

Fig 7.4.1 Extracts from Fenner wedge belt catalogue *continued on following pages*

TABLE 3: SERVICE FACTORS

SPECIAL CASES		TYPES OF PRIME MOVER					
For speed increasing drives of: Speed ratio 1,00 – 1,24 multiply service factor by 1,00 Speed ratio 1,25 – 1,74 multiply service factor by 1,05 Speed ratio 1,75 – 2,49 multiply service factor by 1,11 Speed ratio 2,50 – 3,49 multiply service factor by 1,18 Speed ratio 3,50 and over multiply service factor by 1,25		'Soft' starts Electric Motors: AC – Star Delta start DC – Shunt Wound Internal Combustion Engines with 4 or more cylinders All prime movers fitted with Centrifugal Clutches, Dry or Fluid Couplings or Electronic Soft Start devices			'Heavy' starts Electric Motors: AC – Direct-on-Line start DC – Series & Compound Wound Internal Combustion Engines with less than 4 cylinders Prime movers not fitted with soft start devices		
		Hours per day duty					
	TYPES OF DRIVEN MACHINE	10 and under	Over 10 to 16	Over 16	10 and under	Over 10 to 16	Over 16
Class 1 Light Duty	Agitators (uniform density), Blowers, Exhausters and fans (up to 7,5 kW), Centrifugal compressors and pumps, Belt conveyors (uniformly loaded).	1,0	1,1	1,2	1,1	1,2	1,3
Class 2 Medium Duty	Agitators and mixers (variable density), Blowers, Exhausters and fans (over 7,5 kW), Rotary compressors and pumps (other than centrifugal), Belt conveyors (not uniformly loaded), Generators and exciters, Laundry machinery, Lineshafts, Machine tools, Printing machinery, Sawmill and woodworking machinery, Screens (rotary).	1,1	1,2	1,3	1,2	1,3	1,4
Class 3 Heavy Duty	Brick machinery, Bucket elevators, Compressors and pumps (reciprocating), Conveyors (heavy duty), Hoists, Mills (hammer), Pulverisers, Punches, Presses, Shears, Quarry plant, Rubber machinery, Screens (vibrating), Textile machinery.	1,2	1,3	1,4	1,4	1,5	1,6
Class 4 Extra Heavy Duty	Crushers (gyratory-jaw roll), Mills (ball-rod-tube).	1,3	1,4	1,5	1,5	1,6	1,8

Power Ratings—SPC Wedge Belts

Rev/min of faster shaft	RATED POWER (kW) PER BELT FOR SMALL PULLEY PITCH DIA (mm)																	Belt Speed (m/s)
	224	**236**	**250**	**265**	**280**	**300**	**315**	**335**	**355**	**375**	**400**	**425**	**450**	**475**	**500**	**530**	**560**	
100	1,99	2,20	2,45	2,72	2,99	3,34	3,60	3,96	4,31	4,65	5.09	5,52	5,95	6,38	6,80	7,31	7,82	
200	3,59	4,00	4,48	4,99	5,49	6,17	6,67	7,33	8,00	8,66	9,48	10,30	11,11	11,92	12,72	13,68	14,64	
300	5,04	5,64	6,33	7,07	7,81	8,78	9,51	10,48	11,44	12,39	13,58	14,76	15,93	17,10	18,26	19,64	21,01	
400	6,40	7,17	8,07	9,04	9,99	11,26	12,20	13,45	14,70	15,93	17,47	18,99	20,50	22,00	23,48	25,25	27,00	10
500	7,67	8,62	9,72	10,90	12,06	13,61	14,76	16,28	17,80	19,30	21,16	23,00	24,82	26,63	28,42	30,54	32,64	
600	8,87	9,99	11,28	12,66	14,04	15,85	17,20	18,98	20,75	22,50	24,67	26,80	28,92	31,00	33,06	35,50	37,89	
700	10,00	11,29	12,77	14,35	15,92	17,98	19,52	21,55	23,56	25,54	27,99	30,40	32,77	35,11	37,40	40,11	42,75	20
720	**10,22**	**11,54**	**13,06**	**14,68**	**16,28**	**18,40**	**19,97**	**22,05**	**24,10**	**26,13**	**28,63**	**31,09**	**33,51**	**35,89**	**38,23**	**40,99**	**43,67**	
800	11,08	12,52	14,19	15,95	17,71	20,02	21,73	23,99	26,22	28,42	31,12	33,77	36,37	38,93	41,42	44,34	47,18	
900	12,10	13,69	15,53	17,48	19,41	21,95	23,82	26,30	28,73	31,12	34,05	36,92	39,71	42,44	45,09	48,18	51,15	
960	**12,68**	**14,36**	**16,30**	**18,35**	**20,38**	**23,05**	**25,02**	**27,61**	**30,16**	**32,66**	**35,71**	**38,69**	**41,58**	**44,39**	**47,12**	**50,27**	**53,29**	
1000	13,06	14,79	16,80	18,92	21,02	23,77	25,80	28,47	31,08	33,64	36,77	39,82	42,77	45,63	48,39	51,58	54,61	
1100	13,96	15,83	18,00	20,28	22,53	25,48	27,65	30,49	33,27	35,98	39,27	42,46	45,52	48,47	51,29	54,41	57,53	30
1200	14,81	16,81	19,12	21,55	23,95	27,08	29,38	32,37	35,29	38,12	41,54	44,82	47,96	50,94	53,77	56,82	59,87	
1300	15,59	17,72	20,17	22,74	25,27	28,56	30,97	34,09	37,12	40,05	43,56	46,90	50,05	52,92	55,79	58,69	61,59	40
1400	16,31	18,56	21,14	23,84	26,48	29,92	32,42	35,65	38,77	41,76	45,32	48,55	51,78	54,55	57,32	59,97	–	
1440	**16,59**	**18,88**	**21,50**	**24,25**	**26,94**	**30,43**	**32,96**	**36,23**	**39,37**	**42,38**	**45,91**	**49,11**	**52,32**	**54,92**	**57,73**	**60,25**	–	
1500	16,98	19,33	22,02	24,84	27,59	31,15	33,73	37,04	40,22	43,24	46,80	49,96	53,13	55,73	58,34	–	–	
1600	17,58	20,03	22,83	25,75	28,59	32,24	34,88	38,25	41,45	44,47	47,99	51,03	54,08	56,45	–	–	–	
1700	18,11	20,65	23,55	26,56	29,47	33,20	35,88	39,27	42,47	45,45	48,87	51,73	–	–	–	–	–	
1800	18,57	21,20	24,18	27,26	30,23	34,01	36,71	39,98	43,25	46,16	49,43	–	–	–	–	–	–	
1900	18,97	21,67	24,71	27,85	30,87	34,67	37,36	40,58	43,80	46,59	49,65	–	–	–	–	–	–	
2000	19,29	22,05	25,15	28,34	31,37	35,18	37,84	40,96	44,09	46,72	–	–	–	–	–	–	–	

Rev/min of faster shaft	ADDITIONAL POWER (kW) PER BELT FOR SPEED RATIO									
	1,00 to 1,01	1,02 to 1,05	1,06 to 1,11	1,12 to 1,18	1,19 to 1,26	1,27 to 1,38	1,39 to 1,57	1,58 to 1,94	1,95 to 3,38	3,39 and over
100	0,00	0,02	0,06	0,11	0,14	0,17	0,20	0,23	0,25	0,26
200	0,00	0,04	0,12	0,21	0,29	0,35	0,41	0,46	0,50	0,53
300	0,00	0,07	0,18	0,32	0,43	0,52	0,61	0,69	0,75	0,79
400	0,00	0,09	0,24	0,42	0,57	0,70	0,81	0,92	1,00	1,06
500	0,00	0,11	0,30	0,53	0,72	0,87	1,02	1,15	1,25	1,32
600	0,00	0,13	0,36	0,63	0,86	1,04	1,22	1,37	1,50	1,59
700	0,00	0,15	0,42	0,74	1,00	1,22	1,43	1,60	1,75	1,85
720	**0,00**	**0,16**	**0,43**	**0,76**	**1,03**	**1,25**	**1,46**	**1,65**	**1,80**	**1,90**
800	0,00	0,17	0,48	0,84	1,15	1,39	1,63	1,83	2,00	2,11
900	0,00	0,20	0,54	0,95	1,29	1,56	1,85	2,06	2,25	2,38
960	**0,00**	**0,21**	**0,58**	**1,01**	**1,37**	**1,67**	**1,95**	**2,20**	**2,40**	**2,54**
1000	0,00	0,22	0,60	1,05	1,43	1,74	2,04	2,29	2,50	2,64
1100	0,00	0,24	0,66	1,16	1,57	1,91	2,24	2,52	2,75	2,91
1200	0,00	0,26	0,72	1,26	1,72	2,09	2,44	2,75	3,00	3,17
1300	0,00	0,28	0,78	1,37	1,86	2,26	2,65	2,98	3,25	3,44
1400	0,00	0,31	0,84	1,47	2,00	2,43	2,85	3,21	3,50	3,70
1440	**0,00**	**0,31**	**0,87**	**1,51**	**2,06**	**2,50**	**2,93**	**3,30**	**3,50**	**3,81**
1500	0,00	0,33	0,90	1,58	2,15	2,61	3,05	3,44	3,75	3,96
1600	0,00	0,35	0,96	1,68	2,29	2,78	3,26	3,67	4,00	4,23
1700	0,00	0,37	1,02	1,79	2,43	2,96	3,46	3,90	4,25	4,49
1800	0,00	0,39	1,08	1,89	2,58	3,13	3,66	4,12	4,50	4,76
1900	0,00	0,42	1,14	2,00	2,72	3,30	3,87	4,35	4,75	5,02
2000	0,00	0,44	1,20	2,10	2,86	3,48	4,07	4,58	5,00	5,29

NOTE: Only pulleys of Fenner manufacture should be used where belt speed falls between 30 and 40 m/s.

Fig 7.4.1 Extracts from Fenner wedge belt catalogue *continued from previous page*

Power Ratings—SPB Wedge Belts

Rev/min of faster shaft	RATED POWER (kW) PER BELT FOR SMALL PULLEY PITCH DIA (mm)													Belt Speed (m/s)
	140	150	160	170	180	190	200	212	224	236	250	280	315	
100	0.73	0.82	0.92	1.01	1.10	1.20	1.29	1.40	1.51	1.62	1.74	2.01	2.33	
200	1.33	1.51	1.69	1.87	2.05	2.22	2.40	2.61	2.82	3.02	3.26	3.78	4.37	
300	1.89	2.15	2.41	2.67	2.93	3.18	3.44	3.74	4.04	4.35	4.70	5.44	6.30	
400	2.42	2.76	3.09	3.43	3.77	4.10	4.43	4.83	5.22	5.61	6.07	7.04	8.15	
500	2.92	3.33	3.75	4.16	4.57	4.98	5.39	5.87	6.36	6.84	7.39	8.58	9.94	
600	3.40	3.89	4.38	4.87	5.35	5.83	6.31	6.89	7.45	8.02	8.67	10.06	11.66	
700	3.86	4.43	4.99	5.55	6.11	6.66	7.21	7.87	8.52	9.17	9.92	11.50	13.32	10
720	**3.95**	**4.53**	**5.11**	**5.69**	**6.26**	**6.82**	**7.39**	**8.06**	**8.73**	**9.39**	**10.16**	**11.79**	**13.65**	
800	4.31	4.95	5.59	6.22	6.84	7.47	8.08	8.82	9.55	10.28	11.12	12.90	14.93	
900	4.75	5.46	6.16	6.86	7.56	8.25	8.93	9.75	10.56	11.36	12.29	14.25	16.47	
960	**5.00**	**5.75**	**6.50**	**7.24**	**7.98**	**8.71**	**9.43**	**10.29**	**11.15**	**11.99**	**12.97**	**15.03**	**17.37**	
1000	5.17	5.95	6.72	7.49	8.25	9.01	9.76	10.65	11.53	12.41	13.42	15.55	17.96	
1100	5.58	6.42	7.27	8.10	8.93	9.75	10.56	11.52	12.48	13.43	14.52	16.80	19.39	
1200	5.97	6.89	7.79	8.69	9.58	10.46	11.34	12.37	13.40	14.41	15.57	18.01	20.75	20
1300	6.36	7.34	8.31	9.27	10.22	11.16	12.09	13.19	14.28	15.36	16.59	19.17	22.05	
1400	6.73	7.77	8.81	9.83	10.84	11.84	12.82	13.99	15.14	16.27	17.57	20.28	23.28	
1440	**6.88**	**7.95**	**9.00**	**10.05**	**11.08**	**12.10**	**13.11**	**14.30**	**15.47**	**16.63**	**17.96**	**20.70**	**23.75**	
1500	7.09	8.20	9.29	10.37	11.44	12.49	13.53	14.76	15.97	17.15	18.51	21.33	24.43	
1600	7.44	8.61	9.76	10.90	12.02	13.12	14.21	15.50	16.76	18.00	19.41	22.33	25.51	
1700	7.78	9.01	10.21	11.40	12.58	13.73	14.87	16.21	17.52	18.81	20.27	23.27	26.51	
1800	8.11	9.39	10.65	11.90	13.12	14.32	15.50	16.89	18.25	19.58	21.08	24.15	27.43	30
1900	8.43	9.76	11.08	12.37	13.64	14.88	16.11	17.54	18.94	20.31	21.85	24.97	28.27	
2000	8.73	10.12	11.48	12.82	14.14	15.43	16.69	18.16	19.60	20.99	22.57	25.72	29.01	
2100	9.02	10.46	11.88	13.26	14.62	15.94	17.24	18.75	20.22	21.64	23.23	26.41	29.67	
2200	9.31	10.79	12.25	13.68	15.07	16.44	17.76	19.31	20.80	22.24	23.85	27.03	30.22	
2300	9.57	11.11	12.61	14.08	15.51	16.90	18.26	19.83	21.35	22.80	24.42	27.57	30.68	
2400	9.83	11.41	12.95	14.46	15.92	17.34	18.72	20.32	21.85	23.31	24.93	28.05	31.04	40
2500	10.08	11.70	13.28	14.82	16.31	17.76	19.16	20.77	22.31	23.78	25.38	28.44	–	
2600	10.31	11.97	13.59	15.16	16.68	18.14	19.56	21.19	22.73	24.19	25.78	28.76	–	
2700	10.53	12.23	13.88	15.47	17.02	18.50	19.93	21.56	23.11	24.56	26.12	28.99	–	
2800	10.73	12.47	14.15	15.77	17.33	18.83	20.27	21.90	23.44	24.87	26.40	–	–	
2880	**10.89**	**12.65**	**14.35**	**15.99**	**17.57**	**19.07**	**20.51**	**22.14**	**23.67**	**25.08**	**26.57**	–	–	
2900	10.93	12.69	14.40	16.04	17.62	19.13	20.57	22.20	23.72	25.12	26.61	–	–	
3000	11.10	12.90	14.63	16.30	17.89	19.40	20.84	22.46	23.96	25.33	26.76	–	–	

Rev/min of faster shaft	ADDITIONAL POWER (kW) PER BELT FOR SPEED RATIO									
	1.00 to 1.01	1.02 to 1.05	1.06 to 1.11	1.12 to 1.18	1.19 to 1.26	1.27 to 1.38	1.39 to 1.57	1.58 to 1.94	1.95 to 3.38	3.39 and over
100	0.00	0.01	0.02	0.04	0.04	0.06	0.07	0.07	0.08	0.08
200	0.00	0.01	0.04	0.07	0.09	0.11	0.13	0.15	0.16	0.17
300	0.00	0.02	0.06	0.10	0.14	0.17	0.20	0.22	0.24	0.25
400	0.00	0.03	0.07	0.13	0.19	0.22	0.26	0.29	0.32	0.34
500	0.00	0.04	0.09	0.17	0.23	0.28	0.33	0.37	0.40	0.43
600	0.00	0.04	0.12	0.20	0.28	0.34	0.40	0.45	0.48	0.51
700	0.00	0.05	0.13	0.24	0.33	0.39	0.46	0.52	0.57	0.59
720	**0.00**	**0.05**	**0.14**	**0.25**	**0.33**	**0.41**	**0.48**	**0.54**	**0.59**	**0.62**
800	0.00	0.06	0.16	0.28	0.37	0.45	0.53	0.60	0.65	0.69
900	0.00	0.07	0.18	0.31	0.42	0.51	0.60	0.66	0.72	0.77
960	**0.00**	**0.07**	**0.19**	**0.32**	**0.44**	**0.54**	**0.62**	**0.70**	**0.77**	**0.81**
1000	0.00	0.07	0.19	0.34	0.46	0.56	0.66	0.74	0.81	0.86
1100	0.00	0.08	0.22	0.37	0.51	0.62	0.72	0.81	0.89	0.94
1200	0.00	0.09	0.23	0.41	0.56	0.68	0.79	0.89	0.97	1.03
1300	0.00	0.09	0.25	0.44	0.60	0.73	0.86	0.96	1.05	1.11
1400	0.00	0.10	0.28	0.48	0.65	0.79	0.93	1.04	1.13	1.20
1440	**0.00**	**0.10**	**0.28**	**0.48**	**0.66**	**0.79**	**0.94**	**1.06**	**1.15**	**1.21**
1500	0.00	0.10	0.29	0.51	0.69	0.84	0.99	1.11	1.21	1.28
1600	0.00	0.11	0.31	0.54	0.75	0.90	1.05	1.19	1.29	1.37
1700	0.00	0.12	0.34	0.58	0.79	0.95	1.12	1.26	1.37	1.45
1800	0.00	0.13	0.35	0.61	0.84	1.01	1.19	1.34	1.45	1.54
1900	0.00	0.13	0.37	0.65	0.88	1.07	1.25	1.41	1.54	1.63
2000	0.00	0.14	0.39	0.68	0.93	1.13	1.32	1.48	1.62	1.71
2100	0.00	0.15	0.41	0.72	0.98	1.18	1.39	1.56	1.69	1.79
2200	0.00	0.16	0.43	0.75	1.02	1.24	1.45	1.63	1.78	1.88
2300	0.00	0.16	0.45	0.78	1.07	1.29	1.51	1.71	1.86	1.97
2400	0.00	0.17	0.47	0.82	1.11	1.35	1.58	1.78	1.94	2.05
2500	0.00	0.18	0.49	0.85	1.16	1.41	1.65	1.86	2.02	2.14
2600	0.00	0.19	0.51	0.89	1.21	1.46	1.72	1.92	2.10	2.22
2700	0.00	0.19	0.53	0.92	1.25	1.52	1.78	1.99	2.18	2.31
2800	0.00	0.20	0.54	0.95	1.29	1.57	1.84	2.07	2.26	2.39
2880	**0.00**	**0.20**	**0.56**	**0.97**	**1.32**	**1.60**	**1.88**	**2.11**	**2.31**	**2.44**
2900	0.00	0.21	0.57	0.99	1.34	1.63	1.91	2.15	2.34	2.48
3000	0.00	0.22	0.59	1.02	1.39	1.69	1.98	2.23	2.42	2.57

NOTE: Only pulleys of Fenner manufacture should be used where belt speed falls between 30 and 40 m/s.

Fig 7.4.1 Extracts from Fenner wedge belt catalogue *continued from previous pages*

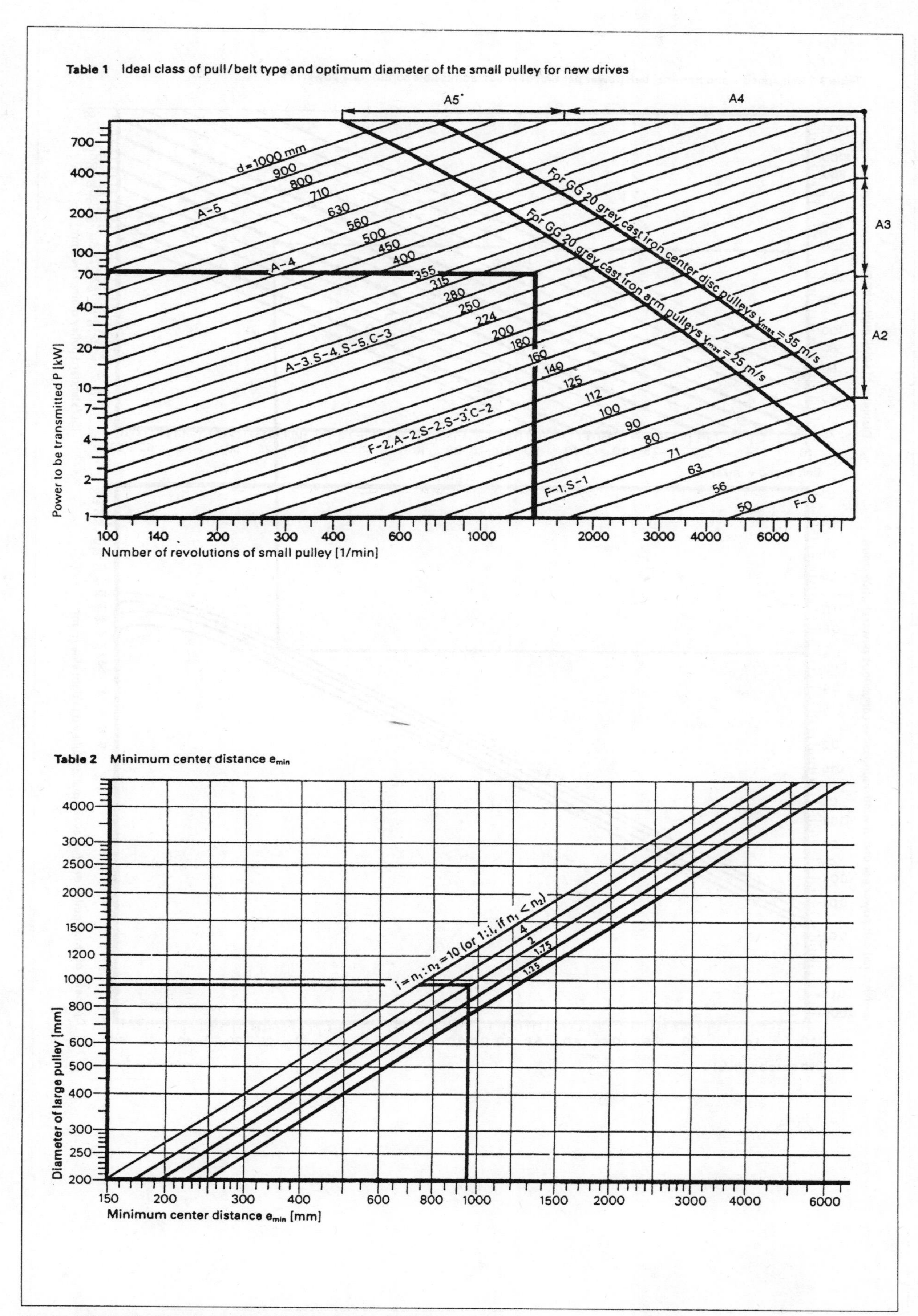

Fig 7.4.2 Extracts from Habasit catalogue *continued on following pages*

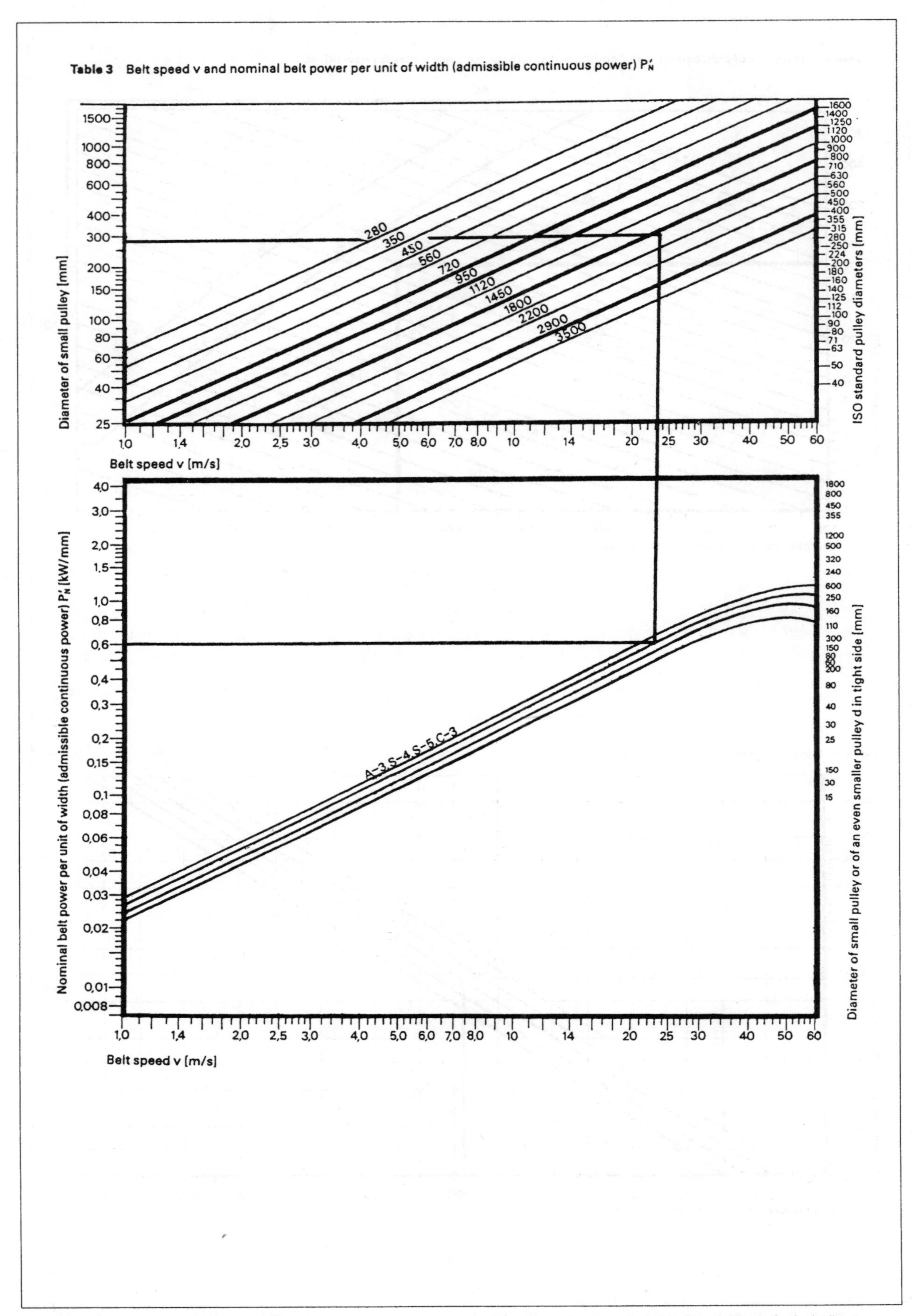

Fig 7.4.2 Extracts from Habasit catalogue *continued from previous page* (only data relevant to examples is included)

Table 4 Arc of contact factor c_1

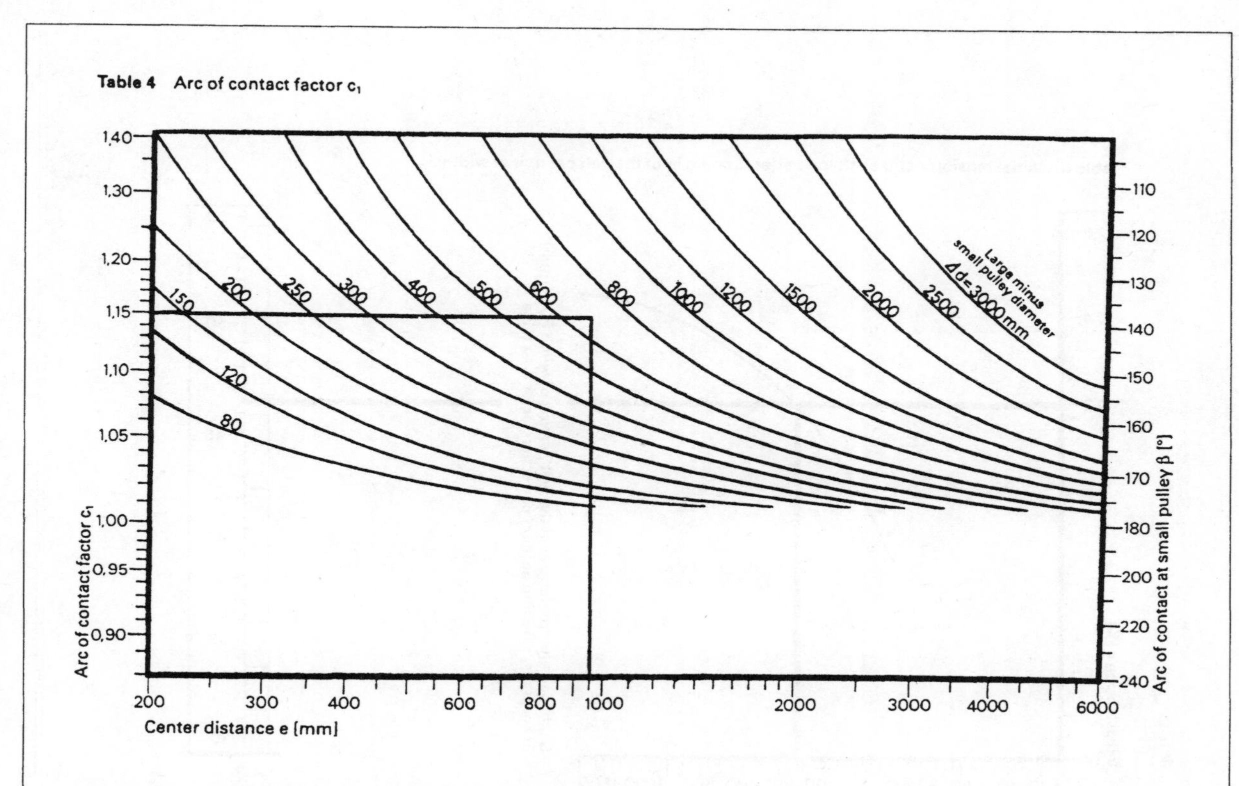

Table 5 Service factor c_2 (also takes starting torques into account)

Operating conditions	Examples	Service factor c_2 for belt group			
		F	A	C	S
Uniform operation. Small inertia forces.	Pumps, blowers, fans, generators, lathes, intermediate gears, conveyors	1.1	1.0	1.0	1.0
Variable operation. Medium inertia forces.	Metal, wood working, textile and printing machines, centrifuges, stirrers, kneading machines, cyclinder mills, elevators	1.2	1.1	1.1	1.1
Variable operation. Considerable inertia forces, shock loads.	Slotting, planing machines, beater rolls, presses, looms, cards, small rolling mills, punching presses	1.4	1.2	1.2	1.2
Variable operation. Great inertia forces, heavy shock loads.	Beating mills, stone crushers, frame saws, rolling mills, pan grinders, calenders, piston compressors	1.6	1.4	no application	no application
Very wet, oily and dusty conditions.	In all cases	add 0.4 to c_2			

Fig 7.4.2 Extracts from Habasit catalogue *continued from previous pages* (only data relevant to example is included)

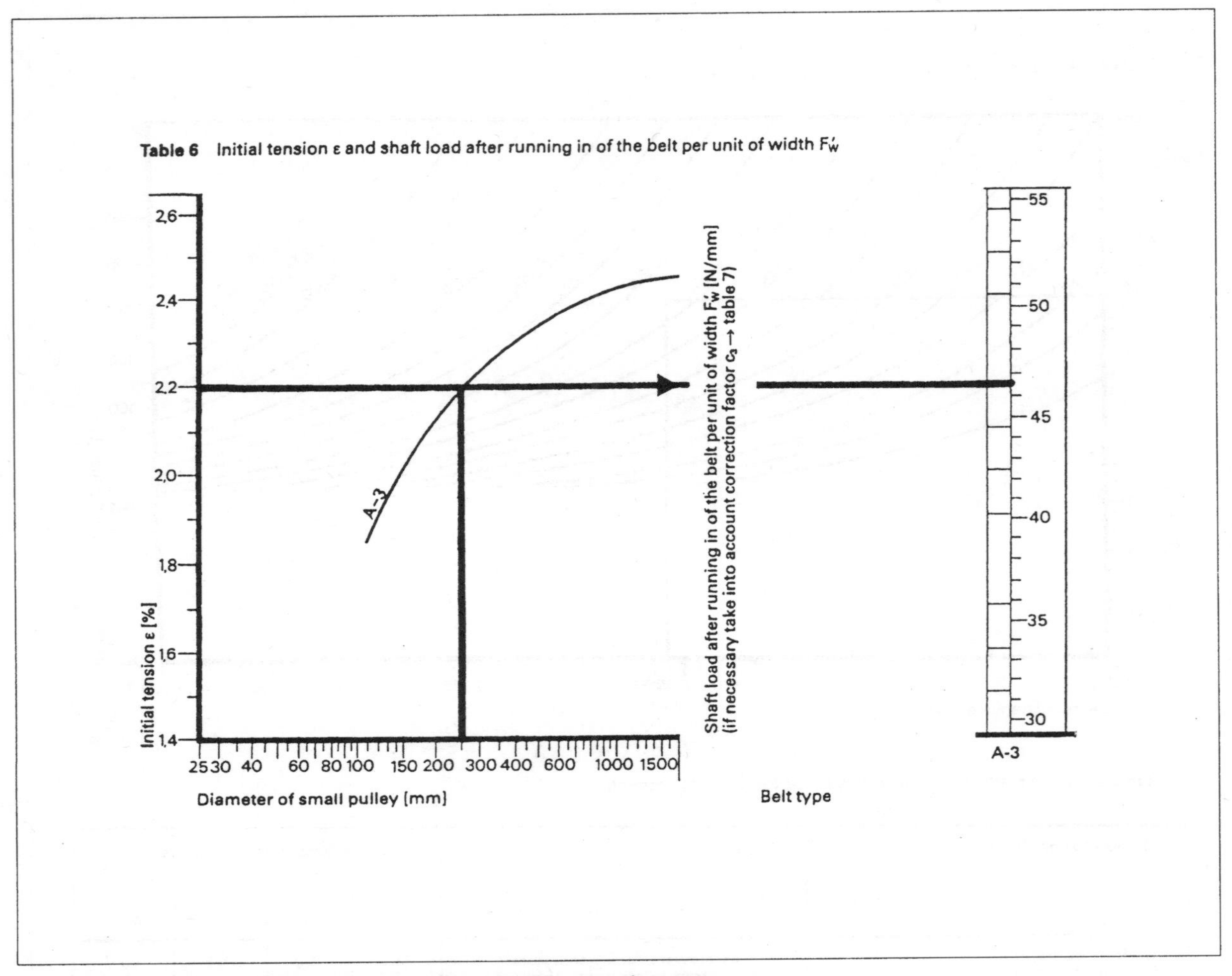

Fig 7.4.2 Extracts from Habasit catalogue *continued from previous pages*
(only data relevant to examples is included. Table 7 is not required, our belt speed v < 30 m/s)

Photo 7.4.1 Flat belt drive on a 62 kW 3-jet Pelton installation (see Fig 7.5.11). (IT/Jeremy Hartley – Sri Lanka)

Selection of wedge belt **Example 7.4.1**

Following the problem posed in Example 7.1.1 and using information from the Fenner belt catalogue:

1 Calculate speed ratio – already known 3.33:1 increasing

2 Obtain service factors – use catalogue Table 3. This is a speed-increasing drive, therefore:
Speed-increasing factor = 1.18
The duty factor should also be obtained from the table, but note that a water turbine is not included in the list of prime movers. The nearest equivalent is an electric motor driving a centrifugal pump. The essential difference is that, on starting, the electric motor generates a very high torque which rapidly accelerates the motor, the drive and the pump. In a micro-hydro scheme the turbine will accelerate slowly as the main valve is opened. It is therefore sensible to use a duty factor between 1 and 1.2 depending on the number of hours per day that the drive is operating.
For this example use: Duty factor = 1.2

3 Design power for belts = Power to be transmitted × Factors = 71 × 1.18 × 1.2 = 100 kW

4 Use Table 2 (Fig 7.4.1) to select belt section.
Belt section: 100 kW at 1500 rpm may be transmitted by either SPB or SPC belts. It is usual to use the small section. For this example use SPB belts. (SPC belts are considered in Example 7.4.3)

5 Minimum pulley diameter – use catalogue Table 1. 100 kW at 1500 rpm gives minimum pulley diameter = 236 mm. Reference to a list of pulleys shows this to be a stock size.

6 Large pulley diameter 236 × 3.33 = 786 mm diameter.
Refer to a list of pulleys and assume 800 is the nearest available.The ratio will then be 800/236 = 3.39
Consider how this will affect the generator speed. 450 turbine speed × 3.39 = 1525 rpm
Conclude that this is acceptable.

7 Belt length and centre distance. An initial estimate of centre distance may be calculated from:
Centre distance (approx) = Dia. of large pulley + Dia. of small pulley = 800 + 236 = 1036 mm.
A better approach is to refer to the table of standard belt lengths in the catalogue (not included in these extracts). Ø 236 and Ø 800 pulleys require a standard belt length of 4000 mm and a centre distance of 1152 mm. Note that there is a correction factor which accounts for the arc of contact.
Correction factor = 0.95

8 Basic power per belt. From power rating table Ø 236 and 1500 rpm gives 17.15 kW/belt

9 Speed ratio power increment. The power determined at (8) should be increased by an increment based on the speed ratio. The lower table shows 1500 rpm and 3.39 ratio as having an extra 1.28 kW/belt.

10 Corrected power per belt
= (Basic power + increment) × Factor = (17.15 + 1.28) × 0.95 = 17.84 kW

11 Number of belts

$$= \frac{\text{Design power}}{\text{power per belt}} = \frac{100}{17.84} = 5.6$$

Use six belts and check that Ø 236 and Ø 800 pulleys are available with six grooves.

12 Belt tension. The catalogue formula simplifies to

T_D = 32PN.
N = number of belts
P = force to depress belt 16 mm per metre span.
Interpolate across the range given depending on pulley diameter.
P = 65
Belt tension = 32 × 65 × 6 = 12480 Newtons.

Note: The running belt tension is also the load applied to the shafts by the belt drive.
Centrifugal tension need not be considered for micro-hydro schemes. It will usually reduce the shaft loads slightly.

Example 7.4.2 **Flat belt drive (modern)**

This example selects a Habasit belt for the transmission specification described in Example 7.1.1.
Following Habasit catalogue No. 1210, the selection recommended is as follows:

1 Belt type – Catalogue indicates Type A belt
2 Belt type – Table 1 Intersection of 71 kW and 1500 rpm show A3 belt on 280 pulley.
3 Pulley dia. – Use 280 mm dia. and 280 × 3.33 = 932 mm dia.
4 Min. centre distance – Table 2 Intersection of Ø932 and 3.33 ratio line yields 950 mm approx.
5 Belt power per unit width – Table 3 Intersection of Ø 280 and 1500 rpm yields 22 m/sec belt speed.
6 Extend into lower chart where intersection of 22 m/sec and Ø 280 for A3 belt yields 0.6 kW/mm width.
7 Arc of contact factor – Table 4 Difference in pulley dia. = 932 – 280 = 652 mm. Intersection of 950 centre distance and 652 diameter difference yields 1.15 factor.
8 Service factor – Table 5 Belt type A driving a generator yields a factor of 1.
9 Belt width = Power × Arc of contact factor × Service factor/Belt power per unit width
= 71 kW × 1.15 × 1/0.6 = 136 mm
10 Initial tension – Table 6 Intersection of Ø 280 with A3 belt yields a strain of 2.2% approx. Extend this to right-hand axis of chart to A3 belt gives 46.5 N/mm width.
11 Using 140 mm belt width (136 rounded up) Initial tension = 140 × 46.5 = 6510 N
12 Shaft load = Initial tension.
13 Consult belt supplier regarding method of setting initial belt tension (2N per millimetre belt width may be used if manufacturer's data is not available)

In order to make a direct comparison between a flat belt and the calculated wedge belt drive it is necessary to repeat the flat belt calculations using the same diameter pulleys as for the wedge belts.

Exercise

Repeat the calculations using a Habasit belt running on pulleys of Ø 236 and Ø 800 but otherwise as in this example.

Example 7.4.3 **Wedge belt drive**

Use Fenner wedge belt drive catalogue.
Use SPC section belts and continue from step 4 from Example 7.4.1.

5 Minimum pulley dia. = 236 mm
6 Large pulley dia. use 800 giving ratio 3.39
7 Belt length and centre distance
Table in catalogue gives belt length = 4000 mm
Centre distance = 1152 mm
Correction factor = 0.9
8 Basic power per belt – use catalogue tables
Basic power = 19.33
9 Additional power = 3.96
Total = 23.29
10 Using factor from step 7.
23.29 × 0.9 = 20.96
11 Number of belts

$$= \frac{100}{20.96} = 4.77 \quad \text{use 5 belts.}$$

12 Belt tension
$T_D = 32\ P \times 5$
For SPC belts P = 120 Newtons
Therefore,
$T_D = 32 \times 120 \times 5 = 18400$ N

Flat belt (traditional) **Example 7.4.4**

Use data given in *Machinery Handbook*, which is American so most data is in imperial units.

71 kW = 95 HP

Ø 236 = 9 inches approx.

9-inch dia. pulley at 1500 rpm gives

$$\text{Belt speed} = \frac{1500 \times 9 \times \pi}{12} = 3534 \text{ ft/min}$$

The relevant table shows that at 3500 ft/min a leather belt of 2 ply and 5/16 inch (8 mm) thickness has a theoretical capacity of 10 HP per inch width.

Three service factors must be obtained from various tables and are:

M = 2.0

P = 0.7

F = 1.35

$$\text{Belt width} = \frac{95 \times 2 \times 1.35}{0.7 \times 10}$$

= 36.6 inches

= 930 mm

Note: This belt is too wide to be a practical proposition.

Comparison of various drive calculations **Table 7.4.2**

	Wedge belt Fenner SPB	**Flat belt Habasit A3**
Small pulley diameter (mm)	236	280
Large pulley diameter (mm)	800	932
Pulley width (mm)	120 (from catalogue)	160 (from Fig 7.3.1)
Shaft load		
Fenner calculation	12480 N	
Habasit calculation		6510 N

Checklist of belt selection **Note 7.4.1**

- What experience of belt drives is available locally?
- What experience of the belt supplier relates to drives similar to that for a micro-hydro scheme?
- Are the belts easily obtained both initially and as replacements in later years?
- Will the belt supplier give advice or check the drive design, including loads imposed on shafts?
- Will the belt supplier provide written guarantees?
- Is the price right? This should include the price of pulleys, fitting them to shafts, fitting belts to pulleys, the belt tensioning system, extra extension shafts and bearings as well as the belts themselves.

7.5
Bearings

All shafts run in bearings. The function of a bearing is to allow a shaft to rotate freely with a low coefficient of friction and at the same time hold the shaft in the correct position against any forces imposed on the shaft.

The two main families of bearings are *sliding* bearings and *rolling element* bearings.

Sliding bearings

Sliding bearings require correct lubrication and regular maintenance. They are not usually used for micro-hydro schemes.

Rolling element bearings

The characteristics of common types of bearings are shown in Table 7.5.1. Rolling element bearings include the very common ball bearings as well as other types. Correctly installed, lubricated and protected from water and dirt they will run trouble free for thousands of hours without the need for any attention.

Detailed information from SKF about deep groove ball bearings (the usual common ball bearings) is given in Fig 7.5.1, for plain roller bearings in Fig 7.5.2 and for spherical roller bearings in Fig 7.5.3.

It should be noted that ball bearings only are available with seals and shields incorporated into the bearing. The use of shields and seals is recommended in micro-hydro applications, as dust and dirt should be prevented from entering the bearing.

When plain or spherical roller bearings are required it is necessary to fit external seals into the bearing housing.

Bearing load calculations

Calculation of the loads applied to bearings is necessary in order to choose the correct type and size of bearings. These calculations will be demonstrated using data from Examples 7.4.1 and 7.4.2.

At this stage we will assume we are intending to drive a generator. Literature from the manufacturer Newage Stanford is included in Fig 7.5.4. We have 71 kW available at the generator shaft. For reasons explained in Chapter 8 we will choose an overrated 100 kW generator with a frame size code C344D. It is necessary to calculate whether the bearings in that machine are adequate for use with a drive arrangement similar to that shown in Fig 7.1.2, that is having an overhung shaft. See Example 7.5.1.

Alternative method

There is an alternative method using graphs or nomograms provided in bearing tables; an example nomogram is given in Fig 7.5.8.

For this method we need:

P = Bearing equivalent load

N = Machine RPM (nominated 'n' in nomogram)

L_{10h} = The required bearing life in hours

Find the machine RPM in the n (r/min) column, and the chosen life in the L_{10h} (operating hours) column, and draw a straight line between them. The ratio of C/P is then read off and C can be easily calculated (C = Bearing dynamic capacity).

If you know the shaft diameter, you have to look through the catalogue for a suitable bearing.

Table 7.5.1 **Common types of bearings**

	Radial load characteristics	Axial load characteristics	Angular misalignment characteristics
Deep groove ball	Medium	Medium	Limited
Angular contact ball	Medium	Good	Limited
Cylindrical roller	Good	Poor	Slight (up to 2 mins)
Spherical roller	Good	Good	Good (up to 2.5°)

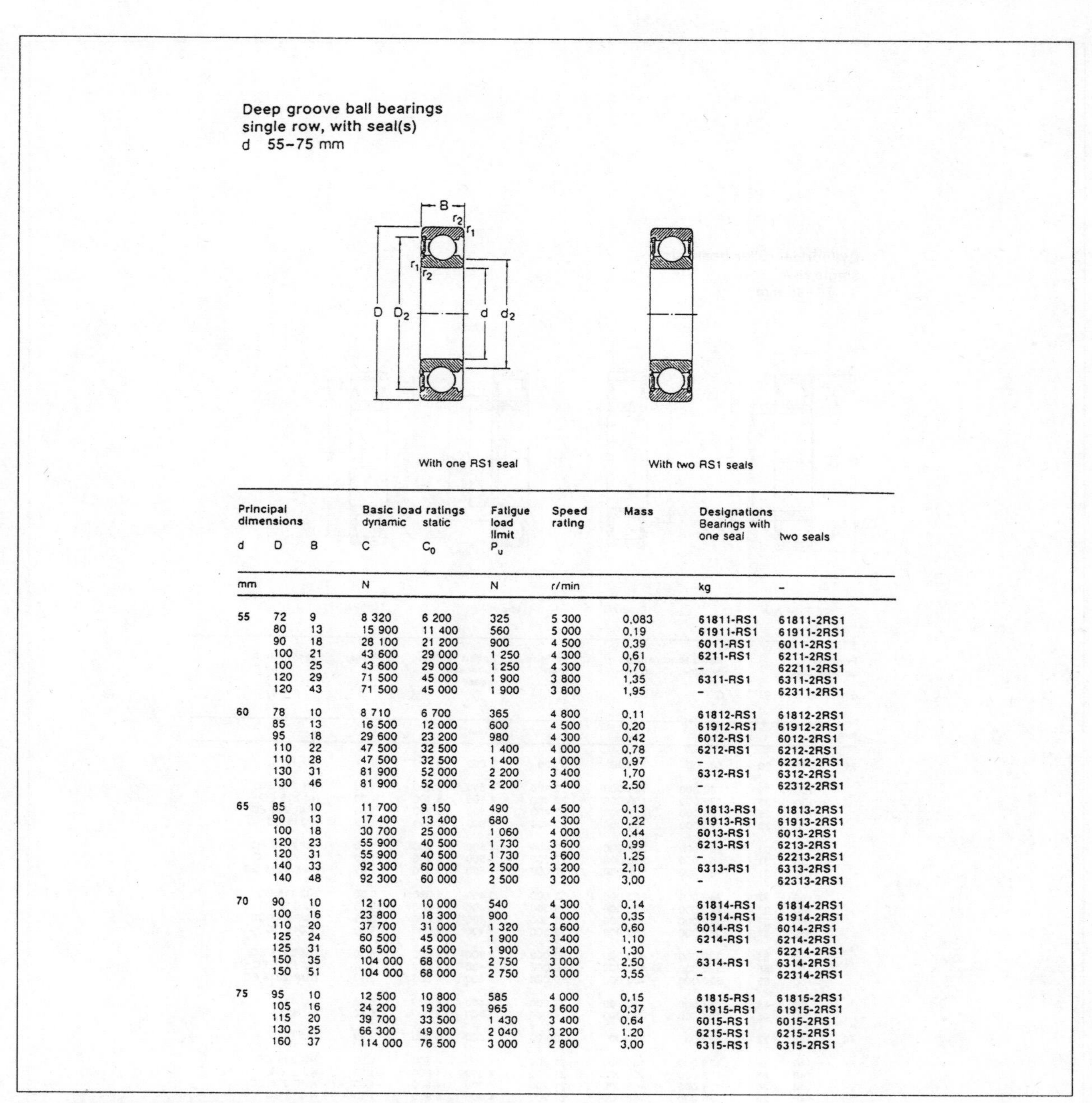
Deep groove ball bearings
single row, with seal(s)
d 55–75 mm

Principal dimensions			Basic load ratings dynamic	static	Fatigue load limit	Speed rating	Mass	Designations Bearings with one seal	two seals
d	D	B	C	C_0	P_u				
mm			N		N	r/min	kg		–
55	72	9	8 320	6 200	325	5 300	0,083	61811-RS1	61811-2RS1
	80	13	15 900	11 400	560	5 000	0,19	61911-RS1	61911-2RS1
	90	18	28 100	21 200	900	4 500	0,39	6011-RS1	6011-2RS1
	100	21	43 600	29 000	1 250	4 300	0,61	6211-RS1	6211-2RS1
	100	25	43 600	29 000	1 250	4 300	0,70	–	62211-2RS1
	120	29	71 500	45 000	1 900	3 800	1,35	6311-RS1	6311-2RS1
	120	43	71 500	45 000	1 900	3 800	1,95	–	62311-2RS1
60	78	10	8 710	6 700	365	4 800	0,11	61812-RS1	61812-2RS1
	85	13	16 500	12 000	600	4 500	0,20	61912-RS1	61912-2RS1
	95	18	29 600	23 200	980	4 300	0,42	6012-RS1	6012-2RS1
	110	22	47 500	32 500	1 400	4 000	0,78	6212-RS1	6212-2RS1
	110	28	47 500	32 500	1 400	4 000	0,97	–	62212-2RS1
	130	31	81 900	52 000	2 200	3 400	1,70	6312-RS1	6312-2RS1
	130	46	81 900	52 000	2 200	3 400	2,50	–	62312-2RS1
65	85	10	11 700	9 150	490	4 500	0,13	61813-RS1	61813-2RS1
	90	13	17 400	13 400	680	4 300	0,22	61913-RS1	61913-2RS1
	100	18	30 700	25 000	1 060	4 000	0,44	6013-RS1	6013-2RS1
	120	23	55 900	40 500	1 730	3 600	0,99	6213-RS1	6213-2RS1
	120	31	55 900	40 500	1 730	3 600	1.25	–	62213-2RS1
	140	33	92 300	60 000	2 500	3 200	2,10	6313-RS1	6313-2RS1
	140	48	92 300	60 000	2 500	3 200	3,00	–	62313-2RS1
70	90	10	12 100	10 000	540	4 300	0.14	61814-RS1	61814-2RS1
	100	16	23 800	18 300	900	4 000	0,35	61914-RS1	61914-2RS1
	110	20	37 700	31 000	1 320	3 600	0,60	6014-RS1	6014-2RS1
	125	24	60 500	45 000	1 900	3 400	1,10	6214-RS1	6214-2RS1
	125	31	60 500	45 000	1 900	3 400	1,30	–	62214-2RS1
	150	35	104 000	68 000	2 750	3 000	2.50	6314-RS1	6314-2RS1
	150	51	104 000	68 000	2 750	3 000	3,55	–	62314-2RS1
75	95	10	12 500	10 800	585	4 000	0,15	61815-RS1	61815-2RS1
	105	16	24 200	19 300	965	3 600	0,37	61915-RS1	61915-2RS1
	115	20	39 700	33 500	1 430	3 400	0.64	6015-RS1	6015-2RS1
	130	25	66 300	49 000	2 040	3 200	1.20	6215-RS1	6215-2RS1
	160	37	114 000	76 500	3 000	2 800	3,00	6315-RS1	6315-2RS1

Fig 7.5.1 Deep groove ball bearings taken from SKF catalogue

Calculation factors X and Y for deep groove ball bearings — Table 7.5.2

The following is taken from SKF catalogue.

F_a / C_0	e	X	Y
0.025	0.22	0.56	2.0
0.04	0.24	0.56	1.8
0.07	0.27	0.56	1.6
0.13	0.31	0.56	1.4
0.25	0.37	0.56	1.2
0.50	0.44	0.56	1.0

The above values for X and Y apply when $F_a / F_r > e$. When $F_a / F_r \leq e$ use X = 1 and Y = 0.

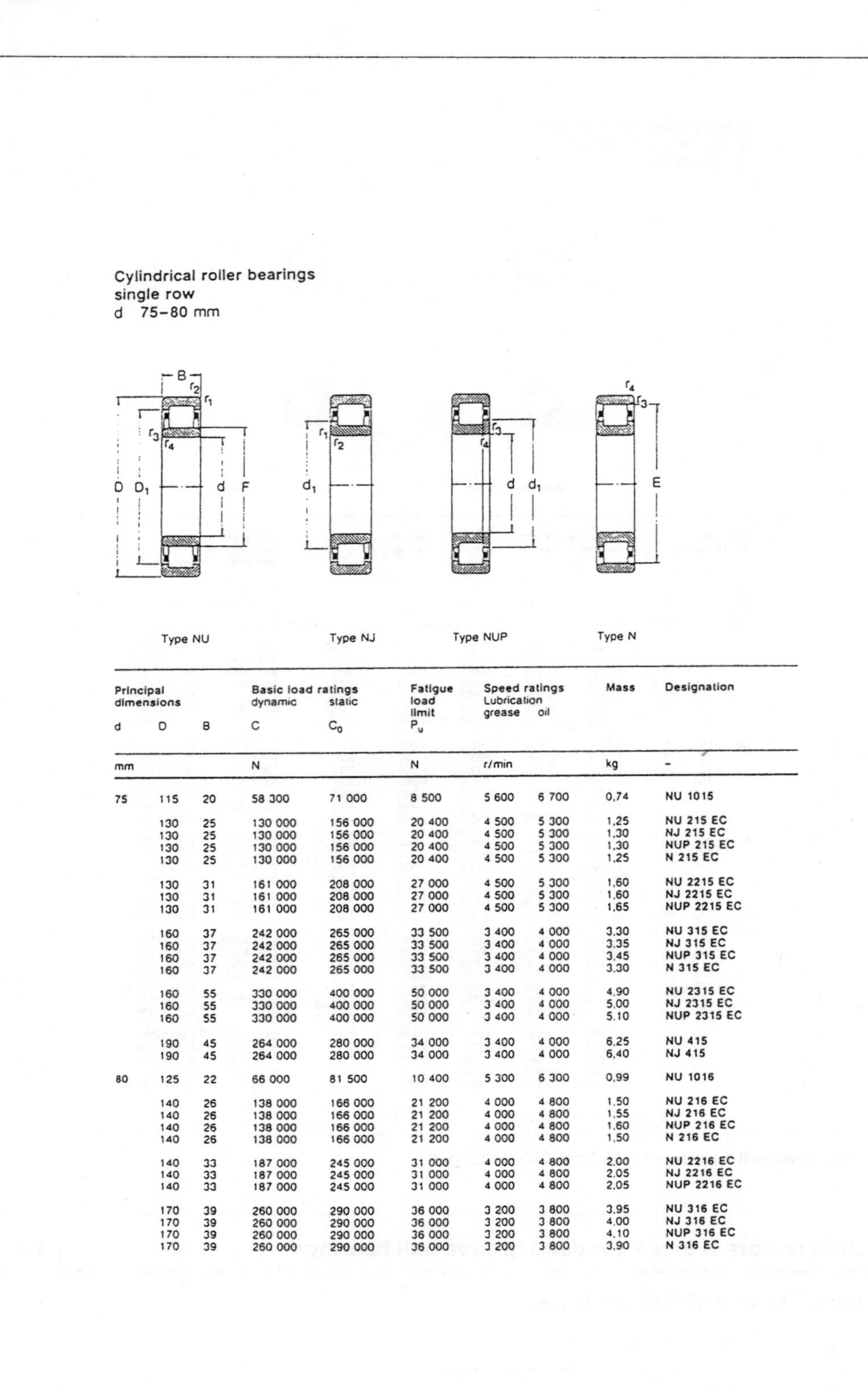
Cylindrical roller bearings
single row
d 75–80 mm

Principal dimensions d	D	B	Basic load ratings dynamic C	static C_0	Fatigue load limit P_u	Speed ratings Lubrication grease	oil	Mass	Designation
mm			N		N	r/min		kg	-
75	115	20	58 300	71 000	8 500	5 600	6 700	0,74	NU 1015
	130	25	130 000	156 000	20 400	4 500	5 300	1,25	NU 215 EC
	130	25	130 000	156 000	20 400	4 500	5 300	1,30	NJ 215 EC
	130	25	130 000	156 000	20 400	4 500	5 300	1,30	NUP 215 EC
	130	25	130 000	156 000	20 400	4 500	5 300	1,25	N 215 EC
	130	31	161 000	208 000	27 000	4 500	5 300	1,60	NU 2215 EC
	130	31	161 000	208 000	27 000	4 500	5 300	1,60	NJ 2215 EC
	130	31	161 000	208 000	27 000	4 500	5 300	1,65	NUP 2215 EC
	160	37	242 000	265 000	33 500	3 400	4 000	3,30	NU 315 EC
	160	37	242 000	265 000	33 500	3 400	4 000	3,35	NJ 315 EC
	160	37	242 000	265 000	33 500	3 400	4 000	3,45	NUP 315 EC
	160	37	242 000	265 000	33 500	3 400	4 000	3,30	N 315 EC
	160	55	330 000	400 000	50 000	3 400	4 000	4,90	NU 2315 EC
	160	55	330 000	400 000	50 000	3 400	4 000	5,00	NJ 2315 EC
	160	55	330 000	400 000	50 000	3 400	4 000	5,10	NUP 2315 EC
	190	45	264 000	280 000	34 000	3 400	4 000	6,25	NU 415
	190	45	264 000	280 000	34 000	3 400	4 000	6,40	NJ 415
80	125	22	66 000	81 500	10 400	5 300	6 300	0,99	NU 1016
	140	26	138 000	166 000	21 200	4 000	4 800	1,50	NU 216 EC
	140	26	138 000	166 000	21 200	4 000	4 800	1,55	NJ 216 EC
	140	26	138 000	166 000	21 200	4 000	4 800	1,60	NUP 216 EC
	140	26	138 000	166 000	21 200	4 000	4 800	1,50	N 216 EC
	140	33	187 000	245 000	31 000	4 000	4 800	2,00	NU 2216 EC
	140	33	187 000	245 000	31 000	4 000	4 800	2,05	NJ 2216 EC
	140	33	187 000	245 000	31 000	4 000	4 800	2,05	NUP 2216 EC
	170	39	260 000	290 000	36 000	3 200	3 800	3,95	NU 316 EC
	170	39	260 000	290 000	36 000	3 200	3 800	4,00	NJ 316 EC
	170	39	260 000	290 000	36 000	3 200	3 800	4,10	NUP 316 EC
	170	39	260 000	290 000	36 000	3 200	3 800	3,90	N 316 EC

Fig 7.5.2 Cylindrical roller bearings taken from SKF catalogue

Spherical roller bearings
d 60–85 mm

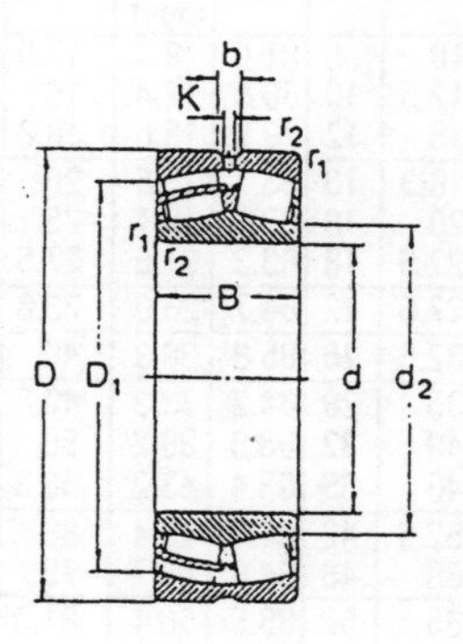

Cylindrical bore

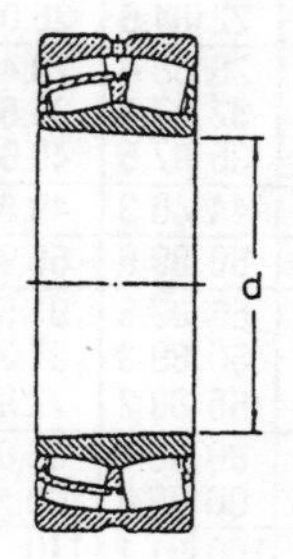

Tapered bore

Principal dimensions			Basic load ratings dynamic	static	Fatigue load limit	Speed ratings Lubrication grease	oil	Mass	Designations Bearings with cylindrical bore	tapered bore	Calculation factors	
d	D	B	C	C_0	P_u						e	Y_1
mm			N		N	r/min		kg	–			
60	110	28	122 000	146 000	16 300	4 000	5 000	1,10	22212 CC	22212 CCK	0,24	2,8
	110	28	140 000	173 000	19 000	4 300	5 300	1,15	22212 E	22212 EK	0,24	2,8
	130	31	161 000	200 000	23 200	3 000	3 800	1,95	21312 CC	21312 CCK	0,24	2,8
	130	46	235 000	280 000	30 000	3 000	3 800	2,95	22312 CC	22312 CCK	0,35	1,9
	130	46	271 000	335 000	36 500	2 800	3 600	2,90	22312 E	22312 EK	0,35	1,9
65	120	31	148 000	183 000	21 200	3 800	4 800	1,45	22213 CC	22213 CCK	0,24	2,8
	120	31	176 000	216 000	24 000	3 800	4 800	1,50	22213 E	22213 EK	0,25	2,7
	140	33	184 000	240 000	27 000	2 800	3 600	2,45	21313 CC	21313 CCK	0,24	2,8
	140	48	253 000	300 000	32 000	2 600	3 400	3,55	22313 CC	22313 CCK	0,35	1,9
	140	48	299 000	360 000	38 000	2 600	3 400	3,55	22313 E	22313 EK	0,35	1,9
70	125	31	148 000	186 000	21 200	3 600	4 500	1,55	22214 CC	22214 CCK	0,23	2,9
	125	31	179 000	228 000	25 500	3 600	4 500	1,55	22214 E	22214 EK	0,23	2,9
	150	35	207 000	260 000	29 000	2 600	3 400	3,00	21314 CC	21314 CCK	0,24	2,8
	150	51	311 000	380 000	40 000	2 400	3 200	4,30	22314 CC/W33	22314 CCK/W33	0,35	1,9
	150	51	345 000	430 000	45 000	2 200	3 000	4,30	22314 E	22314 EK	0,33	2
75	130	31	158 000	208 000	23 600	3 400	4 300	1,65	22215 CC	22215 CCK	0,22	3
	130	31	184 000	240 000	26 500	3 400	4 300	1,70	22215 E	22215 EK	0,22	3
	160	37	235 000	300 000	32 500	2 400	3 200	3,55	21315 CC	21315 CCK	0,23	2,9
	160	55	345 000	430 000	44 000	2 200	3 000	5,25	22315 CC/W33	22315 CCK/W33	0,35	1,9
	160	55	385 000	475 000	48 000	2 200	3 000	5,25	22315 E	22315 EK	0,35	1,9
80	140	33	176 000	228 000	26 000	3 200	4 000	2,05	22216 CC	22216 CCK	0,22	3
	140	33	207 000	270 000	29 000	3 200	4 000	2,10	22216 E	22216 EK	0,22	3
	170	39	258 000	335 000	36 000	2 200	3 000	4,20	21316 CC	21316 CCK	0,23	2,9
	170	58	374 000	455 000	46 500	2 000	2 800	6,20	22316 CC/W33	22316 CCK/W33	0,35	1,9
	170	58	431 000	540 000	54 000	2 000	2 800	6,20	22316 E	22316 EK	0,35	1,9
85	150	36	210 000	270 000	31 000	3 000	3 800	2,55	22217 CC/W33	22217 CCK/W33	0,22	3
	150	36	244 000	325 000	34 500	2 800	3 600	2,65	22217 E	22217 EK	0,22	3
	180	41	293 000	375 000	40 000	2 000	2 800	5,00	21317 CC	21317 CCK	0,23	2,9
	180	60	420 000	520 000	52 000	1 900	2 600	7,25	22317 CC/W33	22317 CCK/W33	0,33	2
	180	60	477 000	620 000	61 000	1 900	2 600	7,25	22317 E	22317 EK	0,33	2

Fig 7.5.3 Spherical roller bearings taken from SKF catalogue

Ratings

3 PHASE

FRAME SIZE	50 Hz 1500 Rev/Min				60 Hz 1800 Rev/Min			
	kVA	kW	Eff%	kW Input	kVA	kW	Eff%	kW Input
C234/244A	15	12	85.2	14.1	18.8	15	85.7	17.5
	20	16	84.1	19.0	25	20	85.0	23.5
	22.5	18	83.4	21.6	27.5	22	84.6	26.0
C234/244B	25	20	88.0	22.7	31.3	25	88.0	28.4
	32.5	26	87.3	29.8	40	32	87.8	36.5
	35	28	87.0	32.2	43.8	35	87.5	40.0
C234/244C	43.8	35	88.0	39.8	55	44	88.3	49.8
C234/244D	50	40	89.2	44.8	62.5	50	89.5	55.9
C334/344A	55	44	89.1	49.4	68.8	55	89.5	61.5
	60	48	88.9	54.0	75	60	89.3	67.2
	65	52	88.7	58.6	81.3	65	89.2	72.9
C334/344B	80	64	89.8	71.3	100	80	90.1	88.8
	90	72	89.5	80.5	112.5	90	89.8	100
C334/344C	100	80	90.8	88.1	125	100	91.1	110
	110	88	90.6	97.1	137.5	110	91.0	121
C334/344D	125	100	91.3	110	156.3	125	91.5	137

1 PHASE

FRAME SIZE	50 Hz 1500 Rev/Min				60 Hz 1800 Rev/Min			
	kVA	kW	Eff%	kW Input	kVA	kW	Eff%	kW Input
C234/244A	10	8	81.0	9.9	12.5	10	81.5	12.3
	12.5	10	80.5	12.4	15	12	81.5	14.7
	15	12	79.9	15.0	18.8	15	80.8	18.6
C234/244B	16.3	13	83.7	15.5	20	16	84.0	19.1
	20	16	83.6	19.1	25	20	84.1	23.8
	22.5	18	83.3	21.6	27.5	22	84.0	26.2
C234/244C	27.5	22	84.7	26.0	33.8	27	85.0	31.8
C234/244D	32.5	26	85.8	30.3	40	32	86.1	37.2
C334/344A	35	28	84.2	33.3	43.8	35	84.6	41.4
	40	32	83.8	38.2	50	40	84.4	47.4
	45	36	83.4	43.2	56.3	45	84.0	53.6
C334/344B	52.5	42	85.1	49.4	65	52	85.6	60.8
	60	48	84.6	56.7	75	60	85.2	70.4
C334/344C	65	52	86.5	60.1	81.3	65	86.9	74.8
	70	56	86.5	64.7	87.5	70	86.7	80.7
C334/344D	80	64	87.2	73.4	100	80	87.5	91.4

Industrial ratings are based on a Class F temperature rise, 40°C ambient and altitude of 1000 metres. As the insulation system is Class H an extended insulation life can be expected.

Dimensions/Weights

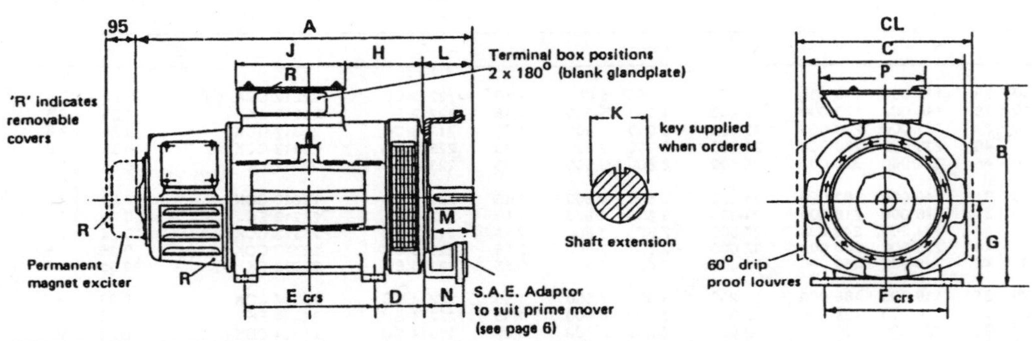

FRAME SIZE	FRAME			FEET				TERMINAL BOX				SHAFT			
	A	C	CL	D	E	F	G	B	H	J	P	K	L	M	N*
2A/B	779	420	470	137	241	356	225	538	100.5	314	274	55	122	110	105
2C/D	849	420	470	137	311	356	225	538	170	314	274	55	122	110	105
3A/B	913.5	506	556	157	291	406	270	645	132.5	340	320	70	151.5	140	126
3C/D	1028.5	506	556	157	406	406	270	645	245	340	320	70	151.5	140	126

All dimensions in mm.
Keys and keyways are to BS 4235 Part 1.
Fully dimensioned general arrangement drawings are available on request.

Approximate Nett Weights

FRAME SIZE	kg	FRAME SIZE	kg
244A	241	234A	254
244B	274	234B	287
244C	308	234C	321
244D	333	234D	346

FRAME SIZE	kg	FRAME SIZE	kg
344A	377	334A	390
344B	424	334B	437
344C	504	334C	517
344D	541	334D	554

Fig 7.5.4 Newage Stamford alternator details taken from literature from:
Newage Engineers Ltd, PO Box 17, Barnack Road, Stamford, PE9 2ND, UK.

Selection of bearings **Example 7.5.1**

Calculate the loads applied to the bearings of a generator with a C344D frame and driven by wedge belts as calculated in Example 7.4.1

Referring to the manufacturers' literature shown in Fig 7.5.4 for frame size specified, the total shaft length is 1028.5 mm and the part of the shaft where the pulley is fixed is 151.5 mm (see Fig 7.5.5).

We can then assume that

$$A = 1028.5 - 151.5 = 0.877 \text{ mm} = 0.877\text{m}$$

and that if the pulley is fixed centrally on the shaft extension, then

$$B = \frac{151.5}{2} = 75.75 \text{ mm} = 0.877 \text{ m}$$

Now we can represent the shaft as shown in Fig 7.5.6.

For the sake of clarity we are not including the self weight of the generator rotor, which can be taken as a force acting halfway between R_1 and R_2. The generator catalogue gives the total weight as 541 kg; assume half this weight is the rotor, giving a force of 2.7 kN. This should be taken into account in a full calculation of bearing loads. The simplest way is to add 2.7/2 kN to each of the two bearing radial loads. Notice that if you do this in the following calculation, the effect of rotor weight is small.

Note that the directions of the reactions at the bearings (R_1 and R_2) are nominal at this stage.

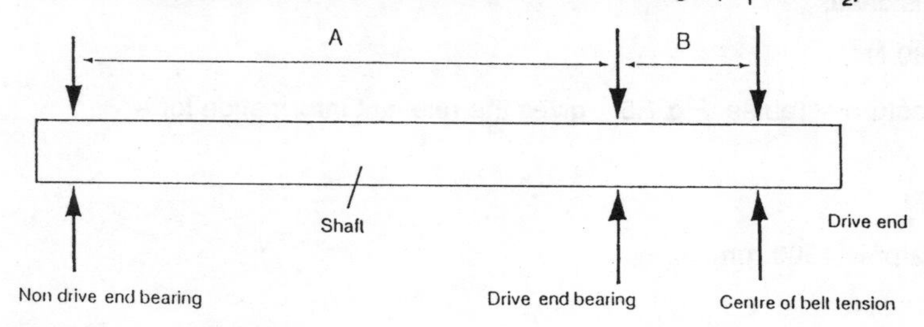

Fig 7.5.5 Schematic of shaft

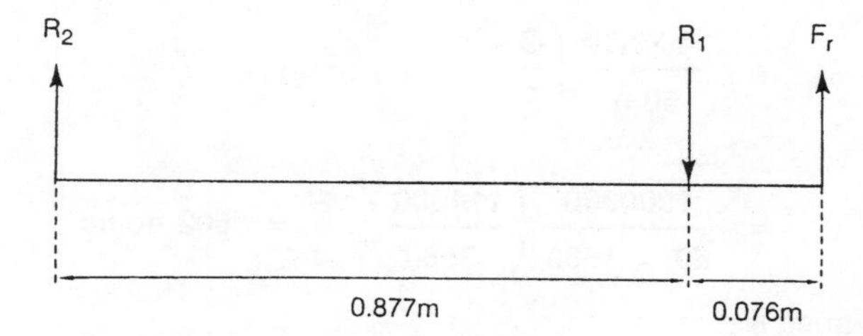

Fig 7.5.6 Force diagram

Example 7.4.1 resulted in a shaft load of:

$F = 12.48$ kN

Taking moments about R_1:

$0.877\ R_2 = 0.076 \times 12.48$ kN

$R_2 = 1.08$ kN

Resolving vertically:

$F - R_1 + R_2 = 0$

$R_1 = F + R_2 = 12.48 + 1.08$

$= 13.56$ kN

Thus, the radial load on the 'drive end' bearing (R_1) is 13.56 kN and on the 'non-drive end' is 1.08 kN.

Let us consider the life of the drive end bearing (R_1). Suppose that the manufacturers have told us this bearing is a deep-groove ball bearing type 6315-2RS1 made by SKF, and the shaft diameter in the bearing is 75 mm (see Fig. 7.5.1).

The bearing life is expressed as

$$L_{10h} = \frac{1000000}{60\ n}\left(\frac{C}{P}\right)^q$$

where

L_{10h}	=	Likely bearing life in hours	n	=	Rotational speed in rpm
C	=	Dynamic load rating in Newtons	P	=	Equivalent dynamic load in Newtons
q	=	Exponent for life equation			

For ball bearings q = 3 and for roller bearings q = 3.3.

continued opposite

Example 7.5.1 *continued* **Selection of bearings**

Equivalent dynamic load

P is calculated as

$P = X F_r + Y F_a$

Where F_r = Radial load = R_1

in this example = 13.56 kN

F_a = Axial load

X and Y are factors: see Table 7.5.2.

These factors take account of the applied forces F_a and F_r, the relationship between them, and the relationship of F_a to the static capacity of the bearing.

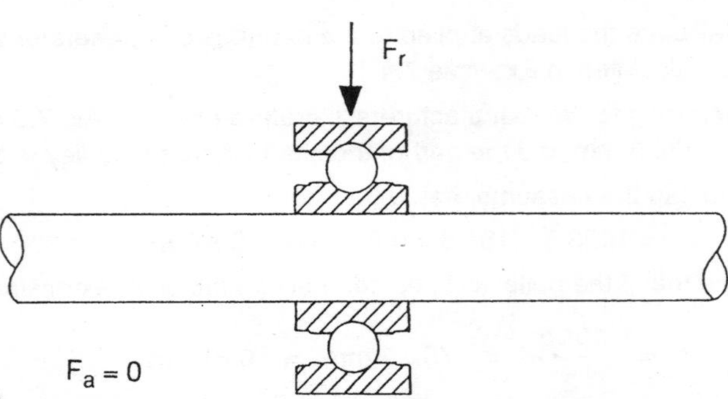

Fig 7.5.7 Loads applied to bearings In our example, since the alternator is mounted horizontally, there will effectively be no axial load on the bearing, so $F_a = 0$.

In our case, $F_a = 0$ so F_a/C_o is 0

Also, $F_a/F_r = 0$ and consequently less than any value of e quoted.

Therefore, X = 1 and Y = 0.

Where there is no axial load, as in this case,

$$P = XF_r = 1 \times 13.56\text{kN} = 13\,560\ \text{N}$$

C may be found from bearing manufacturers' tables. Fig 7.5.1 gives the relevant information for a 6315-2RS1 bearing, where:

C = 114000 N

Rotational speed for the 50Hz alternator is 1500 rpm

Therefore:

$$L_{10h} = \frac{1000000}{60\ n}\left(\frac{C}{P}\right)^q$$

$$= \frac{1000000}{60 \times 1500}\left(\frac{114000}{13560}\right)^3 = 6602 \text{ hours}$$

or bearing life

$$= \frac{6602}{24 \times 30} = 9.2 \text{ months}$$

This is obviously inadequate. A higher duty bearing is required.

Cylindrical roller bearings have better radial load characteristics, but are not suitable for high axial loads. In our example, we have no axial load, so a cylindrical roller bearing can be used.

The outside diameter of the existing bearing is 160 mm, and its thickness is 37 mm.

From the bearing catalogue we find that a cylindrical roller bearing No. NU 315 has a 75 mm bore, 160 mm outside diameter and 37 mm thickness. Hence this bearing will fit directly into the existing housing.

It has a dynamic load rating of 242 kN.

We can now calculate the life of this bearing:

$$L_{10h} = \frac{1000000}{60\ n}\left(\frac{242000}{13560}\right)^{3.3}$$

(Note that the exponent here is 3.3, because we are using a roller bearing.)

L_{10h} = 149931 = 208 months = 17 years

This bearing life is more than satisfactory.

Selection of bearings **Example 7.5.2**

Repeat example 7.5.1 except use the shaft force previously calculated in example 7.4.2.

Namely F = 6510N. Detailed calculations should be made as an exercise but the following may be used as a check.

$0.877R_2 = 0.076 \times 6.51$

$R_2 = 0.56\text{kN}$ $\qquad R_1 = 6.51 + 0.56 = 7.07\text{kN}$

$P = R_1 = 7070\text{N}$ $\qquad C = 114\,000\text{N}$ (for a 6315-2RS1 ball bearing)

$$L_{10h} = \frac{1000000}{60 \times 1500}\left(\frac{114000}{7070}\right)^3 = 46581$$

$$\text{Bearing life} = \frac{46581}{24 \times 30} = 64.7 \text{ months} = 5.39 \text{ years}$$

This could be a satisfactory bearing life and drive arrangement but refer to Section 7.4 before making a decision.

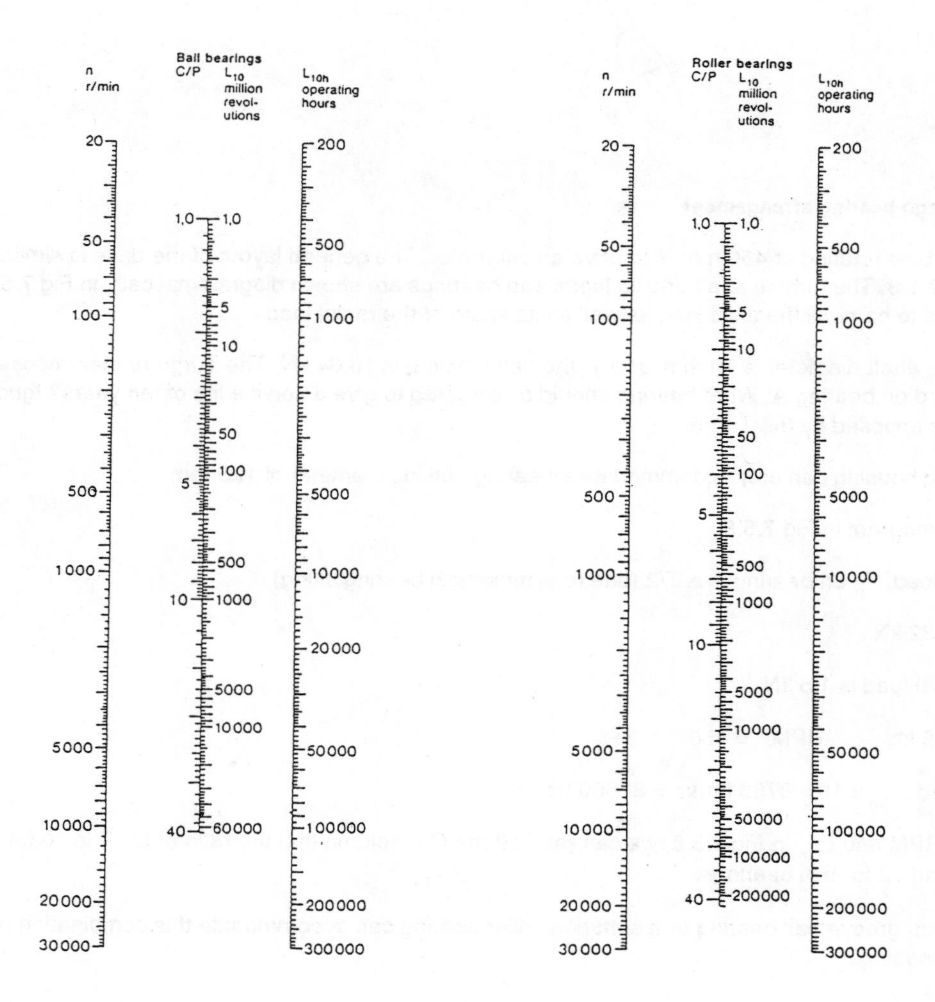

Fig 7.5.8 Bearing life calculation chart

Example 7.5.3 **Bearing selection with axial and radial loads**

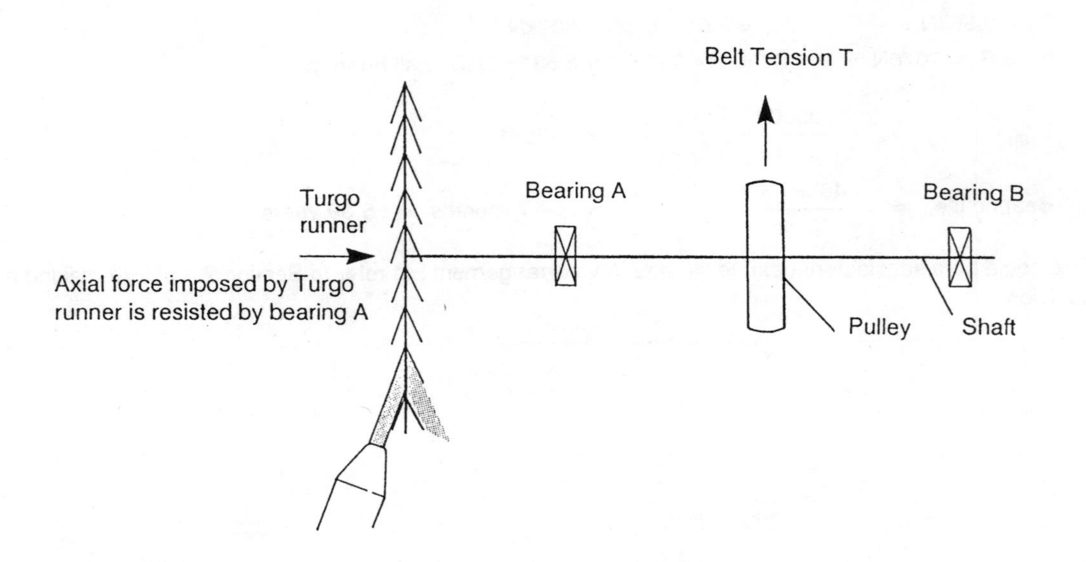

Fig 7.5.9 Turgo bearing arrangement

A Turgo turbine rotating at 450 rpm is to drive an alternator. The general layout of the drive is similar to that in Fig 7.1.3. The turbine shaft and its loads and bearings are shown diagrammatically in Fig 7.5.9. Bearing A is to carry all the axial load as well as its share of the radial load.

Assume the shaft diameter is 70 mm, and T, the belt tension, is 16.64 kN. The Turgo runner imposes a 1.5 kN axial load on bearing A. What bearing should be selected to give a service life of ten years? Ignore radial loads imposed by the Turgo.

The bearing housing can only accommodate a bearing outside diameter of 125 mm.

Use the nomogram in Fig 7.5.8.

The radial load, F_r, on bearing A is T/2 (due to symmetrical bearing siting)

F_r = 8.32 kN

and the axial load is 1.5 kN

F_a = 1.5 kN RPM = 450

Life required L_{10h} = 10 × 8760 hours = 87600 hours.

Using the RPM and L_{10h} in Fig 7.5.8, we can read off the C/P column that the ratio of C / P is 10 for roller bearings and 12 for ball bearings.

Either a deep groove ball bearing or a spherical roller bearing can accommodate this combination of axial and radial loading.

continued overleaf

Bearing selection with axial and radial loads **Example 7.5.3** ***continued***

In the case of deep-groove ball bearings, looking at the bearings with a 70 mm bore, and an outside diameter of 125 mm, we find that a 6214-2RS1 bearing has a C of 60.5 kN.

P, the equivalent load, is calculated using the factors given in Table 7.5.2.

F_r = 8.32 kN

F_a = 1.50 kN

C = 60.50 kN

C_o = 45 kN

$$\frac{F_a}{C_o} = \frac{1.5}{45} = 0.033$$

From Table 7.5.2: e is approximately 0.24

$$\frac{F_a}{F_r} = \frac{1.5}{8.32} = 0.18$$

Therefore

$$\frac{F_a}{F_r} \le e$$

Hence X = 1

and Y = 0

so P = (1 × 8.32) + (0 × 1.5) = 8.32 kN

For a 10-year life C/P must be 12 or greater.

$$\frac{C}{P} = \frac{60.5}{8.32}$$

$$= 7.27$$

Thus, a deep-groove ball bearing cannot be used.

Let's now look at the spherical roller bearing. Figure 7.5.3 gives the details of 70 mm bore spherical roller bearings, and a bearing of outside diameter 125 mm; 22214 CCK has the following characteristics.

C = 148 kN

C_o = 186 kN

e = 0.23, Y_1 = 2.9

$$\frac{F_a}{F_r} = 0.18 \text{ calculated previously}$$

thus

$$\frac{F_a}{F_r} \le e$$

$$P = Fr + Y_1\,Fa$$

$$= 8.32 + (2.9 \times 1.5)$$

$$= 12.67$$

$$\frac{C}{P} = \frac{148}{12.67} = 11.68$$

As this is close to 12 the C/P required for a 10-year life, a 22214 CCK spherical roller bearing is the best choice in this case. See Fig 7.5.9 regarding bearings having tapered bores.

Bearing modifications

When deciding to modify the generator bearings, consider:

a Will the manufacturer do it for you?

b Will you invalidate any clauses in the contract with the supplier if you do the modification yourselves after delivery of the generator?

c Both inner and outer races of cylindrical roller bearings must be fixed axially. Does the generator bearing housing enable this to be done?

d Roller bearings cannot be fitted with internal shields or seals. There is much air flow through a generator for cooling purposes. Consequently lubricant in the bearing is liable to be sucked in or blown out. Does the generator bearing housing enable you to fit seals on BOTH sides of the bearing?

e Deep-grooved ball bearings locate the shaft axially. Cylindrical roller bearings do not. The above modification may be satisfactory if questions a, b, c and d are all answered. A check should also be made that it is a deep-groove ball bearing at the non-drive end and that this bearing is positively located axially on both the outer race and inner race. This check is important because the non-drive end bearing will be the only means of keeping the generator rotor within the stator.

Use of plummer blocks and bearings with adaptor sleeves

A plummer block containing a spherical roller bearing is shown at Fig 7.5.10. Such plummer blocks are ideal for use in micro-hydro schemes because they can eliminate much of the precision work needed on site.

The arrangements shown in figs 7.5.10 and 7.5.11 indicate the simplicity of construction using plummer blocks or flanged housings with bearings with adaptor sleeves.

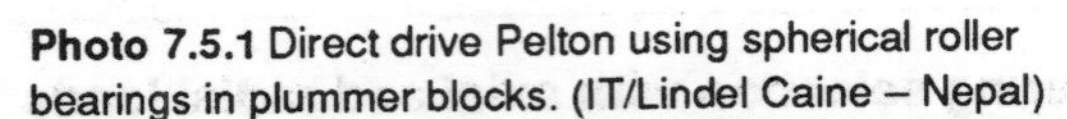

Photo 7.5.1 Direct drive Pelton using spherical roller bearings in plummer blocks. (IT/Lindel Caine – Nepal)

The main features and their advantages are:

- Simple two bolt fixing to a pedestal;
- Fixing bolts go through slots so a little positional adjustment is possible;
- Spherical roller bearings in a plummer block will allow a little angular misalignment but nothing like as much as for spherical roller bearings on their own, because of the seals;
- The plummer blocks have seals which keep lubricant in the bearing but keep dirt, dust and water out;

The bearing shown in Fig 7.5.9 has a tapered bore and an adaptor sleeve. A nut on the adaptor sleeve is tightened against the bearing. This pulls the tapered adaptor sleeve into the tapered bored bearing. The adaptor sleeve is slotted so that it tightens onto the shaft as it is tightened against the tapers. This system enables the use of plain diameter shafts without specially turned bearing seats and shoulders. It is known as the Taper lock method.

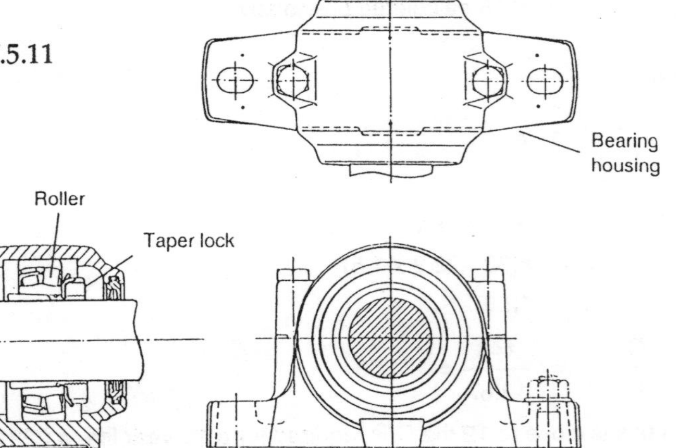

Fig 7.5.10 Plummer block

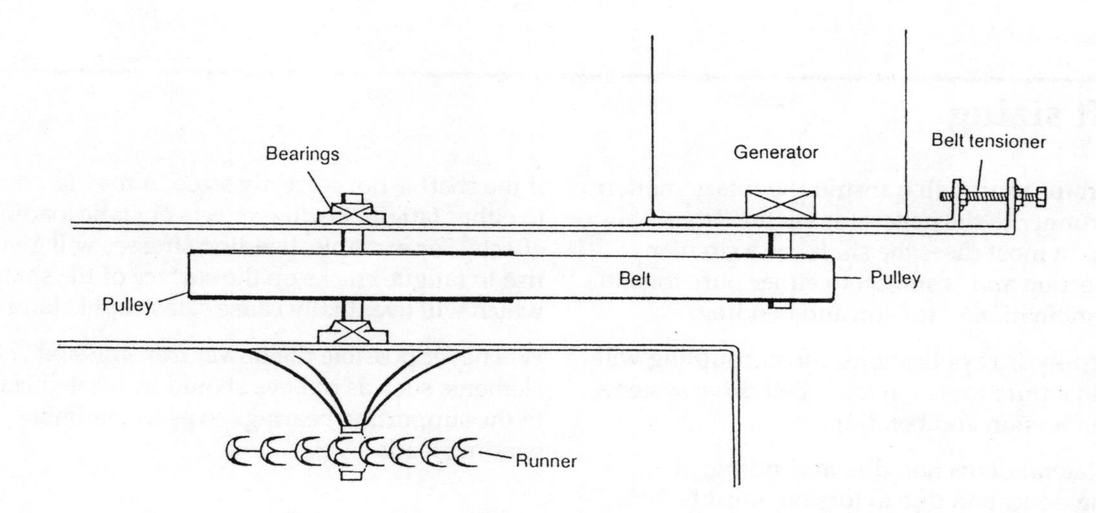

a Vertical turbine and generator showing belt tightening arrangement (the turbine shaft is that shown in **b** below).

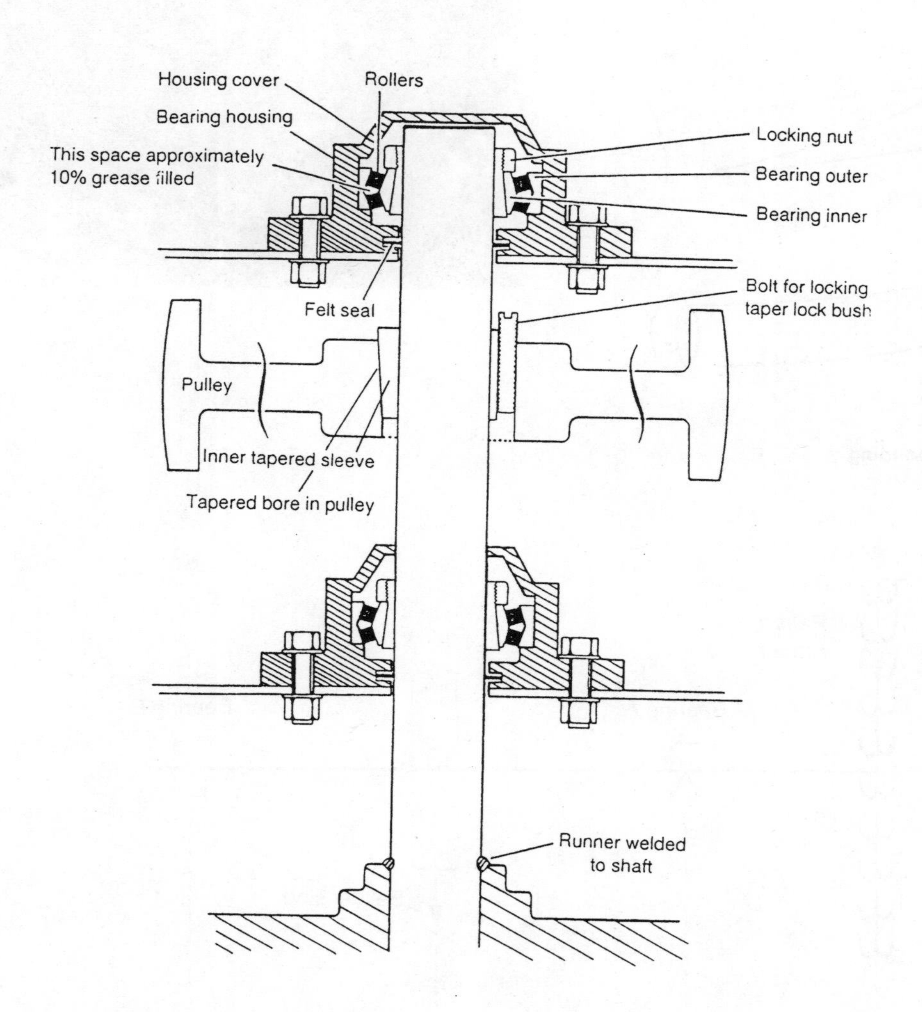

b Vertical turbine shaft mounted in flanged bearing housings (similar to plummer blocks).

Fig 7.5.11 An example of a manufacturer's arrangement for a vertical axis Pelton turbine/generator set.

7.6

Shaft sizing

The turbine shaft will transmit the rotary motion of the runner to the generator via the drive system. In most cases the shaft has a circular cross-section and is subject to either pure torsion or a combination of torsion and bending.

In micro-hydro applications, direct coupling will result in a pure torsional load. Belt drive systems result in torsion and bending.

When lateral deflection due to bending, or twisting deflection due to torsion, must be held to close limits, the shaft should be sized with respect to these deflections. However, the stresses in the shaft must always be analysed to check on the suitability of the size arrived at by a deflection analysis.

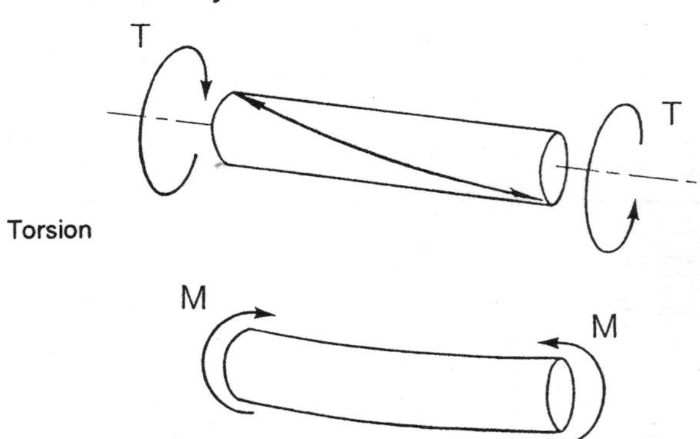

Fig 7.6.1 Torsion and bending

If the shaft is not correctly sized, it may fail due to either fatigue loading effects or static loading effects. For example, bending stresses will give rise to fatigue cracks on the surface of the shaft which will eventually cause catastrophic failure.

Whenever possible the power transmission elements such as pulleys should be located close to the supporting bearings so as to minimize bending loads.

Photo 7.6.1 1935 Boving Pelton with multi-Vee belt drive. (IT/Jeremy Hartley – Sri Lanka)

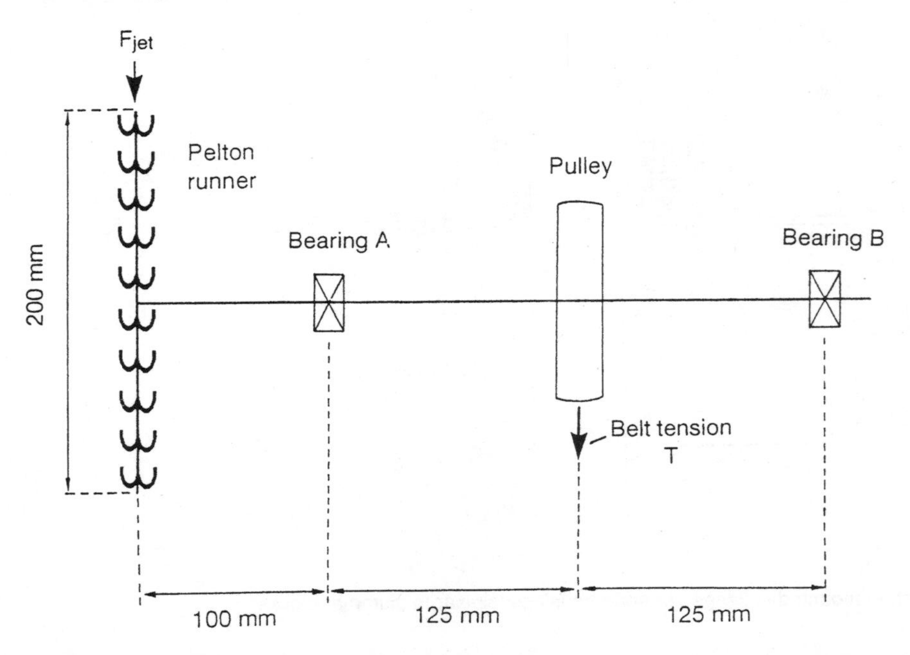

Fig 7.6.2 Pelton turbine and pulley drive. In this example the belt tension and turbine forces act in the same direction.

Calculation of shaft diameter — Example 7.6.1

Consider the Pelton turbine system in Fig 7.6.2.

Suppose that the jet force F_j and belt tension T act in the same direction, and assume that the following is already known:

Belt tension, T = 16.64 kN

Rotational speed of runner = 450 rpm

$= 2\pi \times 450/60$

= 47.12 rad/s

Turbine shaft power = 71 kW

Turbine runner pitch circle diameter (pcd)

= 0.2 m

Thus:

$$\text{Torque} = \frac{71000}{47.12} = 1507 \text{ Nm}$$

$$\text{And } F_j = \frac{\text{Torque}}{0.5\text{ pcd}} = \frac{1507}{0.1} = 15.07 \text{ kN}$$

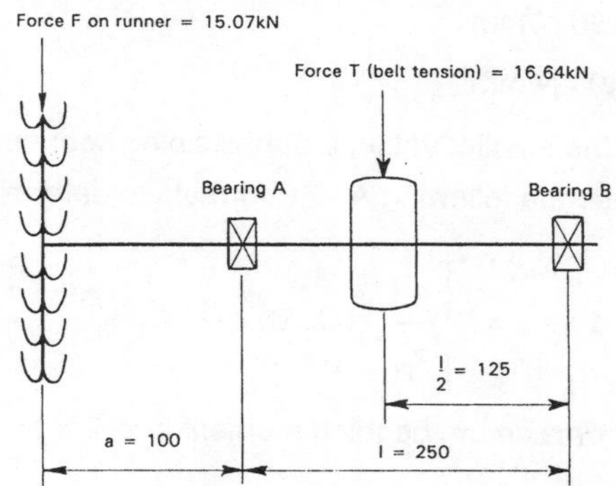

Fig 7.6.3 Force diagram

Note that the self weight of the runner is considered negligible and need not be introduced into the calculation. The force diagram is shown in Fig 7.6.3.

Moments about B:

(remember the sign convention: clockwise positive)

$R_A \times 0.25 - T \times 0.125 - F_j \times 0.35 = 0$

$(R_A \times 0.25) = (16.64 \times 0.125) + (15.07 \times 0.35)$

$R_A = 29.42$ kN

$R_B = 15.07 - 29.42 = -14.35$

The minus sign indicates that R_B is a force acting in the opposite direction from R_A.

We now wish to find the maximum bending moment. Using the following formula:

Bending moment at any point = (The sum of the forces to the left of that point) × (Distance from left to the point):

Moment at runner = 0

Moment at A = $-(15.07 \times 0.1) = -1.507$ kNm

Moment at belt centre = $-(15.07 \times 0.225) + (29.42 \times 0.125) = 0.287$ kNm

Moment at B = $-(16.64 \times 0.125) + (29.42 \times 0.25) - (15.07 \times 0.35) = 0$

Thus, the maximum bending moment is at 'A' and is of magnitude 1.507 kNm.

We can now calculate the necessary shaft diameter. Several formulae are available for this purpose. We will use a historical approach based on a code established by the ASME (American Society of Mechanical Engineers) in 1927. This method is based on maximum-shear-stress theory and will tend to give a conservative result.

The ASME code defines a permissible shear stress which is the smaller of the following values:

a $t_p = 0.30\ S_y$ or

b $t_p = 0.18\ S_{ut}$

where S_y is the yield stress of the shaft material and S_{ut} is the ultimate tensile stress.

These stresses should be reduced by 25% if there is a possibility of stress concentration caused by, for instance, a shoulder fillet or keyway.

continued overleaf

Example 7.6.1 *continued* **Calculation of shaft diameter**

For a high quality steel the yield strength S_y is approximately 300 N/mm² and the ultimate tensile strength S_{ut} is about 550 N/mm². Thus,

a $t_p = 90$ N/mm²

b $t_p = 99$ N/mm²

We must take the smaller value, a permissible shear stress of $t_p = 90$ N/mm².

We can now use the following ASME formula to determine shaft diameter:

$$d = \left[\frac{5.1}{t_p}\left\{(C_m M)^2 + (C_t T)^2\right\}^{0.5}\right]^{0.33}$$

Where M is our maximum bending moment and T is the torque we have calculated.

This formula may be rewritten:

$$d^3 = \frac{5.1}{t_p}\left\{(C_m M)^2 + (C_t T)^2\right\}^{0.5}$$

This form is preferred by many people because it looks simpler.

(NOTE that the units must be consistent: if t_p is N/mm² then M and T must be Nmm.)

The shock factor C_m and fatigue factor C_t are given in Table 7.6.1. Their values depend on the expected conditions of service of the shaft.

For belt drives, the shaft is rotating and the load is considered steady. Thus,

$$C_m = 1.5 \quad \text{and } C_t = 1$$

Hence

$$d^3 = \frac{5.1}{90}\left\{(1.5 \times 1.507\text{ E6})^2 + (1 \times 1.567\text{ E6})^2\right\}^{0.5}$$

$$d^3 = 155900$$

$$d = 53.6 \text{ mm}$$

(Note that units have been kept in Newtons and millimetres to give an answer direct in millimetres).

Thus, we choose a shaft of 55 mm diameter as the nearest standard size above our calculated value. It is important to err on the side of caution in shaft selection, but remember that there is a very steep relationship between shaft strength and diameter. A shaft with an extra 5 mm of diameter will be very much stronger; the increase in strength will be approximately proportional to d to the power of 3.

Table 7.6.1 Values of bending-moment C_m and torsional moment factor C_t

	Type of loading	C_m	C_t
Stationary shaft	Load applied gradually	1.0	1.0
	Load applied suddenly	1.5 - 2.0	1.5 - 2.0
Rotating shaft	Load applied gradually	1.5	1.0
	Steady load:	1.5	1.0
	Load applied suddenly, minor shocks	1.5 - 2.0	1.0 - 1.5
	Load applied suddenly, heavy shocks	2.0 - 3.0	1.5 - 3.0

7.7 Deflection of shafts

It is normal to calculate the size of shafts from strength considerations as shown in Example 7.6.1. It is advisable to check that such a shaft will deflect only within allowable limits.

The maximum allowable deflection for machine shafts may be taken as 0.0005 times the distance between bearings. No position on the shaft must deflect more than that amount irrespective of whether the position is between or outside the bearing centres.

Shaft deflections are calculated by the method used for beams. Most engineering handbooks contain diagrams and formulae for common arrangements.

A common way of calculating deflections for arrangements where there is more than one load is to calculate deflections separately for each load in turn and then to add the deflections algebraically. Care must be taken to use + and – signs correctly to signify the direction of the deflection.

Shaft deflection calculation

The arrangement of bearings and loads used in the previous example is repeated in Fig 7.7.1(a). Fig 7.7.1(b) shows that arrangement as it would deflect under the belt tension load only. Fig 7.7.1(c) shows the same arrangement as it would deflect under the turbine runner force only. The deflection under both loads acting together is the algebraic sum of the separately obtained deflections.

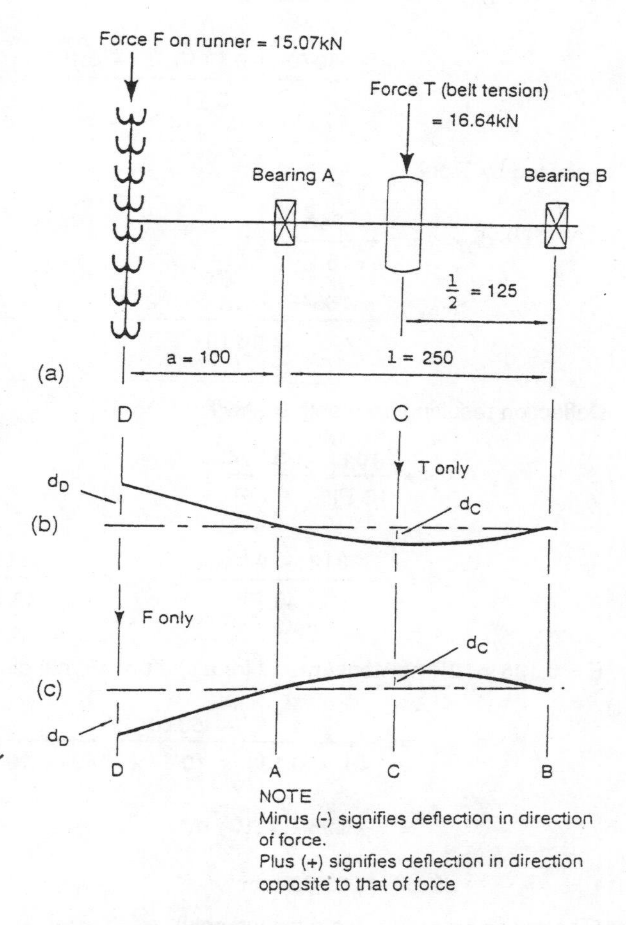

Fig 7.7.1 Shaft deflections

Shaft deflection — Note 7.7.1

E is the modulus of elasticity of the shaft material $= 0.196 \times 10^{12}$ N/m².

I is the moment of inertia of the shaft and is given by $I = \dfrac{\pi D^4}{64}$

Formulae for calculating the deflections in Fig 7.7.1 are:

$$d_{CT} = -\frac{Tl^3}{48EI} \qquad d_{DT} = +\frac{Tal^2}{16EI}$$

$$d_{CF} = +\frac{Fal^2}{16EI} \qquad d_{DF} = -\frac{Fa^2(l + a)}{3EI}$$

Example 7.7.1 **Shaft deflection**

Determine the deflection (d) at the turbine runner for the arrangement shown in Fig 7.7.1. Refer to Note 7.7.1 for symbols.

d caused by F only:

$$d_{DF} = -\frac{Fa^2(l + a)}{3EI}$$

$$= -\frac{15070 \times 0.1^2\ (0.25 + 0.1)}{3\ EI} = -\frac{52.74}{3\ EI}$$

d caused by T only:

$$d_{DT} = +\frac{Tal^2}{16\ EI}$$

$$= +\frac{16640 \times 0.1 \times 0.25^2}{16\ EI} = +\frac{104}{16\ EI}$$

Deflection resulting from both F and T:

$$+\frac{104}{16\ EI} - \frac{52.74}{3\ EI}$$

$$= +\frac{312 - 843.84}{48\ EI} = -\frac{531.84}{48\ EI}$$

$E = 0.196 \times 10^{12}$ newtons/m^2. I for a shaft of 55 mm dia. = 4.49×10^{-7} m^4.

$$d = \frac{-531.84}{48 \times 0.196 \times 10^{12} \times 4.49 \times 10^{7}}$$

$$= -1.259 \times 10^{-4}\ m$$

$$= -0.126\ mm$$

(The minus sign indicates direction only)

This deflection is not acceptable as allowable deflection is $0.0005 \times 250 = 0.125$ mm.
It would be necessary, in this case, to redesign using a reduced 'e', or a larger shaft, for example.

Exercise

Check that the deflection at C is less than at D.

7.8

Achieving alignments

The turbine casing will be rigidly connected to the penstock. It is therefore not possible to move the turbine, so all adjustments must be done on the other components.

Alignment of pulleys (see Fig 7.8.1)

First, sight along one side of the pulleys as close to the centre as the shaft will allow. Adjust the position of the pulley by moving the pulley or pulley and generator until all four points, a b c d, are in line. Usually, a straight edge or taut string will be needed to check alignment.

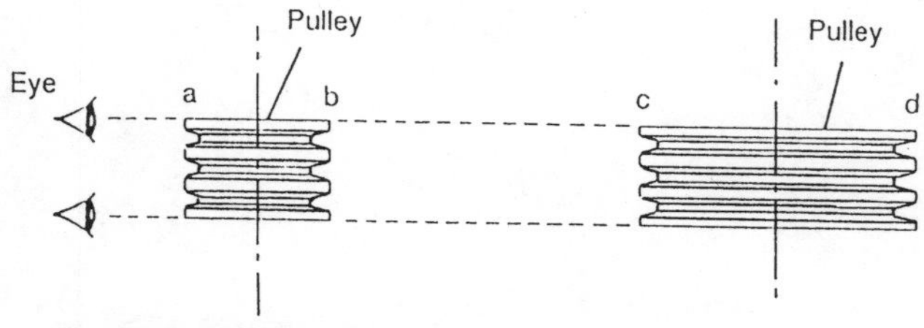

Fig 7.8.1 Alignment of pulleys

When this is satisfactory, check on the other side of the pulley. Rotate each shaft half a revolution and check again. If points a, b, c and d are no longer in line, it is possible that one or both of the pulleys are not assembled correctly onto the shafts.

With care an accuracy of 0.25 mm is possible.

Alignment of shafts

Shafts are connected by flexible couplings and these have to be aligned to within the specification set by the coupling supplier. Typical tolerances are 0.05 mm axially and less than one degree of angular tolerance.

These tolerances are difficult to achieve without the proper equipment, and if they are not achieved, the results will be premature failure of one or more components. Stresses will be set up in bearings and fatigue stresses set up in shafts, causing cracks to develop. Vibrations set up may cause mountings to fail.

A dial gauge is normally required, and procedures for setting up are laid out in suppliers' catalogues.

Once set up correctly, a direct drive system is almost maintenance-free.

7.9

Plinths and frames

Every force has an equal and opposite reaction. The turbine casing must be fixed so that the rotation and power is transferred to the generator shaft. The generator stator and casing must also be fixed; otherwise they would rotate with the rotor.

It therefore follows that the turbine and the generator, and any extra shafts and bearings, must be bolted down. The actual positions in which they are bolted must have adjustments to correct alignments and to permit assembly and tensioning of belts.

The simplest system is the direct coupled drive shown in Fig 7.1.1. Such a system may be bolted onto a concrete floor which has a plinth to compensate for the different heights of components. Such an arrangement is shown in Fig 7.9.1.

No concrete plinth is required for the installation of the machines where belts are used but there must be a good concrete floor to which the items may be bolted.

Floors and plinths should be made with good quality concrete. It is recommended that you use:

1 part cement

2-3 parts coarse sand

4-6 parts gravel or crushed stone between 6 mm and 25 mm.

Typically, the concrete floor should be 150 mm thick on a well compacted bed of rubble.

The diameter of the holding-down bolt may be ascertained from the literature on the machine which is being installed.

Holding-down bolts which can be made locally and used for bolting machines to concrete floors, are shown in Fig 7.9.2. Their method of use is given in Note 7.9.1. Many types of manufactured holding-down bolts can be purchased but they offer little advantage over those illustrated.

An alternative to the welded washer and plates shown is to cut a slot in the bottom of the bolt and carefully splay the end.

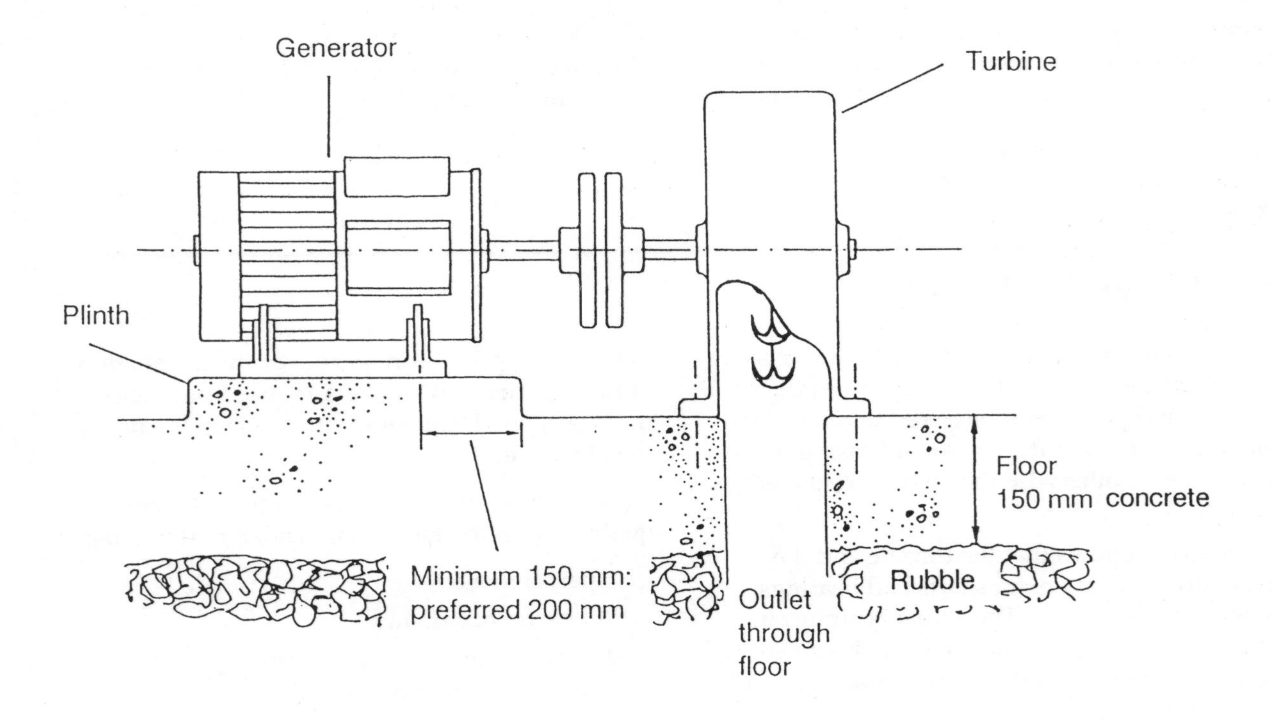

Fig 7.9.1 Concrete floor and plinth The floor and plinth should be made with good quality concrete. It is recommended that you use:1 part cement, 2-3 parts coarse sand, 4-6 parts gravel or crushed stone. The concrete floor should be 150 mm thick on a well compacted bed of rubble.

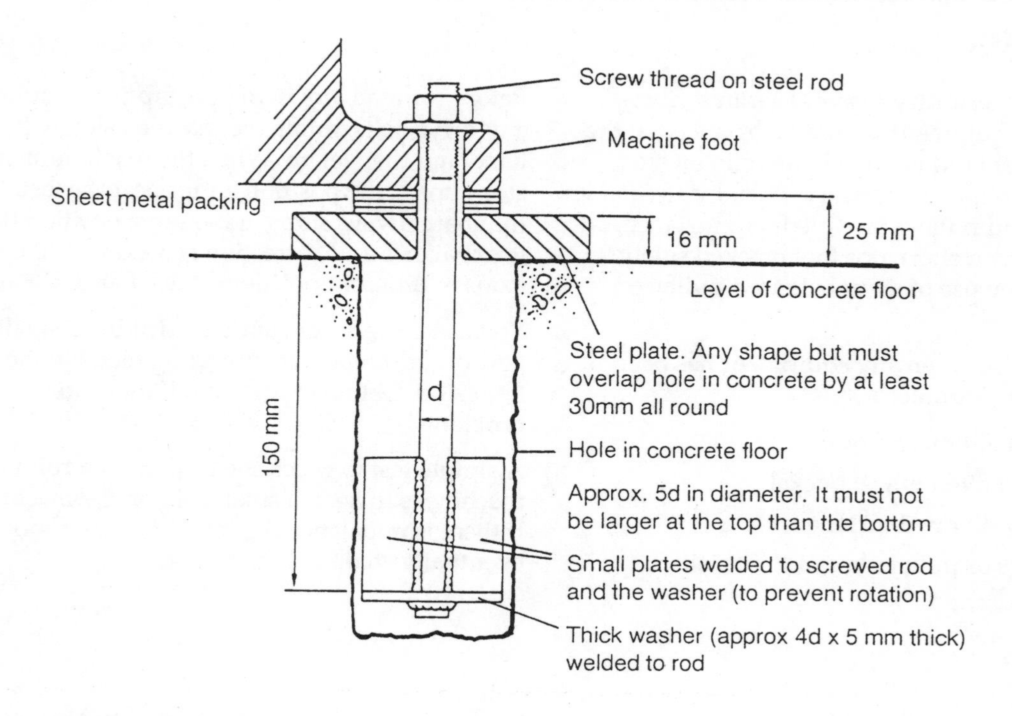

Fig 7.9.2 Holding down bolts The diameter of the holding-down bolt may be ascertained from literature on the machine which is being installed. Alternative to the welded washer and plates shown is to cut a slot in the bottom of the bolt and carefully splay the end.

Use of holding-down bolts — Note 7.9.1

1. Make holes in the concrete to about 170 mm depth. The holes must be in the correct positions.
2. Position the holding-down bolts in each hole using small stones to obtain the correct height.
3. Position the machine accurately in its correct setting relative to other machines or structures. Use temporary packing and wedges so arranged that access may be obtained to the machine feet and the holes.
4. Assemble nuts onto the bolts loosely.
5. Work a concrete mix into the holes ensuring that there are no voids.
6. Leave to cure.
7. Undo the nuts, lift machine and remove temporary supports.
8. Install the steel plates and the sheet metal packing to the designed thickness.
9. Lower the machine. Reassemble nuts and tighten.
10. Check position of machine and alignments.
11. Adjust thickness of sheet metal packing and use slots in machine feet to obtain correct position and alignment.
12. Finally tighten down and recheck alignment. This is the arrangement shown in Fig 7.1.2.
The turbine must connect to the penstock and water outlet so its position is fixed. The generator must consequently be able to slide in order to install the belts and tension them. The most convenient way as previously mentioned is to use motor slide rails under the generator. These slide rails are bolted to the concrete floor.

7.10
Balancing

Balancing is necessary to avoid serious vibrations. New equipment should be balanced at the factory and should not need attention on site. Special care may be necessary when locally made couplings and pulleys are fitted. For instance, on one occasion an alternator foot cracked solely because of the use of an unbalanced pulley on the shaft.

Problems arise when any equipment has been repaired; for example:

- rewound alternator rotor
- replaced Pelton wheel bucket
- any reconditioned turbine rotor
- couplings or pulleys bored to fit a shaft.

Before running any item of equipment, remove any drive belts, rotate the machine slowly by hand and then let go. When the machine shaft stops, mark the position and repeat the test. If the shaft always stops in the same position there is imbalance and correction is needed. If the shaft stops in different positions it is balanced.

You can correct imbalance by drilling a small hole on the *heavy* side or adding metal to the light side. Obtain expert help if there are serious problems.

A simple test to check the balance of a running machine is to keep a small coin on the machine. If there is imbalance the coin will move and will eventually shake off the machine.

7.11
Safety and guards

Most countries have laws relating to the safe installation and use of equipment. Clearly it is prudent to find out about the law in your own country and obey it to the full. In any case it is good design practice to guard machinery so that no person can possibly touch any moving parts. You should consider the old or infirm who may fall or faint and the very young who could crawl under or even into equipment.

You should also consider safety during maintenance.

Photo 7.11.1 Expanded mesh guard. (Sri Lanka)

Note 7.11.1 **Safety check**

A test to show that guards are probably adequate is to:

a Switch off all electricity and turn off water.

b Positively lock all moving parts so they cannot move.

c You yourself must deliberately try and twist and stretch to touch any parts that would normally move.

If you are satisfied that you cannot get close to anything that could move:

d Ask someone else to try to touch any parts that would normally move as a check.

You must improve safety if you or anyone else can reach close to any parts or material which is likely to move.

Electrical power

8

8.1 Introduction

Machinery in many grain mills is driven directly from a water turbine. It would be less efficient and far more costly to generate electrical power first and use that power to drive a motor which would then drive the mill. However, there are real advantages in generating electrical power, for instance:

- readily available electrical equipment (such as lights, radios, refrigerators) can be used;
- different appliances can be situated in different, more convenient places;
- the control of electrical equipment is so easy and effective that the ability to exercise such control very often offsets the inefficiency of generating electrical power and then using a motor to convert back to mechanical power where required.

No attempt is made here, however, to provide guidance as to the type of drive, mechanical or electrical, which should be used with a concentrated load near a water turbine. Each case must be considered on its merits. The present chapter concerns situations where the electrical system has been chosen because of the advantages above.

It is normally necessary to contract in a fully qualified and experienced electrical engineer to design and supervise the installation of the electrical system. At the same time, it is equally important that the local engineer responsible for the overall design of the scheme has a full appreciation of the specification and purchase of all the electrical and safety equipment used. Accordingly this chapter describes a full range of design choices and relevant calculation procedures. The chapter is divided into the following sections:

- Basic electricity
- Choosing the supply system: ac or dc? (this includes a section on battery charging)
- ac systems (choice of single phase or 3-phase; sizing of the generator from power factor calculations)
- Synchronous generators (how to specify and purchase; types of voltage regulators; use of electronic load controllers)
- Stand-alone induction generators (how they work; how to select a motor and convert it to a generator)
- Switch gear and protection (choosing and sizing)
- Transmission (sizing of transmission cables)

8.2
Basic electricity

Electricity is a convenient way of transporting energy. At the delivery end the energy is converted into a useful form, such as light, or heat, or rotating shafts. The driving force which causes the electricity to flow is known as the 'potential difference' or voltage, measured in volts (V). The flow itself is current, which has the symbol I, and is measured in amps (A). The rate at which energy is transported and converted is the power, which is a combination of both voltage and current; this is very similar to the way in which the power derived from falling water in a penstock is a combination of the head and flow.

The quantity of current flowing in a wire or a device such as a light bulb depends on two things: the voltage driving it, and the resistance of the material through which it is conducted. Any conductor of electricity has a resistance (symbol R) which is measured in ohms (Ω). The current is the voltage divided by the resistance:

$$I = \frac{V}{R} \text{ amps (A)}$$

Electrical fields and magnetic fields often occur in electrical circuits. These absorb energy. An electrical field is created by a capacitor, and the extent to which energy is absorbed in this way is known as capacitance (symbol C), measured in farads (F). Magnetic fields are caused by coils or 'inductors' and the extent of absorption of energy in this case is inductance (symbol L) measured in henrys (H).

The energy absorbed in electrical and magnetic fields circulates within the electrical system but it does no useful work. The power triangle shows how the electrical system accomodates this circulation of energy, known as the "reactive power" of the circuit, as well as useful power, or "real" power. The practical effect of this is that circuit components, such as the generator, must be large enough to accomodate the resultant power, known as "apparent" power, which is measured not in watts but in volt-amps (VA).

The apparent power is simply the voltage across the circuit multiplied by the current flowing:

Apparent power
$= V \times I$ volt-amps (VA)

The real power is the apparent power reduced by a factor:

Real power
$= V \times I \times \text{Power factor}$
$= V \times I \times Pf$ watts (W)

Not all the real power in a circuit arrives for use by the consumer. Some will be lost in inefficiencies like heating of wires. For instance, if there is a long transmission cable, it will have a large resistance, and some voltage is needed to overcome this. This is known as the 'volt drop' across the cable. Heat is generated by the flow of current (I) overcoming the resistance of the cable:

Real power loss in cable
$= V_{drop} \times I$ watts (W)

If the resistance of the cable (R_{cable}) is known, then it is not necessary to calculate the volt drop, since we know that for the cable $R = V/I$, or $V = IR$ (This is 'Ohm's law').
Therefore substituting:

Real power loss in cable
$= I \times I \times R_{cable} = I^2 R_{cable}$ watts (W)

"I^2R" losses like this are common for most components of an electrical system. At the end of the day the electrical power system must be designed with both real power losses and reactive power capacity in mind.

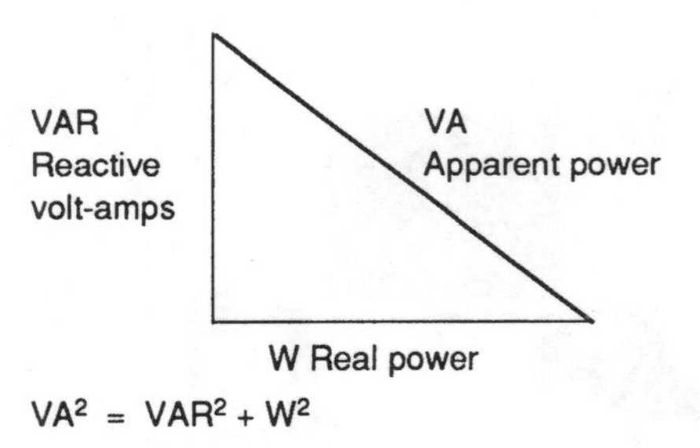

$VA^2 = VAR^2 + W^2$

ac and dc

Electrical generators can produce either alternating current (ac) or direct current (dc). In the case of ac, a voltage cycles sinusoidally with time, from a positive peak value to a negative peak value. Because the voltage changes its sign, the current resulting also reverses direction in a cyclic pattern. dc current flows in a single direction as the result of a steady voltage.

Alternating voltage is produced in a coil which is rotated in a magnetic field. This is shown by Figs 8.5.1 and 8.5.2. The magnetic field can either be produced by a permanent magnet or by another coil (an electro-magnet), known as a field coil, which is fed by direct current. A machine or generator supplying alternating current is known as an alternator, to distinguish it from a machine designed to supply dc current which is known as a dc generator or dynamo.

Phase angle

Current flows when a voltage difference is placed across a conducting body. In ac circuits the magnitude and the timing of the current cycle relative to the voltage cycle will depend on whether the body is resistive, inductive, capacitive, or some combination of these elements.

Fig 8.2.1 shows how in the first three cases respectively the current cycle either stays in phase with the voltage, runs behind ('lags') by a phase angle of 90°, or leads the voltage by a phase angle of 90°. Circuits in which the load causes the current and voltage to be out of phase are said to be 'reactive circuits'. Generally a load is a combination of resistance, capacitance, and inductance, described by the term *impedance* (symbol Z), and causes a phase difference between current and voltage of angle ϕ.

Power factor

Fig 8.2.1**d** shows the current-voltage characteristic of a circuit where the load causes the current (I) to lag the voltage (V) by an angle ϕ. The load in this case may consist of both capacitive and inductive components but predominantly inductive. Because the current and voltage are not perfectly in phase the useful power available is reduced by the power factor. The power factor is the cosine of the phase angle ϕ:

Power factor $= \cos \phi$
Real power $= VI \cos \phi$

The power usefully consumed by the load is VI cos ϕ, although the power supplied is VI.
A typical electrical system might consist of a generator supplying a workshop with 10 kW of useful power for lights and motors. The power factor of this kind of load could be about 0.8, implying a current lag behind the voltage of 37°

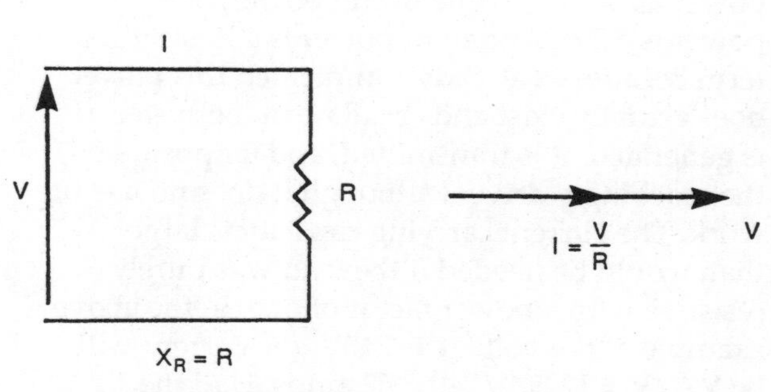

a Resistance (V and I in phase)

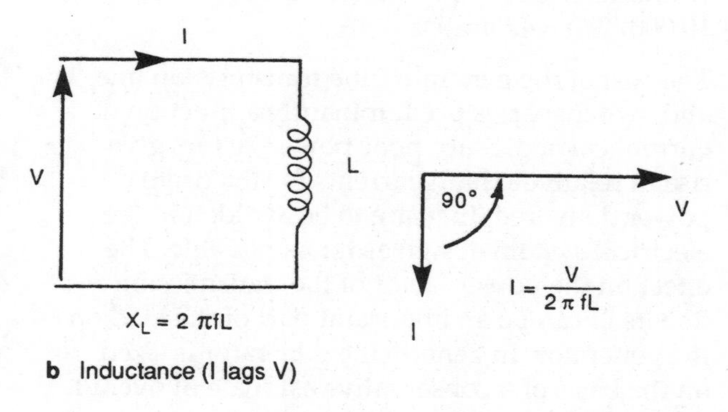

b Inductance (I lags V)

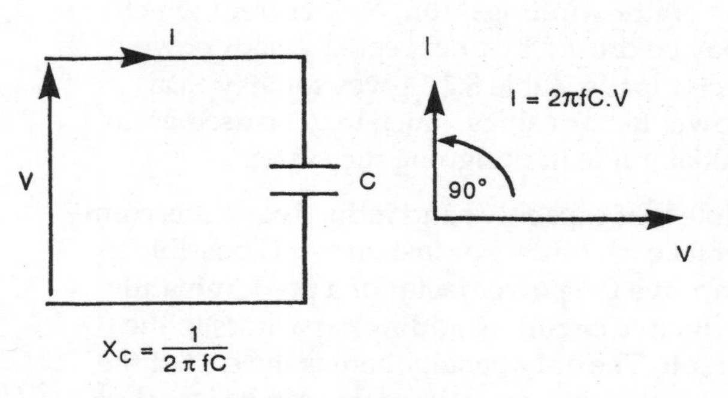

c Capacitance (I leads V)

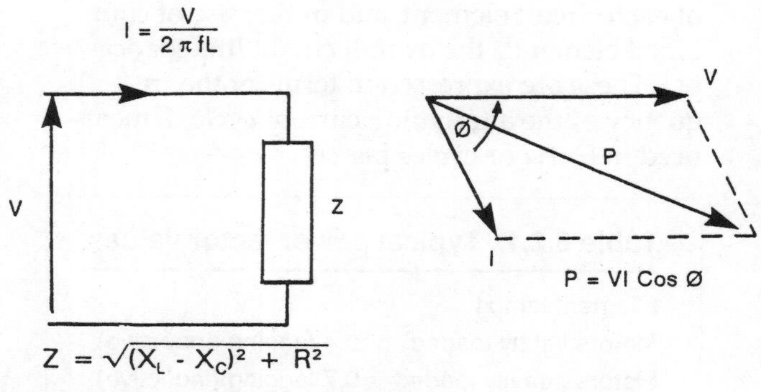

d Reactance (combined effect)

Fig 8.2.1 Phase angles

(cos 37° = 0.8). Since the useful power is VI cos ϕ, or VI × 0.8, the apparent power is 10 kW/0.8 = 12.5 kVA. The convention is to refer to apparent power as 'kVA', while useful power, or 'real' power is 'kW'. Apparent power is a misleading term because as already commented this power does actually exist and circulate in the system; it is generated, it is transmitted, and it appears in the machinery driven, although it does no useful work. The current carrying capacity is larger than would be needed if the load was purely resistive with a power factor of one. In the above example, if the voltage is 240V, the current will be VA/V = 12,500/240 = 52 amps. Had the 10 kW load been purely resistive, for instance all filament lights and no motors, the power factor would have been one and the current equal to 10,000/240 = 42 amps.

The cost of the generator, the transmission line, and switchgear, is predominantly a function of current carried. Since poor power factors give rise to relatively high currents for the useful power delivered, they are to be avoided in the electrical system design as far as possible. The effect on the power factor of the transmission line itself can be an important part of the load on the generator. In general the generator is sized on the basis of a conservative estimate of overall power factor, and care is taken to protect the generator windings from over current which may be caused by unexpectedly poor power factor loads. Table 8.2.1 gives some typical power factor values which may be used as an initial guide in designing the system.

Note that capacitive and inductive effects counteract each other. For instance, it is possible to improve the power factor of a predominantly inductive circuit by adding capacitors to the circuit. The only penalty here is the cost of the capacitors, but sometimes this can be less than the cost of increasing the supply current.

Fig 8.2.1 gives expressions for the reactance (X) of each circuit element, and in the case of combined elements, the overall circuit impedance (Z). These are expressed in terms of the frequency of the alternating current cycle, f, measured in Hertz or cycles per sec.

Table 8.2.1 Typical power factor values

Filament lamps	1
Motors lightly loaded	0.4 lagging (inductive)
Motors heavily loaded	0.7 lagging (inductive)
Fluorescent lamps	0.5 – 0.7 lagging (inductive)
Overhead line	0.9 lagging (inductive)

3-phase, star and delta

So far alternating current has been described as induced by a single rotating coil in a magnetic field. This results in the single waveform shown in Fig 8.2.2. An alternator operating in this way is a single phase machine. Three rotating coils evenly spaced on the armature of an alternator produce three such waveforms displaced from each other by a phase angle of 120° – this is a 3-phase alternator (see Fig 8.5.3).

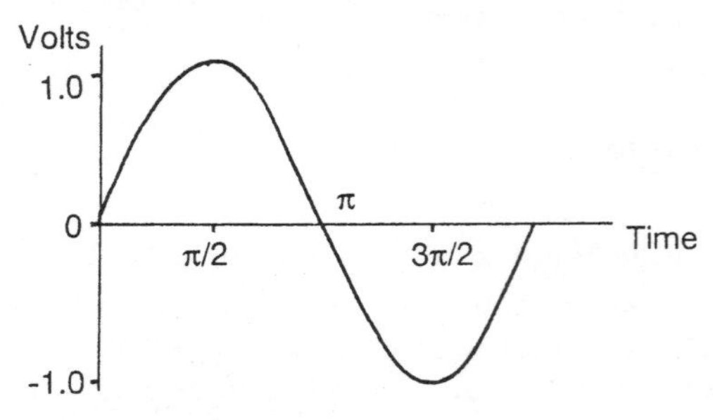

a Single phase ac waveform

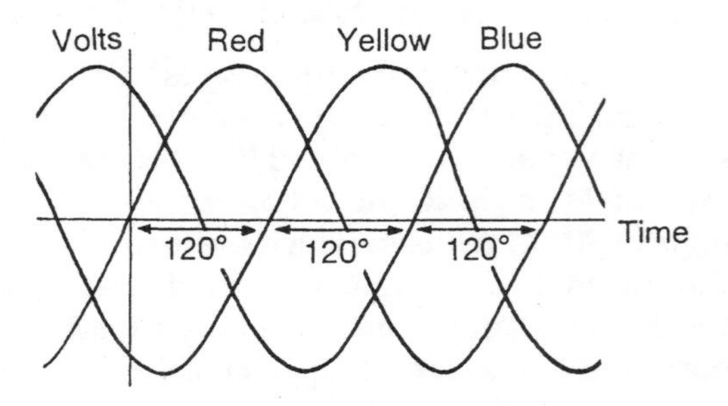

b 3-phase ac waveforms. Each wave is displaced by 120° (or 2π/3 radians) from the next.

Fig 8.2.2 ac waveforms

Fig 8.2.3 shows that the three coils can be connected in one of two ways, either the 'star' configuration, or the 'delta' configuration. The star system as shown is known as the '4 wire star connection' since it has a neutral wire in addition to the wires carrying each of the three phases. In contrast the delta system is always a 3-wire system.

Fig 8.2.3 also shows that the load in a 3-phase system is connected either in the 4-wire star mode or in delta mode. The load in both cases is split into three parts, each connected to one of the alternator phases. In the star system the current drawn by each part of the load (I_{phase}) is clearly equal to the current in the connected line (I_{line}).

If each of the three loads have exactly the same impedance, then phasor diagram analysis shows that the voltage between two lines (V_{line}) is larger than the voltage across a phase (V_{phase}) by a factor of $\sqrt{3}$. The reverse holds for the delta system, as summarized in Fig 8.2.3 below. The kVA ('apparent' power) drawn by the complete load is the same in both the star and the delta systems. It is calculated as the phase voltage and phase current product, multiplied by 3 for the three phases. The real, or useful, power, in kW, is this power multiplied by the load power factor.

The advantage of 3-phase systems is that power is delivered more efficiently. A 3-phase transmission line is cheaper per unit power delivered. The disadvantage is that the load must be split into three balanced parts. Imbalances in the load cause larger currents to flow. For this reason 3-phase systems are rarely used in electrical systems where the full generated power is below 10 kW, since the switching in and out of individual loads will tend to unbalance the phases.

A further factor is that switchgear and control equipment (for instance, over-current protection) tends to be more expensive in a 3-phase system since it is more complex. This last factor is offset by the distinct economic advantages of a 3-phase system:

1. A 3-phase transmission line requires less copper to transmit the same power, and so is less costly.
2. Motors rated over 5 kW are generally only available as 3-phase; they are wired internally to make use of each of the 3 phases.
3. 3-phase motors are usually smaller, cheaper and more easily available than motors of equivalent power wound for single phase.
4. 3-phase alternators are also cheaper than single phase ones.

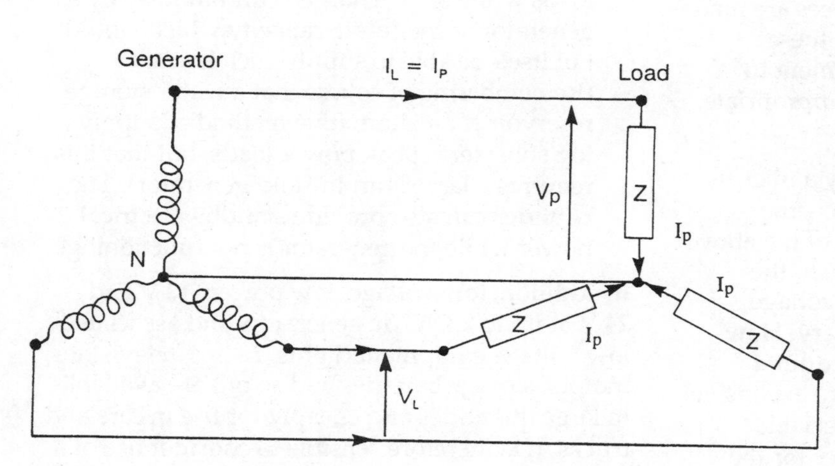

Line voltage $= V_L$

Phase voltage $V_p = \dfrac{V_L}{\sqrt{3}}$

Phase current $I_p =$ line current I_L

Useful power drawn $P = 3\,V_p I_p \cos\phi$

$= \sqrt{3}\,V_L I_L \cos\phi$

a Star system: balanced load

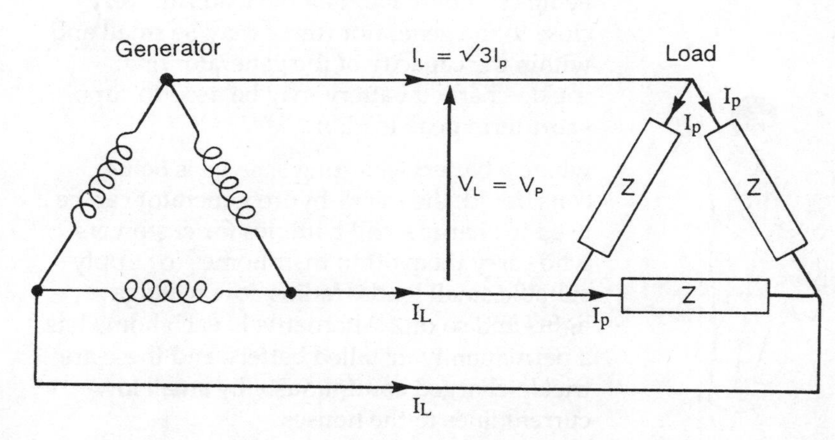

Line voltage $= V_L$

Phase voltage $V_p =$ line voltage V_L

Phase current $I_p = \dfrac{I_L}{\sqrt{3}}$

Useful power drawn $P = 3\,V_p I_p \cos\phi$

$= \sqrt{3}\,V_L I_L \cos\phi$

b Delta system: balanced load

Fig 8.2.3 Alternative 3-phase connections

8.3

Choosing the supply system: ac or dc?

In industrialized countries local distribution systems, usually fed from a countrywide 'grid' electricity supply network, normally operate at a standard voltage – 110 V, 200 V, 220 V or 240 V. They are usually alternating current (ac) at 50 or 60 cycles per sec (Hz). Such electricity supply networks have developed as ac systems because the voltage can be changed easily and cheaply using transformers. By changing to high voltage the sizes of the transmission wires are reduced and costs lowered. So bulk supply from large power stations over considerable distances in the grid system is at very high voltage (in the region of 400 kV) and this is 'transformed' to the standard voltages for distribution.

As electrical equipment and appliances are made in large numbers and so cheaply for these frequencies and voltages, it is convenient to adopt one of these ac standards, as appropriate, in a micro-hydro scheme.

Theoretically, there is no reason why a micro-hydro generation and distribution scheme should not be dc and operate at one of the above voltages, since in a limited area close to the generator (say within 1000 metres), voltage transformation would not be necessary. However, suitable dc generators of the required voltage and size (say above 2 kW) are expensive and have brush gear requiring appreciable maintenance. In addition, dc switches for the voltages and currents concerned are more expensive than their ac equivalents and many ac appliances will not operate satisfactorily on dc.

Fig 8.3.1 Battery charging as a business

So, for these and other reasons, a dc micro-hydro system at the above standard voltages should not normally be considered; an ac system should be used.

Despite this, a dc micro-hydro system has two distinct advantages over an ac system:

- The exact speed at which the dc generator is driven does not matter providing that its automatic voltage regulator (AVR) causes it to produce the required voltage. As a result, no governor or load controller is required and there may be a considerable saving in cost.
- Storage batteries may be used to store electrical energy and so supply substantial peak loads while being charged continuously by a generator of moderate capacity which would not itself be able to supply such loads. (Remembering of course that a water storage reservoir is an alternative method of supplying short term peak power loads, but that this requires a larger turbine and generator). The batteries can also provide standby electrical power while the generator is not functioning.

In addition, low voltage, low power (12 V and 24 V at up to 2 kW) dc generators and associated low voltage equipment (lights, radios, televisions, motors, storage batteries and so on) are available in large quantities and cheaply for use in cars and trucks. It is therefore sensible to consider use of a low voltage, low power dc micro-hydro system in the following circumstances:

- where a small scheme (less than 2 kW) is being considered and all the loads are very close to the generator (these may be small and within the capacity of the generator or a trickle-charged battery may be used to supply short-term peak loads);
- where a battery-charging scheme is being considered: the micro-hydro generator can be used to charge small batteries for customers who carry them off to their homes to supply suitable small loads (radios, low wattage lights and so on). Alternatively each home has a permanently installed battery and these are trickle-charged continuously by small low-current lines to the houses;
- where the turbine is one of many energy sources providing dc to a common battery store. This can be linked to remote loads, but this is an expensive arrangement (Fig 8.3.2).

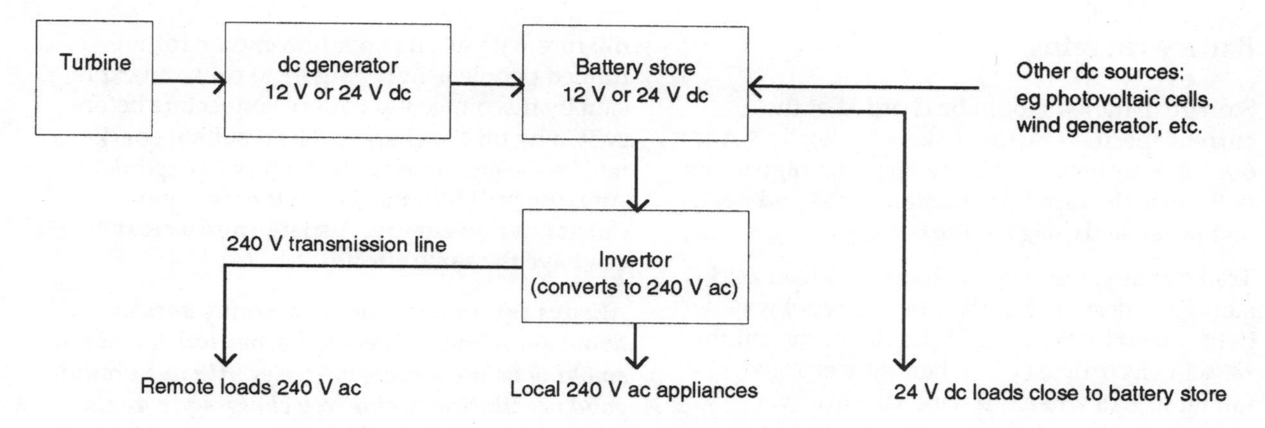

Fig 8.3.2 dc system using an invertor to supply remote ac loads

ac systems versus dc battery-charging systems — Note 8.3.1

ac

Advantages

- Virtually all national electrical supply systems are ac and therefore electrical appliances of all types are produced in large numbers and cheaply for those systems.
- Synchronous generators (Section 8.5) and induction generators (Section 8.6) are manufactured in large numbers, and are readily available and inexpensive.
- Transformers can be used to change the supply voltage by a fixed ratio easily and cheaply. This means that transmission losses over appreciable distances can be minimized by using a high-voltage transmission line (Section 8.8).
- Current interruption is easier than with dc and so switches and circuit breakers are cheaper (Section 8.7).
- Most of the advantages of dc battery-charging systems can be enjoyed by an ac system which delivers power to batteries through transformer/rectifier units ('charging units'). This is an *ac-driven combined system* and is often the best choice. It is usually loaded with both ac loads and either one battery charger for people to carry batteries to for recharging, or many battery chargers which trickle-charge permanently installed batteries.

Disadvantages

- The system frequency must be kept constant within limits, especially if appliances using induction motors are used. So some form of generator speed governing control must be used in addition to a voltage regulator.
- Direct energy storage is not possible as in the case with the use of storage batteries in dc systems.

dc battery-charging systems

Advantages

- Good returns on investment are possible if the system is well designed and looked after. This is because a high load factor can be obtained if the demand is correctly assessed and, in addition, there are no distribution costs.
- The units will have a reasonable secondhand value and their small size will make them easily movable. This is particularly relevant for villages where grid connection is anticipated or for communities that may be required to move.
- Site survey requirements are less rigorous since the turbine/generator speed can be varied within certain limits and under low flow conditions the system can be turned on and off, providing that a small amount of water storage can be created.
- This cycling approach would enable the system always to be run near its best efficiency. This is especially advantageous if turbines with poor part-flow efficiency are used, such as pumps as turbines, or generators with poor part-load efficiency.
- Villages where there is insufficient potential for a standard scheme can be supplied using this approach since the batteries are charged at very low wattages (see Example 8.3.1).

Disadvantages

- 12 V dc systems are restrictive regarding the range of electrical loads that can be used. Productive uses of electricity are less likely.
- Batteries are relatively expensive and have a fairly short life, particularly if they are heavily discharged between recharging operations.

Battery charging

Storage batteries should be charged at the current specified by the makers (typically 4 A to 6 A for a car battery). Charging at too high a rate will cause the rapid production of gas and heat, and possible damage to the battery.

Traditionally, the state of charge of a lead acid storage battery (the most common type) was determined by the specific gravity of the sulphuric acid electrolyte (1.21 when fully charged, falling to 1.18 when discharged). However, modern motor vehicle batteries are sealed and the only method of assessing the state of charge is by measuring the voltage (2.1 V per cell when fully charged to about 1.85 V per cell when discharged). Note that the voltage does not decrease linearly with discharge.

In order to measure both the rate and state of charge of batteries therefore, an appropriate ammeter and voltmeter are required.

The ampere-hour efficiency of a lead acid cell (Ampere-hours discharging/Ampere-hours charging) is about 90%.

The watt-hour efficiency of a lead acid cell (Watt-hours discharging/Watt-hours charging) taking voltage into account, is about 75%.

Attention must be given to human safety. ***Great care must be taken to avoid short-circuiting any storage battery.*** The currents which flow can be very high and sufficient to melt metal, causing a short circuit. Apart from the damage caused, there is the risk of very serious burns.

When lead acid (car) batteries are charged, hydrogen is produced and forms a cloud of gas over the batteries. This forms a highly explosive mixture with air. It is not uncommon for untrained people using chargers to cause an explosion by disconnecting battery connectors before switching off the charger. The resultant spark ignites the gas causing the battery to explode, spraying acid into the face of the user and causing serious injury. A spark from a cigarette can have the same effect.

Always train users and use warning notices about switching off before disconnecting and not smoking near charging batteries. Always provide good ventilation for battery charging to avoid the build-up of explosive gas.

Either motor vehicle generators or 220/240 ac generators can be used. In the latter case battery chargers must be used. Commercial battery chargers, that will charge several batteries at the same time, may be available. Although more expensive, a 220/240 V system is more versatile as the power can be used for a wide range of other loads and lower transmission costs mean that there is less restriction on the location of the battery charger(s).

With a motor vehicle generator batteries will usually be charged in parallel. For example, a system may be intended to charge several standard 12 V batteries in parallel from a generator designed to produce about 15 V. If this is the case, care must be taken to avoid any battery nearly fully charged supplying charging current to others that are discharged. This may occur if the supply from the generator is switched off before disconnecting each battery or if one battery has a 'dead' cell and thus has a low terminal voltage and so appears to be entirely charged. The scheme shown in Fig 8.3.3 could be used to avoid this difficulty.

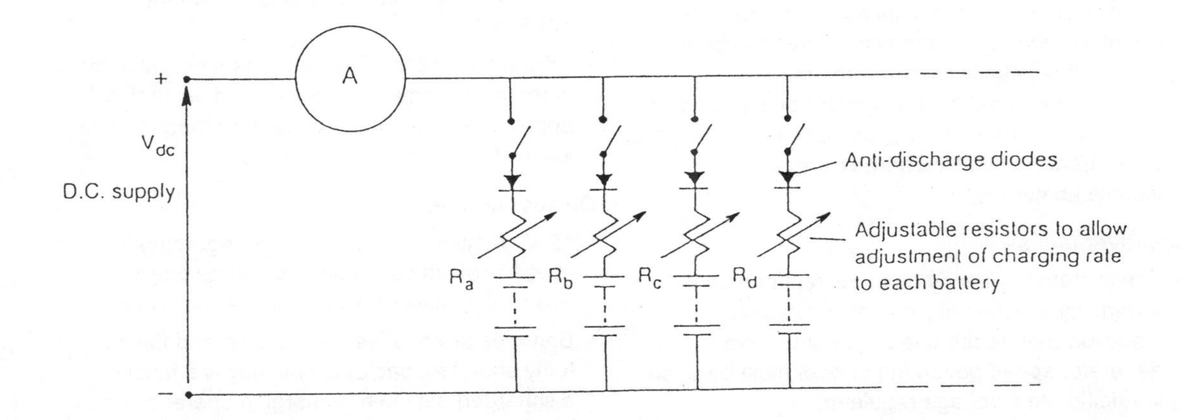

Fig 8.3.3 Battery-charging circuit with single dc supply and anti-discharge diodes

Batteries or ac — Example 8.3.1

A comparison is to be made between two possible schemes for supplying a village: a normal ac system and a battery-charging scheme. There are 100 households, each of which is likely to require 160 watt-hours per day (two 20 watt fluorescent tubes on for four hours). Consider the costs and advantages of each scheme:

Based on an estimate of 160 watt-hours per day per household a 1 kW scheme charging batteries for 20 hours a day at an efficiency of 80% could supply electricity to 100 houses. This is equivalent to a 4 kW standard scheme, when used in the evening for lighting. In terms of economics the battery-charging system will be cheaper initially (at a rough guess, £1000 for the hydro and £2000 for 100 batteries) but, with the cost of replacing/reconditioning the batteries every 2 or 3 years, more expensive in the long run. In addition the limitation in terms of electrical end uses will become more of a disadvantage with larger installations. The main size range for this approach is likely to be the 250 - 1000 watt unit.

When considering this approach it is worth thinking about its effect on the introduction of larger, more productive micro-hydro schemes. Will it encourage larger installations by making people aware of the potential available or stifle them by reducing their market?

Battery-charging circuit design — Example 8.3.2

A simple battery-charging circuit is required which will give controlled charging of 12 V car batteries. The charging rate specified by the battery supplier is 5 A. It is assumed that the minimum voltage per cell when discharged is 1.85 V and the maximum when fully charged and still on charge is 2.4 V (see Fig 8.3.3).

Required dc supply voltage

Maximum dc supply voltage at battery must be: 6 × 2.4 = 14.4 V. If the circuit in Fig 8.3.3 is used, then allowance must be made for the diode volt drop; say 0.7 V for a silicon diode.

Therefore required V_{dc} = 14.4 + 0.7 = 15.1 V, say 15.5 V.

Diode rating

The diode should be rated for maximum charging current (5 A), say 8 A giving a safety margin.

Charging resistors

The minimum value of the charging resistors should give the maximum charging rate (5 A) at the minimum battery voltage. Therefore, taking diode volt drops into account:

$$R_{min} = \frac{15.5 - 0.7 - (6 \times 1.85)}{5} = 0.74 \text{ ohm } (\Omega)$$

Therefore use a resistance rated at least 5 A whose maximum value (if variable) is at least 0.74 Ω.

Vehicle alternators — Note 8.3.2

Advantages

- Low cost and easily available spare parts, repair services.
- Designed for battery charging.
- A voltage regulator is an integral part of an alternator.

Disadvantages

- Designed for high speed (2000 rev/min and above) so a belt drive is required.
- Poor long-term reliability.
- Poor efficiency (approximately 60% maximum at about 25% full load and 55% at full load ignoring mechanical losses).
- Open construction to aid cooling and may overheat if covered to prevent entry of dust or damp.

8.4

ac systems

As discussed earlier, an ac system has many advantages which make it the obvious choice for all but small and very specialized schemes.

The designer of an ac system must first decide between 3-phase and single phase. Guidance on this choice is given below. An important parameter in the sizing of the generator is the power factor of the load and the effect of motor starting currents. These topics are covered subsequently. This section then discusses voltage and frequency standards, and finally introduces the two main types of ac generators.

3-phase and single phase

An ac supply system can be either single phase, as shown in Fig 8.4.1, or 3-phase as in Fig 8.4.2.

As will be seen from Fig 8.4.1, a single phase system has essentially the same circuit as a dc system with a 'go' and a 'return' conductor. A 3-phase system on the other hand essentially consists of three single phase systems with a common return or 'neutral' conductor as in Fig 8.4.2. The three voltages are generated at 120° to each other by equally spacing three coils round the circumference of the generator. The relation of the 'line-to-line' (l – l) voltages to the 'line-to-neutral' (l – n) voltages are shown in Fig 8.4.2 and it should be noted that the system is described by its line-to-line voltage.

Individual 3-phase loads may be either star or delta connected as shown in Fig 8.4.3, and in the case of 3-phase motors will be 'balanced'; that is will present the same load impedance in each phase. The relation of balanced delta phase currents and the resulting line currents is shown in Fig 8.4.3.

It should be noted that no current flows down the neutral line from a star-connected balanced load and so if all the loads are balanced or are delta-connected no neutral is required. This means that such a system would, at best, require only a quarter as much copper as a single phase system supplying the same power in order to give the same voltage drop to the load. While there may be a few such balanced loads in a small, isolated, generation and supply system, there are likely to be many more different single phase loads each connected between one of the three lines and neutral. Single phase appliances are so designed to operate on the line-to-neutral voltage. They are connected between different lines and neutral in such a way as to form as nearly an overall balanced load as possible. The nearer this balance is achieved, the less is the neutral current. So, with care a neutral conductor of reduced cross-section can be used.

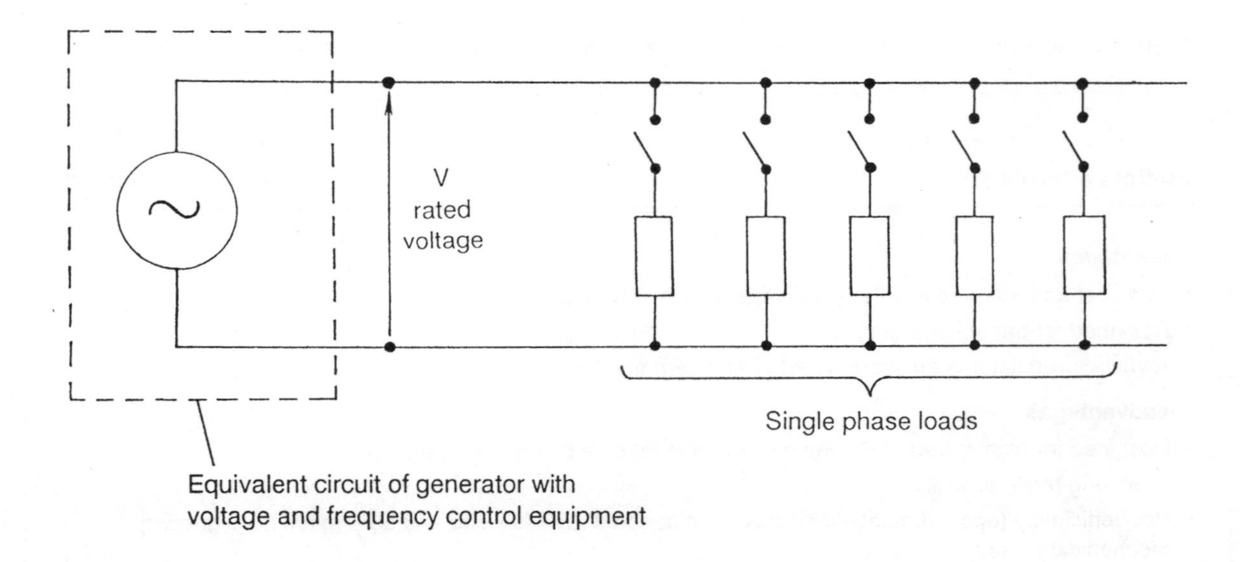

Fig 8.4.1 Single phase supply system

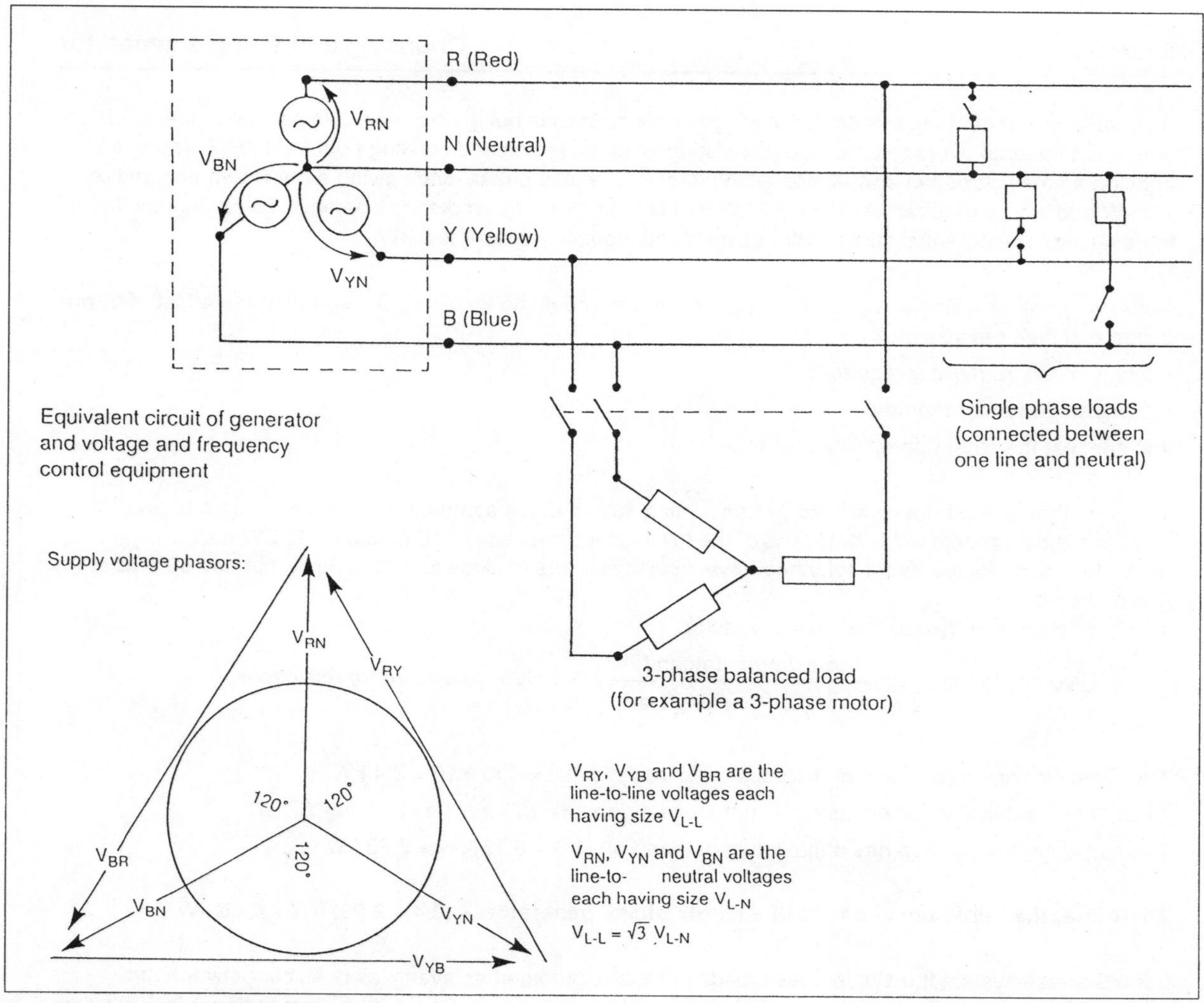

Fig 8.4.2 3-phase supply system

The advantages of single and 3-phase systems relative to each other may be summarized as follows:

Advantages of 3-phase

- A copper conductor saving of up to 75% compared to single phase and dc systems having a voltage equal to the line-to-neutral value. This depends on the degree of load balance achievable.
- Cheaper generators and motors; 3-phase generators and induction motors are more common and cheaper than their single phase equivalents above about 5 kW. Note that 1-phase ac/dc 'universal' commutator motors used in some appliances (for example in electric drills) are not being considered in this comparison.
- 3-phase generators and induction motors are smaller for a given power output than their single phase equivalents.
- Both single phase and 3-phase loads can be supplied.

Advantages of single phase

- Less complicated wiring, insulation arrangements and switches.
- Single phase ELCs are less expensive.
- Generator size determined by maximum load demand. In a 3-phase system it would normally be necessary to install a generator of higher capacity than the total maximum load demand since it is unlikely that the load would be balanced. See Example 8.4.1.

In practice, the decision as to whether to adopt a single phase or a 3-phase system depends on an estimate of the relative costs including the saving in copper and machine costs bearing in mind the degree of load balance that is likely to be achieved. It is clear, however, that the larger the system and the larger the number of individual loads there are, the better the load balance that is possible. As an approximate rule, therefore,

Example 8.4.1 **Choosing and sizing a generator**

A micro-hydro generating system for a village of ten houses is being designed. It is estimated that each house will have an average total load, consisting of lights, radios and cooking pots, of 400 W. There is in addition a small carpenter's workshop which has small single phase loads giving a maximum demand of 1.5 kW and also a requirement for a 0.5 kW induction motor. The workshop operates during the day and in the evening. The potential micro-hydro power continuously available is 6 kW.

Consider whether a single phase or a 3-phase system should be installed. Disregarding the effect of power factor, estimate the size of:

- single phase generator required;
- 3-phase generator required;

and suggest the most appropriate system.

It may be that all the houses will not be consuming 400 W at the same time. You may be able to predict that the maximum demand is only 60% of the full installed capacity of 10 houses × 400 W each.
(Note that many village hydro schemes have diversity factors of close to 100%, so do not use 60% as a rule of thumb.)
This prediction is known as the 'diversity factor':

$$\text{Diversity factor} = \frac{\text{maximum demand}}{\text{sum of all possible loads}} = 60\% \text{ (assumed for this case)}$$

Therefore, maximum demand of the household loads is: $0.6 \times 400 \times 10 = 2.4$ kW

Bearing in mind motor losses, assume motor power input is 0.75 kW (ie efficiency 67%).

The carpenter's workshop has a maximum demand of: (1.5 + 0.75)kW = 2.25 kW

Therefore, the minimum capacity of a single phase generator = (2.4 + 2.25) kW = 4.65 kW

For a 3-phase system the single phase loads must be distributed as evenly as possible between the phases. The motor in the carpenter's workshop can be a 3-phase machine, thus demanding 0.25 kW from each phase.

If the maximum demand of each of the household loads is assumed to be $0.6 \times 400 = 0.24$ kW as above, then again, assuming all the maximum demands occur at the same time, the best phase balance of the loads is:

Red Phase: Workshop single phase loads plus one phase of induction motor: (1.5 + 0.25) kW =1.75 kW

Yellow Phase: Five houses plus one phase of the induction motor: $(0.24 \times 5 + 0.25)$ kW = 1.45 kW

Blue Phase: Five houses plus one phase of the induction motor: = 1.45 kW

Minimum capacity of 3-phase generator required: (3×1.75) kW = 5.25 kW

In this case, with such a small scheme (11 consumers) there is unlikely to be much saving in conductor costs by adopting 3-phase. Moreover, the saving in cost obtained by using a 3-phase generator would be offset by the larger size required.

Note that 25% additional capacity above the minimum is usually required to ensure good motor starting and increase alternator service life in line with other system components. If an ELC is used this allowance must be increased to 60% (see Section 8.5).

Although the above calculations are valid, the situation is simplified as in practice load power factor should be considered and the generator will be rated in kVA. See section following on 'ac loads'.

Note also that an ELC would be very suitable in such a scheme since motors are being started. A single phase ELC is much less expensive than a 3-phase ELC.

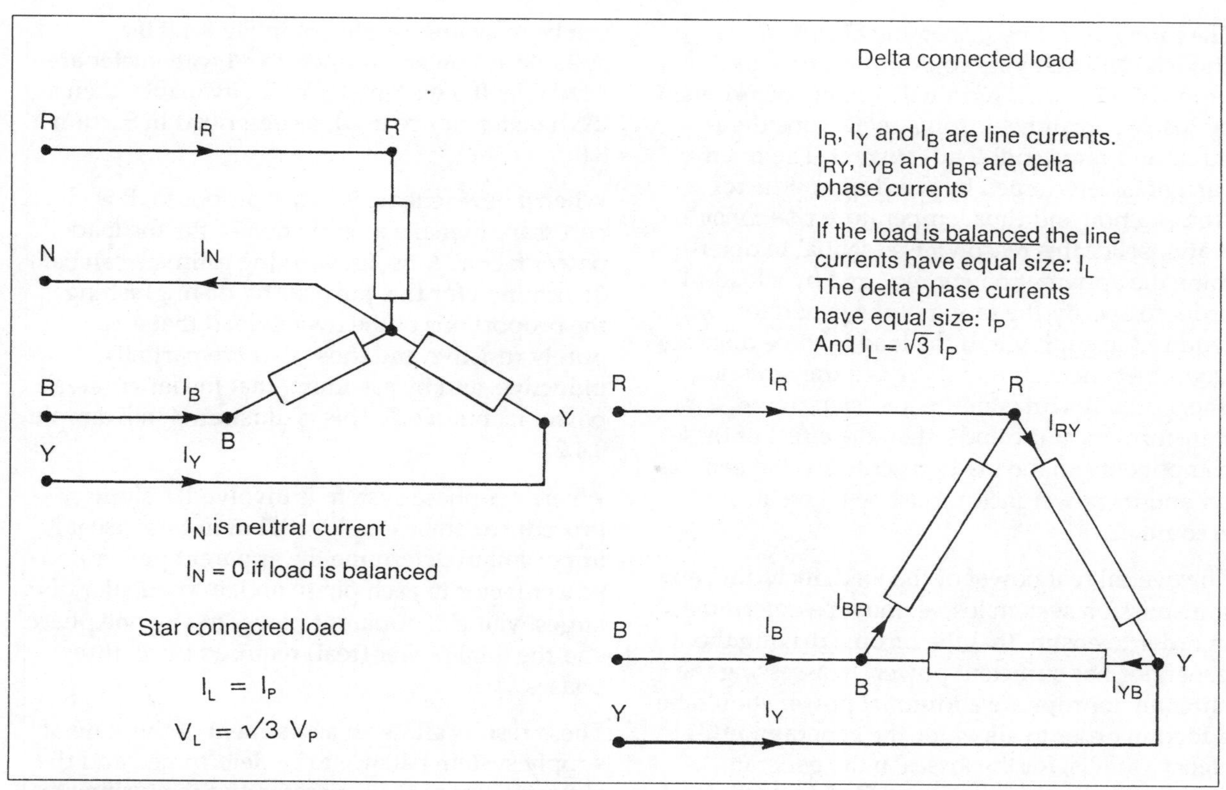

Fig 8.4.3 Star and Delta loads

systems up to 10 kW may be single phase, while systems over 5 kW may be 3-phase. Again see Example 8.4.1.

If a 3-phase system is adopted it should be noted that, for safety reasons, there are limitations on the way single phase loads can be connected to achieve balance. So, all the single phase loads in a house should be connected to one phase so that there is no chance of the line-to-line voltage appearing between any line conductors which might be thought to be at the same voltage. For the same reason, even where a small factory has a 3-phase supply for induction motors, the single phase loads should be connected to one phase.

This is also demonstrated in Examples 8.4.1 and 8.4.2.

ac loads

Resistive and reactive loads: power factor

Section 8.2 introduced the distinction between real power, expressed in watts (W), and apparent power, expressed in volt-amps (VA). The two quantities differ if the current is out of phase with the voltage (current usually lagging because of circuit inductance). The 'real power' is thus the product of the power factor and the 'apparent power':

$$\text{real power (W)} = \text{Pf} \times \text{apparent power (VA)}$$

where power factor (Pf) is the cosine of the angle between the voltage and the current.

A third quantity, reactive volt-amps (VAR), represents the 'reactive power' associated with the component of current completely out-of-phase with the voltage. Following the power triangle in Fig 8.4.4 this is related to real (W)and apparent power by the expression:

$$\text{Apparent power (VA)} = \sqrt{(W)^2 + (VAR)^2}$$

$$= \sqrt{(\text{real power})^2 + (\text{reactive volt-amps})^2}$$

This quantity, reactive volt-amps, is useful in the determination of the apparent power and power factor of a multiple load as in Example 8.4.2.

Fig 8.4.4 The power triangle

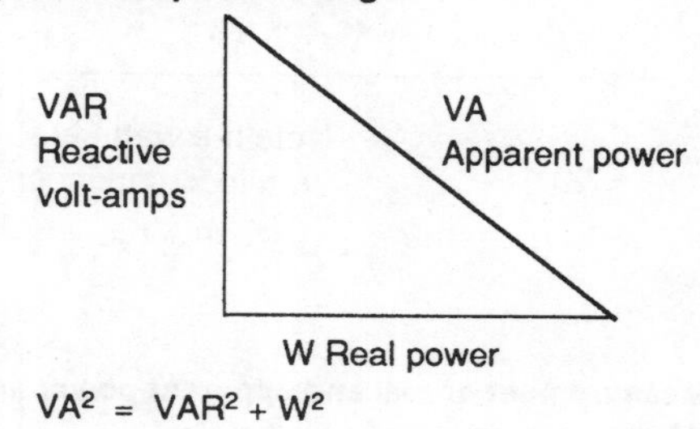

The rating and thus size of any electrical machine is defined by its volt-amp rating – the product of its rated terminal voltage (determined by insulation and electromagnetic considerations) and rated (full-load) current. The rated current is determined by winding conductor cross-section and thus temperature rise considerations. For this reason it is essential to determine the apparent power required by a load in order to specify the rating of the generator required to supply it. If the load is some distance from the generator and there is a transmission line, possibly with step-up and step-down transformers at the ends, then the effect of these components on the load presented to the generator and its power factor must be taken into account.

The overall real power of the load, including any transmission system losses, must be determined in order to ensure that the turbine driving the generator has sufficient power. In assessing the latter an appropriate additional power should be added in order to allow for the generator efficiency (that is, for the losses in the generator). This is discussed in Sections 8.5 and 8.6.

If a load consists of purely resistive appliances such as heaters and ordinary light bulbs (incandescent filament lights), the power factor will be one. On the other hand, if an appreciable portion consists of partially inductive components such as induction motors, appliances incorporating motors, or fluorescent lights, then the power factor will be less than one. The overall power factor and the true and apparent power of a load can be measured as shown in Fig 8.4.5 if a suitable ammeter, voltmeter and watt-meter are available. If a watt-meter is not available, then a kWh meter can be used, as described in Section 8.8.

When a new scheme is being proposed, it is necessary to make a prediction of the the load power factor. A useful working estimate can be determined for this purpose by distinguishing the proportions of the overall load that are purely resistive and those that are partially inductive, and by assuming that the latter have a power factor of 0.7. This is illustrated in Example 8.4.2.

Where a 3-phase system is involved, a similar procedure can be employed, but in this case it is important to determine the apparent power and power factor in each phase and, in particular, the largest value of apparent power in any one phase and the total power (real) required by all three phases.

These results allow an assessment of the state of supply system balance to be determined and the required size of the generator and the turbine to be calculated. Again, if a transmission line is involved, its effects must be considered. Example 8.4.3 illustrates the procedure which may be used to estimate the overall load quantities in a 3-phase scheme.

In all cases an allowance or safety margin of 25% is made on kVA. This means that a 125 kVA alternator would be selected for a calculated 100 kVA load.

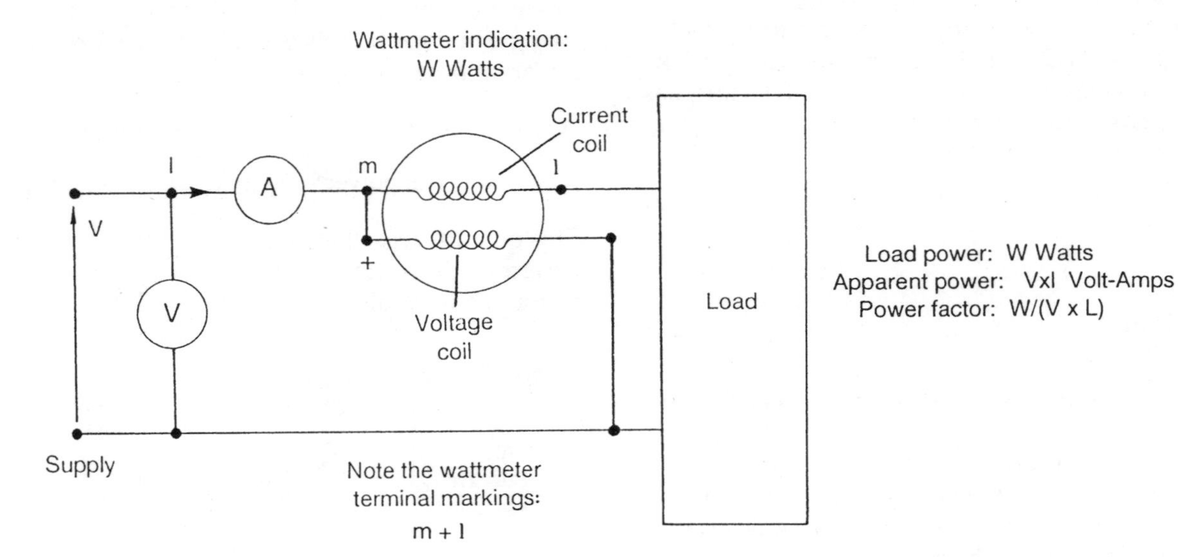

Figure 8.4.5 Measurement of real and apparent power and power factor See also Section 8.8 for measurement of power factor using a kWh meter instead of a watt-meter.

Maximum power rating of motors **Table 8.4.1**

In some circumstances these values can be exceeded

Motor type	Fraction of the rating of the generator supplying the system			
	3-phase		single phase	
	Synchronous	Induction	Synchronous	Induction
3-phase induction	0.5	0.1		
single phase induction	0.3	0.1	0.3	0.1
Universal	0.5	0.3	0.4	0.2

Motor loads: starting

All motors, whether induction motors or 'universal' commutator motors, present a low impedance to a supply until their speed builds up. Thus if they are started by being connected 'direct-on-line' as is often the case, there will be a significant current surge when they are switched on. This may be sufficient to depress the system voltage for a short period or, in the worst circumstances, may depress it to a level where the motor will not start. In order to avoid this:

1 The power rating (kW) of 3-phase motors started 'direct-on-line' should be limited to the fractions of the supply generator ratings shown in Table 8.4.1. The values shown in the table are very conservative. Where high inertia turbines (such as Francis turbines) or flywheels are fitted, bigger motors can be used. Often it is best to consult the alternator manufacturer. If an ELC is used, bigger motors can be accommodated because of faster system response.

2 A 'soft starting' unit may be used; this reduces the starting current (possibly by connecting a large capacitor across the supply, or across each phase of a 3-phase supply, and thus supplying the necessary reactive current) and allows the limits specified in Table 8.4.1 to be relaxed.

3 A 'star-delta' starter may be used for a 3-phase induction motor. This is an arrangement which connects the motor in star when it is first connected to the supply and then reconnects the motor in delta when it is approaching its normal running speed. The system may be manual but is normally automatic in which case the change from star to delta occurs either when the motor speed reaches a specified level, or when the line current becomes less than a preset level.

As the starting torque with a star-delta starter is approximately one third of the direct-on-line starting torque, this system can only be used where the load torque on the motor is low at starting.

It should be noted that the above applies also to appliances containing motors.

Voltage and frequency standards

Voltage and frequency tolerances are discussed in Section 6.3 in connection with turbine governing.

For the reasons given in that section, the voltage and frequency tolerances recommended for an all-purpose micro-hydro electrical supply system are:

Voltage: + 7% to –7% of the rated value.

Frequency: up to 5% above, but not below the rated value.

For a micro-hydro system where there are no transformers or induction motors, that is where there are only lighting (incandescent or fluorescent lamps), heating and universal (commutator) motor loads, the more liberal voltage and frequency limits are possible as discussed in Section 6.3. The limits recommended in this case are:

Voltage: + 7% to –10% of the rated value.

Frequency: + 10% to –5% of the rated value.

The chosen values of rated voltage and frequency are most likely to be determined by the voltage and frequency ratings of the most easily and cheaply obtainable electrical equipment. In practice these ratings are most likely to be determined by the nearest mains supply (grid-connected) system in the country concerned.

Example 8.4.2 **Single phase generator and turbine sizing**

The estimated load to be supplied by a micro-hydro single phase generating scheme consists of 20 houses, and an associated metal fabrication workshop and dairy. It is expected that each house will be equipped with three 60 W incandescent lamps and have a 150 W cooking pot. In addition, six of the houses are likely to have a 200 W compressor type (as opposed to absorption type) refrigerator.
The workshop is to be equipped with two machines each driven by a 0.5 kW induction motor and lighting is to be provided by six 75 W fluorescent tubes. The dairy will use a 1kW induction motor/pump and a 2 kW milk cooler which works on the absorption principle. In this case lighting will be provided by four 75 W fluorescent tubes. It is assumed that all the loads may be connected at once; that is, the 'diversity factor' (see Example 8.4.1) is 1.0.

Calculate the total power (real), the apparent power and the power factor which the overall load will impose on the generator.

Purely resistive loads:

20 × 3 × 60 W incandescent lamps	3600 W
20 × 150 W cooking pots	3000 W
1 × 2 kW absorption type milk cooler	2000 W
Total	**8600 W**

Partially inductive loads:

6 × 200 W refrigerators	1200 W
2 × 0.5 kW induction motors having, say, 67% efficiency	1500 W
1 × 1 kW induction motor pump, having, say, 75% efficiency	1330 W
(6 + 4) × 75 W fluorescent tubes	750 W
Total	**4780 W**

Total reactive volt-amps can be calculated by applying the power triangle (Fig 8.4.4) to the partially inductive part of the load, and then estimating that it will have an overall average power factor of 0.7 (as indicated in Table 8.2.1):

$$\sqrt{VA^2 - W^2} = \sqrt{(4780/0.7)^2 - 4780^2} = 4876.6 \text{ VAR}$$

Overall load:

Total power (real)	= 8600 + 4780	= 13380 W	= 13.38 kW
Total apparent power	$= \sqrt{13380^2 + 4876.6^2}$	= 14240.91	= 14.24 kVA
Power factor	= 13380 / 14240.91		= 0.94 lagging

This shows that because a significant portion of the load is purely resistive, the overall power factor is still reasonably close to 1.0 and the apparent power load is thus only a little greater than the real power load.

The total apparent power determines the size of the generator: it must be rated at more than 14.24 kVA. An additional 25% capacity on kVA should be made on this calculated figure to allow for motor starting currents and to extend the life of the alternator to be more in line with the other system components. The appropriate standard frame size is therefore 15 kW, 18.75 kVA which gives an additional capacity of 32% on kVA. This then is the correct generator size. Note that if an ELC is used an allowance of 60% on capacity is necessary.

The real power load, and the generator efficiency, determine the size of the turbine. The manufacturer of the generator may quote for instance 80% as the generator efficiency. The turbine must therefore be capable of providing 13.38/0.8 = 17 kW.

Note that the generator efficiency will reduce if it is only partly loaded. For machines of more than 10 kW capacity efficiency is typically as high as 70% or greater at 25% of full load, and drops to zero at around 5% of full load – nevertheless, ask the manufacturer for a part-load efficiency curve. For machines of less than 10 kW capacity part-load can become a problem (see Example 5.1.1). For small machines efficiency at 25% of full load can be very low. For a machine over-sized by 130% as recommended here, this can impose severe constraints on part-flow operation.

3-phase generator and turbine sizing — Example 8.4.3

Consideration is also being given to the use of a 3-phase system for the micro-hydro supply system described in Example 8.4.2.

Determine the best degree of load balance which can be arranged and, using the connection scheme devised to achieve this, calculate the real and apparent power supplied by each phase. From this, estimate the volt-amp rating of the required generator and the maximum power it will be called on to deliver.

	Purely resistive	Partially inductive
1 Classify the various types of consumer load patterns		
14 households with purely resistive loads: 3 × 60 W lamps	180 W	
150 W cooking pot	150 W	
Total resistive load	330 W	
6 households with refrigerators:		
Total resistive load, as above	330 W	
Total partially inductive load		200 W
Workshop:		
Total partially inductive load from fluorescent lamps is 6 × 75 =		450 W
The induction motors would be 3-phase machines so that, at 67% efficiency, the resulting total partially inductive load per phase is (2 × 750)/3 =		500 W
Total partially inductive load on one phase (motor and fluorescents)		950 W
Total partially inductive load on each of the other two phases		500 W
Dairy:		
Total resistive load from the absorption type cooler (one phase load);	2000 W	
Total partially inductive load from fluorescent lamps is: 4 × 75 =		300 W
The induction motor of the motor/pump will be a 3-phase machine at 75% efficiency, the partially inductive load per phase is: 1330/3 =		444 W
Total partially inductive load on one phase (motor and fluorescents)		744 W
Total purely resistive load (cooler) on that phase	2000 W	
Total partially inductive load on each of the other two phases		444 W

2 Distribute the loads between phases to achieve the best balance at Maximum Demand. Distribute the large loads first, then balance the partially inductive loads before achieving a final balance with the small loads.

	Purely resistive	Partially inductive
Red phase		
Dairy:	2000 W	744 W
Workshop:		500 W
2 houses with refrigerators:	660 W	400 W
Totals:	**2660 W**	**1644 W**
Yellow phase		
Dairy:		444 W
Workshop:		950 W
1 house with a refrigerator:	330 W	200 W
8 houses with purely resistive loads:	2640 W	1594 W
Totals:	**2970 W**	**990 W**
Blue phase		
Dairy:		444 W
Workshop:		500 W
3 houses with refrigerators:		600 W
6 houses with purely resistive loads:	1980 W	
Totals:	**2970 W**	**1544 W**

Example 8.4.3 *continued* **3-phase generator and turbine sizing**

3 Calculate the loading for each phase and thus overall apparent power loading

Red phase

The reactive volt-amps are:	$\sqrt{(1644/0.7)^2 - 1644^2}$	=	1677.2 VAR
The total power (real) required by the phase is:	2660 + 1644	=	4304 W
Thus, the total apparent power for the phase is:	$\sqrt{4304^2 + 1677.2^2}$	=	4619.2 VA
The total power factor of the phase is then:	4304/4619.2	=	0.93 lagging

Yellow phase

The reactive volt-amps are:	$\sqrt{(1594/0.7)^2 - 1594^2}$	=	1626.2 VAR
The total power (real) required by the phase is:	2970 + 1594	=	4564 W
Thus, the total apparent power for the phase is:	$\sqrt{4564^2 + 1626.2^2}$	=	4845 VA
The power factor of the phase is then:	4564/4845	=	0.94 lagging

Blue phase

The reactive volt-amps are:	$\sqrt{(1544/0.7)^2 - 1544^2}$	=	1575.2 VAR
The total power (real) required by the phase is:	2970 + 1544	=	4514 W
Thus, the total apparent power for the phase is:	$\sqrt{4514^2 + 1575.2^2}$	=	4781 VA
The power factor of the phase is then:	4514/4781	=	0.94 lagging

The largest phase apparent power is 4845 VA (yellow phase). Thus, all the phases of the generator must be capable of supplying this and the rating of the generator must then be at least 3 times 4845 VA, or 14.50 kVA. The next standard frame size is 15 kW, 18.75 kVA, which gives a little more than the required 25% allowance on kVA for motor starting and long life. This is then the correct generator choice.

The total power (real) that the generator must supply is 4304 + 4564 + 4514 W, or 13.38 kW, and thus the turbine should be capable of supplying this plus the losses in the generator.

If the generator is 80% efficient, the required turbine output power is: 13.38/0.8 = 16.73 kW, say 17 kW.

ac generators

There are two types of ac generator suitable for use in a micro-hydro electricity supply scheme. These are synchronous generators (or 'alternators') and induction generators. Induction generators (in which an induction motor is used as a generator) are less common but are being used increasingly in small micro-hydro schemes. Their advantages are that they are easily and cheaply available as motors, simply constructed and repaired, reliable, rugged, require little maintenance and can withstand 100% overspeed.

Induction generators are easily used as generators when connected to an existing supply system (grid) and have been used in this way for many years. When used in a stand-alone application such as an isolated micro-hydro scheme they require fitting with excitation capacitors. They can then be used to power single fixed resistive loads, for instance a complete set of village lights, either all switched on, or all switched off. If fitted additionally with a voltage regulation system (or 'controller') they can be used as all-purpose ac supply generators.

This type of stand-alone self-excited induction generator is now being used increasingly in schemes of less than 50 kW size as a result of the development, since 1985, of an inexpensive voltage and frequency controller – the IGC (induction generator controller). Although strictly these kind of generators should be referred to as 'stand-alone' or 'self-excited' induction generators we usually refer to them simply as 'induction generators' or 'motor generators' (MGs). In some cases they are known as IMAGs – 'Induction motors as generators'.

Synchronous generators are so called because the frequency generated is directly related to the shaft speed. Induction generators are known as asynchronous generators because the frequency generated, though dependent on shaft speed, is not directly proportional to it and does change slightly with load changes. Both machines may have the same stator design but they use quite different rotors, as shown in Fig 8.4.6.

Section 8.5 following provides detailed discussion of synchronous generators and section 8.6 covers induction generators.

Photo 8.4 1 An ac generator of the induction type.

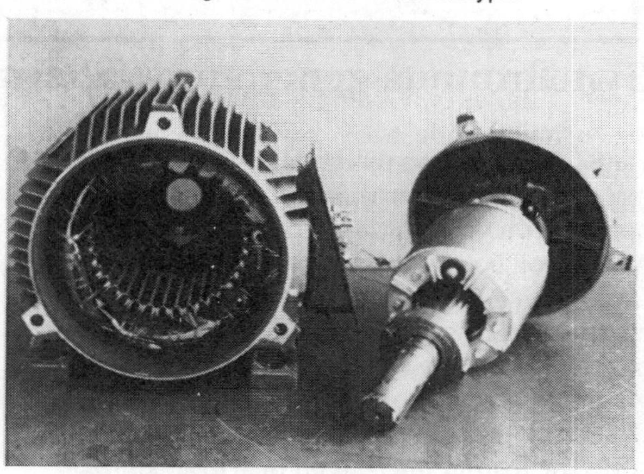

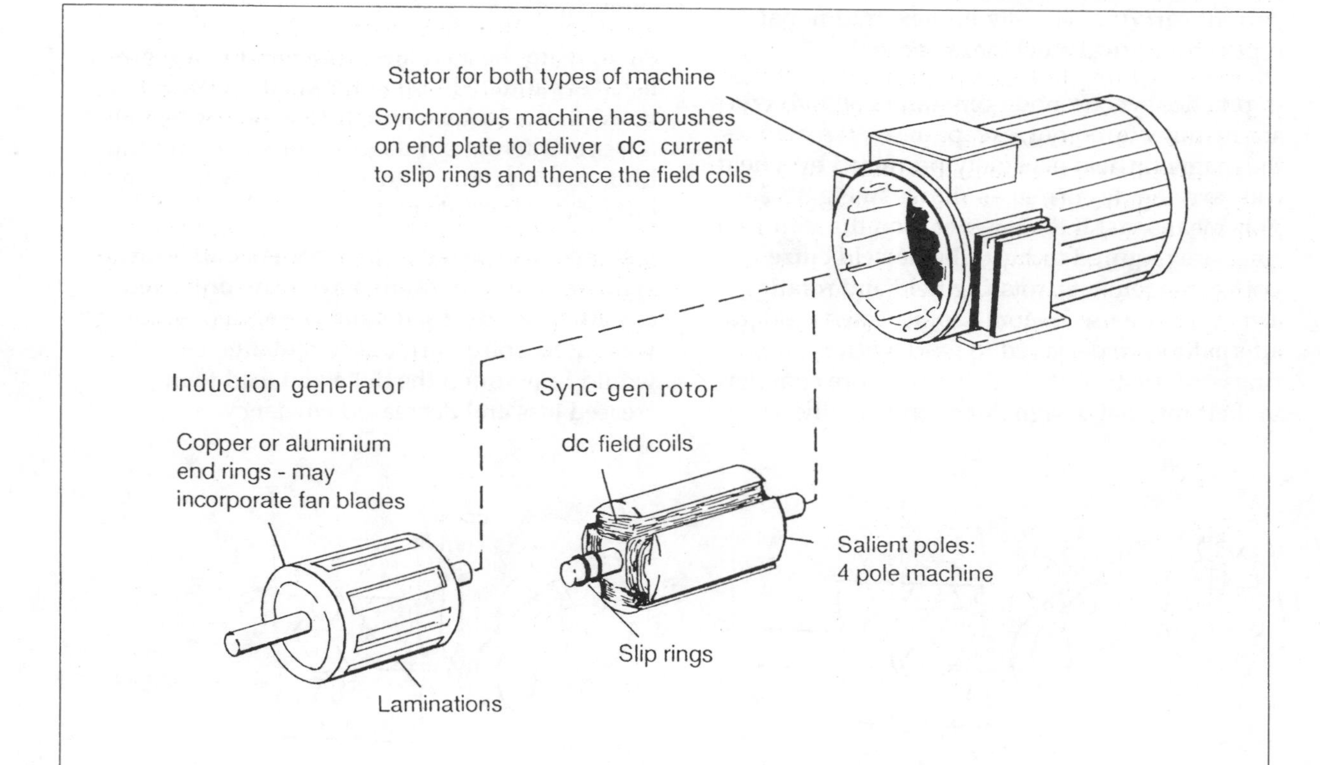

Induction generator:

$$rpm = \frac{120 \times freq}{p}(1 - s)$$

where s = slip and p = number of poles

The slip will be between 0 and 10% (depending on the load and the size of the machine).

It is negative for a generator (positive for a motor). The value of s may for instance be –0.05 for a generator, and (1 – s) will be 1.05.

For instance, a 4 pole induction generator on a 50 Hz supply runs at between 1500 and 1575 rpm, depending on the load.

Synchronous generator (wound rotor):

$$rpm = \frac{120 \times freq}{p}$$

where p = number of poles

For instance, a 4 pole synchronous generator at 50 Hz will run at a speed of 1500 rpm.

$$rpm = \frac{120 \times 50}{4} = 1500 \text{ rpm}$$

Figure 8.4.6 Generator types Induction (asynchronous) generator & synchronous generator (alternator) of the rotating field type

8.5

Synchronous generators (alternators)

Fig 8.5.1 shows two methods by which alternating current (ac) electricity can be generated. In both cases a voltage is generated in a load or output coil because it cuts or is cut by a magnetic field. If a load is connected to the output coil a current will be supplied to it. This requires direct connections to the rotating coil in the first case, whereas slip-ring connections must be made to the stationary coil in the second case. In both cases the load or output coil or winding is also called the 'armature', a name given to the load-current-carrying winding in most traditional types of electrical machine, ac or dc.

In practice synchronous generators of both types are manufactured but, except in small machines, the magnetic field is usually produced by a field coil carrying dc current as shown in Fig 8.5.2. This means that if the magnet is on the rotor, slip rings are required to carry the dc field current. Both arrangements, rotating field and rotating armature, are used. More sophisticated modern alternators are designed to work with no slip rings or 'brushes' at all; these are more complex in their internal design but operate on the same principles (see Fig 8.5.11). They generally have built-in electronic voltage regulators.

3-phase synchronous generators

The synchronous generators described in Figs 8.5.1 and 8.5.2 are all single phase machines. A 3-phase synchronous generator is effectively three single phase generators in one, using the same magnetic field, as illustrated in Fig 8.5.3. The waveform produced by such a machine is shown in Fig 8.5.4.

Apart from the advantages of 3-phase systems listed separately in Section 8.4.3 it should be noted that a 3-phase armature winding uses all the circumference of the rotor or stator and thus leads to a machine with a higher power/weight ratio than for single phase.

It should also be noted that virtually **all 3-phase synchronous generators have star-connected armature windings.** If delta-connected windings were used, third harmonic circulating currents would flow round the delta and lead to increased loss and decreased efficiency.

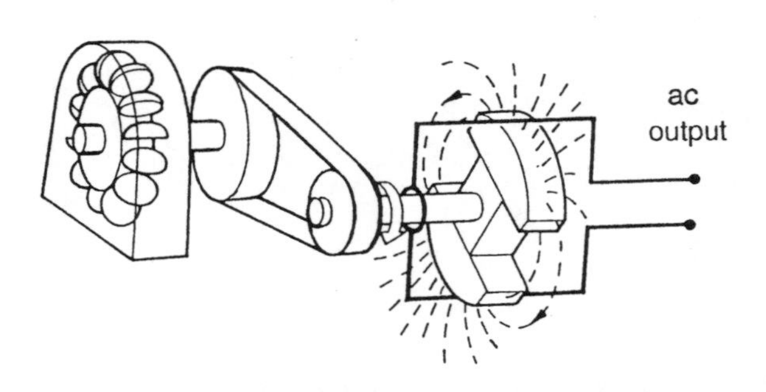

Rotating field (permanent magnet)

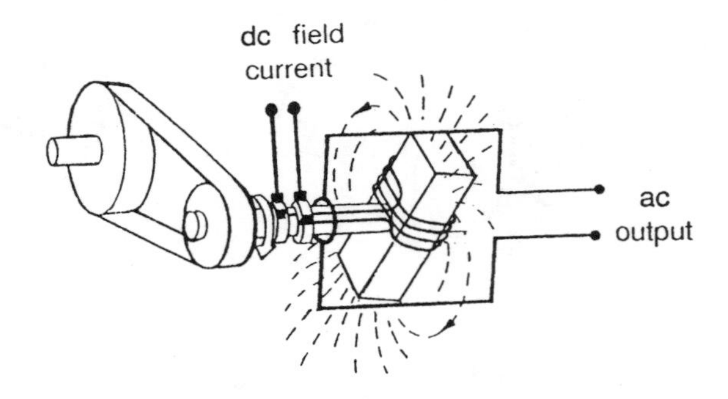

Rotating field (wound field)

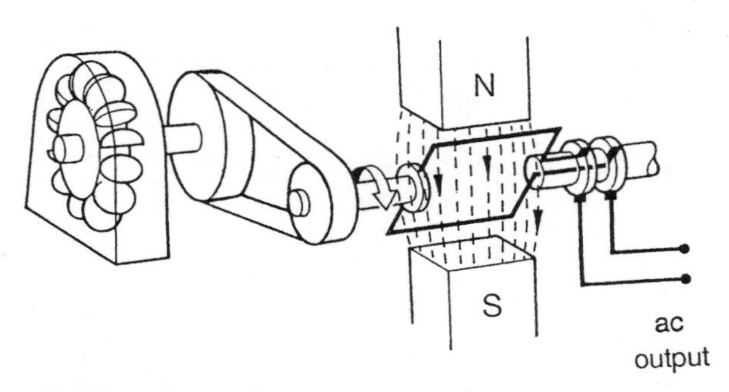

Rotating armature (permanent magnet)

Fig 8.5.1 Synchronous generator (design variations)

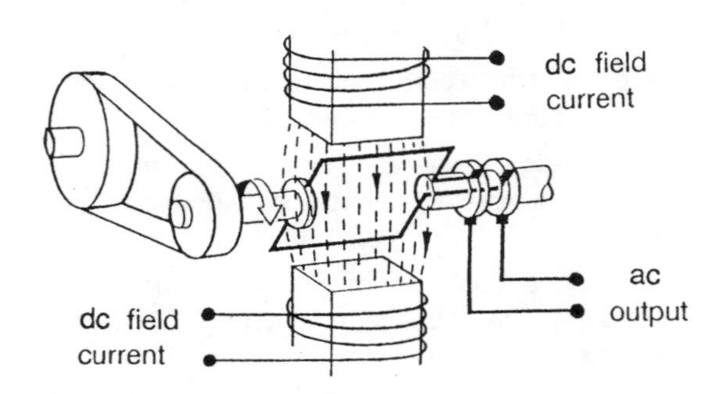

Rotating armature (wound field)

Fig 8.5.2 Synchronous generator (design variations)

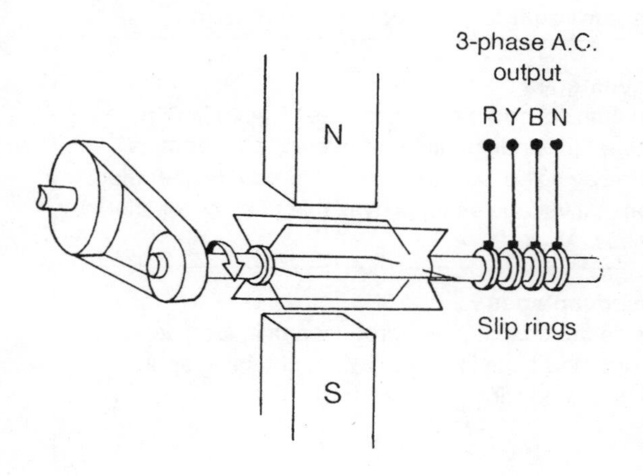

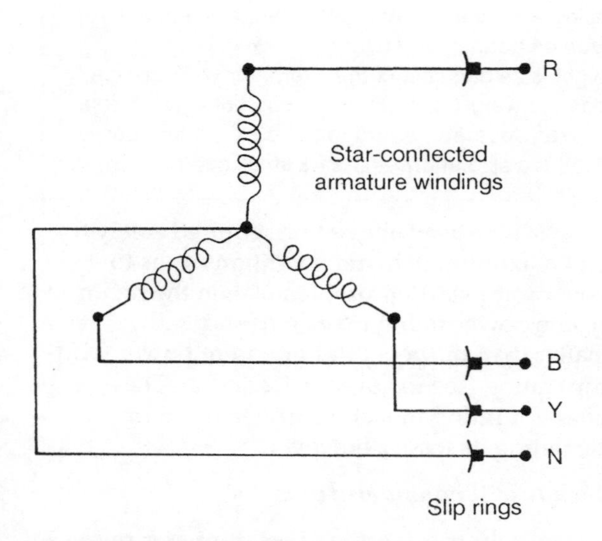

Fig 8.5.3 3-phase synchronous generator Rotating armature permanent magnet type.(3-phase synchronous generators usually have wound fields and armature windings on the stator as in Fig 8.5.2.

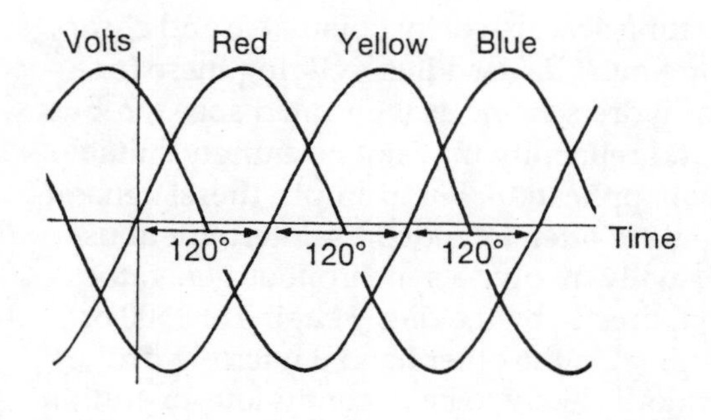

Fig 8.5.4 Ideal waveform Ideal sinusoidal voltage waveform produced by a 3-phase synchronous generator. Note that each wave is displaced 120 degrees from the next.

Single phase synchronous generators (alternators)

Small (up to 10 kW) micro-hydro schemes would normally be of the single phase type. A single phase generator is often called an alternator and is normally larger than a 3-phase generator of the same rating since only a portion of the stator circumference is used for the load winding. The same factors and considerations regarding rating, robustness and drive arrangements should be taken into account as for 3-phase synchronous generators.

Specification of synchronous generators

Electrical specifications

The capacity, ie size, of a generator is described by its volt-amp rating. Examples 8.4.2 and 8.4.3 illustrate how the best load balance may be achieved. If there is some sort of transmission system (for example a transmission line and step-up and step-down transformers) between the generator and the loads, then allowance must be made for the extra volt-amps that will be required.

It is clear from Section 8.4 that the generator must have a volt-amp rating sufficient to supply the total required volt-amps in conditions of maximum demand, and that it is wise to employ a generator having a considerably larger volt-amp rating because:

- This allows for possible expansion of the loads and supply systems.
- This minimizes momentary voltage reductions which occur when induction motors are started. Use Table 8.4.1 or the approximate rule that the volt-amp rating of a generator should be more than five times the kW rating of any induction motor (including those built into electrical appliances) which may be started 'direct on line'.
- Over-rated generators will run cooler.
- Over-rated generators are necessary when using an ELC (see below).

Always add an extra 25% (or more) to kVA capacity for these factors, ie select a 125 kVA alternator for a 100 kVA load. For an ELC, increase this to 60% extra kVA capacity.

In practice, most synchronous generators are rated in power output (kW) at a specified power factor; for example 5 kW at 0.8 power factor. This allows the engine or other unit (turbine) driving

Note 8.5.1 **Comparison of off-the-shelf and custom-built alternators**

Off-the-shelf	Custom-built to suit hydro specification
Advantages Easily obtained; about half the cost of custom-made unit; replacement parts easily found. Generally will be high speed so relatively small and light: A 1500 rpm, 100 kW alternator can be moved with small manual hoist.	**Advantages** Mechanical and electrical duties as specified; no danger of winding failure at overspeed, bearings correctly sized, so will not wear rapidly. Higher than nominal voltage settings available to allow for line losses.
Disadvantages Maximum safe speed rating often not known; therefore risk of mechanical failure at turbine overspeed. Bearings usually not sufficiently strong, will wear rapidly. Voltage adjustment may be limited.	**Disadvantages** Twice the cost of mass-produced unit; long delays in ordering quite likely; delays in obtaining spare parts are possible.
Precautions Assess quality of workmanship and adhesive strength. Request bracing of windings if necessary to protect from mechanical failure at overspeed. If necessary, request replacement of bearings; check that bearing housings can accommodate heavier bearings.	**Precautions** If a low-speed alternator requested, savings will be made on gearing and bearing-loading, but the weight, size and cost of the alternator will increase, since the weight of copper is proportional to inverse of speed squared. (A 500 rpm, 100 kW alternator will be the size and weight of a small car.)

the generator to be specified for the power rating of the generator and assumes that in general the electrical loads to be supplied have a power factor similar to that at which the generator is rated. As we have seen in Section 8.4, while the generator shaft will be designed to transmit the rated power, its electrical design will be dependent on the volt-amp rating:

$$\text{Volt-amp rating} = \frac{\text{Power rating}}{\text{Power factor}}$$

So, the 5 kW, 0.8 power factor generator has a volt-amp rating of:

$$\frac{5000}{0.8} = 6.25 \text{ kVA}$$

The total power required by the load and transmission lines will, together with the losses in the generator, determine the power output required from the micro-hydro turbine and will thus determine the size of the turbine. The efficiency of a synchronous generator generally increases with its size and, typically, may vary from 65% for a 1 kVA machine to 90% for a 20 kVA machine. A knowledge or estimate of this may be used to determine the generator losses and thus the required turbine power. Again, the total power required must be determined for maximum demand conditions and it may be wise to allow for possible future increases in required load power. If, on the other hand, the generator rating is increased above that required solely to meet maximum demand conditions so as to avoid motor starting problems, then the required turbine power rating may be considerably smaller than the generator apparent power (volt-amp) rating (see Example 8.4.2 and 3). The generator rating must be borne in mind in calculating its losses, however.

Mechanical considerations

Once the electrical rating of an alternator required for a micro-hydro scheme has been determined, some thought must be given to the mechanical requirements in such an application. 'Off-the-shelf' synchronous generators (alternators) intended for use in diesel and other motor generating sets are easily obtainable and cheap, but care must be used in specifying these for a micro-hydro scheme as their rated speed, robustness and reliability may not be entirely suitable for that application. For example, diesel generating sets are often intended for intermittent use and usually involve a synchronous generator driven directly by the diesel engine at 1500 or 3000 rpm. On the other hand a micro-hydro generator is likely to be in continuous operation and the water turbine speed is likely to be in the range 200 to 1000 rpm. Thus, it is worth giving careful consideration at an early stage to the alternative possibility of ordering a custom-made machine from a manufacturer. The two approaches are considered in Note 8.5.1.

In particular, if an off-the-shelf unit is to be used, special consideration must be given to the bearings, overspeed and insulation, as follows:

The bearings

These are likely to be undersized, being intended for a direct drive from a diesel engine in line with the alternator. If a pulley and belt arrangement is to be used to achieve the correct speed, then the manufacturer should be consulted. Alternatively, a flexible coupling should be used, and the pulley mounted on separate bearings. (see Chapter 7).

Overspeed

A second condition experienced in hydro applications but not usually with diesel engine drives is overspeed. If this occurs, windings on the rotor of a synchronous generator intended for a diesel drive may be pulled out of position by the centripetal force occurring at excessive speeds. This will cause severe damage followed by extensive delay and expense in rewinding. The manufacturers of low-cost machines may not be willing to conduct full tests of their machines at overspeed, so may not guarantee the security of the windings.

If you are unable to obtain a machine guaranteed for overspeed, your options are:

1 Incorporate an overspeed trip which disconnects the generator mechanically.

2 Adopt high turbine speed for normal operation, as described in Chapter 6, to reduce the overspeed value.

Fig 8.5.5 Overspeed occurs when load is removed or reduced.

3 Incorporate a spoiler into your turbine so that it cannot reach runaway speeds. **But be warned** spoiler designs often do not work reliably and they can cut down turbine efficiency under normal operating conditions.

4 Take the imported rotor to an experienced local rewind shop and ask them to install wedges and extra binding.

5 Risk it. This is very unwise except in special cases and when you are dealing with a reputable company which has assured you that the machine is 'probably all right' at overspeed. A very small company may not be able to give any real assurance and may not have access to good adhesives for retaining the windings, and may not exercise full quality control over its products. It may also be the case that the machine is made in a small workshop which is simply borrowing the name of a large more reputable company.

If you are forced into this option, test the alternator at overspeed in safe conditions before commissioning.

Operating temperature and insulation class

Western manufacturers usually make use of advanced insulation materials which can withstand temperatures in the order of 200°C. If these are used in the alternator windings, the machine can be allowed to run very hot; an alternator made by Newage (for instance) can run at a winding temperature of 180°C. Stator windings produced by Western manufacturers are normally designed to have a continuous life of 20,000 hours (2.3 years) if run at the rated temperature of the class of insulation used. Operation at 8 to 10°C above the rated temperature is likely to halve the life, whereas the life may be doubled by operating at 8 to 10°C below the rated temperature. This should be borne in mind when planning a micro-hydro installation and choosing the site for the generator.

The copper wire in an alternator is a dominant factor in its cost; if wire of a smaller cross-section can be used to carry the alternator current, it will stabilize at a higher temperature, but less overall weight of copper is required, so reducing the cost and bulk of the alternator, which will be more economic even though the heat losses are greater.

A compact generator which runs hot will conduct heat to its bearings, which will also run hotter than those of a large alternator. This will reduce the bearing grease viscosity. Careful consideration should be given to the choice and sizing of the bearings, taking the expected running temperature into account.

Note 8.5.2 **Alternator specification: Checklist**

This checklist is useful when consulting a supplier of alternators or a catalogue. An example of a letter to a supplier is given in Note 8.5.3 below.

1 The turbine type (eg Pelton, Francis, etc) and power entering the turbine.
2 An assessment of turbine efficiency and drive system efficiency. State how sure you are of this. Refer to turbine manufacturers for more information.
3 Desired power delivery in kW to consumers.
4 Estimate of transmission line efficiency voltage drops, and how sure you are of this. Describe length of transmission, type of transmission.
5 Nature of load, whether motors, lights, heaters, etc. Estimate power factor, including transmission line power factor.
6 Duty: state that alternator will run continuously.
7 Speed – ask whether the alternator can handle overspeeds up to 180%.
8 Overspeed – turbine: give overspeed characteristic of your turbine.
9 Voltage desired – types of appliances envisaged.
10 Nominal voltage adjustment range – a high nominal voltage is often preferred with high loss transmission lines.
11 Phases – appliances envisaged.
12 Frequency – standard for that region (50 or 60 Hz).
13 Environment – dust? flour dust?
14 Ambient temperature, humidity, altitude.
15 Monitoring equipment (frequency meter, voltage meter, current meter, power meter) if available from alternator manufacturer.
16 Automatic Voltage Regulator (AVR) – ask what is available.
17 Governing system – state what type of system you are proposing.
18 Switchgear (see Section 8.8.7 below).

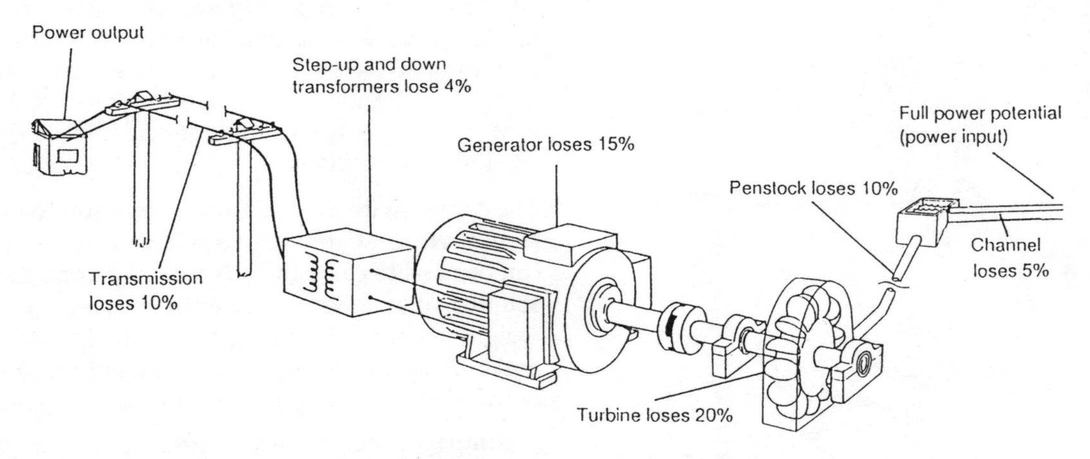

$$\begin{aligned}\text{Power output} &= e_{\text{civil works}} \times e_{\text{penstock}} \times e_{\text{turbine}} \times e_{\text{generator}} \times e_{\text{line}} \times \text{power input}\\ &= e_o \times \text{power input}\\ &= 0.95 \times 0.9 \times 0.8 \times 0.85 \times 0.96 \times 0.9 \times \text{power input}\\ &= 0.5 \times \text{power input}\end{aligned}$$

Fig 8.5.6 Typical system efficiencies for a scheme running at full design flow. For part-flow efficiencies see Note 5.1.1.

When a decision has been taken as to whether an 'off-the-shelf' or 'custom-built' synchronous generator is to be purchased, then manufacturers can be approached in the appropriate way. The basic procedure involved in each case is covered in the following sections.

Standard ('off-the shelf') synchronous generators – the manufacturer's catalogue

If you are considering purchase of an 'off-the-shelf' low-cost alternator, the manufacturers will send you catalogues listing equipment within the power range you are interested in.

Catalogues specify alternators in terms of:

1 Brushed or unbrushed: rotating armature or rotating field
2 3-phase or single phase
3 Insulation class (class F, class H, etc.), operating temperature
4 Voltage regulation and excitation (voltage band eg $\pm$ 1.5%), number of poles
5 Frequency/voltage characteristics, speed (rpm)
6 kVA, kW, power factor, efficiency
7 Built-in protection (underspeed, overload, short circuit)
8 Compliance to engineering standards/codes of practice
9 Vulnerability to dust; effect of fitting air filters
10 Response to thyristor loads
11 Length of shaft extension; option of extended length
12 Life expectancy of components, eg replacement of bearings with heavier duty

Photo 8.5.1 A synchronous generator and crossflow turbine. (Sri Lanka)

13 Recommended spare parts
14 Frame size, weight.

Confronted with such a catalogue it is best to make a provisional choice of unit, then write to the manufacturers stating the conditions of use for the alternator, following the guidance given in Note 8.5.2. This is a list of factors which must be considered in each application. The manufacturer will then give advice as to whether the provisional choice is suitable.

Custom-built synchronous generator

The list of factors given in Note 8.5.2 should be considered and used to prepare an initial letter of enquiry to manufacturers. In the light of the replies, or if the requirements are sufficiently clear at an early stage, a letter of specification requesting a tender quotation of the form shown in Note 8.5.3 may then be sent.

Excitation and voltage regulation

Voltage regulation

Unlike a dc generator, the output voltage of a simple synchronous generator falls very rapidly as the load current it supplies increases (see Fig 8.5.7b). This is due to an effect called 'armature reaction', a phenomenon which occurs in all traditional electrical machines and which is caused by the load current 'de-magnetizing' the main field. In a dc machine interpoles are used to nullify the effect of the armature reaction field, but there is no way in which such nullification can be achieved in a synchronous generator. As a result, it is essential that the main field dc magnetizing current is adjusted as the load current changes so that the output voltage remains the same for all loads. The necessary adjustment is not proportional to load current and thus cannot be achieved in a simple way. An 'automatic voltage regulator' (AVR) must therefore be provided to control the dc field current. In the case of small and moderate-sized modern synchronous generators (up to 50 kVA) this is usually an electronic unit built into the machine. Older machines may have electro-mechanical AVRs.

Apart from the armature reaction effect, a good AVR will also keep the generator output voltage constant when the speed changes. However, any change in speed means a corresponding change in frequency and it is important to avoid all but small changes in frequency.

Note 8.5.3 **Detailed specification for a generator and request for tender**

GENERAL SUPPLIES LTD

For the attention of the Sales Department

We would be pleased if you would tender us for the generator detailed below by Friday 30th February 1999.

Type of turbine, drive system	Pelton wheel, belt drive
Number off	One
Turbine input power	45 kW
Generator rating	50 kVA
Duty	Maximum continuous rated
Speed	1500 rev/min
Overspeed	180% continuously
Voltage, phases	415 V, 3-phase
Frequency	50 Hz
Power factor	0.8 lagging
Mounting	Vertical shaft, skirt mounted
Type	Salient pole, synchronous
Bearings	Grease lubricated, suitable for radial thrust of 400 kg/continuously
Excitation	Shunt with forcing current transformers (CTs)
AVR	Mounted internally, voltage regulation = ± 2.5% with under/overvoltage and under/over-frequency protection
Electronic switching	ELC; Thyristor action; (how many switches per sec?)
Ambient temperature	40° C
Altitude	500 metres asl
Humidity	90%

The generator should be designed and rated in accordance with BS4999/BS5000, or equivalent, and should be provided with a Class F insulation system. The generator will be located in a small powerhouse, measuring 6 by 5 metres, and there will be a manually operated hoist. Please advise the weight of your proposed generator, and provide an outline drawing with your tender. Please advise us of suitable monitoring equipment (voltage, current and frequency meters) and switchgear. The following accessories should be included in the price of the generator: slide rails, holding down bolts, operating and maintenance manuals; spares for 5 years operation. Spares should include rectifier sets, diodes, bearings and advice on other parts with limited life. Please show additional prices for: anti-condensation heaters; winding temperature detectors; tropical paint finish; shaft-mounted speed switch.

Your tender should be submitted subject to the following commercial conditions:

Prices: Your prices should be firm and fixed for the validity and despatch periods stated and include for works testing, packing for shipment, and delivery FOB Port in the country of origin. Your prices should be quoted in US$ or UK £

Validity: 120 days from the date of tender.

Conditions of sale: Your tender should be based on I Mech E/IEE Model form B2 conditions of contract.

Terms of payment: 100% of contract value on delivery FOB

Liquidated damages: Liquidated damages would apply at a rate of 0.5% of the contract value per week of delay up to a maximum of 5.0% of the contract value.

Please do not hesitate to contact us if you require any further assistance.

Yours faithfully,
Purchasing Manager

Simple excitation systems and AVR principles

Fig 8.5.7a shows a simple synchronous generator field excitation system with manual ac output voltage control.

The machine should be driven at the correct constant speed to give the required frequency. The field current is then adjusted, using the variable resistance R, to cause the required ac voltage to be generated on no load. If a purely resistive load is then connected to the generator with the machine running at constant speed and no further adjustment to R, the ac output voltage will fall as in the load/voltage characteristic at Fig 8.5.7 b. The fall would be greater with an inductive load.

Manual ac output voltage regulation could be achieved by continuously adjusting R so that the ac voltage is kept constant whatever the load. However, an AVR is clearly preferable to this and in modern units consists of an ac voltage-sensing element which continuously checks the ac voltage and adjusts the field current (usually electronically) so that the correct ac voltage is generated. See Fig 8.5.8.

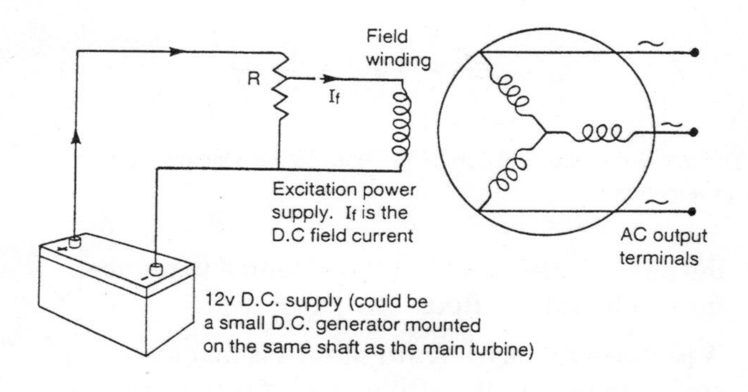

a Separately excited field and manual voltage control, by variable resistor R.

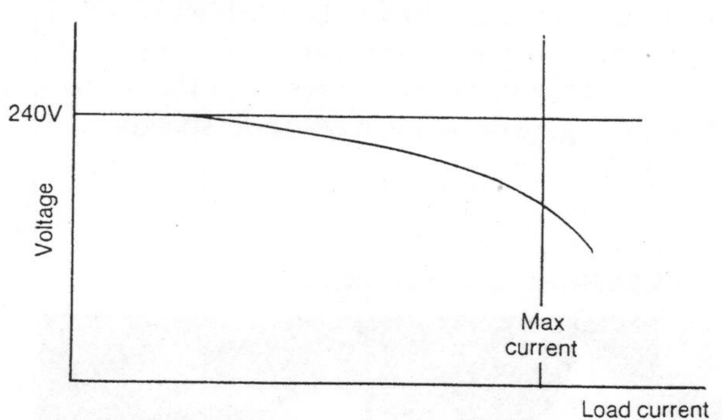

b Load/voltage characteristic of an unregulated synchronous generator – that is, driven at constant speed with constant field current

Fig. 8.5.7 Manual voltage control and effect of absence of control.

The source of the dc field current supply might be a battery, but it is much more likely to be a dc generator mounted and driven on the same shaft as the main synchronous generator. Alternatively, especially in small machines, the dc field supply may be obtained by rectifying the ac output voltage. Such a machine is then a 'self-excited' synchronous generator and Fig. 8.5.9 illustrates the arrangement.

It should be noted that the ability of a self-excited synchronous generator to build up its field excitation when the machine is started depends on the residual magnetism in the iron core and this must not be lost. This matter is dealt with in more detail in Section 8.6, on induction generators.

It should also be noted that an AVR is even more necessary in a self-excited synchronous generator than in the separately excited type as the dc field current supply voltage is directly dependent on, and thus decreases with, the ac output voltage.

There are many other types of synchronous generator field excitation circuits, including brushless alternators, and many AVR arrangements, and it is important to recognize the system used on an existing alternator. A range of excitation and AVR arrangements are shown in Fig 8.5.12.

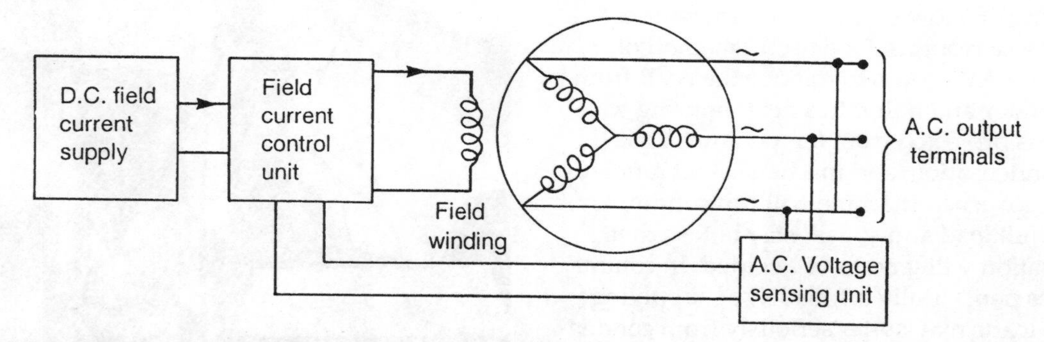

Figure 8.5.8 Automatic Voltage Regulator (AVR) – principal elements. Separately excited generator with automatic voltage control.

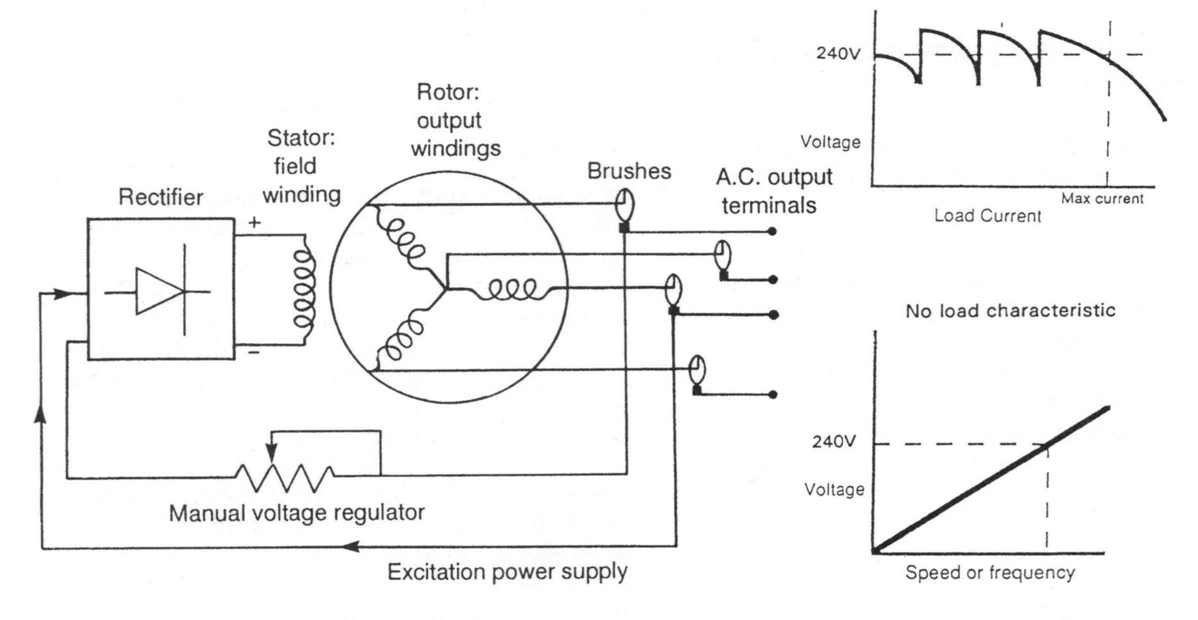

Fig 8.5.9 Self-excited synchronous generator: stationary field and rotor armature configuration without an AVR supplying a unity power factor load. Fig 8.5.12 shows developments from this arrangement.

If it has been decided to install a synchronous generator (alternator) in a new micro-hydro scheme, special care should be taken in selecting an appropriate AVR.

The AVR: practical considerations

AVRs designed for diesel-driven generators with lighter duties have tended to be highly unreliable in micro-hydro installations, although the situation is improving. Continuous use under conditions of varying load or shaft speed has caused endless AVR failures, and the engineer is well advised to keep a stock of spare parts, especially rectifier sets and diodes. (A diode expected to have a two-year life in a diesel driven generator may only have an eight month life in a continuously running micro-hydro generator.) The use of a thyristor type ELC (electronic load controller) will impose electrical and mechanical loading on the AVR which will shorten its life.

Oversizing of a low-cost type of generator can help in these respects. One useful method of prolonging AVR life is to remove the AVR from the generator and bolt it to a neighbouring wall where it is protected from temperature variations, condensation, and mechanical vibrations. An ELC governed machine will run continuously at full load and so will stay hot, so that condensation will tend to be avoided. In contrast machines periodically off-loaded or supplying variable loads may suffer seriously from condensation, especially if situated in damp conditions or in a humid environment. Proper ventilation of the powerhouse at all times is essential to reduce the condensation effect.

A problem in micro-hydro set-ups is that low-speed running of the turbine and generator may occur for prolonged periods. The AVR may respond by boosting the excitation in order to raise output voltage but will be unable to do this continuously without overheating. The high field currents forced in this way have been known to destroy the generator, even on underspeeds of only 5%.

Photo 8.5.2 Repairs to an AVR. (Nepal)

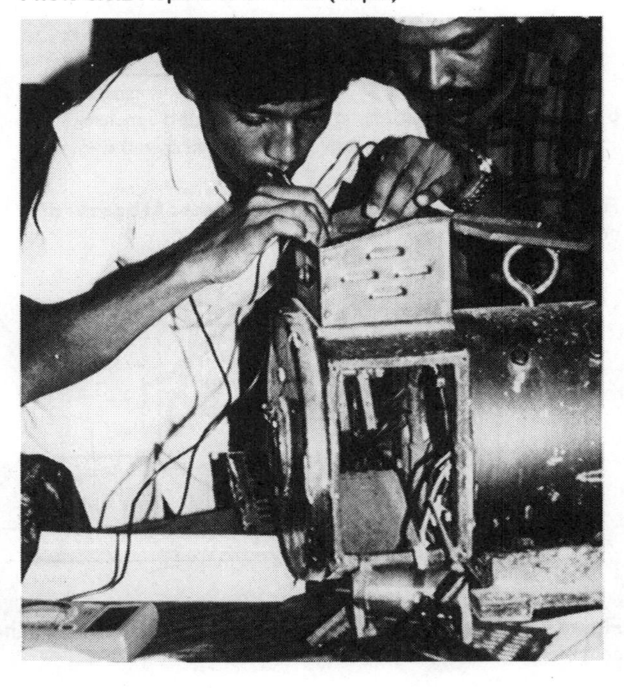

Common methods of protection from underspeed are:

1 Connection of AVR to a shaft-mounted speed switch ensures that excitation is only applied when shaft speed is about 80% or 90% of normal running speed.

2 Purchase of an 'intelligent' AVR which for instance gives a voltage frequency characteristic which reduces excitation with speed. (Low speed will cause low frequencies at the generator output, which will cause the AVR to reduce excitation. This is called 'frequency roll-off' and is illustrated in Fig. 8.5.10.)

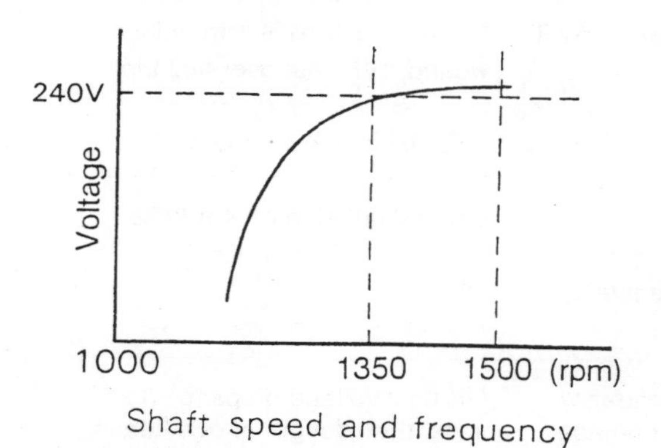

Fig 8.5.10 Frequency roll-off

In general this is the recommended method of protection. When purchasing a generator it is best to specify an AVR capable of full protection, from under/overspeed, under/over voltage, under/over-current, and under/over-temperature.

3 An alternative to **2** is to add a frequency trip to the circuit. A time delay on underspeed is needed to avoid unnecessary tripping on motor starts.

These methods can also be used to trip out the AVR excitation at higher than normal shaft speeds; this will protect the AVR from damage at turbine runaway conditions, and protect the load from over-frequency.

The engineer should be warned against 'add-on trips' since these tend to push the final generator cost to uneconomic levels. Usually it is cheaper to buy a more expensive sophisticated generator with an 'intelligent' AVR than to buy a simple one and add trips – a low frequency trip, for instance, might add almost 10% to cost.

Some AVRs will tend to give a voltage 'droop' on increasing load. This effect is then accentuated by increased transmission line losses, resulting in insufficient voltage levels at the load end. When buying low-cost alternators, the best solution to this problem is to buy an alternator rated to supply at least 125% of the expected full load kVA. The extra cost involved is minor in view of gains made in greater reliability, longer service, and better voltage regulation.

Careful contingency planning against revenue losses caused by generator breakdown is worthwhile. The failure of an AVR could result in long delays before the system is back in operation. Normally a plentiful supply of spare parts will be sufficient, with immediate re-ordering of the stock when any of the spare parts are put to use. It may in some cases be economic to purchase two low-cost generators when setting up the scheme, allowing immediate replacement of a failed generator with a good one at any time. Repairs on the failed generator can then be made without withdrawal of the electrical power supply.

The type of voltage regulation used in a generator will determine the frequencies/voltage combinations fed to the appliances used. Together with the type of governing system in use, this will affect the life span and reliability of the appliances, and so has important maintenance and cost implications. The engineer should categorize the proposed system along the lines shown in Table 8.5.1, and make sure that the governing/voltage regulation system matches the type of loads in use.

Photo 8.5.3 Bench run for an alternator. (Peru)

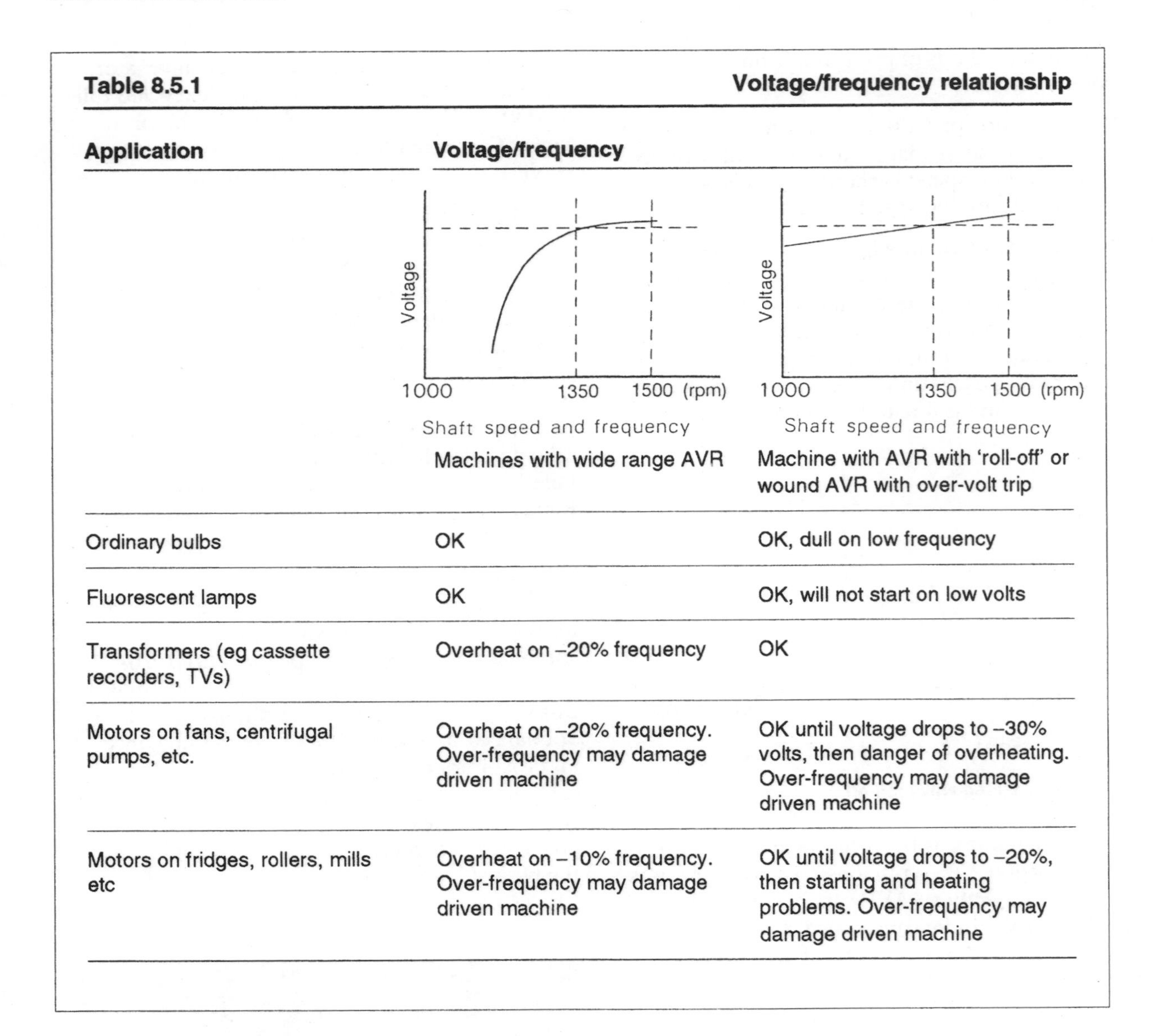

Table 8.5.1 **Voltage/frequency relationship**

Application	Voltage/frequency	
	Machines with wide range AVR	Machine with AVR with 'roll-off' or wound AVR with over-volt trip
Ordinary bulbs	OK	OK, dull on low frequency
Fluorescent lamps	OK	OK, will not start on low volts
Transformers (eg cassette recorders, TVs)	Overheat on –20% frequency	OK
Motors on fans, centrifugal pumps, etc.	Overheat on –20% frequency. Over-frequency may damage driven machine	OK until voltage drops to –30% volts, then danger of overheating. Over-frequency may damage driven machine
Motors on fridges, rollers, mills etc	Overheat on –10% frequency. Over-frequency may damage driven machine	OK until voltage drops to –20%, then starting and heating problems. Over-frequency may damage driven machine

Photo 8.5.4 An alternator driven by a Francis turbine. (Sri Lanka)

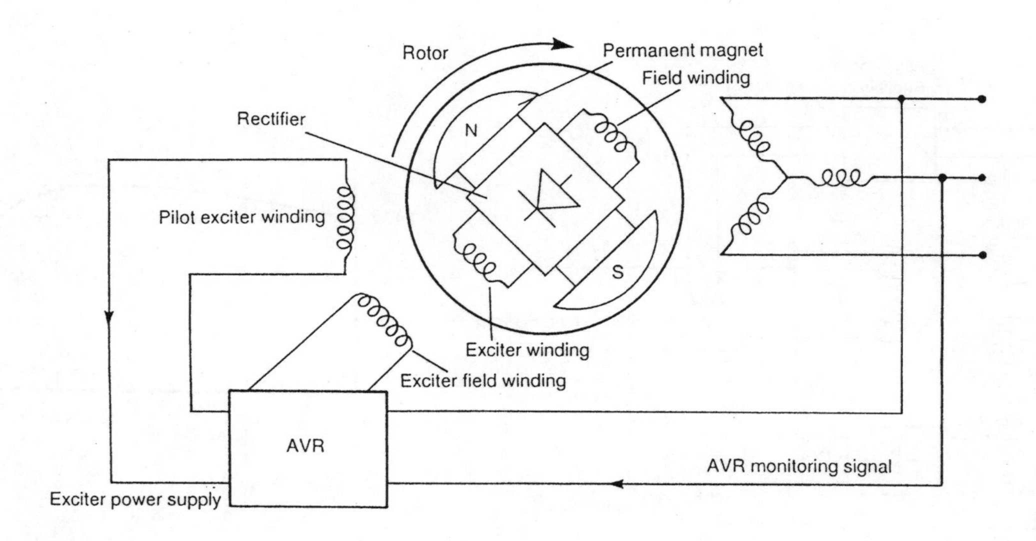

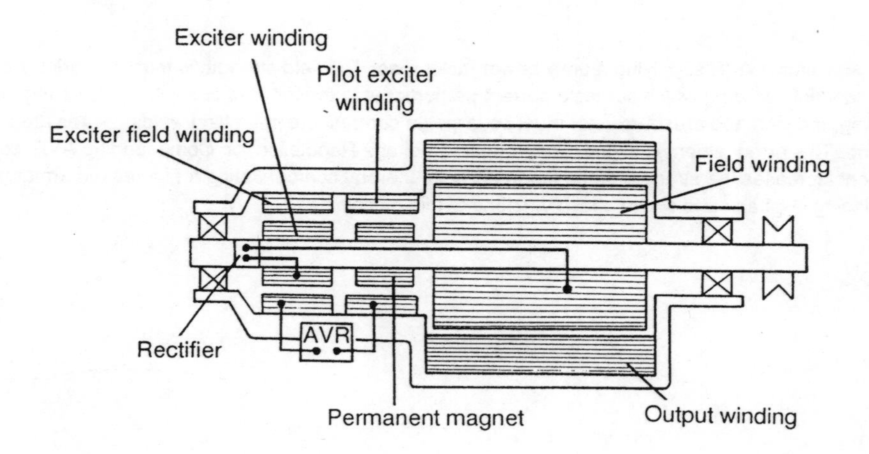

a In this arrangement power for the exciter field winding is drawn from a permanent magnet and pilot exciter winding.

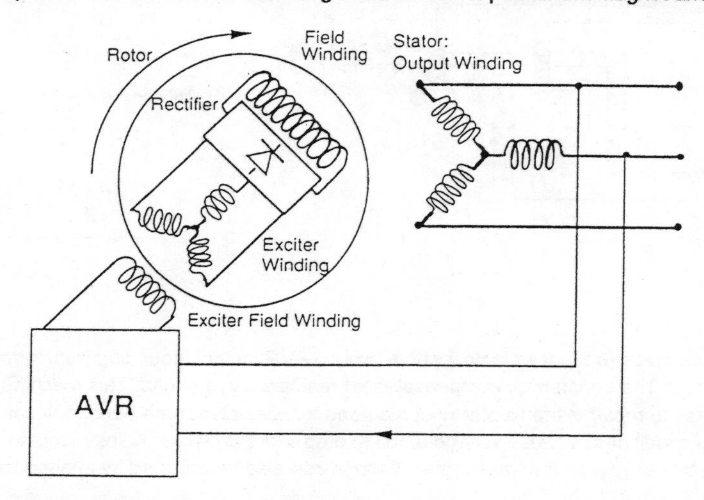

b In this arrangement power for the exciter field winding is drawn from the output of the generator itself.

Fig 8.5.11 Modern type brushless alternator

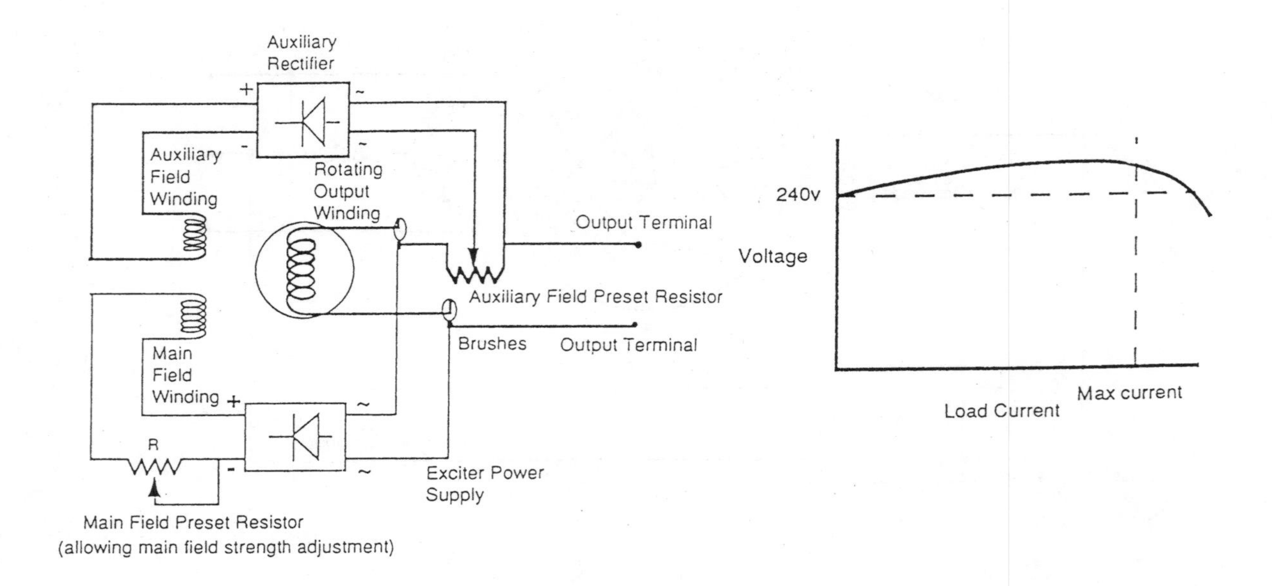

a A self-excited machine, with simple AVR supplying a unity power factor load. The field strength is manually adjusted. The AVR consists of an auxiliary field winding which carries a current proportional to the full load current. Because this current is large, the auxiliary winding, rectifier, and preset resistor must be large. In contrast the main field winding wires are thin, carrying only about 1 amp. This arrangement is called 'Compounding Voltage Regulation' or 'Compounding AVR' because an increase in load current increases the strength of the magnetic field, thereby compensating for increased armature reaction in the output winding. 3-phase and single phase versions of this machine exist.

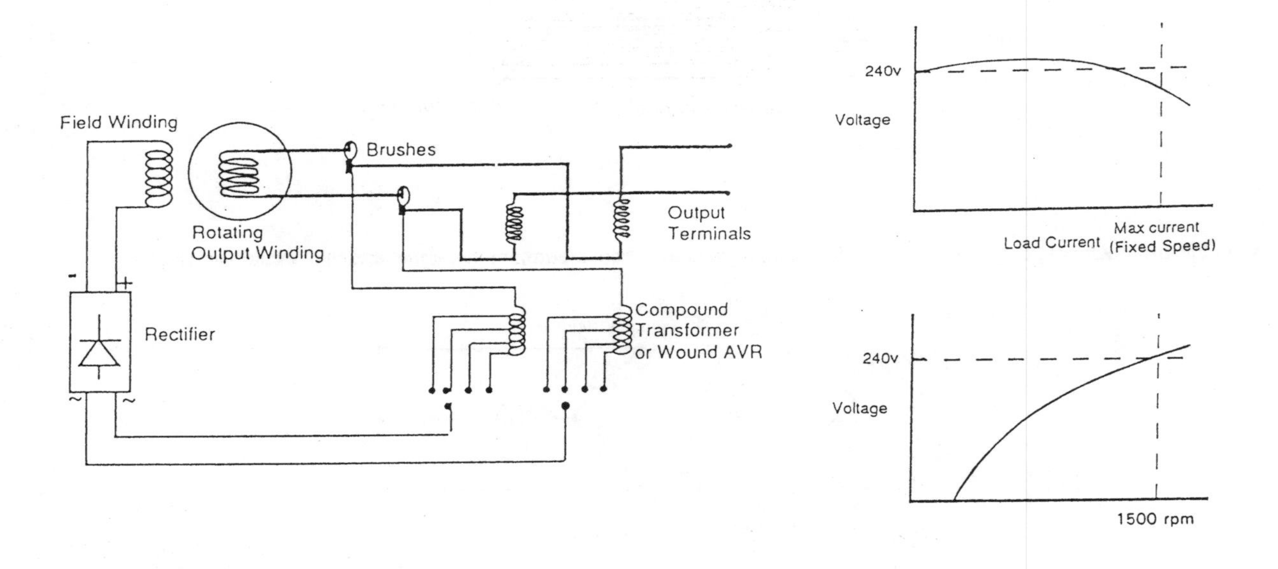

b Self-excited machine for reactive loads (0.8 power factor) with a 'wound AVR' or 'compounding transformer'. Both single phase and 3-phase versions exist. The transformer compensates for reactance in the load. This machine has proved to be highly reliable and relatively easy to repair in the field without the need for specialist spare parts replacements. It is safe to use with motors in underspeed conditions because voltage tends to drop with frequency. Adjustments to no load and full load voltage are made by changing the air gap on the transformer. Voltage can also be adjusted by altering the connections to tappings on the transformer.

Fig 8.5.12 Various AVR/excitation designs These are becoming outdated but are still found on older installations and in some cases are still manufactured. The wound AVR is a very robust device for remote locations.

The electronic load controller (ELC)

The electronic load controller is essentially a convenient, reliable and responsive type of turbine governor intended, as with other types of turbine governors, to keep the turbine speed constant. Its operation was described in Chapter 6. Both 3-phase and single phase load controllers are available.

Need for an AVR with a load controller

The load controller ensures that the synchronous generator (alternator) runs at constant frequency, and thus speed, by imposing a constant total resistive load. It might thus be thought that the field excitation of the generator could be fixed and the expense and problems associated with an AVR might be avoided. However, if the power factor of the load changes – for example, decreases – then the required generator current will increase even though the load controller ensures that the power generated (and speed) remains the same. As a result, the ac output voltage would decrease if the excitation were fixed and, except with purely resistive loads, an AVR must be provided.

Generator size, load controller installation, and ballast sizing

Details of the ratings and types of electronic load controllers should be obtained from the manufacturer. Installation should be as advised by the instruction manual for the controller obtained.

The manufacturer should be consulted with respect to the size of the generator. Strictly the generator should be sized following the "add 60%" rule given below but certain conditions may allow the use of a smaller generator.

The ELC varies the current flow to the ballast load by adjusting the phase angle at which conduction starts in each cycle. As a result, under conditions when the conduction is retarded by 90° (half of a half cycle), the ballast load appears to the generator and supply system as a load having a lagging power factor of about 0.75. The same effect occurs at lesser switching phase angles (greater power to the ballast load and higher effective power factor) and greater switching angles (less power to the ballast load and lower effective power factor). However, the worst condition as far as the reactive volt-amp requirement is concerned is a switching angle of

Photo 8.5.5 Assembly of an ELC. (IT/Jeremy Hartley – Nepal)

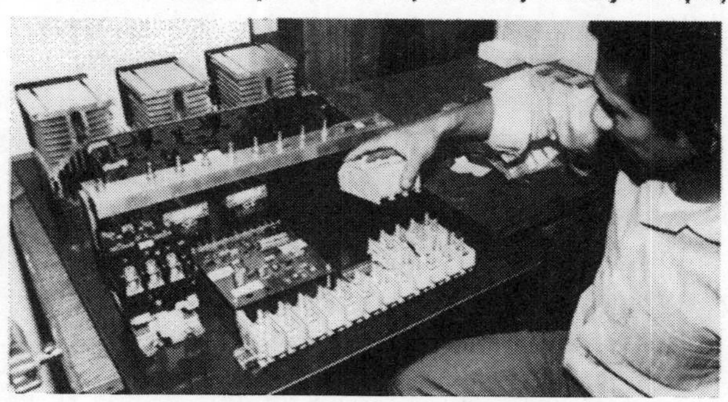

about 90°. This is the condition when the ballast load is being used to about half its rated power dissipation capacity.

In this condition a user load of low power factor will combine with the low power factor effect of the ballast, so demanding from the generator excessive reactive power. This will create large currents capable of damaging the AVR or generator windings. If for instance a user load of 0.8 power factor occurs while the ballast is drawing half the generated power, then only a generator sized according to the following "add 60%"rule can provide sufficient reactive power:

Generator KVA = (maximum kW load/Pf) x 1.6

For a user power factor of 0.8 this implies a KVA rating of twice the maximum kW demand. This allowance will be satisfactory if the ballast is not oversized. It is recommended that the power dissipation capacity of the ballast load should be between 5% and 15% greater than the usual expected maximum power output of the generator.

Water-cooled ballast

The water-cooled dump load is the most common type in existing micro-hydro ELC installations. Its great advantage is that it is compact and can provide a source of heated water. The cooling water may be obtained from a small outlet from the turbine penstock or, preferably, from a secondary local source of water if this is available. It would normally consist of a correctly rated insulated water heater immersed in water flowing through a suitable tank (ballast tank) arranged as shown in Fig 8.5.13. The important features of the installation are as follows:

1 If the cooling water is taken from the turbine penstock, its flow is likely to be controlled by a valve feeding a relatively small pipe leading to the ballast tank. Since the water in the penstock will be at high pressure and the

cooling water flow is likely to be moderate, the valve will normally only be slightly open. Steps must therefore be taken to avoid grit reaching the valve from the penstock and thus blocking the flow through the valve. Careful choice of the position and angle of the water 'take off' from the penstock will help in this.

2 Whatever the source of cooling water, the ballast tank should be fitted with a thermostat. An immersion heater thermostat is suitable and its contacts should be arranged so that if the water temperature rises above 50°C, they:

- cause a suitably placed warning bell to ring indicating dump load overheat,
- and if necessary they also cause any jet deflectors (if provided in the turbine) to operate and thus prevent the turbine from overspeeding.

An alternative or an additional failsafe device is a water level switch which rings a separate alarm bell. *In all cases the ballast must not be disconnected either manually or automatically.*

3 Several water heating elements connected in parallel should be used rather than one element of the required total rating. This will increase reliability since failure of one element will not be disastrous. If only one large element were used, its failure would lead to turbine overspeed when the supply system load became small.

4 If the ballast tank is metal it should be earthed for safety.

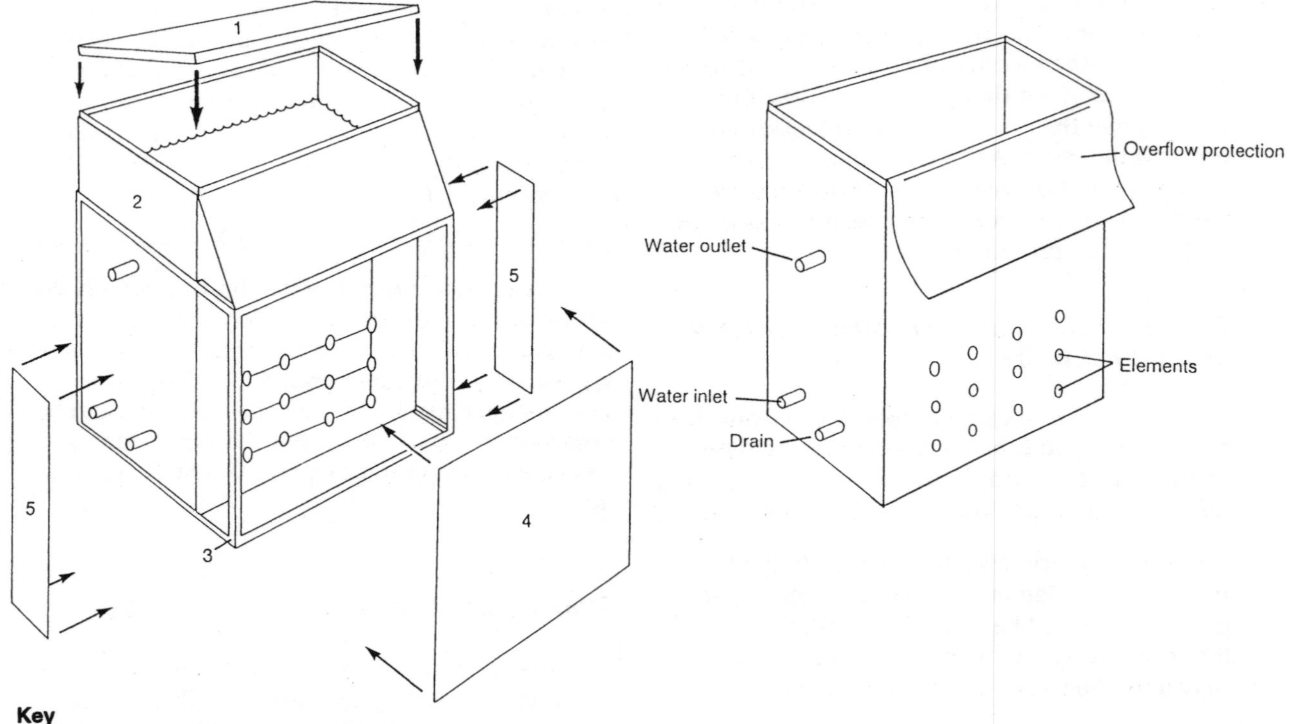

Key

1 Top cover or lid. This should fit onto a ledge *inside* the tank so that condensation drips remain inside.

2 Ballast tank made from sheet copper or stainless steel – for instance, dimensions 1000 × 300 × 650 mm.
Fitted with kettle elements on one side and pipe fittings on another side. The tank is also fited with a thermostat that is wired into an alarm, and ideally also a low water level switch which sounds a second alarm. Cold water enters the tank near the bottom and exits near the top. The kettle elements are fitted towards the bottom of the tank. The elements are wired into a terminal box which is on the same side as the elements. The sloping top cover above the elements prevents water falling onto them if an overflow occurs. (Alternatively put large overflow pipes on the side opposite the electrics).

3 An angle-iron frame which supports the ballast tank and cover plates.

4+5 The side cover plates protect the wiring and electrical connections.

Plumbing: Note the water outlet is above the elements to ensure they are immersed. The drain valve should have its handle removed to protect from accidental opening.

Inlet pipe: It is safer to enter the inlet at the top and then take it down inside the tank. Perforate with small holes to prevent siphoning.

Fig 8.5.13 Water-cooled ballast tank for an ELC

Air-cooled ballast

Suitable domestic or industrial air convector heaters may also be used as dump loads. The important features of this type of installation are as follows:

1. As in the case of water-cooled dump loads, several convector heaters connected in parallel should be used rather than one large heater of the rating required. The number required may be calculated as for water-cooled dump loads. At least two convector heaters should be used.
2. It is essential that convector heaters are located in a dry, well ventilated building with adequate space for dry air to enter and leave the heater *as specified by the heater manufacturer.*
3. The heaters may be used for practical purposes such as room heating, providing their position fulfils 2 above and there is no chance of the airflow through them being limited. Again, *the manufacturer's instructions as to location and use must be followed.*
4. A check should be made to ensure that the heaters used are fitted with over-temperature trips. These will interrupt the current if the heater temperature becomes too great, perhaps because the airflow has been impeded. An appropriate fuse should be fitted in series with the high voltage end of each heater as an additional safeguard.

   Note that if either the over-temperature trip operates or the fuse blows, the dump load will be reduced by the power rating of the heater concerned. Thus, it is important to know it has happened (fit an alarm) and investigate the cause of the failure as soon as possible and remedy it so as to avoid possible turbine overspeeding.
5. Convector heaters with adjustable thermostats should be avoided. They are more expensive than the simple type with only an over-heat trip, and inadvertent reduction of the thermostat setting could lead to the generator drive turbine overspeeding.

Photo 8.5.6 An ELC is used to govern a rehabilitated Pelton turbine. Note the water-cooled ballast tank in the far corner. The original mechanical governor is still connected to the Pelton spear valve but is now out of use. (IT/Jeremy Hartley – Sri Lanka)

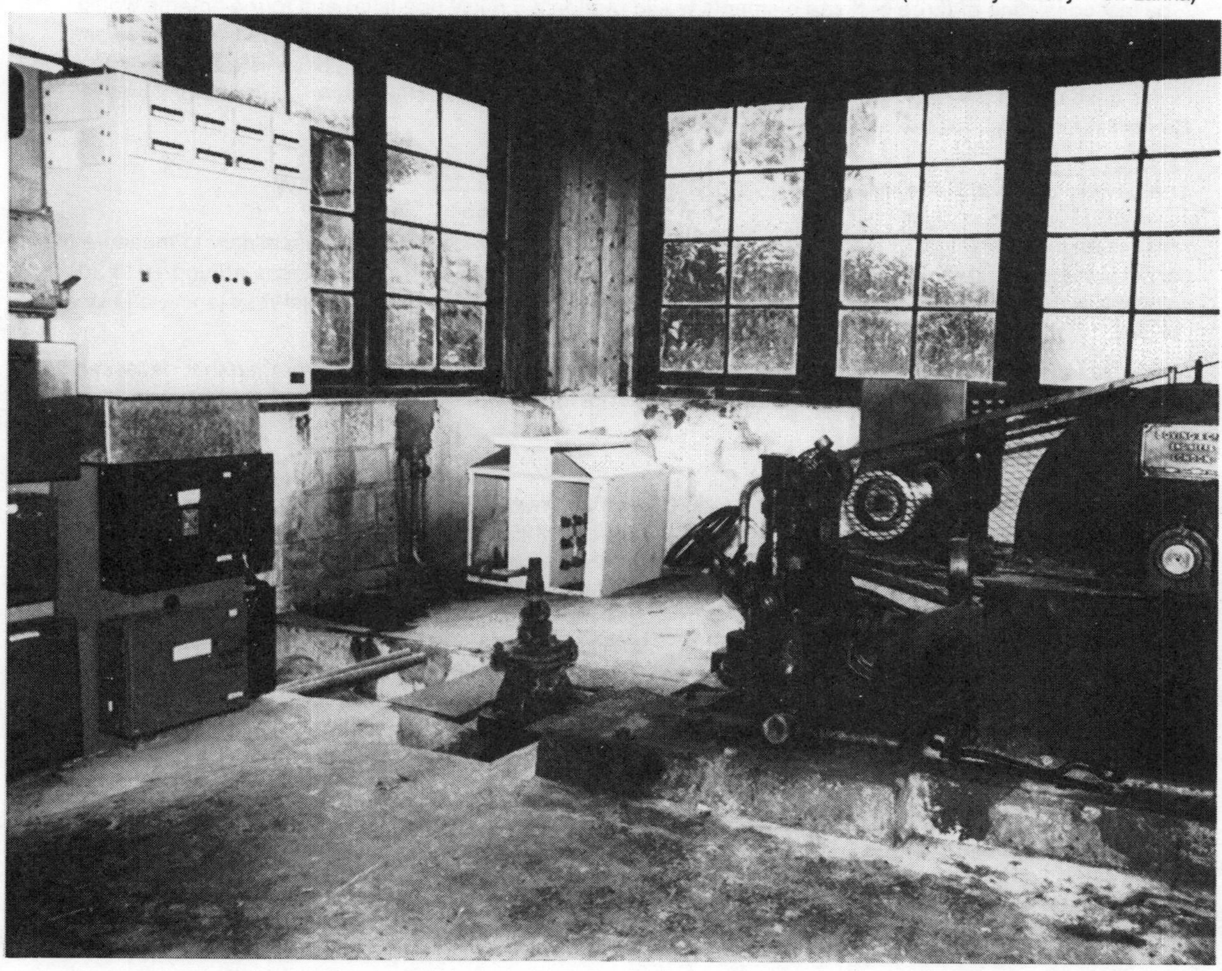

Example 8.5.1 **Sizing of electronic load controller ballast (water-cooled)**

A micro-hydro system is being designed which must be capable of supplying a single phase, 230 V, 50 Hz, electrical distribution system having a predicted maximum demand of 6.0 kW at 0.8 power factor lagging.

Although the maximum expected demand ('m.e.d.') is 6.0 kW at 0.8 pf (implying 7.5 kVA) it is necessary to add an extra 60% to the kVA capacity to ensure generator and AVR reliability, and accommodate the effect of the ELC thyristor action which would be damaging to a generator running at full output. In this case choose a generator of 12 kVA capacity.

For a water-cooled ballast it is possible to use ordinary kettle elements so long as they are proven to be reliable. These elements should provide sufficient load on the generator such that when the user load is off, and the turbine is providing full power to meet the 6 kW m.e.d, all the power can be safely absorbed by the ballast without turbine and generator speed increase. The voltage across the ballast should then read between 210 V and 220 V. The generator output voltage would normally be around 230 V, about 10 volts greater than the ballast voltage. The ballast voltage indicates the amount of power absorbed by the ballast, because it reflects the effect of the chopped waveform. At full ballast power the waveform cannot return completely to normal, so there is still a small voltage reduction.

To estimate in advance the necessary resistance calculate the power dissipated by the kettle elements when run about 10 volts under rated voltage. Suppose the only elements available are rated 2 kW/230 V and each one consists of two 1 kW sub-elements. Since Power = V^2/R this implies a sub-element resistance of

$$\frac{V^2}{\text{power}} = \frac{230 \times 230}{2000} = 26.45 \text{ ohms } (\Omega)$$

At 220 V the power dissipated is $\frac{V^2}{R} = \frac{220 \times 220}{26.45} = 1830$ watts

A ballast consisting of three of these elements would dissipate 5.5 kW (too little) and four elements would dissipate 7.32 kW, which is 7.3/6 = 22% overballasting. In practice 10% overballasting is the target to aim for, and overballasting above 15% should be avoided as the ELC will be loaded above its rated capacity. In this case one of the sub-elements could be removed, so that the ballast effectively consists of 7 sub-elements in parallel, a total resistance of $(26.45 \times 2)/7 = 7.56\ \Omega$.

The power which would be dissipated at 220 V is $\frac{V^2}{R} = \frac{220 \times 220}{7.56} = 6.4$ kW

which is an overballasting of 6.4/6 = 6% which is acceptable. In practice it will be necessary to measure the ballast voltage with the main load off and adjust the ballast resistance until the reading is around 10 to 20 volts below the generator voltage. If the ballast voltage reads low, increase the ballast resistance, and vice versa.

Notice that the ELC and ballast will be overloaded if the turbine produces more power than that necessary to meet the 6 kW m.e.d. The turbine should not be sized not to allow this or should have flow limiting controls.

Water flow rate

If a water-cooled ballast load were chosen, using seven 1 kW kettle sub-elements, then it would be wise to ensure that the cooling water flowing over the elements is sufficient to dissipate the maximum power ever likely to be produced by the generator. (If a still water tank is used, then there must be a failsafe method of replacing water continuously which is boiled away. Tail water can be used for this).

With a generator output of 6 kW and the maximum temperature of the water entering the ballast tank 25° C, then assuming the ballast tank outlet water temperature is not to exceed 50° C (just too hot to keep one's hands in continuously), then the rate of heat dissipation must be 6000 W, that is, 6000 Joules/second, and, as the specific heat of water is 4200 J/kg K (Joules per kilogram per degree K) for the temperature range concerned,

$$\text{Flow required} = \frac{\text{maximum demand}}{4200 \times \text{temp difference}} = \frac{6000}{4200 \times (50 - 25)} = 0.057 \text{ kg/sec of water.}$$

1 litre of water weighs 1 kg at the temperature concerned. Thus, the required water flow rate must be greater than 0.057 litres/second or, say, **0.1 litres of cooling water per second**.

8.6 Induction generators

The induction generator is a simpler and more reliable machine than the synchronous generator. It contains fewer parts, is less expensive, and is more easily available from electrical suppliers. It can withstand 200% runaway speeds without harm, and has no brushes or other parts which require maintenance. These factors all make the induction generator an attractive choice for micro-hydro installations.

An induction generator can be designed as such but normally an induction motor would be used. Because induction motors are the most common type of electrical machine, their production on a large scale makes their use as induction generators attractive. Not only are they simple, robust machines, they are also cheap and plentiful.

Induction generators may operate on a stand-alone basis or they may be connected to an existing supply. In considering their operation it is easiest to start by dealing with the latter type.

The practical problems of grid connect are not covered here. Nevertheless, it is useful to introduce the topic of induction generators by first considering a grid connect system.

Induction motor operation

The stator winding is a 3-phase distributed winding of the same type as a synchronous generator armature (ac load) winding. When this is connected to a 3-phase supply, lagging magnetizing current flows from the supply at a level dependent on the supply voltage and creates a rotating magnetic field in the machine at a synchronous speed directly determined by the supply frequency and the number of stator winding 'poles' as in a synchronous generator. The field produced is just the same as if an energized synchronous generator rotor were rotated at the same speed.

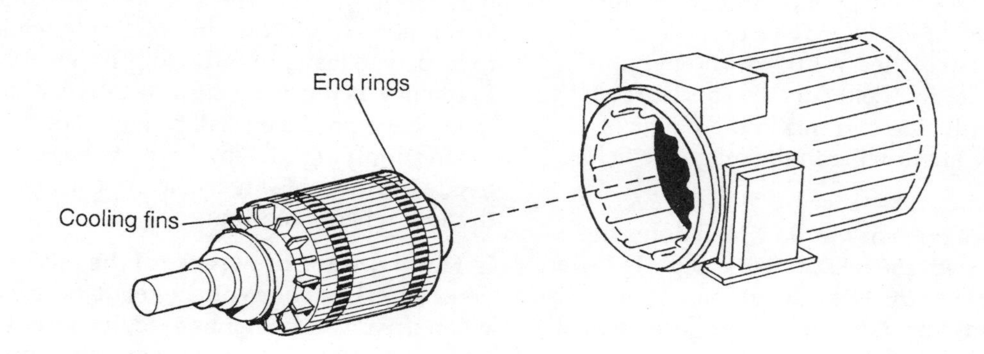

Rotor: 'Squirrel Cage'

The rotor speed (N) is:

$$N = \frac{120 \times freq}{p}(1 - s) \text{ rpm}$$

where s = slip (negative for a generator)
and p = number of poles

For instance, a 4-pole induction generator producing 50 Hz supply has negative slip and runs at between 1500 and 1575 rpm depending on the load. For a very small generator the maximum slip will be around 2% rising to 5% for sizes approaching 50 kW.

A motor has positive slip so with 4 poles on a 50 Hz supply it will produce a shaft speed of between 1425 and 5000 depending on load.

Stator: Balanced 3-phase winding.

The speed of the rotating field due to the magnetizing component of the current in the stator windings is called the *synchronous speed* (N_s).

This is the same as the speed of the rotor in a synchronous generator with the same number of poles and producing the same frequency. That is:

$$N_s = \frac{120 \times freq}{p} \text{ rpm}$$

where p is the number of poles

The slip is defined as:

$$s = \frac{(N_s - N)}{N_s}$$

Fig 8.6.1 Induction generator or motor

Photo 8.6.1 Induction motor (or generator) showing the 'squirrel cage' rotor.

The simplest and most common type of induction machine rotor is the 'squirrel cage' or 'cage' type (see Fig 8.6.1) which consists of a set of copper or aluminium bars laid, or cast, in many evenly spaced slots in the rotor lamination stack. These are short-circuited at both ends of the lamination stack by copper or cast aluminium end rings, one of which may be designed to incorporate fan blades or fins for cooling purposes. The rotating field cuts the short-circuited rotor bars, inducing currents in them which, because they are flowing in the magnetic field, react with it.

The result is a force or torque which drags the rotor round with the field, but at a slightly lower speed, otherwise the effect would not occur – no currents would be induced if the rotor turned at the same speed as the field.

The difference between the speed of the rotor and the speed of the rotating field is called the 'slip' (s) and is defined as:

$$\text{Slip, s} = \frac{(N_s - N)}{N_s}$$

where N_s is the synchronous speed of the rotating field.

N is the rotor speed

On no-load the slip of an induction motor may be very small indeed. It increases, however, as the load is increased. For a small motor, say 1 kW, the full-load slip may be 0.05 (that is 5%). For a large machine, say 100 kW, the full load slip may be less than 0.01 (that is, 1%) – see Example 8.6.1.

Supply-connected induction generators

Imagine that an induction motor, still connected to its supply of electrical power, is running; imagine that its shaft is then coupled to the drive shaft of a turbine which is rotating above synchronous speed. The slip now becomes negative and a current flows *to* the supply.

However, it still takes a lagging magnetizing current from the supply in order to create the rotating field, just as though it were a motor. Again, full-load power output is achieved at a slip of similar value (but negative) to the full-load motoring slip.

Stand-alone induction generators

The lagging reactive magnetizing current of an induction machine can be supplied by capacitors and in the case of an induction motor this means that the capacitors, if of the correct value, correct the power factor to unity. Ideally, the same value of capacitors connected across the terminals of the same induction machine will cause it to generate at the same voltage as that supplied to it as a motor if it is driven at the same speed in stand-alone conditions. In practice it would have to be driven faster in order to produce the same frequency as the slip is now negative. Moreover, the voltage produced will be less, especially when supplying current, as the voltage drops inside the machine are in the opposite direction to those when motoring.

In fact one of the drawbacks of the induction generator is its poor voltage regulation. Even when driven at constant speed, its output voltage drops rapidly with increasing load (see Fig 8.6.2). On the other hand, the voltage is

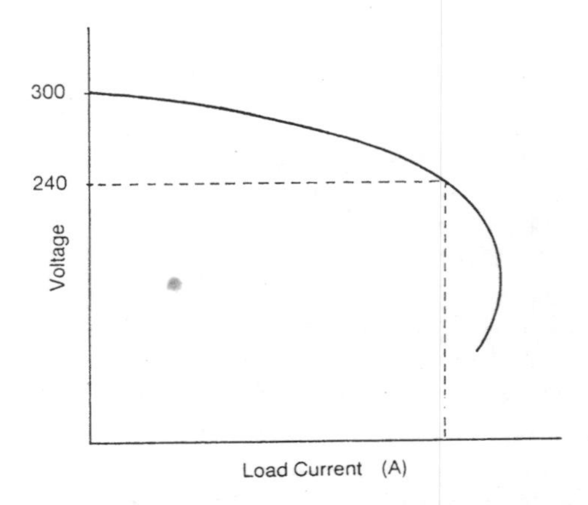

Fig 8.6.2 Voltage regulation of a self-excited (stand-alone) induction generator when driven at constant speed

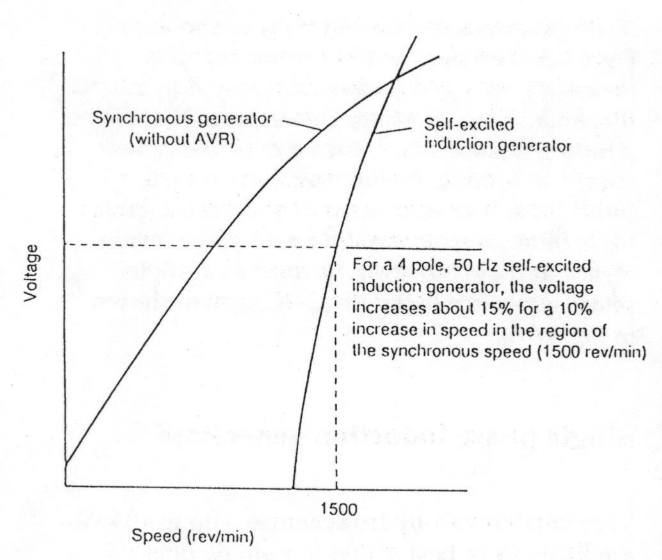
Synchronous generator
(without AVR)
Self-excited
induction generator
Voltage
For a 4 pole, 50 Hz self-excited
induction generator, the voltage
increases about 15% for a 10%
increase in speed in the region of
the synchronous speed (1500 rev/min)
1500
Speed (rev/min)

much more dependent on the speed with which it is driven than is the synchronous generator (see Fig 8.6.3). The water turbine driven induction generator can in fact be used to operate multiple fixed resistive loads, such as village lights, with only the addition of capacitors, if some voltage variation and de-excitation problems are tolerated for the sake of very low cost. The development, in recent years, of a simple voltage and speed controller, the induction generator controller (IGC), which has overcome these problems and made the stand-alone induction generator an increasingly attractive proposition for micro-hydro schemes. The IGC is described and its operation is explained later.

Fig 8.6.3 Variation of voltage with speed: comparison of synchronous with self-excited (stand-alone) induction generator

Selection of turbine speed for an induction generator — Example 8.6.1

Slip can be defined as:

$$s = \frac{N_s - N}{N_s} \times 100$$

where s is slip (%), N_s is the speed of the field, and N is the speed of the rotor. This is easily turned round to give an expression for motor shaft speed (in revs per second):

$$N = N_s \left(1 - \frac{s}{100}\right)$$

The relationship between field speed (N_s), frequency and the number of poles is exactly the same as with a synchronous generator, so that:

$$N_s = \frac{2 \times f}{p} \text{ revs per second}$$

where f in frequency in Hertz and p is the number of poles. The shaft speed (expressed in rpm) of the induction motor is then:

$$N_{rpm} = \frac{2 \times f \times 60}{p} \left(1 - \frac{s}{100}\right)$$

As described in the text above, the induction generator differs from the induction motor only in that the slip is reversed and is thus negative, so that the same equation for speed is used.

Suppose that you have an induction motor with four poles and a mains frequency of 50 Hz.

1 What is its shaft speed if its full load slip is 6%?

2 What speed should the turbine provide to run this motor, as an induction generator, to achieve 50 Hz at full load?

1 As a motor running at full load:

$$N = \frac{2 \times 50 \times 60}{4} \left(1 - \frac{6}{100}\right) = 1410 \text{ rpm}$$

2 As a generator, at full load, the slip is – 6%, and

$$N = \frac{2 \times 50 \times 60}{4} \left(1 - \frac{-6}{100}\right) = 1590 \text{ rpm}$$

3-phase induction generators may be either star or delta-connected and the excitation capacitors connected across the terminals may, theoretically, also be in star or delta as shown in Fig 8.6.4. If the generator is star-connected, the capacitors may be connected in either star or delta. If they are connected in star, the capacitor star point should not be connected to the generator and system neutral as waveform distortion occurs. Moreover, three times as much capacitance is required in comparison to the delta connection, though lower voltage and thus, theoretically, cheaper capacitors may be used. If the capacitors are connected in delta then less capacitance is required as indicated and, because of the standard voltages for which capacitors are manufactured, the cost per microfarad may not be greater than for the star-connected case. Delta-connected excitation capacitors are thus almost always cheaper and to be preferred.

If the generator is *delta-connected* as shown in Fig 8.6.5, then the third harmonic currents necessary for a good waveform may flow round the delta of the winding. However, as there is no winding neutral point, such a machine cannot supply a 3-phase, 4-wire distribution system (with three lines and neutral) and such a generator is thus *not recommended for a 3-phase supply system*. It may, however, be used as a single phase generator using the C-2C system shown in Fig 8.6.6.

Single phase induction generators

Very small micro-hydro schemes – up to 10 kW – are likely to be best suited to a single phase supply and thus require a single phase induction generator. 'Capacitor-run' single phase induction

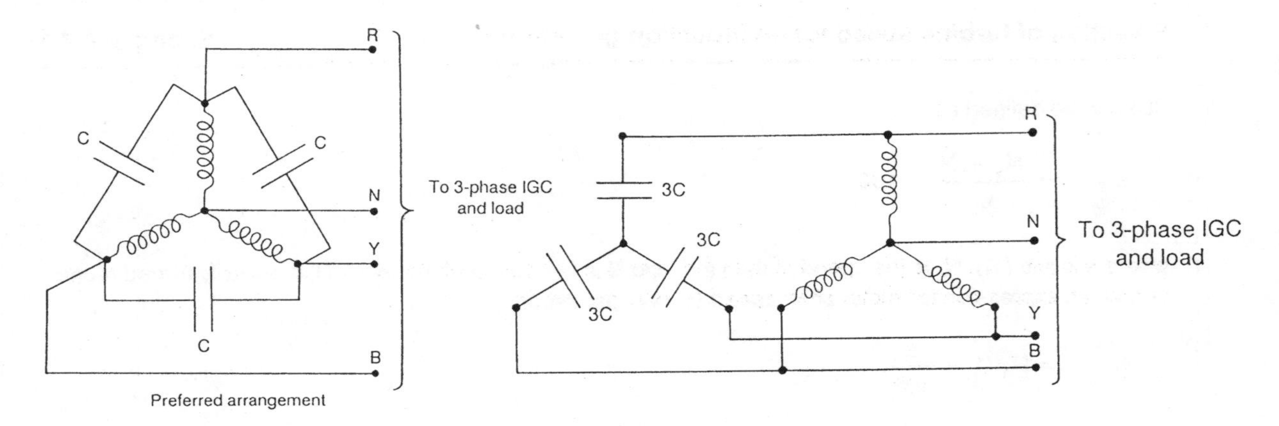

Fig 8.6.4 Star-connected induction generator with delta or star-connected excitation capacitors

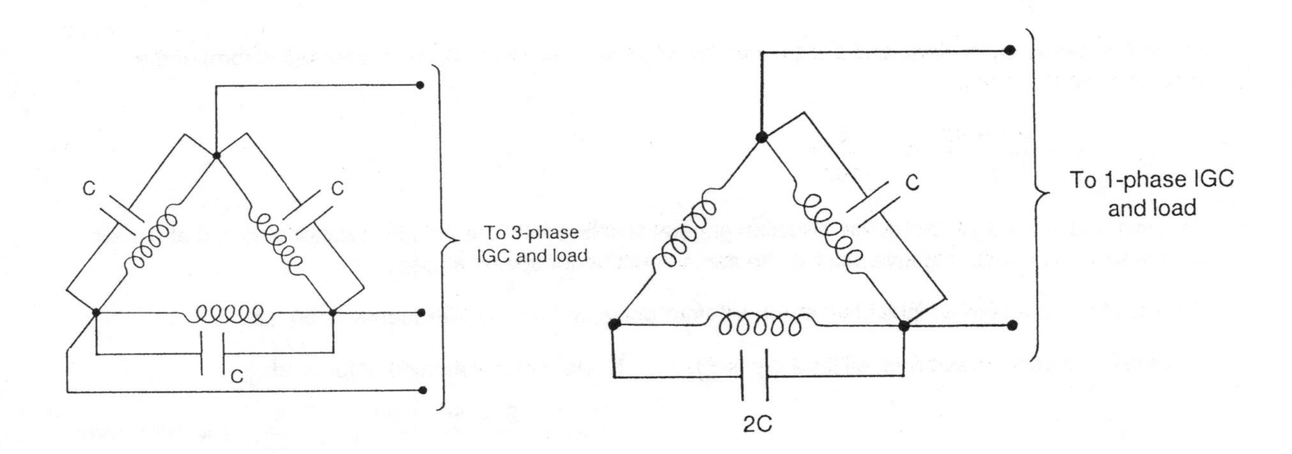

Fig 8.6.5 Delta-connected induction generator with excitation capacitors

Fig 8.6.6 Single phase supply system using a 3-phase induction generator and excitation capacitors connected in the C-2C arrangement. A delta (as shown) or star-connected generator may be used

motors may be used for this purpose, but problems can be experienced in achieving easy excitation and in deciding the size and arrangement of the capacitors required. Moreover, machines of this type are not normally manufactured in sizes above 2 kW. As a result, such machines are not recommended at present. Some work is in progress, however, to determine the best excitation arrangements and, more important, to find the optimum way in which larger, 3-phase induction machines may be rewound for single phase operation.

For the present, a very satisfactory alternative is to use a 3-phase induction generator with excitation capacitors connected as shown in Fig 8.6.6, where the value of the capacitance C is that which would be used between lines for a delta excitation capacitor arrangement (see Fig 8.6.5) using the same induction machine as a 3-phase generator. The C-2C arrangement was developed in New Zealand and allows the generator to supply single phase power at 80% of the machine's rating as a 3-phase generator.

In using the C-2C excitation arrangement it is essential to ensure that the direction of rotation of the machine rotor is correct in relation to the phases to which the C and 2C values of capacitance are connected. The easiest way to achieve this is probably a practical load test.

If the capacitors are arranged correctly the 'C phase' will produce its peak voltage before the '2C phase'. If the opposite occurs, the generator will run inefficiently and overheat.

The following load tests can be used to determine the correct capacitor arrangement. A constant input power must be used for both tests; this is usually best achieved by running the turbine at maximum power output. Connect C to one phase of the machine and 2C to another phase. Increase the turbine output power until 240 volts is produced across the 'C phase' then add resistive load to the 'C phase' and further increase the turbine power output to maintain 240 volts. When turbine maximum power output is reached, measure the load current. Repeat the test with the 2C capacitance connected to the phase that previously had no capacitance across it. A higher load current reading will be obtained for the correct capacitance arrangement because the generator will run more efficiently. A 10% efficiency improvement is typical.

This test is very easy to perform if an IGC is used because the controller will maintain a constant voltage and the ballast meter will indicate which connection produces the maximum power output.

Induction generator specification

The required volt-amp (or power) and power factor output capacity of the induction generator to be used for a micro-hydro scheme is calculated from the expected loads in the same way as for a synchronous generator (see Examples 8.4.1, 8.4.2 and 8.4.3). Providing that an IGC is to be employed and the load power factor is not, at worst, less than 0.8 lagging, then an induction generator of the output rating determined in this way may be used. If the worst reactive volt-amp requirement of the load is such that a full load power factor of less than 0.8 lagging is necessary, then the largest, more inductive loads should have their power factor corrected (using capacitors permanently connected across them) so that the overall maximum demand load has a power factor of at least 0.8.

Selection of an induction motor for use as an induction generator

Manufacturers do not make induction machines specifically as generators in the small sizes likely to be required for micro-hydro schemes. It is thus not appropriate to write asking a manufacturer for induction generators or even induction motors for use as induction generators in the same terms as one would for a synchronous generator. Instead, as much technical data as possible should be obtained from as many manufacturers as possible. Fig 8.6.7 shows typical full induction motor data published by a manufacturer. Rating for rating, induction motors tend to cost less than synchronous generators up to about 25 kW capacity. Larger sizes of induction motor cost more than a synchronous generator of the same rating, but may still be selected in preference to the latter because of the other advantages of induction machines.

To choose a motor to act as a generator, simply divide the generator rating that you require by a derating factor of 0.8. In other words if you wish to supply 10 kW choose a motor of 10/0.8 = 12.5 kW rating. This allowance compensates for possible load imbalance in the case of 3-phase machines and winding imbalance in the case of single phase machines.

This simple electrical derating rule is sufficient in almost all cases. Nevertheless, it is worthwhile being aware that mechanical limitations also exist.The power rating (kW) of an induction motor is the rated full-load (and thus maximum) mechanical power output the machine is designed to supply continuously. The mechanical parts of the motor, principally the shaft and bearings, will have been designed to transmit this power. The electrical power input to the motor at full-load will be greater than the rated mechanical output power by the amount of the losses in the machine. Thus, if an induction motor of known rating is to be used as an induction generator, the maximum mechanical input power must not be greater than the rated output power of the machine as motor, otherwise the shaft and bearings may be overloaded. As the losses in the generating mode will be supplied from the mechanical input power, the rated output electrical power of the machine as a generator will be significantly less than its rating as a motor. The full-load efficiency of an induction machine is approximately the same in both the motoring and generating modes. Therefore when choosing a motor to act as a generator make sure it is not rated less than the generator rating you are looking for divided by its full-load efficiency as a motor. In practice this is normally the case anyway when you follow the simple electrical derating factor of 0.8 prescribed above.

When an induction machine is driven as a generator it will produce a terminal voltage approximately 10% lower than the supply voltage which would be necessary when operating on the same frequency, with the same magnetizing current, as a motor. Thus, in designing a micro-hydro system the decision may be taken to supply electric power at a voltage a little lower than the motor rated voltage of the induction machine used as a generator. For example, a 415 V (240 V line-to-neutral), 50 Hz, 3-phase, star-connected induction motor might be used as a generator to supply a 380 V (220 V line-to-neutral) 3-phase, 4-wire supply system. Alternatively induction motor induction manufacturers will, at little or no extra cost, wind standard induction motors for operation at a voltage specified by the purchaser. Thus, any standard supply system voltage can be obtained by specifying correctly the voltage rating of the motor to be used as a generator.

Alternatively, a 415 V (240 V line to neutral) 3-phase, 50 Hz, induction motor could be used as a generator to supply a system of the same voltage rating by increasing the generated frequency from 50 to about 53 Hz. This would be acceptable using an IGC, providing that the load does not contain induction motors rated at 50 Hz which drive fans or pumps. The torque of such loads increases as the square of speed and might lead to the motors overheating at the slightly higher frequencies that the IGC may allow. It is worth noting that a 10% increase in frequency will increase the generated voltage by about 15%.

The selection of a standard induction motor for use as an induction generator is illustrated in Example 8.6.2.

Mechanical considerations

Induction motors with ratings up to 50 kW usually have bearings and shafts which are designed to drive through pulleys and belts. However, it is important to check with the manufacturer's data, or directly with the manufacturer, that a particular machine is capable of this. It is essential to check this in the case of large machines. In both cases, a manufacturer may on request fit heavy duty bearings to a particular motor if there is some doubt whether the normal bearings will withstand belt drive forces. Such a modified machine will, of course, be more expensive than the same machine in standard form.

While most small induction motors are designed to deal with belt-driven loads, the same considerations as in the case of synchronous generators apply to such drives when an induction motor is used as a generator. In particular, if a drive pulley is mounted on the machine shaft, it should be placed as close as possible to the adjacent bearing.

Off-the-shelf synchronous generators (alternators) designed for a diesel engine drive will normally be 4-pole machines which will thus produce 50 Hz at a speed of 1500 rev/min (rpm). Slower speed synchronous generators (with more poles) suitable for micro-hydro applications may be more difficult to obtain or may have to be specially made at considerable expense.

Induction motors, on the other hand, are commonly manufactured as 2, 4, or 6-pole machines for all but the smallest ratings, as will be seen from Figure 8.6.7. The corresponding speeds, neglecting slip, for the production of 50 Hz when used as a generator are 3000 rpm (2-pole), 1500 rpm (4-pole) and 1000 rpm (6-pole). It is thus much easier to select a machine with a speed close to the maximum power output speed of the turbine. A direct drive arrangement thus becomes a possibility with the resultant advantage of longer turbine and generator bearing life. However, it should be noted that 6 and 8-pole motors are normally more expensive and may be more difficult to obtain (longer delivery) than 2 or 4-pole machines of the same rating. Belt drive speed increase pulley ratios of up to 3:1 are usually satisfactory.

Ratings and performance

The performance data listed is given in good faith as indicative values obtained from serial manufacture and may be changed from time to time without notice, as it is subject to continuing improvements in manufacture, materials and design.

Specific data will be given against specific enquiries requiring guarantee, noting that tolerances on values as specified in BS 4999 : Part 69 apply to all performance data.

The following notes and definitions covering the ratings and performance figures are based on IEC 50 (411) 1973 where applicable.

Rating

The whole of the numerical values of the electrical and mechanical quantities with their duration and sequences assigned to the machine by the manufacturer and stated on the rating plate, the machine complying with the specified conditions.

Column 1 **Rated output**

The value of the output included in the rating, expressed in kW.

Column 2 **Frame size**

Assigned to a frame having the same fixing dimensions as specified in clause 6 of IEC 72 and BS 4999 : Part 10.

Column 3 **Rated speed**

The value of the rotational speed included in the rating expressed in r/min.

Columns 4 & 5 **Rated current**

The value of the root mean square line current of the motor included in the rating when delivering rated output at rated speed with rated voltage and frequency applied, expressed in Amperes.

Columns 6,7,& 8 **Efficiency**

The ratio of output to input, expressed as a percentage. The values quoted are in line with IEC 34-2 (1960) BS269/1927 and amendments or BS4999 : Part 33

Columns 9, 10 & 11 **Power-factor**

The value of the ratio input kW to input kVA.

Column 12 **Locked rotor current**

The measured steady state root mean square current taken from the line with the motor at rest with rated voltage and frequency applied, expressed as the ratio locked rotor current to rated current.

Column 13 **Rated torque**

The value of the shaft torque of a motor corresponding to rated output and speed, expressed in Newton metres.

1 lbf ft = 1.3558 Nm
1 kgfm = 1 kpm = 9.814 Nm
1 Nm = 0.102 kpm = 0.738 lbf ft

Column 14 **Locked rotor torque**

The minimum measured torque which the motor will develop with the rotor locked and rated voltage applied at rated frequency, expressed as the ratio locked rotor torque to rated torque.

Column 15 **Pull-out torque**

The highest torque that the motor can develop whilst running at rated voltage and frequency, expressed as the ratio pull-out torque to rated torque.

Column 16 **Pull-up torque**

The smallest torque developed by the motor between zero speed and the speed which corresponds to the pull-out torque when the motor is supplied at the rated voltage and frequency, expressed as the ratio pull-up torque to rated torque.

Column 17 **Maximum locked rotor time**

The number of seconds for any part of the motor to attain the maximum temperature permitted for that part. (Commencing at the temperature attained at rated output in the maximum specified ambient) when supplied at rated voltage and frequency with the rotor locked.

Column 18 **Maximum 'starting seconds'**

Starting seconds is the calculated number of seconds for any part of a motor to attain the maximum temperature permitted for that part (commencing at the temperature attained at rated output in the maximum specified ambient) and subjected to a heating rate equal to the root mean square heating rate when accelerating the load from standstill to rated speed.

Column 19 **Maximum load inertia**

The load inertia referred to rated speed which the motor will accelerate to rated speed within the stated maximum starting seconds with rated voltage and frequency applied.

Column 20 **Rotor inertia**

The moment of inertia of a rotor with respect to the longitudinal axis of the shaft is the sum of the products obtained by multiplying the mass of each component of the rotor by the square of its distance from the axis expressed in kg m^2.

Column 21 **Rotor mass**

Rotor mass in kilograms, for a standard machine.

Columns 22 & 23 **Noise level**

The mean sound pressure level measured at one metre from the surface of the motor is expressed in dB(A) referred to a base pressure of 2×10^{-5} N/m^2.

The sound power level is expressed in dB (A) referred to a base power of 10^{-12} Watt.

Fig 8.6.7 Example of manufacturer's data for induction motors. From GEC Small Machines, England B65 0QT

Cooling form IC0141
50Hz, 3-phase. Class F insulation
Temperature rise not greater than 80K by resistance
Direct-on-line starting

Poles	Rated output	Frame size	Performance at rated output									Locked rotor current ratios	Torque ratios				Max locked rotor time	Max starting seconds	Moment of inertia		Rotor mass	Noise level	
			Speed	Current		Efficiency %			Power-factor (cos ø)				Rated torque	Locked rotor	Pull-out	Pull-up			mr^2 Load max	Rotor		Mean sound pressure level at 1 metre	Sound power level
				at 415V	at 380V	1/1	3/4	1/2	1/1	3/4	1/2												
	kW		r/min	A	A	x rated output			x rated output			x rated current	Nm	x rated torque			s		kg m²	kg m²	kg	dB(A)	dB(A)
	1	2	3	4	5	6	7	8	9	10	11	12	13	14	15	16	17	18	19	20	21	22	23
2	0.25	D63	2800	0.63	0.68	67.0	66.0	60.0	0.83	0.76	0.63	4.5	0.85	2.0	2.1	1.4	15	18	0.07	0.000363	0.96	56	64
	0.37	D71	2800	0.88	0.96	69.0	68.0	63.0	0.85	0.77	0.65	4.9	1.25	2.0	2.1	1.4	8	10	0.05	0.000533	1.42	58	66
	0.55	D71	2800	1.27	1.38	71.0	71.0	68.0	0.85	0.77	0.65	4.9	1.88	2.0	2.4	1.5	8	10	0.08	0.000533	1.42	58	·66
	0.75	D80	2800	1.70	1.86	72.0	72.0	70,0	0.85	0.79	0.68	5.0	2.56	2.4	2.4	1.7	14	17	0.20	0.000980	1.96	62	70
	1.1	D80	2800	2.35	2.60	76.0	76.0	74.0	0.85	0.79	0.68	5.2	3.75	2.4	2.6	1.7	14	17	0.30	0.00114	1.28	62	70
	1.5	D90S	2820	3.10	3.35	77.0	77.0	75.0	0.88	0.83	0.74	6.0	5.08	2.5	2.6	1.7	11	13	0.37	0.00161	3.04	65	73
	2.2	D90L	2820	4.45	4.85	78.0	78.0	75.0	0.88	0.83	0.74	6.0	7.45	2.5	2.6	1.7	11	13	0.45	0.00199	3.29	65	73
	3.0	D100L	2820	5.9	6.5	79.0	79.0	77.0	0.89	0.86	0.81	6.0	10.2	2.4	2.6	1.7	11	13	0.62	0.00643	6.08	70	78
	4.0	D112M	2820	7.7	8.4	81.0	81.0	79.0	0.89	0.86	0.81	6.0	13.6	2.4	2.6	1.7	11	13	0.82	0.00735	7.00	70	78
	5.5	D132S	2870	10.5	11.4	82.0	82.0	80.0	0.89	0.86	0.81	6.0	18.3	2.4	2.6	1.7	16	19	1 6	0.0165	7.80	80	89
	7.5	D132S	2870	14.0	15.2	84.0	84.0	82.0	0.89	0.85	0.82	6.5	25.0	2.4	2.6	1.7	14	17	1.9	0.0190	12.4	80	89
	11.0	D160M	2910	21.0	22.5	86.5	86.0	83.5	0.85	0.82	0.74	6.3	36.1	2.0	2.5	1.4	25	31	5.6	0.038	20.9	81	90
	15.0	D160M	2900	27.5	30.0	88.0	87.5	85.0	0.87	0.84	0.77	6.5	49.4	2.0	2.5	1.4	25	31	7.7	0.0463	20.9	81	90
	18.5	D160L	2920	33.5	36.5	89.0	89.0	88.0	0.87	0.84	0.77	7.0	60.5	2.0	2.5	1.4	25	31	9.4	0.0520	23.6	81	90
	22.0	D180M	2925	41.5	45.5	89.5	88.5	86.0	0.82	0.80	0.68	7.5	71.8	2.5	3.0	1.8	18	22	9.4	0.0600	24.0	81	90
	30.0	D200L	2945	58.5	64.0	89.0	87.5	83.5	0.80	0.76	0.67	5.0	97.2	1.9	2.0	1.4	25	31	12.2	0.171	46.0	93	103
	37.0	D200L	2945	70.0	76.5	89.5	90.0	87.5	0.82	0.78	0.71	5.0	120	2.1	2.0	1.5	25	31	15.0	0.187	51.0	93	103
	45.0	D225M	2945	83.0	90.5	91.0	90.0	87.5	0.83	0.80	0.74	5.5	146	2.1	2.0	1.5	25	31	18.0	0.204	58.0	94	104
	55.0	D250S	2955	95.5	104.0	91.0	90.0	87.0	0.88	0.85	0.78	6.5	178	1.7	2.6	1.5	18	22	17.0	0.442	72.0	89	99
	75.0	D250M	2960	126.0	138.0	91.5	90.5	87.5	0.90	0.88	0.81	6.5	242	1.7	2.8	1.5	14	17	21.0	0.590	90.0	89	99
4	0.18	D63	1400	0.57	0.62	64.0	60.0	52.0	0.69	0.57	0.44	4.0	1.23	2.1	2.8	1.5	20	24	0.27	0.000365	0.97	50	58
	0.25	D71	1400	0.76	0.83	66.0	63.0	56.0	0.69	0.57	0.44	4.5	1.71	2.1	2.8	1.5	14	17	0.27	0.000543	1.44	50	58
	0.37	D71	1400	1.05	1.15	69.0	67.0	59.0	0.71	0.60	0.48	4.5	2.52	2.1	2.8	1.5	14	17	0.40	0.000543	1.44	50	58
	0.55	D80	1400	1.44	1.57	70.0	70.0	65.0	0.76	0.65	0.50	4.5	3.75	2.0	2.5	1.4	12	14	0.50	0.00131	2.17	52	60
	0.75	D80	1400	1.90	2.10	72.0	72.0	65.0	0.76	0.65	0.50	5.0	5.10	2.1	2.6	1.5	12	14	0.70	0.00156	2.58	52	60
	1.1	D90S	1410	2.50	2.75	75.0	75.0	72.0	0.81	0.73	0.60	5.5	7.45	2.1	2.5	1.5	8	10	0.65	0.00343	4.06	54	62
	1.5	D90L	1420	3.45	3.75	75.0	75.0	74.0	0.81	0.73	0.60	6.0	10.1	2.1	2.5	1.5	7	8	0.78	0.00393	4.65	54	62
	2.2	D100L	1420	4.70	5.1	80.0	80.0	78.0	0.82	0.75	0.63	6.0	14.8	2.1	2.7	1.5	9	11	1.5	0.00980	7.18	54	62
	3.0	D100L	1420	6.2	6.8	81.0	81.0	79.0	0.83	0.78	0.66	6.0	20.2	2.0	2.7	1.4	9	11	2.0	0.0115	8.65	54	62
	4.0	D112M	1420	8.1	8.8	83.0	82.0	81.0	0.83	0.78	0.66	6.5	27.0	2.4	2.8	1.7	7	8	2.1	0.0135	9.95	54	62
	5.5	D132S	1440	10.9	11.8	85.0	84.0	80.0	0.83	0.78	0.68	6.0	36.4	2.2	2.5	1.6	14	17	5.5	0.0310	15.5	64	72
	7.5	D132M	1440	14.8	16.1	85.0	84.0	80.0	0.83	0.78	0.68	6.0	49.5	2.2	2.5	1.6	14	17	7.6	0.0338	16.9	64	72
	11.0	D160M	1450	21.5	23.5	87.0	87.0	85.5	0.81	0.75	0.63	5.5	72.4	2.0	2.4	1.4	18	22	15	0.064	18.1	67	75
	15.0	D160L	1450	29.0	32.0	88.5	88.5	87.0	0.81	0.75	0.63	6.0	98.7	2.3	2.4	1.7	18	22	20	0.081	26.0	67	75
	18.5	D180M	1455	35.5	39.0	89.5	89.5	87.5	0.81	0.75	0.63	6.0	121	2.5	2.6	1.8	12	15	19	0.0983	31.0	71	79
	22.0	D180L	1450	41.0	44.5	90.0	89.5	88.5	0.83	0.78	0.68	6.5	145	2.5	2.7	2.0	20	25	29	0.116	40.5	71	79
	30.0	D200L	1460	56.0	61.5	90.5	90.5	89.5	0.82	0.76	0.65	6.0	196	2.3	2.4	1.7	20	25	48	0.188	50.0	75	81
	37.0	D225S	1470	67.5	73.0	92.0	92.0	91.0	0.83	0.81	0.75	6.5	240	2.5	2.4	1.8	20	25	59	0.343	72.0	75	85
	45.0	D225M	1460	80.5	88.0	92.5	92.5	91.5	0.84	0.82	0.75	6.5	294	2.5	2.4	1.8	20	25	59	0.378	87.0	80	90
	55.0	D250S	1480	98.0	107.0	93.0	92.5	91.5	0.84	0.78	0.68	6.5	355	1.9	2.6	1.5	15	30	52	0.625	82.0	78	88
	75.0	D250M	1480	133.0	145.0	93.5	93.5	92.5	0.84	0.78	0.68	6.5	484	2.0	2.8	1.5	11	24	62	0.750	100.0	78	88
6	0.37	D80	910	1.15	1.25	66.0	64.0	58.0	0.68	0.53	0.44	4.0	3.89	1.9	2.1	1.4	25	31	1.6	0.00161	2.66	50	58
	0.55	D80	920	1.66	1.80	67.0	65.0	60.0	0.69	0.59	0.46	4.0	5.72	1.9	2.1	1.4	25	31	2.4	0.00161	2.66	50	58
	0.75	D90S	930	1.90	2.05	73.0	71.0	56.0	0.75	0.65	0.48	4.5	7.7	2.0	2.2	1.4	25	31	3.3	0.00340	4.03	55	63
	1.1	D90L	930	2.90	3.15	73.0	71.0	66.0	0.72	0.62	0.45	4.5	11.3	2.0	2.2	1.4	25	31	4.8	0.00388	4.60	55	63
	1.5	D100L	930	3.45	3.75	78.0	77.0	74.0	0.78	0.70	0.57	5.5	15.4	2.1	2.3	1.5	10	12	2.6	0.0116	8.54	58	66
	2.2	D112M	930	4.90	5.4	79.0	75.0	75.0	0.78	0.70	0.57	5.5	22.6	2.1	2.3	1.5	8	10	3.0	0.0138	10.0	58	66
	3.0	D132S	950	6.6	7.2	83.0	82.0	79.0	0.76	0.70	0.56	5.5	30.2	2.1	2.3	1.5	15	18	7.5	0.0335	15.3	58	66
	4.0	D132M	950	8.7	9.5	83.0	83.0	81:0	0.77	0.70	0.56	5.5	40.3	2.1	2.3	1.5	15	18	10	0.0340	15.4	58	66
	5.5	D132M	950	11.8	12.9	84.0	84.0	82.0	0.77	0.70	0.56	6.0	55.3	2.1	2.3	1.5	15	18	14	0.0415	18.6	58	66
	7.5	D160M	965	15.9	17.4	86.5	86.5	84.5	0.76	0.70	0.60	5.5	74.2	2.0	2.3	1.6	15	18	18	0.092	24.9	62	71
	11.0	D160L	960	23.5	25.5	85.5	85.5	84.5	0.76	0.71	0.60	5.5	109	2.0	2.3	1.6	15	18	26	0.114	32.1	62	71
	15.0	D180L	960	30.5	33.0	87.0	86.5	85.0	0.79	0.74	0.61	5.5	149	2.0	2.3	1.6	20	25	49	0.146	45.9	68	78
	18.5	D200L	970	36.0	39.5	89.5	89.5	88.0	0.80	0.76	0.67	6.0	182	2.3	2.3	1.8	25	31	75	0.249	57.4	68	78
	22.0	D200L	970	43.0	47.0	89.0	89.0	87.5	0.80	0.72	0.62	6.0	216	2.3	2.3	1.8	25	31	89	0.359	65.2	68	78
	30.0	D225M	975	56.5	62.0	91.0	91.0	90.0	0.81	0.77	0.67	6.0	294	2.3	2.3	1.8	25	31	100	0.471	87.0	71	81
	37.0	D250S	985	65.0	71.0	93.0	93.0	92.5	0.85	0.82	0.74	6.5	358	1.8	3.0	1.4	15	18	102	1.01	92.0	74	85
	45.0	D250M	985	78.0	85.0	93.5	93.5	93.0	0.86	0.82	0.74	7.0	436	2.0	3.0	1.6	14	17	117	1.17	110.0	74	85
8	0.37	D90S	700	1.16	1.27	67.0	65.0	58.0	0.66	0.56	0.45	4.0	5,85	2.0	2.2	1.4	30	36	3.0	0.00340	4.03	47	55
	0.55	D90L	700	1.73	1.89	67.0	65.0	57.0	0.66	0.56	0.45	4.0	7.5	2.0	2.2	1.4	30	36	5.0	0.00388	4.60	47	55
	0.75	D100L	700	2.10	2.30	71.0	70.0	65.0	0.70	0.61	0.50	4.5	10.2	2.0	2.2	1.4	20	24	4.5	0.0116	8.54	60	69
	1.1	D100L	700	3.05	3.30	71.0	70.0	65.0	0.71	0.61	0.50	4.5	15.0	2.0	2.2	1.4	20	24	7.2	0.0116	8.54	60	69
	1.5	D112M	700	4.00	4.40	72.0	72.0	70.0	0.72	0.62	0.52	4.5	20.4	2.0	2.2	1.4	15	18	7.4	0.0138	10.0	60	69
	2.2	D132S	710	5.70	6.20	80.0	80.0	75.0	0.67	0.60	0.49	5.0	29.6	2.0	2.2	1.4	30	36	19.0	0.0335	15.3	60	69
	3.0	D132M	710	7.5	8.2	80.0	80.0	75.0	0.69	0.65	0.53	5.0	40.3	2.0	2.2	1.4	30	36	27	0.0415	18.6	60	69
	4.0	D160M	710	9.8	10.7	82.0	81.5	79.5	0.69	0.60	0.48	4.5	54.0	2.0	2.2	1.6	35	43	39	0.092	24.9	64	73
	5.5	D160M	710	13.4	14.6	83.0	82.5	80.5	0.69	0.60	0.48	4.5	73.9	2.0	2.2	1.6	35	43	53	0.092	24.9	64	73
	7.5	D160L	710	18.2	19.9	83.0	83.0	81.0	0.69	0.62	0.50	4.5	101	1.8	2.2	1.4	30	38	64	0.114	32.1	64	73
	11.0	D180L	720	24.0	26.0	86.0	85.5	84.0	0.74	0.68	0.55	5.0	146	1.8	2.0	1.6	30	38	80	0.146	45.9	66	76
	15.0	D200L	725	32.5	36.0	87.5	87.5	87.0	0.73	0.69	0.62	4.5	197	1.9	1.8	1.5	40	50	140	0.249	57.4	66	76
	18.5	D225S	725	43.5	47.5	87.0	87.0	86.0	0.68	0.62	0.50	4.5	243	2.0	2.0	1.6	40	50	170	0.416	71.0	66	76
	22.0	D225M	730	48.0	52.0	89.0	89.0	88.0	0.72	0.68	0.59	4.5	288	2.0	2.0	1.6	40	50	200	0.471	87.0	70	80
	30.0	D250S	735	58.0	63.5	92.0	92.0	91.5	0.78	0.73	0.61	6.0	389	1.8	2.9	1.4	19	24	200	1.40	122	68	80
	37.0	D250M	740	71.0	77.5	93.0	93.0	92.0	0.78	0.73	0.61	6.5	477	1.9	2.6	1.5	20	25	226	1.67	142	68	80

Fig 8.6.7 Example of manufacturer's data (continued) for induction motors. From GEC Small Machines, England B65 0QT

Selecting a motor as generator — Example 8.6.2

Using the data provided in Fig 8.6.7, suitable 3-phase induction motors can be selected for the schemes described in Examples 8.4.3 and 8.4.1.

Scheme in Example 8.4.3

The consumer load is 13.4 kW at a power factor of 0.93 lagging. Given a turbine speed of 500 rpm and a required voltage of 380 V (220 V line-to-neutral) at 50 Hz select the correct motor.

1 Fix the nominal motor speed

First try to match motor and turbine for a direct drive system. If this is not possible, use the fastest motor possible using a speed increasing drive of not more than 3 : 1 (as recommended in Section 5.1).

The slowest motor in Fig 8.6.7 is the 8-pole range, which runs at a nominal speed of

$$\frac{120 \times 50}{8} = 750 \text{ rpm}$$

This is still too fast for the turbine, so direct drive is not possible. The fastest motor we can select is then 3 × turbine speed = 1500 rpm. This corresponds to a 4-pole machine.

2 Select motor power

Use a derating factor of 0.8 and column 1 in the data. $\frac{13.4}{0.8} = 16.75 \text{ kW}$

From column 1 the next size is then 18.5 kW with a D180M frame size.

3 Check required generator input power

Use the efficiency data to predict input power at design load.

Use the rated power of the motor and the load to predict generator efficiency from motor data. At the design load the generator will be running at

$$\frac{13.4}{18.5} = 72\% \text{ of output.}$$

The relevant efficiency will then lie close to column 7 on the data sheet. This gives 89.5%, giving an input power of $\frac{13.4}{0.895} = 15.0 \text{ kW}$

Dividing by drive efficiency will then give required turbine output power.

Scheme as in Example 8.4.1

The consumer load is 4.65 kW. Given a power factor of 0.8, turbine speed of 300 rpm and a required voltage of 220 V, single phase at 50 Hz, select the correct motor.

1 Fix the motor speed

The slowest motor is 750 rpm, so direct drive is not possible. Maximum speed increase of 3 : 1 gives a motor speed of 300 × 3 = 900 rpm. The next lowest speed is 750 rpm, corresponding to an 8-pole machine.

2 Select motor power

Derating by 0.8, we need at least $\frac{4.65}{0.8} = 5.81 \text{ kW}$

From the data sheet the next size up from this figure is 7.5 kW, frame size D160L.

3 Check required input power

At design load the generator will be running at $\frac{4.65}{7.5} = 62\% \text{ of output.}$

The relevant efficiency is then between column 7 and 8, 82%, giving a required generator input power of

$$\frac{4.65}{0.82} = 5.67 \text{ kW}$$

It should be noted that, in general, induction generators (and most other electrical machines) have a power/weight ratio that varies approximately as the inverse of shaft speed. Thus, a 6-pole, 50 Hz induction generator (speed nominally 1000 rpm) will have approximately 1.5 × the weight and size of a 4-pole, 50 Hz induction generator (speed nominally 1500 rpm) of the same power rating. The exception to this rule concerns the nominally 1500 and 3000 rpm (4-pole and 2-pole) machines which are approximately the same size, probably because of the need for the stator coil ends to extend round half the stator circumference in the 2-pole machine. This matter of machine size and speed is important when planning the transport and the housing of a micro-hydro generator.

Calculation of induction generator excitation capacitance

A precise calculation of the capacitance required to generate a given voltage under specified load conditions is only possible with an accurate knowledge of the electrical parameters of the induction machine, including the variation of the parameters with voltage. The parameters can be obtained by standard tests, but expensive equipment is required.

In practice it is sufficient to calculate an approximate value of excitation capacitance and adjust the turbine speed until the required system voltage is obtained (see the previous section, *Selection of an induction motor*). Providing that the resulting frequency of operation is not below or not more than about 5% above the rated system frequency there should be no problems.

Two methods are recommended for the approximate calculation of the excitation capacitance required: an electrical test method and a technique requiring only the induction machine manufacturer's data. The electrical test method involves a 'no-load test' in which the line current and supply voltage are measured when the induction machine concerned is run as a motor without a mechanical load. A voltmeter and ammeter are required as well as a supply of rated voltage and frequency which has sufficient capacity to start the machine (the starting current will be five or six times the rated full load current for a direct on-line start). The 'magnetizing reactance' calculated from the no-load voltage and current is the capacitive reactance which must be provided by the excitation capacitors. A calculation using this method is given in Example 8.6.3.

Photo 8.6.2 An IGC in Nepal.

The second method, using the manufacturer's full data, involves an estimation of the magnetizing current component of the full-load current and power factor. Again, the magnetizing reactance and hence the capacitive reactance of the excitation capacitors can be obtained direct from this and the rated voltage. Example 8.6.4 illustrates this calculation.

As in the case of most electrical components, capacitors are available in standard or 'preferred' sizes. In making up a calculated value of excitation capacitance such standard values should be used and it is unnecessary to use precisely the predicted value. Under normal circumstances, a total value within about 10% of the calculated value will be adequate.

It is important that the excitation capacitors should have a substantial margin of rated voltage. Capacitor life is very dependent on the voltage of operation in relation to the rated voltage. For example, if a capacitor will last a year at its rated voltage, its life at half its rated voltage may be 30 years! Care should also be taken in observing whether the voltage ratings are for 'peak' or 'rms' voltage and the correct rating selected accordingly.

Capacitor life is also affected by frequency. Increasing the frequency of a micro-hydro system will reduce capacitor life, so it is wise to keep increases to a minimum.

The induction generator controller (IGC)

As shown in Fig 8.6.3 a stand-alone, capacitor-excited induction generator shows considerable voltage regulation with increasing load and also an output voltage which is dependent on speed. It is therefore very important, as in the case of a synchronous generator, that there is some form of voltage control (AVR for a synchronous

Sizing the capacitors — Example 8.6.3

A 5.5 kW 920 rpm, 50 Hz, 415 V (line-to-line) 3-phase induction motor draws a line current of 6 Amps when the machine is supplied at its rated voltage (line-to-line) and frequency and run as a motor with no mechanical load. Calculate the capacitance of the excitation capacitors which must be connected in delta between lines in order to make the machine generate at approximately its rated voltage when driven at slightly above its rated speed.

The delta-connected excitation capacitive reactance must equal the equivalent delta magnetizing (that is, approximately, no-load) reactance of the induction machine. Thus:

$$X_{C\,(delta)} = \frac{415}{\left(\frac{6}{\sqrt{3}}\right)} = 119.8 \text{ ohms } (\Omega)$$

The required capacitance between lines for a frequency of 50 Hz is then:

$$C = \frac{1}{(2 \times \pi \times 50 \times 119.8)} = 26.6 \text{ micro-farads } (\mu F)$$

These capacitors will experience a minimum peak voltage of 415 V × √2 = 586.9 V, and should be rated at a voltage giving a considerable margin above this voltage; say, at least 800 V (peak), that is 566 V (rms). It is normally acceptable to use capacitors of, or made up of, standard values to within 10% of the calculated value.

Note that if the excitation capacitors are to be connected in star between the lines, then the capacitive reactance required will be equal to the induction motor equivalent star magnetizing reactance, that is:

$$X_{C\,(delta)} = \frac{415}{\sqrt{3} \times 6} = 39.9\ \Omega$$

The required capacitance between each line and the excitation capacitance network star point will then be:

$$\frac{1}{(2 \times \pi \times 50 \times 39.9)} = 79.9\ (\mu F) = 3 \times C \text{ (see above)}$$

Again, capacitors with standard values should be used and their voltage rating should leave a considerable margin of safety. A rating of 400/450 V (rms) is recommended in this case.

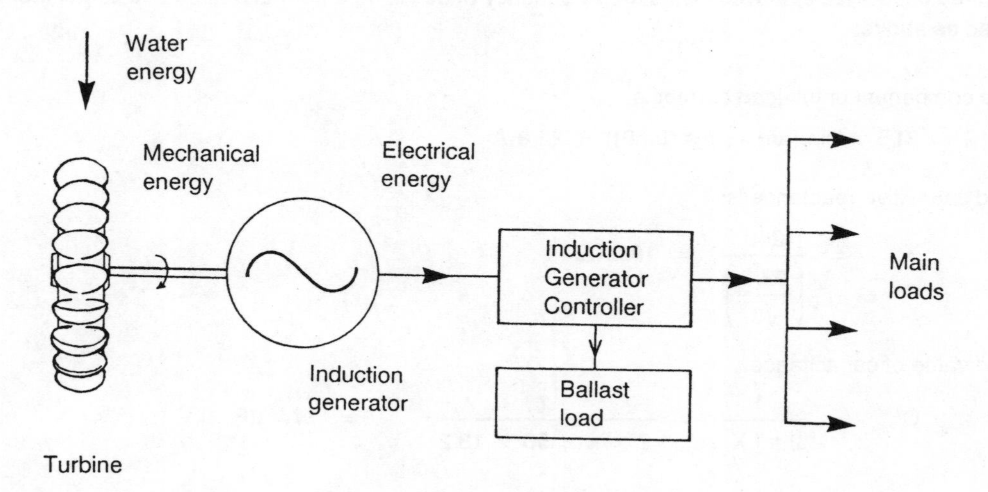

Fig 8.6.8 The induction generator controller (IGC)

Example 8.6.4 **Sizing the capacitors**

Calculate from the manufacturer's data shown in Fig 8.6.7 the values of excitation capacitance required for the generators chosen in Example 8.6.2 for the two micro-hydro schemes concerned.

18.5 kW 4-pole machine for scheme in Example 8.4.3:

Chosen induction machine: 415 V, 18.5 kW, Frame size D180M.

From Fig 8.6.7, the full-load current for a 415 V machine is seen to be 35.5 A (column 4) and the full-load power factor is 0.81 (column 9). The reactive component of this current can be taken as the magnetizing current and this is:

$$35.5 \times \sin(\text{arc} - \cos(0.81)) = 20.8 \text{ A}$$

The required delta-connected capacitive reactance is thus:

$$X_c = \frac{415}{\left(\frac{20.8}{\sqrt{3}}\right)} = 34.5 \text{ ohms } (\Omega)$$

The required capacitance between lines for a frequency of 50 Hz is then:

$$C = \frac{1}{2\pi f X_c} = \frac{1}{2 \times \pi \times 50 \times 34.5} = 92 \ \mu\text{F}$$

The notes in Example 8.6.3 regarding standard capacitor sizes and voltage rating also apply in this case.

7.5 kW 8-pole machine for scheme in Example 8.4.1:

Chosen induction machine: 240 V, 7.5 kW, Frame size D160L.

From Fig 8.6.7, the full-load current for a 415 V machine is 18.2 A (column 4) and the full-load power factor is 0.69 (column 9). If wound for 240 V (line-to-line) the full-load current will be:

$$\frac{415}{240} \times 18.2 = 31.5 \text{ A}$$

The required capacitance between lines for a frequency of 50 Hz on a normal 3-phase basis will then be calculated as above:

Reactive component of full-load current is:

$$31.5 \times \sin(\text{arc} - \cos(0.69)) = 22.8 \text{ A}$$

Required capacitive reactance is:

$$X_c = \frac{240}{\left(\frac{22.8}{\sqrt{3}}\right)} = 18.2 \ \Omega$$

Required value of capacitance:

$$C = \frac{1}{2\pi f X_c} = \frac{1}{2 \times \pi \times 50 \times 18.2} = 174 \ \mu\text{F}$$

Again, the notes in Example 8.6.3 regarding standard capacitor sizes and voltage rating apply.

Photo 8.6.3 An induction generator coupled directly to a 4-jet Pelton. An IGC can be seen above, and to its right a standard heating element serves as an air-cooled ballast. (Sri Lanka)

generator) and speed and so frequency control (electronic load controller or turbine governor for a synchronous generator). The induction generator controller which has been developed by ITDG and Nottingham Polytechnic performs both these functions at once by making use of the load/speed characteristics of the turbine driving the generator, together with the generator voltage/speed characteristic. Essentially, the IGC senses the generator output voltage, which it then controls by diverting varying amounts of power to a ballast load.

If the voltage rises, due perhaps to a reduction in user load, the turbine speed and generator voltage rise, and the IGC senses this and increases the power sent to the ballast load.

This in turn increases the generator and turbine load and reduces the speed, frequency and voltage to the desired level. There is thus some similarity to electronic load controller (ELC) operation, but voltage rather than frequency is sensed and thus the need for an AVR is avoided.

The IGC allows reactive loads to be supplied providing the load power factor is not less than 0.8 lagging. This is achieved by allowing a limited frequency rise when such loads are supplied and thus effectively increasing the capacitive reactance available from the generator excitation capacitors. The increased capacitive reactance then supplies the magnetization requirements (excitation) of the generator and also provides the necessary capacitive reactance to correct the lagging load power factor.

When an inductive load, such as an induction motor, is added to the user load, the voltage falls more than would be the case if a resistive load taking the same true power as the motor were added. The IGC then responds by reducing the power sent to the ballast load until the voltage returns to the desired level. As the voltage change is more than would be the case for the true power change, the ballast load is reduced by more than the true power change, and the turbine speed and generator speed – and thus frequency – increase. The load power factor is thus corrected as mentioned above.

These effects which take place with lagging power factor loads normally involve a frequency rise of only 5% or at most 10%. The frequency variation can of course be reduced by correcting the power factor of the larger, low power factor user loads so that the overall load power factor is increased.

Fig 8.6.8 is a schematic diagram illustrating IGC operation and applies to both 3-phase and single phase micro-hydro systems. Both 3-phase and single phase IGCs are available and both use power electronic switching devices to control the power flow to a single ballast load circuit. The design of the ballast loads is exactly the same as for an ELC and Section 8.5 should be consulted for this.

Finally, Table 8.6.1 illustrates the advantages (and some disadvantages) of using an induction generator and an IGC in comparison with other possible generator systems. The induction generator voltage controller mentioned in the table normally adjusts the value of generator excitation capacitance. These units tend to be complex and may have to be manufactured specially.

Photo 8.6.4 Induction generator belt-driven from a crossflow. A water-cooled ballast for the IGC is on the left. (Indonesia)

Table 8.6.1 *see notes opposite* **Comparing the induction generator with other systems**

Advantages	Disadvantages	Approx cost for 10 kW	
1 Synchronous generator and mechanical governor			
Saves water in the case of hydro schemes which are not run-of-the-river. Some local repairs may be possible. Large motors can be started.[2]	Poor long-term reliability. Routine maintenance required Slow acting governor. Cost savings by local manufacture only possible in well-equipped workshops. Water hammer. Not inherently overload safe.[1]	**Generator** **Frequency control** **Total**	$ 1530 $ 3570 $ 5100
2 Synchronous generator and electronic load controller			
Almost constant V & f. Large motors can be started.[2]	Not inherently overload safe.[1] 60% oversizing of the generator adds cost and may reduce part-load performance.	**Generator** **Frequency control** **Total**	$ 1530 $ 1190 $ 2720
3 Induction generator, electronic load controller and voltage control			
Almost constant V & f. Generator both cheap and reliable.	V controller adds complexity. Not inherently overload safe.[1] Added expense required for large motor loads.[2]	**Generator** **Frequency control** **Voltage control** **Total**	$ 510 $ 1190 $ 1020 $ 2720
4 Induction generator and induction generator controller			
Half the cost of 1 & 2. Very reliable. Inherently overload safe.	5-10% frequency variation.[3] Extra modules required for large loads.[2] Slow response for large inductive load changes.[2] Large motor starting is more limited than other options	**Generator** **IGC** **Total**	$ 510 $ 510 $ 1020

Photo 8.6.5 A 3-jet Pelton directly drives an induction generator. Note use of standard glued PVC and standard valves for the manifold. (Sri Lanka)

Notes to Table 8.6.1

1 Overload safety

Synchronous generators incorporating AVRs and induction generators with separate voltage controllers are designed to maintain the output voltage irrespective of the load on the generator. Extra hardware is available, either separately or as part of the controller, that will sense an overload and protect the generator provided that it functions correctly. With an induction generator using an IGC no additional protection hardware is required since an overload will cause the frequency, and so the voltage, to fall. The voltage against frequency characteristic of an induction generator with fixed excitation is always such that an overload will not cause overheating, hence the IGC system is inherently overload-safe.

The only generator protection that is required is for an overvoltage that may arise from an underload condition due to failure of the IGC or ballast loads. This can be achieved by fusing the excitation capacitors. This is a cheaper and simpler solution to the standard overfrequency trip. It is fast-acting since an increase in frequency decreases capacitor impedance as well as increasing the voltage.

2 Induction motor starts

With a synchronous generator incorporating an AVR the increase in excitation required to start an induction motor is easily achieved. For an isolated induction generator power factor correction is required. The amount of correction capacitance required to achieve a unity power factor direct-on-line start of an induction motor is typically eight times that required for full load operation. Hence, for an induction generator with voltage controller considerably more power factor correction must be available if large motors, in comparison to the generator, are to be started. With the IGC power factor correction is achieved by allowing the turbine speed to increase. For large motor loads, compared to the generator rating, the voltage will drop initially whilst the frequency increases. Too large a motor will cause excitation collapse. Star-delta starters (for 3-phase systems) and power factor correction should be used to help start motors.

In the event of the excitation collapsing, or sometimes at the installation of an induction generator it will be found that the residual magnetism in the rotor has been lost and the machine will fail to excite even with the correct or added capacitance across the stator terminals. However, a satisfactory level of residual magnetism can be restored by connecting a 6 or 12 V vehicle battery across two stator terminals for one to two seconds.

3 Effects of frequency variations

The matter of system frequency variation has been considered previously, but it is included again here in connection with Table 8.6.1 for the sake of completeness.

Provided that the minimum operating frequency is the rated frequency, 5% frequency regulation will have no detrimental effect upon all common types of load. Such regulation is similar to that of some frequency controllers and many mechanical governors. For higher frequency operation induction motor overheating will occur where highly speed-dependent loads, such as fans and pumps, are used. Also the reduction in starting torque, due to increased frequency, may be significant if accompanied by a large drop in line voltage. Power factor correction of significant inductive loads will prevent too large a frequency variation.

8.7

Switchgear and protection

Switchgear

Almost all micro-hydro installations which generate electricity will have some form of switchgear. The purpose of the switchgear is to isolate the power supply when necessary and also to have some control over the electrical power flow.

The protection equipment will work along with the switches and will be in a position to isolate the power supply when a fault occurs.

Given below are some of the common switches used on micro-hydro installations and their functions. These are available in 3-phase and single phase.

1 Isolators – these are manually operated switches which have the basic function of isolating the load from the supply. These are very cheap but they are not used very often because they do not offer any form of protection to the power supply.

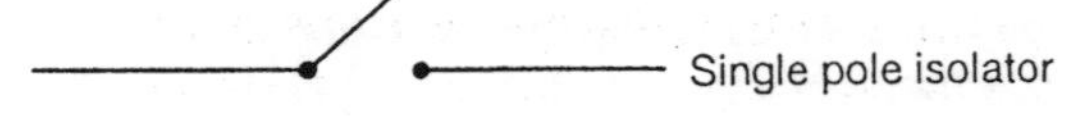

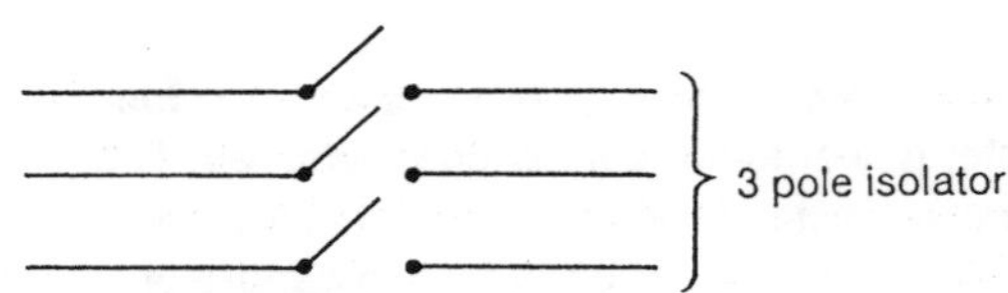

Fig 8.7.1 Symbols for isolators

2 Switch fuses or main switches – these are again manually operated switches which are similar to the isolator but with an addition of a fuse on the phase conductor. The addition of a fuse allows this switch to be used where current limiting is required.

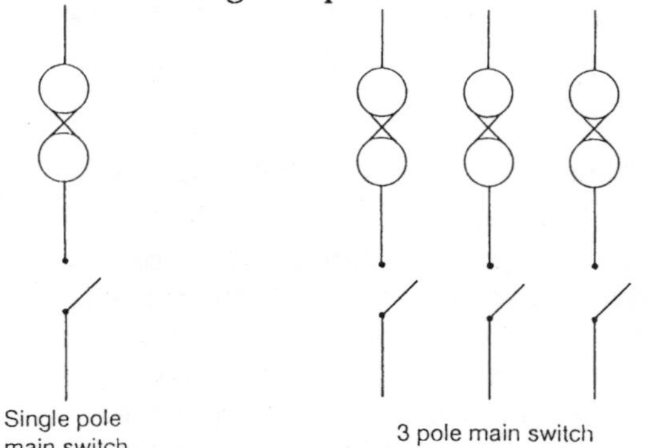

Fig 8.7.2 Symbol for switch fuse of 'main switch'

3 Moulded Case Circuit Breaker (MCCB) – these replace the switch fuse as they are cheaper in most cases. The over-current function in the MCCB is handled by bimetallic strips which isolate the load.

In addition it is possible to have the following as extra on the MCCB: control trip coils to trip the circuit breaker, magnetic trips to protect against short circuits, earth fault relays to be connected to the trip coils, motor drives to switch the circuit breaker ON and OFF, etc.

Fig 8.7.3 Symbol for MCCB

4 Oil and Air Circuit Breakers – Oil circuit breakers are being replaced by the MCCB. They are very costly and offer the same facilities as the MCCB. ACBs were used at currents of over 600 amps and are also being replaced by the MCCB.

5 Earth Leakage Circuit Breaker (ELCB) – this is a special type of MCCB which can protect the circuit against earth leakage.

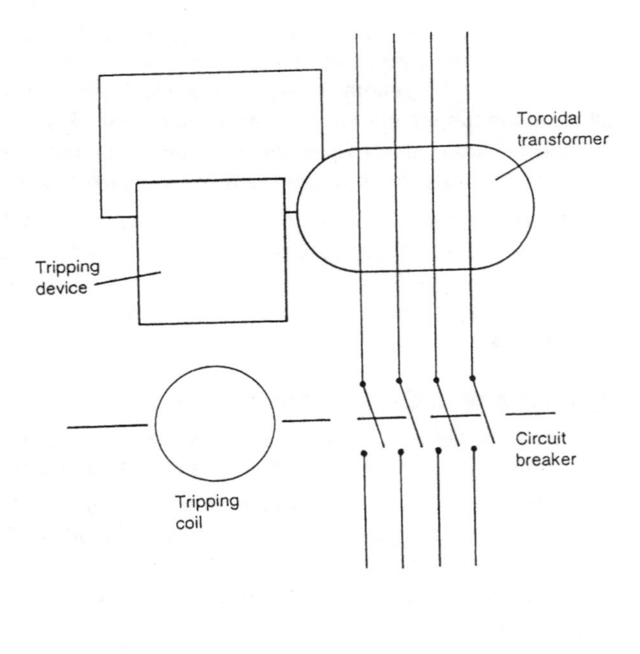

Fig 8.7.4 Symbol for ELCB

6 Contactor – these are generally used in control circuits as in motor starters etc. However they could be used as circuit breakers along with the relevant trips. This makes them very expensive.

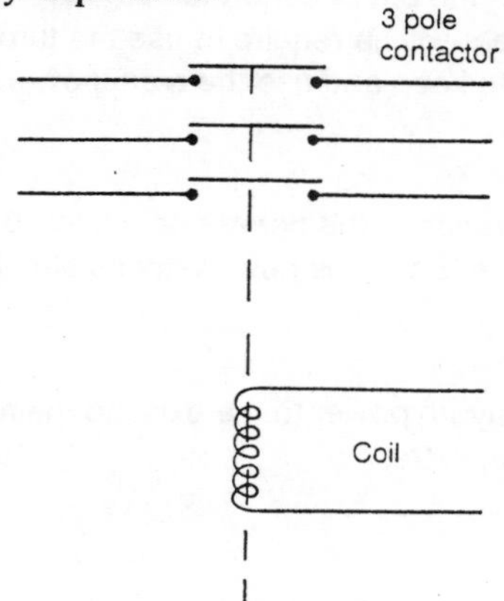

Fig 8.7.5 Symbol for contactor

7 Changeovers – these are used to select a source of power to work a machine. In micro-hydro systems a changeover may be used to select power either from the main grid or from the micro-hydro. The changeover switch allows the load to be connected to the main grid in one position and the hydro supply to be connected in the other position.

Protection

Two types of protection are considered; one is to protect the machines from faults and the second is to protect people. In a micro-hydro installation the electrical generation and distribution is protected with the following trips in conjunction with either an MCCB or a contactor to electrically isolate the circuit.

1 Under-voltage trip – this will trip the circuit breaker if the voltage drops below a pre-set value, normally adjustable down to -15%. Under-voltage for long periods could damage electric motors or the alternator itself; it may occur in a micro-hydro installation if:

   a The automatic voltage regulator in the generator is defective.
   b The machine is overloaded
   c The load has a bad power factor
   d Bad regulation in a transmission line.

2 Over-voltage trip – this will trip the circuit breaker if the voltage rises over a preset value. Over-voltage could damage loads such as lamps, heating elements etc. and if it is very high could cause a breakdown of certain insulating materials in the circuit. It is generally caused by:
   a a defective AVR
   b a leading power factor load
   c bad regulation on a transmission line
   d generator overspeed.

3 Over-current trips – current limiting is the most common form of protection used in electric systems. Over-current can cause damage to the generator windings, switches, cables, and other equipment due to the excess heat generated in the conductors.

The simplest form of over-current protection is to use a fuse on the current-carrying conductor. The next form of protection is to use an MCCB, and the third form of protection is to use an over-current trip. These forms of current limiters will isolate the load from the generator if the load is taking current in excess of the pre-set value. It is also possible to use a positive temperature coefficient (PTC) thermistor as a current trip, or as a form of current limiter.

Over-current is caused by:
a excess load being connected to the generator
b faulty equipment being connected to the generator
c lagging power factor
d short circuit
e incorrect frequency

It is important to protect the three phases of a 3-phase system when protecting against over-current.

A fuse will have a range of protection up to a value of 1.6 times its rating, and an MCCB will have a value of 1.2 times its rating when protecting against over-current.

4 Frequency trips – these will disconnect the load if the supply frequency is over or under the preset value. If supply frequency is not the rated value or not within its limits, it could cause damage to the electric motors and also the alternator.

Under-frequency may be caused by:
a overloading the machine
b insufficient water
c defective governor
d belt slip

Over-frequency may be caused by:
a excess water
b defective governor
c defective alternator or AVR

Example 8.7.1 **Selection and sizing of switches**

1 A person whose house is connected to the mains supply now acquires a small turbine with a single phase 240 volts 50 Hz synchronous generator. The output power of the 0.8 pf generator is 5 kW. The turbine set is installed very close to the house. What extra switchgear will he require to use the turbine, a) if the main power supply is removed completely b) if he intends to keep and use the main power during dry periods?

First calculate the current to be handled by the switches. The kVA rating of the generator is 5 kW/0.8 = 6.25 kVA. The current of the generator is 6.25 kVA/240 V = 26 amps. So all his new switches should be 30 amps.

a If the mains power is removed completely, he can connect the hydro power to the existing main switch and use it as it is. If no ELCB is available he can have that as an extra.

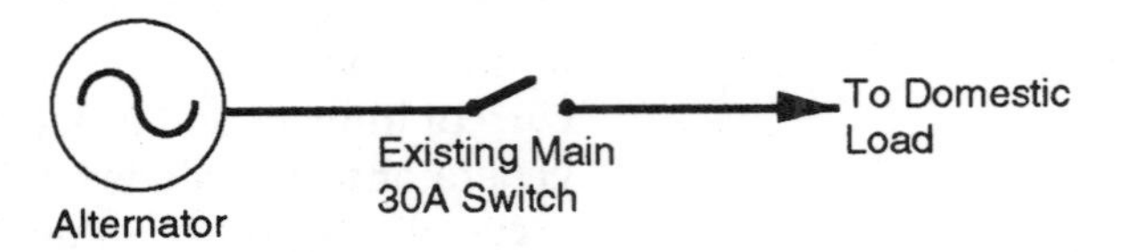

Fig 8.7.6

If he wants to install trips eg under-volts, frequency etc. he needs a 30 amp contactor with coil voltage of 240 volts. This contactor can be wired as shown below.

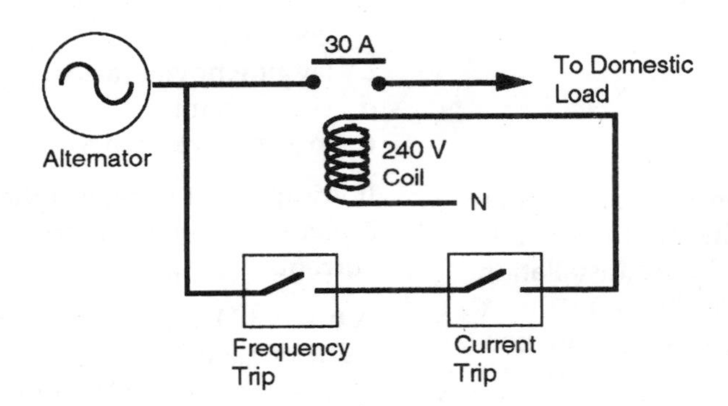

Fig 8.7.7

b If he wants to keep the mains supply along with the hydro supply he needs a manual changeover of 30 amps rating, and also a 30 amp main switch. If he hopes to use trips then he requires a contactor as mentioned above. See Fig 8.7.8

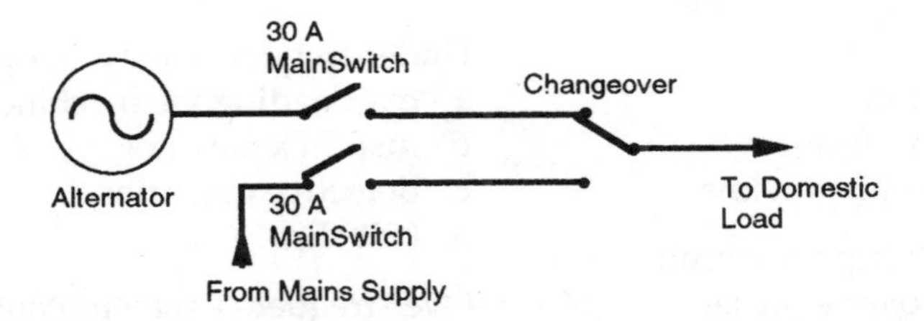

Fig 8.7.8

5 Temperature trips – alternators can be ordered with temperature sensing probes embedded in the windings. These probes can be wired to trips or to alarms to indicate when the winding temperature has risen over the safe value.

The winding temperature may rise due to:

a overloading the alternator

b incorrect frequency

c excess current due to poor power factor

d waveform distortion due to the load

e defective bearings

f bad ventilation

g ambient temperature too high

h altitude not within rated value

Note: all the above trips have a time delay so that they will not trip under transient conditions.

6 Lightning protection – where power from a micro-hydro plant is transmitted from the powerhouse to a load by means of a transmission line, the transmission line must be protected against direct and indirect lightning strikes. Direct hits are rare and can only be protected by installing an earth wire over the transmission line.

Lightning discharge in the vicinity of the transmission line can induce high voltages in the transmission line which could damage electrical and electronic equipment connected to the line. To protect against this, lightning arrestors are installed on the line to divert these high voltage spikes to earth. The lightning arrestors will have a very high resistance at normal voltage and if the voltage rises over the safe value the insulating material breaks down and conducts to earth. Once the high voltage is removed the arrestor returns back to its original state.

Metal oxide varistors work in a similar manner but can dissipate less energy. They are used to protect the ac line against high voltage spikes which may damage electronic components.

The undermentioned trips are not widely used on micro-hydro installations:

a phase failure; to protect against loss of one or two phases

b phase sequence; to monitor if the sequence of the alternator is correct – used when one alternator is synchronized with another

c reverse power relay; to prevent power from one alternator flowing into the other

7 Enclosures are necessary to keep all live components out of reach of people and also to keep all switchgear and control equipment away from dust and water sprays.

Earthing

It is necessary to earth all equipment as this will give the zero voltage reference for all other voltages. Generally all metal work is earthed by electrically connecting to an earth electrode.

The earth electrode consists of a GI pipe or a copper plate buried in the ground. In the past it was believed that the copper plate provided a better earth due to the surface area but this is not necessarily true. A long conductor (about 2 metres) buried in the ground can provide a better earth. The earth resistance of this electrode should be less than one ohm. Resistance can be measured with an earth megger.

Earthing systems

There are three common earthing systems that are used in conjunction with earth fault protection. One system (used in Sri Lanka) is the TT system in which case the neutral at the source is connected to the earth, that is: at the supply transformer the star point which is the neutral is connected to the earth. Similarly for micro-hydro alternators the neutral point of the generator is earthed.

Earth fault protection

Live wires and live parts can be protected from people with the help of barriers. However, if metal parts accidentally become live due to faulty insulation they must be isolated from the source. The earth fault relay (EFR) and the earth leakage circuit breaker (ELCB) will disconnect the power supply if a fault is detected.

The commercially available EFRs have variable setting for leakage currents and are used to disconnect the main circuit breaker if a fault is detected. The ELCBs which have a fixed setting can be used down the line to disconnect distribution circuits during smaller faults and thereby allow a continuous supply to the unaffected loads.

Wiring between alternator and switchboard

The general practice of wiring is to follow the local standards for distribution of electricity within buildings and factories, which are basically an extension or an extract from the IEE regulations. The wire sizes are taken from the relevant tables and these wires are drawn through a suitable steel or PVC conduit. If the alternator is belt-driven a flexible conduit is used between the alternator and the rigid conduit.

At the ends of the cables, sockets are installed for the purpose of connecting the cables on to the terminals.

In general the cross-sectional area of the neutral cable can be a third or a half of the size of the main cables. If an electronic load controller with continuous voltage control is used the neutral cables between the alternator and the ELC and the ELC and the ballast tank should be double the area of the phase cables. This is because the neutral cable carries a fairly high current between the alternator and the ballast load.

The output from the alternator is wired directly into the main circuit breaker which controls the flow of power into the hydro busbars. The protection circuit trips are wired to control this circuit breaker and switch it off if there is a fault.

In some cases there are a few control wires from the alternator also coming into the switch board, eg for voltage adjustment and temperature sensing.

The consumer electrical loads are connected to the hydro busbar. If no other source of power is available then the loads are connected directly to the hydro busbar through a set of switch fuses or MCCBs. If another source of power is available, eg main grid or diesel power, then the two supplies are brought into a changeover switch and then fed through a switch fuse or MCCB into the load. The changeover allows the use of either the hydro power or the other source.

Metering equipment

Often it is necessary to know what is happening to the turbine with regard to the load and for this purpose meters are installed. A revolution meter (RPM) and a pressure gauge will be part of the turbine. The switchboard should at least have a voltmeter to indicate the voltage of the alternator and an ammeter to indicate the amperes drawn from the alternator.

Example 8.7.2 **Switchboard design**

The person in Example 8.7.1 wants to use a grinding mill on the turbine only during the day, and some electrical equipment in his house. What will he need extra as switch gear and protection equipment to see that the alternator is not overloaded during the day? The grinding mill operation will require a 3.5 kW induction motor and his total domestic consumption is 3 kW.

As shown the rating of the switches will be 30 amps.

He will need the following items:

- Manual or contactor type changeover for the domestic line
- Contactor to switch ON the hydro supply
- A main switch and a starter for the 3.5 kW motor
- Frequency trip
- Over-current trip in starter

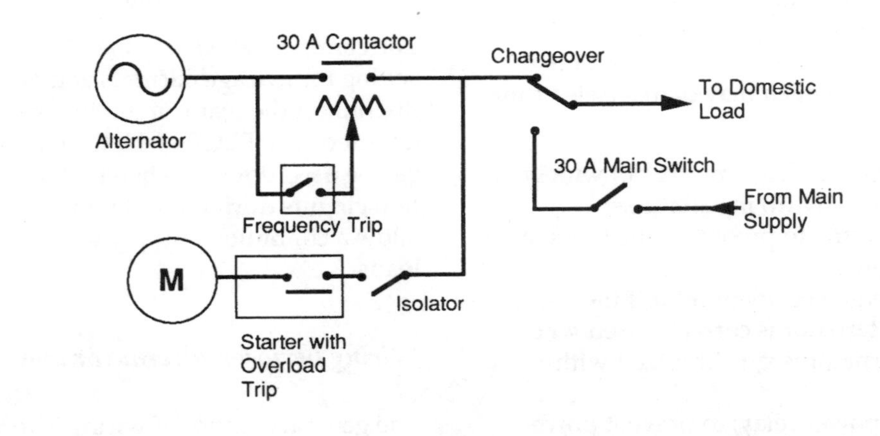

Fig 8.7.9

On more expensive plants it is possible to have three ammeters and one voltmeter with a selector switch to read the voltage between phases and the line voltage, a frequency meter, a power factor meter, a kilowatt meter, a kilowatt hour meter, a dc ammeter to read the exciter amps, etc. Meters are available in various sizes, shapes, classes of accuracy, and also prices.

Other equipment

Sometimes it is necessary to install a heater with temperature control within the switchboard to keep it dry.

Switchboard design

Example 8.7.3

A tea factory has an old turbine which has been rehabilitated and its output now is 30 kVA. Design a switchboard so that the main grid and the hydro power can always be used. The tea factory load is normally induction motors which run on the grid supply. The load is as follows:

One main switch controlling four troughs, each 5.5 kW

Factory lighting on single phase 3 kW

Security lighting on 3-phase 4.5 kW

20 hp dryer motor

Three 15 hp rollers

Available on the estate are two 30 amp 4 pole changeovers and one 100 amp main switch.

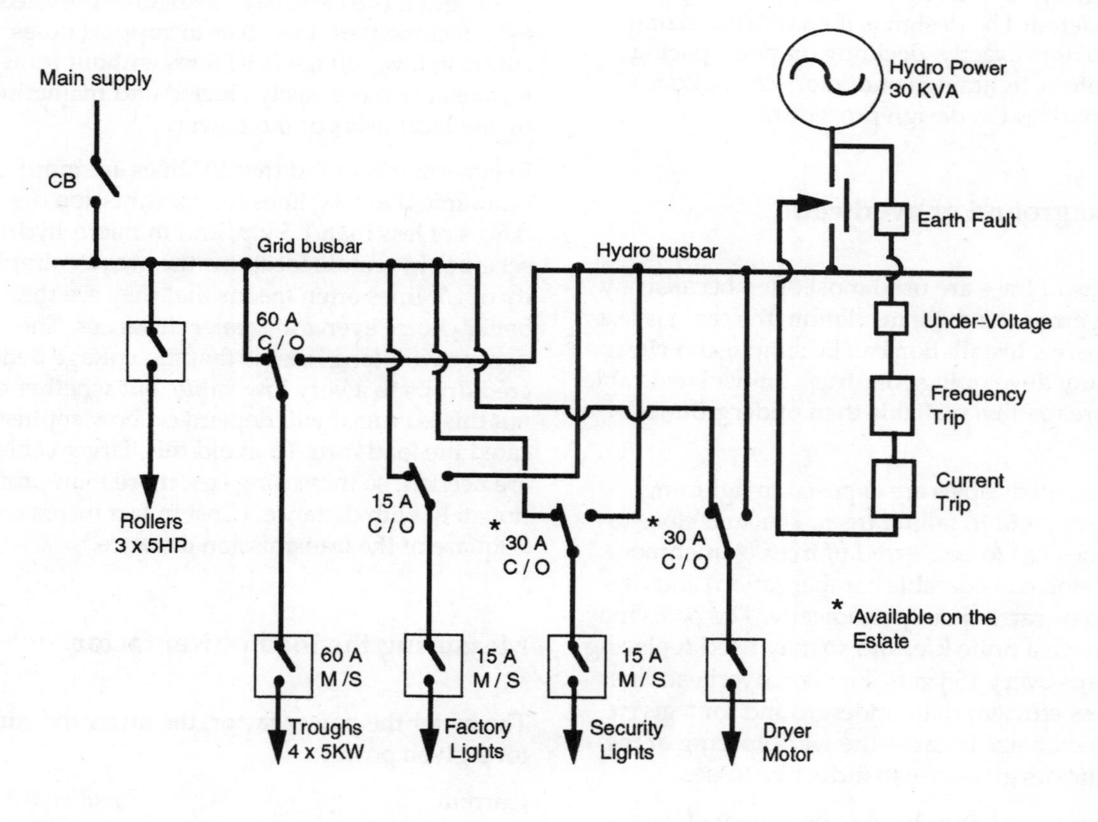

Fig 8.7.10

8.8 Transmission lines

If you are lucky, the turbine set may be very close to the load, as is often the case for a mill or a tea factory. But usually the load is remote from the best powerhouse site, and a transmission line is needed. This section is concerned with the sizing of the cables of such a line.

The main design criteria to consider are:

- the maximum allowable voltage variation from no load to full load;
- the maximum economic power loss;
- protection from lightning and other damage;
- structural stability in high winds (or, in temperate areas, in ice and snow);
- safety for people living and working near the lines.

The main aim of the design procedure is to work out the conductor size, and hence the power lost in transmission and the approximate total cost. If this is too high, then a higher volt drop, or a high voltage system (using transformers) may be considered. The design is then refined, sizing conductors exactly, deciding on pole spacing, insulators, lightning conductors etc. Note 8.8.1 summarizes the design procedure.

Underground or overhead?

Overhead lines are used most often because, by using air as the cable insulation, the cable is less expensive. Installation can be simple and cheap. In many developing countries, uninsulated cable is more readily available than underground cable.

Uninsulated cables are exposed to lightning, however, and to falling trees. The land close to the lines has to be cleared of trees (sometimes involving considerable compensation) and this has to be carried out periodically. The poles may also have a finite life, and so may need replacing, perhaps every 15 years. Further, overhead lines are less efficient than underground for a given conductor size because the wide spacing of the conductors gives rise to inductive losses.

Underground lines have to be insulated, and protected against ground movement, ploughing, new buildings etc. Once installed, however, the line should run without maintenance until the insulating material deteriorates – usually longer than 50 years.

The calculations for overhead or underground lines are the same, but the different cost and maintenance implications must be carefully assessed for any given site.

High voltage (HV) or low voltage (LV) lines

If transformers are used to step up the voltage to high values, the currents in the conductors are smaller and the cables are smaller. The lower cost of the cable is offset by the cost of the two transformers needed, one at the start of the transmission line, and one at the end, to step the voltage back down to the standard value. The purchase cost of the transformers is not the only additional cost associated with high voltage (HV) lines: the transformers need maintenance, for instance checking of ventilation or cooling oil level, and more expensive insulators are needed for attachment of the cables to support poles. By contrast, low voltage (LV) lines without transformers are more easily erected and maintained by the local users of the power.

In general it is found that LV lines are more economic than HV lines for transmission distances of less than 1.5 km; and in micro-hydro schemes in remote locations the greater simplicity of LV lines often means that they are the better choice even for greater distances. The danger with long lines is that the voltage delivered drops to a very low value, but whether or not this is crucial will depend on how sophisticated the loads are. To avoid this, larger cables are needed, so increasing cost more than proportionately with distance. (Cost in fact increases as a square of the transmission distance.)

Measuring the load power factor

The better the power factor, the lower the current for a given power :

Current *(Equation 8.8.1)*

$$= \frac{\text{Power}}{\text{voltage} \times \text{power factor}}$$

for single phase

Sizing of transmission cables

Note 8.8.1

Summary of method

1 Volt limits. Decide on allowable voltage drop. Consider upper and lower voltage limits.

2 Power factor. Estimate worst power factor of load.

3 Spacing. Decide arrangement and spacing of conductors.

4 Conductor specification. Choose a conductor material, specification and size from manufacturers tables. Consider feasibilities of both underground and overhead transmission.

5 Resistance. Calculate resistance of the line (R) either from manufacturer's tables or from the cross-sectional area of conductor and conductivity valve of the material.

6 Reactance. Calculate induction reactance (X_L). In the case of underground cables calculate also capacitive reactance (X_c).

7 Impedance. Calculate impedance.

8 Volt drop. Calculate volt drop and check with upper and lower voltage limits (assessed in step 1 above).

9 If within acceptable limits record cable details on a table like Table 8.8.1. Return to step 4 to choose another cable option, for instance one size smaller or bigger.

10 Power loss. Calculate power lost in transmission for each option.

11 HV or LV? Consider use of transformers for the design of a high voltage (HV) line and the associated cost and maintenance implications.

12 Additional benefit. Calculate differences in power loss. Choose one of the options as a reference. Attribute zero additional benefit to the reference option, and enter the additional power provided by the other options.

13 Line cost. Calculate the cost of each option. To choose the best cable size quickly, do not include associated costs which are shared by the various options (for instance, in some cases you can assume that pole and erection costs will be similar for all overhead options). Include associated costs not shared by the options, for instance if one option is strengthened copper and the other is unreinforced aluminium then pole spacing will vary and erection costs may vary significantly.

14 Additional cost. Calculate the differences in cost of the options and enter these under 'additional costs' in the table, attributing zero additional cost to the reference option.

15 Peak power and comment. Consider whether peak power has a high value or low value in the particular scheme considered. This is discussed in Section 3.11 with respect to penstock sizing. Often this consideration is more important in choosing the best option than calculating the cost/benefit ratio. Enter under 'comment' on the table. Consider other restrictions to choice, such as budget limits.

16 Total cost and benefit. Enter the nominal power for the scheme under 'total benefit' and nominal total cost under 'total cost' for the reference option. One way to arrive at nominal power and cost is to assume a realistic cost/kW target, say $2,000/kW and knowing the scheme is designed to produce 70 kW nominal power, this implies a nominal total cost of $140,000.

17 Compare total cost/benefit ratios for each option.

or, current *(Equation 8.8.2)*

$$= \frac{\text{Power}}{3 \times \text{voltage} \times \text{power factor}}$$

for 3-phase (where voltage is line-to-neutral)

An inductive or lagging power factor means that the voltage lags the current. A capacitive or leading power factor means that voltage leads current. If the voltage and current are in phase, the load is resistive and the power factor is 1.0. Some explanation is given in Section 8.2 and typical power factors are listed in Table 8.2.1.

Motors have lagging power factors – typically 0.7 – but much worse (up to 0.4) if they are running lightly loaded.

Transformers have lagging power factors, typically 0.8 to 0.95.

Fluorescent lamps have lagging power factors of between 0.5 and 0.7.

Lagging power factors are sometimes 'corrected' that is, improved from say 0.8 to 0.95 by the use of power factor correction capacitors. This is usually a good investment but often outside the control of the hydro engineer.

Power factors of existing loads can be measured by using a kWh meter and an ammeter, or a special power factor meter (Fig 8.4.5) which is rather expensive but quicker. To find the power factor with a kWh meter, simply run on a steady typical load for, say, 10 minutes, record the kWh units, and divide by the time (in hours) to give the power in kW:

$$\text{Power} = \frac{\text{units (kWh)}}{\text{time (hours)}}$$

Also measure current and voltage and calculate kVA :

$$\text{kVA} = \frac{V \times I}{1000} \quad \text{(single phase)}$$

$$\text{or} = \frac{3 \times V \times I}{1000} \quad \text{(3-phase, where V is measured line-to-neutral)}$$

$$\text{Now power} = \text{kVA} \times \text{power factor}$$

$$\text{Power factor} = \frac{\text{power (kW)}}{\text{kVA}}$$

This will give an accurate answer so long as the load is steady and (if 3-phase) balanced, and so long as the kWh meter is accurate. If there is uncertainty, run the test with a resistive load and check that kVA = kW (ie power factor = 1).

Allowable volt drop

This is a very critical design parameter – a slight change here can alter costs greatly. The figure used in the UK for allowable volt drop is 6%. This is 14 V on a 230 V supply. 6% is a good starting point for allowable volt drop along a typical hydro transmission line. In order to avoid delivered voltage being too low, it is sensible to raise the generated voltage in the powerhouse as much as possible.

Maximum voltage in the powerhouse is limited by the following considerations:

1. The alternator AVR may not like a high voltage, because it increases field power and heating which can cause failure. It is best to specify 'normal plus 10%' on new alternators, and to check with makers when using existing machines.
2. Tungsten (filament) light bulbs have a shorter life at higher voltage. Fluorescents are not so sensitive. Powerhouse lights are often a problem whenever the voltage is more than 4% above the rated light bulb value.
3. Coils and transformers have shorter lives on voltages more than 4% above rated. This means protective relays using coils can burn out more frequently than expected.

On the other hand, delivery to the consumer appliances of low voltages is also a source of trouble.

Limits to minimum voltage are:

1. Motors start to draw more current, and run hotter. Heavily loaded motors will have a shorter life at low voltage. Significant at 6% below rated voltage.
2. Lights and heaters will have reduced power. Tungsten lamp brightness changes with the fourth power of voltage. Fluorescents are less sensitive.

This means that it is best to run the alternator about 4% higher than the nominal voltage of bulbs and motors, and design line loss to give drops between 4% and 11%, depending on the economics.

Choice of conductor material

The choice is between aluminium and copper, and usually it is dictated simply by local cost and availability. Copper is stronger, and cables with steel strands incorporated, or even galvanized steel, can be considered where strength is required. The advantage of a strong conductor is that support poles can be spaced more widely, so saving on pole costs. The relative price of poles depends on local conditions, but often is significant on transmission lines of over 500 m length.

Calculation

Example 8.8.1 illustrates the calculation method described in Note 8.8.1. Note that this method ignores some factors, such as the effect of line impedance on the total power factor, and losses incurred by connecting cables. The method therefore gives a slightly optimistic result.

First of all, let's put some symbols on the variables :

R = resistive loss (Ω)

X_L = reactance due to inductive effect (Ω)

X_C = reactance due to capacitive effect (Ω)

Z = impedance (Ω)

C = capacitance (F)

L = inductance (H)

d = effective conductor spacing (m)

d_1, d_2, d_3 = distance between conductors (m)

r = overall radius of conductor (m) (see Note 8.8.2)

Θ = phase angle

A = effective line length (m)

f = frequency (Hz)

V = volt drop (V)

I = line current (A)

The formulae are :

$$d = \sqrt[3]{d_1\, d_2\, d_3} \quad \text{(m)}$$

$$L = A \times (5 + 46 \, \text{Log}_{10} \frac{d}{r}) \times 10^{-8} \quad \text{(H per phase)}$$

$$C = \frac{A \times 2.41\ 10^{11}}{\text{Log}_{10} (\frac{d}{r})} \quad \text{(F per phase)}$$

$$X_L = 2 \pi f L \quad (\Omega \text{ per phase})$$

$$X_C = \frac{1}{2 \pi f C} \quad (\Omega \text{ per phase})$$

$$V = Z \times I \quad \text{volts}$$

$$Z = \sqrt{(X_L - X_C)^2 + R^2}$$

$$\text{Power loss} = I^2 \times R \quad \text{watts}$$

For overhead lines X_C can be ignored.

For single phase lines, the length 'A' is twice the length of the line, as the current has to travel down two wires. For 3-phase lines, the analysis is done assuming a balanced load, when there is no neutral current and so 'A' is simply the length of the line.

Photo 8.8.1 Transmission pole. (IT/Lindel Caine – Nepal)

Photo 8.8.2 1 kV to 220 V transformer. (Nepal)

Computer program

Designing lines like this is tedious and it is very easy to make mistakes. A computer program sorts out the units once and for all, and allows you to experiment with different conductor spacings, different voltages etc., and to do sensitivity checks, without getting too bored. The program 'OHL4' is listed and explained in Appendix 2.

Pole spacing

See *Kempe's* listed in Appendix 3, bibliography, for this calculation. In many countries, for powers over 20 kW and relatively cheap poles, pole cost is not significant and a spacing of 30 m is usually chosen.

> **Overall radius r** — **Note 8.8.2**
>
> r
>
> For seven-strand cable, draw a diagram as above and observe that the overall radius is 3 × strand radius.
>
> For other numbers of strands follow the same procedure.

Table 8.8.1 **Comparison of transmission cable options**

Step			Material: Aluminium overhead	Material: Aluminium overhead	
	Conductor spacing		0.3 m vertical	0.3 m vertical	
1	**Allowable voltage variation**		line to line 420 - 380 V 40 volt drop	line to line 420 - 380 V 40 volt drop	
2	**Power factor** Pf		0.5	0.5	
4	**Conductor specification**		7 / 4.90	7 / 4.18	
5	**Resistance** R	Ω	0.076	0.105	
7	**Impedance** Z	Ω	0.119	0.142	
	Current I	amps	165	165	
8	**Voltage drop** V	volts	line to line 34.11 (within limits)	line to line 40.7 (approx. limits)	
10	**Power loss**	kW	6.0	8.4	
12	**Additional benefit**	kW	0	-2.4	
13	**Line cost**	$	3656	2636	
14	**Additional cost**	$	0	- 1020	
15	**Comment** ie peak power value		In this scheme every extra kW keeps a consistent value so $/kW calc. below is relevant	The volt drop just exceeds limits which implies no safety factor in voltage variation	
16	**Total cost**	$	140.000	138.980	
16	**Total benefit**	kW	70	67.6	
17	**Comparative index**	$/kW	2000	2056	

Sizing of transmission cables — Example 8.8.1

Let us suppose a 60 kW hydro set powers a tea factory requiring 3-phase, 50 Hz supply, and the factory is 350 m distance from the generator and turbine.

The power factor was measured using a kWh meter and volt and ammeters, and the worst power factor measured was 0.65, with an estimated accuracy of ±20%.

The alternator manufacturer says the maximum voltage for the alternator is 420 V phase/phase. The tea factory owner says he doesn't want the voltage to drop below 380 V phase/phase.

Assuming an overhead line of aluminium, what size should we choose? (see Table 8.8.2)

Step 1

Decide on volt drop. The limits given above are 420 V to 380 V – a drop of 40 volts which is 9.5%. Decide to use this as a starting point.

Step 2

Decide on power factor. The measurement was 0.65, with an accuracy of 20%. The worst limit would then be 0.54. Say 0.5 to be safe.

Step 3

Decide on conductor spacing. Wooden poles are available and the cheapest way is to mount the conductors vertically. A safe spacing is 0.3 m between conductors.

Step 4

Guess a conductor size from the manufacturer's table (for example Table 8.8.2), say 7/4.90 – the metric notation for 7 strand aluminium cable with each strand being 4.9 mm in diameter. This is overhead cable, not insulated or armoured.

Step 5

Calculate resistance. The manufacturer gives 0.217 Ω/km, so for 350 m resistance is

$$R = \frac{0.217 \times 350}{1000} = 0.076\ \Omega$$

continued overleaf

Example of values and costs for Al conductors — Table 8.8.2

Nominal area mm²	Metric notation	Resistance Ω/km	Cost $/km
22	7/2.06	1.227	515
30	7/2.44	0.875	680
35	7/2.59	0.777	770
40	7/2.79	0.669	871
60	7/3.40	0.4505	1256
70	7/3.66	0.389	1436
75	7/3.79	0.365	1548
90	7/4.18	0.300	1883
100	7/4.29	0.2702	2093
125	7/4.90	0.217	2610

Example 8.8.1 *continued* **Sizing of transmission cables**

Step 6 Calculate inductive reactance, X_L:

$$d = \sqrt[3]{d_1 d_2 d_3} = \sqrt[3]{(0.3 \times 0.3 \times 0.6)} = 0.378 \text{ m}$$

From table, strand diameter = 4.9 mm, number of strands = 7

For seven strand cable, overall radius

$$r = \text{strand radius} \times 3 = 0.0049 \times \frac{3}{2} = 0.00735 \text{ m}$$

$$L = A \times (5 + 46 \text{ Log}_{10} (\frac{d}{r})) \times 10^{-8} = 350 \times (5 + 46 \text{ Log}_{10} \frac{0.378}{0.00735}) \times 10^{-8}$$

$$= 350 \times (5 + (46 \times 1.71)) \times 10^{-8} = 2.93 \times 10^{-4} \text{ H per phase}$$

$$X_L = 2 \pi f L = 2 \times \pi \times 50 \times 2.93 \times 10^{-4} = 0.0920 \ \Omega \text{ per phase}$$

Capacitive reactance, X_C, is not significant.

Step 7 Calculate impedance (X_C can often be neglected)

$$Z = \sqrt{X_L^2 + R^2} = \sqrt{(0.0920)^2 + (0.076)^2} = 0.119 \ \Omega \text{ per phase}$$

Step 8 Calculate volt drop (V): $V = Z \times I$

Current, $I = \dfrac{\text{power}}{3 \times \text{voltage} \times \text{power factor}}$ Voltage (line-to-neutral) is $\dfrac{420}{\sqrt{3}} = 242 \text{ V}$

$$\therefore \quad I = \frac{60000}{3 \times 242 \times 0.5} = 165 \text{ A}$$

Volt drop, $V = 165 \times 0.119 = 19.7 \text{ V}$

This is the line-to-neutral volt drop. The line-to-line volt drop will be : $19.7 \times \sqrt{3} = 34.1 \text{ V}$

giving a minimum voltage of $420 - 34.1 = 385 \text{ V}$

This is within the requirement (380 V min)

Step 9

Enter the above option on Table 8.8.1 since it is within acceptable volt drop limits. Check to see if a smaller diameter will do. Return to Step 4, try 7/4.18.

Step 5, *new cable* Resistance $= 0.3 \dfrac{\Omega}{\text{km}} \times 0.35 \text{ km} = 0.105 \ \Omega$

Step 6, *new cable, 7/4.18*

radius, $r = \text{strand diameter} = 4.18 \text{ m} \quad \therefore \quad r = 0.00627 \text{ m}$

$$L = 350 \times (5 + 46 \text{ Log}_{10} \frac{0.378}{0.00627}) \times 10^{-8} = 3.04 \times 10^{-4} \text{ H per phase}$$

$$X_L = 2 \times \pi \times 50 \times 3.04 \times 10^{-4} = 0.0955 \ \Omega \text{ per phase}$$

Step 7, *new cable* $Z = \sqrt{(0.096)^2 + (0.105)^2} = 0.142 \ \Omega$

Step 8, *new cable* $V = Z \times I = 0.142 \times 165 = 23.5 \text{ V}$

This gives 40.7 V line to line, or 420/379 V – just outside requirements.

continued opposite

Sizing of transmission cables **Example 8.8.1** *continued*

Step 9

Include the 7/4.18 cable in Table 8.8.1, since it is close enough to acceptable limits for an economic analysis including this option to be useful.

Step 10

Power loss = $I^2 R$ watts

which for the 7/4.90 is $(165)^2 \times 0.076$ = 2069 W = 2 kW per phase (6 kW total)

and for the 7/4.18 is $(1652)^2 \times 0.0105$ = 2859 W = 2.8 kW per phase (8.4 kW total)

Step 11

For a 350 m transmission distance an LV line is appropriate.

Step 12

Choose the 7/4.9 cable as a reference option. Additional benefit for this is therefore 0, and for the 7/4.18 option it is 6 – 8.4 = – 2.4 kW

Step13

Cost of 7/4.18 cable = \$1883 × 0.35 km = \$659 × 4 lengths (3 phases and neutral)

Cost of 7/4.9 cable = \$2610 × 0.35 km = \$914 × 4 lengths (3 phases and neutral)

Ancilliary costs such as erection of poles and maintenance will be similar in both cases.

Step 14

The 7/4.9 cable is the reference option so has zero additional cost whereas the 7/4.18 cable has an additional cost of \$914 – \$659 = – \$255 × 4 lengths = – \$1020

Step 15

In this example every extra kW supplied has the same consistent value, because the hydro electricity is used to reduce the quantity of electricity bought by the tea factory from a mains supply grid.

In another scheme it could be that the extra 2.4 kW delivered through use of the bigger cable would have been surplus to actual requirements, such that its value is low and the bigger cable would be an uneconomic proposition regardless of the calculation in Step 17. Alternatively, it may be the case that the extra peak power produced by the bigger cable is essential to the project requirement and the bigger cable will then be economic regardless of the results of Step 17.

Step 16

Assume that the scheme fitted with the 7/4.9 cable provides a total of 60 kW at a total cost of \$140,000. Consequently, the other cable will provide 57.6 kW at a cost of \$140,000 – \$1020 = \$138,980

Step 17

Compare the total cost/total benefit ratios as illustrated in Table 8.8.1. The bigger cable is more economic, but only given the assumption made as to peak power value in Step 15. In this case it is also important to avoid use of the smaller cable because the large voltage drop will put strain on the appliances and allows less tolerance for inaccurate assumptions, substandard cable, load factor changes and other such factors which could worsen the volt drop in the future. It is worthwhile entering some other cable options in the table (use the computer program for speed).

Financial evaluation

9

9.1 Introduction

A micro-hydro scheme is expensive. It is also a risk because most of the cost must be met at the start of the project. The investor (a private individual, prospective owner, funding agency, or rural development bank) will need to be convinced that such a major investment is safe.

The designer of the hydro scheme has the job of convincing the investor that the scheme will produce financially viable results, and must identify which proposed schemes are likely to fail and warn the investor of poor financial potential.

To do this, the designer must be familiar with the language of finance. This chapter introduces some of the terminology of finance and some methods of financial projection. It is highly recommended that the design engineer learns these methods and includes them in feasibility studies.

It is likely that the investor will not be an engineer. He will not be interested in engineering design studies. But he will understand very well the need for good management, training schedules for the operators of the scheme, good planning of maintenance, and so on. The investor will also understand certain financial indicators, and certain types of financial projections.

Two indicators have already been discussed in Chapter 1. These are:

- Plant factor
- Unit energy cost

These two indicators carry a lot of information and are important and useful. Before continuing to read this chapter, it is worthwhile to review Chapter 1. The usefulness of the first two indicators is that they can be quickly calculated at a very early stage of planning a scheme (the 'pre-feasibility' stage) to predict financial viability. Most importantly, they can be used to make initial comparisons with alternative power sources. Often a proposed hydro scheme must be compared with a diesel installation or grid extension. An essential part of a financial study is to provide analysis of rival power sources using standard comparative indicators.

This chapter introduces two further comparative indicators, net present value (NPV) and internal rate of return (IRR). It also describes the simple and discounted payback periods.

These various indicators will help the investor decide whether the project is worth doing; the motive may be to find the most profitable project, or the safest, most financially self-sustaining, use of rural development funds. When presenting these indicators to the investor you are presenting a 'profitability analysis'. This term is useful to distinguish another stage of financial evaluation which is usually left until later, the 'cash flow analysis'. The cash flow analysis is a more detailed future projection of the likely financial performance of the scheme on a year-by-year or month-by-month basis. For instance, it may be that bank loans will be required, and repayments will be needed in future years. Or it may be that the demand for the services provided by the hydro is expected to grow over the first five years of operation, and that in the first five years the demand will be too small for the scheme to be immediately viable. In this case external funds will be needed both for initial capital costs and to cover operational costs during these five years. The investor will also want to see whether in later years sufficient funds are generated to repay loans, replace worn equipment and cover for contingencies such as the replacement of equipment unexpectedly damaged (by landslide, flood, etc). A year-by-year or month-by-month prediction of money flows into the project account, and out of it, is an essential planning activity for the designer as well as an important indicator for the investor.

Whereas a profitability analysis must always take place in the earliest planning stages (in 'pre-feasibility' or even 'back-of-the-envelope' studies) the cash flow tables are not usually needed until later, when a full feasibility study is prepared.

When you have read this chapter and tried a few exercises, you will be able quite easily to calculate NPV, IRR, payback periods, and cash flow tables. But how accurate will these predictions be? The investor will only be impressed by your conclusions (whether they predict financial success or failure for proposed schemes) if you can also present at the same time the following:

- clear statements of all assumptions made while doing the calculations;
- methods used to collect *base information* as explained below;
- evidence as to the *reliability* of the base information.

In Chapter 1 it is emphasized that a hydro scheme must be designed on the basis of a feasibility study which contains sound demand survey data and management capability survey data. These surveys provide you with essential base information for your financial study. Typical key questions are:

- When is the energy needed (by season, by hour, day or night etc)?
- At what price rate can the users really afford to pay for it, leaving aside their aspirations?
- Do the activities of the users leave them sufficient incentives and time to make use of the energy as planned?
- Are the users sufficiently well organized to sustain operation and management tasks into the future?
- Is the time pattern of supply of power well matched with the time pattern of use of power?

The investor will require answers to these questions, and an indication of how much care has been taken to ensure that correct and realistic assessments have been made of future sustained demand. For instance, the survey can be compared with other similar surveys in villages nearby; some of these may have been verified or called into question by subsequent experience. It may be that grain-milling operations are already starting nearby and a hydro scheme which is aimed at providing the same demand for milling is therefore a questionable financial venture; the feasibility study should take this into account. The investor will be interested to see what cross-checks and verification you have used, to make sure the base information is accurate.

The emphasis in this chapter is on financial evaluation of a project, not on social or economic evaluation. It is quite possible that you may wish to promote one particular project which is less financially viable than another, in order to achieve certain social aims. If this is the case then the feasibility study should contain the relevant arguments. A financial evaluation will nevertheless be needed at the same time.

It is also useful to distinguish an 'economic evaluation' of the project from the financial evaluation. This is because the government takes a wider view of which projects are likely to bring benefit to the nation as a whole, or to parts of the rural population. An economic evaluation is addressed to government planners who will be formulating policy as to subsidies for rural energy supplies, privatization, regulations governing power supply systems, grid extension and so on. Although economic evaluation is an essential part of micro-hydro planning, it is not introduced in this chapter.

9.2

The time value of money

"A bird in the hand is worth two in the bush". This proverb expresses the concept of discounting. A project is expected to achieve financial benefits in the future. But these benefits are not worth the same as if they were securely in our hands now. Consequently we discount them; we may for instance say that $1000 dollars expected in one year's time is worth $800 in the hand now. A decision like this can be entirely personal. One person may be content to accept $800 now as equivalent to $1000 in a year's time (but will not accept anything less than $800). A second person may be happy to accept as little as $600 now as equivalent to $1000 in a year's time. The second person has a higher personal discount rate.

The discount rate can be expressed as a fraction. Taking the first example above, $1000 divided by 1.25 is $800. The discount rate is 0.25, or 25%. The future benefit of $1000 is discounted at 25% per year to become a present value (PV) of $800. If we are thinking of a future benefit of $1000 arising not in one year's time, but in two years' time, then it will be discounted twice:

$$\text{Present value (PV)} = \frac{\$1000}{1.25 \times 1.25} = \$640$$

The general formula for discounting is:

$$PV = \frac{\text{Future value}}{(1 + m)^n}$$

Where n is the number of time periods (usually years – but sometimes months) into the future that the future value is expected to arise, and m is the discount rate over the period (years or months); 0.25 per year in the above example. Table 9.2.1 (following page) lists values of the single sum discount factor.

There are several reasons why we discount future sums of money. One of them is that we cannot be sure that some event will not intervene in the meantime which will stop us enjoying our future sum of money. Another is that we generally expect to get some financial compensation if someone else is using money which we might otherwise be using: that is, we could say:

"I won't have the $800 now, someone else
can use it for a year, but in compensation
I expect to have an extra $200 in a year's time".

A third reason is inflation. We know very well that $1000 in a year's time is worth less than $1000 now, because prices tend to go up at a general inflation rate. For instance, suppose the general inflation rate is 10%, and you are offered $1000 dollars in one year's time. You could calculate that this would be equivalent to receiving ($1000/1.1 =) $909, now. The general formula for inflation (f) is:

$$PV = \frac{\text{Future value}}{(1 + f)^n}$$

In the first example above $1000 in one year's time was considered to have the same value as $800, because the discount rate applied was 25%.

The discount rate would have allowed for the inflation rate of 10%, that is, it took into account that the $1000 dollars would only be worth $909. The 'real' discount rate was therefore the conversion of $909 then to $800 dollars now, which is a real discount rate of 13.6% (since $909/1.136 = $800).

The symbol used for real discount rate is 'r' in this chapter. This distinguishes the real rate from 'm', the market discount rate, which includes an allowance for inflation. For most purposes it is permissible to convert quickly by subtracting inflation from the market rate, that is, $r = m - f$. Strictly the real rate is found from the formula:

$$1 + r = \frac{1 + m}{1 + f}$$

Indicators such as NPV and IRR are usually undertaken using real discount rate 'r' since this greatly simplifies the calculations. See Sections 9.5 and 9.6 below.

Table 9.2.1 Discount factors for single sums

Discount rate (r) $\frac{1}{(1+r)^n}$

Period (n)	1%	2%	3%	4%	5%	6%	7%	8%	9%	10%	12%	14%	15%	16%	18%	20%	24%	28%	32%	36%
1	0.9901	0.9804	0.9709	0.9615	0.9524	0.9434	0.9346	0.9259	0.9174	0.9091	0.8929	0.8772	0.8696	0.8621	0.8475	0.8333	0.8065	0.7813	0.7576	0.7353
2	0.9803	0.9612	0.9426	0.9246	0.9070	0.8900	0.8734	0.8573	0.8417	0.8264	0.7972	0.7695	0.7561	0.7432	0.7182	0.6944	0.6504	0.6104	0.5739	0.5407
3	0.9706	0.9423	0.9151	0.8890	0.8638	0.8396	0.8163	0.7938	0.7722	0.7513	0.7118	0.6750	0.6575	0.6407	0.6086	0.5787	0.5245	0.4768	0.4348	0.3975
4	0.9610	0.9238	0.8885	0.8548	0.8227	0.7921	0.7629	0.7350	0.7084	0.6830	0.6355	0.5921	0.5718	0.5523	0.5158	0.4823	0.4230	0.3725	0.3294	0.2923
5	0.9515	0.9057	0.8626	0.8219	0.7835	0.7473	0.7130	0.6806	0.6499	0.6209	0.5674	0.5194	0.4972	0.4761	0.4371	0.4019	0.3411	0.2910	0.2495	0.2149
6	0.9420	0.8880	0.8375	0.7903	0.7462	0.7050	0.6663	0.6302	0.5963	0.5645	0.5066	0.4556	0.4323	0.4104	0.3704	0.3349	0.2751	0.2274	0.1890	0.1580
7	0.9327	0.8706	0.8131	0.7599	0.7107	0.6651	0.6227	0.5835	0.5470	0.5132	0.4523	0.3996	0.3759	0.3538	0.3139	0.2791	0.2218	0.1776	0.1432	0.1162
8	0.9235	0.8535	0.7894	0.7307	0.6768	0.6274	0.5820	0.5403	0.5019	0.4665	0.4039	0.3506	0.3269	0.3050	0.2660	0.2326	0.1789	0.1388	0.1085	0.0854
9	0.9143	0.8368	0.7664	0.7026	0.6446	0.5919	0.5439	0.5002	0.4604	0.4241	0.3606	0.3075	0.2843	0.2630	0.2255	0.1938	0.1443	0.1084	0.0822	0.0628
10	0.9053	0.8203	0.7441	0.6756	0.6139	0.5584	0.5083	0.4632	0.4224	0.3855	0.3220	0.2697	0.2472	0.2267	0.1911	0.1615	0.1164	0.0847	0.0623	0.0462
11	0.8963	0.8043	0.7224	0.6496	0.5847	0.5268	0.4751	0.4289	0.3875	0.3505	0.2875	0.2366	0.2149	0.1954	0.1619	0.1346	0.0938	0.0662	0.0472	0.0340
12	0.8874	0.7885	0.7014	0.6246	0.5568	0.4970	0.4440	0.3971	0.3555	0.3186	0.2567	0.2076	0.1869	0.1685	0.1372	0.1122	0.0757	0.0517	0.0357	0.0250
13	0.8787	0.7730	0.6810	0.6006	0.5303	0.4688	0.4150	0.3677	0.3262	0.2897	0.2292	0.1821	0.1625	0.1452	0.1163	0.0935	0.0610	0.0404	0.0271	0.0184
14	0.8700	0.7579	0.6611	0.5775	0.5051	0.4423	0.3878	0.3405	0.2992	0.2633	0.2046	0.1597	0.1413	0.1252	0.0985	0.0779	0.0492	0.0316	0.0205	0.0135
15	0.8613	0.7430	0.6419	0.5553	0.4810	0.4173	0.3624	0.3152	0.2745	0.2394	0.1827	0.1401	0.1229	0.1079	0.0835	0.0649	0.0397	0.0247	0.0155	0.0099
16	0.8528	0.7284	0.6232	0.5339	0.4581	0.3936	0.3387	0.2919	0.2519	0.2176	0.1631	0.1229	0.1069	0.0930	0.0708	0.0541	0.0320	0.0193	0.0118	0.0073
17	0.8444	0.7142	0.6050	0.5134	0.4363	0.3714	0.3166	0.2703	0.2311	0.1978	0.1456	0.1078	0.0929	0.0802	0.0600	0.0451	0.0258	0.0150	0.0089	0.0054
18	0.8360	0.7002	0.5874	0.4936	0.4155	0.3503	0.2959	0.2502	0.2120	0.1799	0.1300	0.0946	0.0808	0.0691	0.0508	0.0376	0.0208	0.0118	0.0068	0.0039
19	0.8277	0.6864	0.5703	0.4746	0.3957	0.3305	0.2765	0.2317	0.1945	0.1635	0.1161	0.0829	0.0703	0.0596	0.0431	0.0313	0.0168	0.0092	0.0051	0.0029
20	0.8195	0.6730	0.5537	0.4564	0.3769	0.3118	0.2584	0.2145	0.1784	0.1486	0.1037	0.0728	0.0611	0.0514	0.0365	0.0261	0.0135	0.0072	0.0039	0.0021

To find the present value (PV) of a sum of money arising 'n' years in the future, given a discount or interest rate (r), multiply the future sum (FV) by the discount factor found in the table.

$$PV = FV \times \text{discount factor} = \frac{FV}{(1+r)^n}$$

Table 9.3.1 Discount factors for annuities

Discount rate (r) $\frac{(1+r)^n - 1}{r(1+r)^n}$

Period (n)	1%	2%	3%	4%	5%	6%	7%	8%	9%	10%	12%	14%	15%	16%	18%	20%	24%	28%	32%
1	0.9901	0.9804	0.9709	0.9615	0.9524	0.9434	0.9346	0.9259	0.9174	0.9091	0.8929	0.8772	0.8696	0.8621	0.8475	0.8333	0.8065	0.7813	0.7576
2	1.9704	1.9416	1.9135	1.8861	1.8594	1.8334	1.8080	1.7833	1.7591	1.7355	1.6901	1.6467	1.6257	1.6052	1.5656	1.5278	1.4568	1.3916	1.3315
3	2.9410	2.8839	2.8286	2.7751	2.7232	2.6730	2.6243	2.5771	2.5313	2.4869	2.4018	2.3216	2.2832	2.2459	2.1743	2.1065	1.9813	1.8684	1.7663
4	3.9020	3.8077	3.7171	3.6299	3.5460	3.4651	3.3872	3.3121	3.2397	3.1699	3.0373	2.9137	2.8550	2.7982	2.6901	2.5887	2.4043	2.2410	2.0957
5	4.8534	4.7135	4.5797	4.4518	4.3295	4.2124	4.1002	3.9927	3.8897	3.7908	3.6048	3.4331	3.3522	3.2743	3.1272	2.9906	2.7454	2.5320	2.3452
6	5.7955	5.6014	5.4172	5.2421	5.0757	4.9173	4.7665	4.6229	4.4859	4.3553	4.1114	3.8887	3.7845	3.6847	3.4976	3.3255	3.0205	2.7594	2.5342
7	6.7282	6.4720	6.2303	6.0021	5.7864	5.5824	5.3893	5.2064	5.0330	4.8684	4.5638	4.2883	4.1604	4.0386	3.8115	3.6046	3.2423	2.9370	2.6775
8	7.6517	7.3255	7.0197	6.7327	6.4632	6.2098	5.9713	5.7466	5.5348	5.3349	4.9676	4.6389	4.4873	4.3436	4.0776	3.8372	3.4212	3.0758	2.7860
9	8.5660	8.1622	7.7861	7.4353	7.1078	6.8017	6.5152	6.2469	5.9952	5.7590	5.3282	4.9464	4.7716	4.6065	4.3030	4.0310	3.5655	3.1842	2.8681
10	9.4713	8.9826	8.5302	8.1109	7.7217	7.3601	7.0236	6.7101	6.4177	6.1446	5.6502	5.2161	5.0188	4.8332	4.4941	4.1925	3.6819	3.2689	2.9304
11	10.3676	9.7868	9.2526	8.7605	8.3064	7.8869	7.4987	7.1390	6.8052	6.4951	5.9377	5.4527	5.2337	5.0286	4.6560	4.3271	3.7757	3.3351	2.9776
12	11.2551	10.5753	9.9540	9.3851	8.8633	8.3838	7.9427	7.5361	7.1607	6.8137	6.1944	5.6603	5.4206	5.1971	4.7932	4.4392	3.8514	3.3868	3.0133
13	12.1337	11.3484	10.6350	9.9856	9.3936	8.8527	8.3577	7.9038	7.4869	7.1034	6.4235	5.8424	5.5831	5.3423	4.9095	4.5327	3.9124	3.4272	3.0404
14	13.0037	12.1062	11.2961	10.5631	9.8986	9.2950	8.7455	8.2442	7.7862	7.3667	6.6282	6.0021	5.7245	5.4675	5.0081	4.6106	3.9616	3.4587	3.0609
15	13.8651	12.8493	11.9379	11.1184	10.3797	9.7122	9.1079	8.5595	8.0607	7.6061	6.8109	6.1422	5.8474	5.5755	5.0916	4.6755	4.0013	3.4834	3.0764
16	14.7179	13.5777	12.5611	11.6523	10.8378	10.1059	9.4466	8.8514	8.3126	7.8237	6.9740	6.2651	5.9542	5.6685	5.1624	4.7296	4.0333	3.5026	3.0882
17	15.5623	14.2919	13.1661	12.1657	11.2741	10.4773	9.7632	9.1216	8.5436	8.0216	7.1196	6.3729	6.0472	5.7487	5.2223	4.7746	4.0591	3.5177	3.0971
18	16.3983	14.9920	13.7535	12.6593	11.6896	10.8276	10.0591	9.3719	8.7556	8.2014	7.2497	6.4674	6.1280	5.8178	5.2732	4.8122	4.0799	3.5294	3.1039
19	17.2260	15.6785	14.3238	13.1339	12.0853	11.1581	10.3356	9.6036	8.9501	8.3649	7.3658	6.5504	6.1982	5.8775	5.3162	4.8435	4.0967	3.5386	3.1090
20	18.0456	16.3514	14.8775	13.5903	12.4622	11.4699	10.5940	9.8181	9.1285	8.5136	7.4694	6.6231	6.2593	5.9288	5.3527	4.8696	4.1103	3.5458	3.1129

To find the present value of constant annual sums arising 'n' years in the future, given a discount or interest rate (r), multiply the repeated future sum (the annuity) by the discount factor found in the table.

$$PV = \text{annuity} \times \text{discount factor} = \text{annuity} \times \frac{(1+r)^n - 1}{r(1+r)^n}$$

9.3 The annuity equation

So far we have considered only one single sum of money arising on a future date. Usually, however, we are interested in projects involving repeated annual sums. A micro-hydro project, for instance, might bring in revenue in the form of regular payments (or tariffs) for domestic electricity connections. If for example a constant annual sum of $1000 is collected each year for fifteen years into the future, then we can discount each of these sums to its present value. The formula for discounting is as given in the previous section, except that we use the 'real' discount rate (r). Suppose this rate is 12%:

$$\text{Present value} = \frac{\text{Future value}}{(1 + r)^n} = \frac{\text{Future value}}{(1 + 1.12)^n}$$

The first collection of $1000 after one year has a present value (PV) of 1000/(1.12) = $893. The second has a PV of $1000/(1.12)^2$ = $797 and so on. Finally it is necessary to add all the PVs to calculate the total present value of all the revenue over fifteen years.

A short cut is the annuity equation. An 'annuity' is a constant annual sum. For instance, the annual revenue of $1000 in this example is an annuity. The total present value can be found easily from the annuity equation as follows:

If we call the constant annual sum, or annuity, 'A' then

Total Present Value (PV)

$$= A \times \frac{(1 + r)^n - 1}{r\,(1 + r)^n}$$

$$= A \times \text{discount factor}$$

In this example, with a discount rate of 12%, and an annuity of $1000 for fifteen years:

$$PV = 1000 \times \frac{(1.12)^{15} - 1}{0.12\ (1.12)^{15}} = 1000 \times 6.8109$$

$$= \$6800$$

We have calculated that at a discount rate of 12%, an annual revenue of $1000 over the next 15 years is equivalent to $6800 in hand at this moment.

Notice in the above calculations the term discount factor was used. This is the factor by which a simple future recurring sum or annuity is multiplied in order to convert it to the present value. Since it is tedious to work out the discount factor on a calculator, a full range of values is given in Table 9.3.1.

9.4 Unit energy cost and net income

National electricity supply authorities often give a price for electricity in terms of a unit of energy, the kilowatt-hour (kWh). In Chapter 1 we saw that this can be calculated for a hydro project so long as we have some way of expressing the cost of supplying the electricity (or power) as an annual sum. The annuity equation provides a simple way of converting the initial cost of installing a scheme into an annual cost. We can say that the initial outlay or capital cost (C) is equivalent to a fixed sum outlay (A) each year for the life of the project.

The annuity equation is reversed:

$$A = C \times \frac{r\,(1 + r)^n}{(1 + r)^n - 1} = C \times \frac{1}{\text{discount factor}}$$

Example 9.4.1 (following page) illustrates how to use this procedure together with Table 9.3.1 to calculate unit energy cost.

Some micro-hydro schemes may be for the provision of mechanical power only. In such circumstances the net annual income, say from grinding, after all operating costs have been paid, must be sufficient to meet the fixed sum annual outlay (A). Example 9.4.2 should be studied.

Unit costs and annual revenues from a projected scheme are valuable indicators for an investor. He will need to know the discount rate used in the costings and the estimated plant life. Two further important indicators of the value of a project are the NPV and the IRR. These are explained in following sections.

Example 9.4.1 **Unit energy cost from annuity equation**

A 70kW micro-hydro project is proposed which has a start-up capital cost (C) of $120000. Annual O+M expenses are expected to be 2% of the start-up cost. The plant factor has been estimated, following energy surveys as discussed in Chapter 1, as being 0.5. The project is expected to have a fifteen year life. The real discount rate is 12%. Calculate the unit cost.

First express the original capital cost in terms of a constant annual sum throughout the life of the project.

$$A = C \times \frac{r(1+r)^n}{(1+r)^n - 1}$$

$$= 120000 \times \frac{0.12\,(1+0.12)^{15}}{(1+0.12)^{15} - 1}$$

$$= 120000 \times 0.147$$

$$= \$17619$$

The unit energy cost of a project is discussed in Chapter 1 where the term plant factor (PF) is defined.

$$\text{Unit energy cost } (r = 12\%) = \frac{\text{total annual cost}}{\text{energy consumed usefully per year}}$$

$$= \frac{17619 + (0.02 \times 120000)}{70 \times 8760 \times 0.5}$$

$$= 0.065 \text{ \$/kWh}$$

Note that the factor 0.147 calculated above can be found either with a calculator, or simply by taking the reciprocal of the discount factor given in Table 9.3.1.

Example 9.4.2 **Net income indicator**

A 12kW micro-hydro project for grain-milling is proposed. It has a start-up cost of $20000. The discount rate is 20%.

An energy survey relating to the project established that the grain-milling operation will bring in annual earnings of $7000. The operating and maintenance costs are expected to be $1400 per year. What will be the income of the project, if the costs and earnings are imagined as spread out over 12 years? First calculate the capital cost expressed as an annual cost:

$$A = 20000 \times \frac{0.2\,(1+0.2)^{12}}{(1+0.2)^{12} - 1} = 20000 \times \frac{1}{4.4392} = \$4500$$

$$\text{Net income } (r = 20\%) = \text{annual revenue} - \text{annual O+M} - \text{annual capital cost}$$

$$= 7000 - 1400 - 4500 = \$1100$$

The project may be considered a viable proposition by an investor on the basis of this net income. If the net income had been negative, or even just a bit lower than this, the investor would have thought it too risky a project. The 'profit' of $1100 is a safety margin against unexpected operational or marketing problems.

9.5

Net present value: NPV (r%)

A micro-hydro project is expected to bring in revenue in future years, and also to incur running costs. The NPV is simply the PV of all revenues minus the PV of all running and capital costs. It should always be expressed together with the discount rate 'r' (eg NPV (10%) = $1300). Usually the real discount rate 'r' is used rather than the market rate 'm' since this allows present day prices and projected revenues to be used throughout the life of the project. So general inflation 'f' is removed from the analysis right at the start through the choice of r. Only in the exceptional cases where a particular item is inflating at a rate over and above general inflation will adjustment be required.

If for instance the price of wages in a rural area is increasing at 20% a year when general inflation is 10%, then the extra 10% increase must be included in the analysis. In practice, in a micro-hydro project, exceptional inflation is rarely an important issue; it more often effects other kinds of projects, for instance diesel generating plant can sometimes suffer from exceptional rates of inflation on diesel fuel.

In a particular micro-hydro project the annual revenues and the running costs are estimated to be, respectively, $2400 and $1000 per year. The net annual revenue would then be $1400 per year.

This can now be discounted for each year to its present value, and all the present values added together. The result is the total present value of the project net earnings. If the project life is expected to be fifteen years and the discount factor 12% then, in this example, the total PV would be:

PV (r = 12%) = A × discount factor

= 1400 × 6.8109 (Table 9.3.2)

= $9500

The project will create future wealth equivalent to having $9500 in hand at the present moment, given a 12% discount rate. But how much will it cost to set up the project? It will only be a worthwhile project if it earns more than it costs. The net present value (NPV) is the present value of net earnings (PV) with the present value of the project cost (C) deducted.

NPV = PV – C

If we suppose that the project in this example cost $8000 to implement, then:

NPV(r = 12%)

= PV – C = 9500 – 8000 = $1500

The prospective investor now knows that at the stated discount rate of 12%, the scheme will earn more than it costs. Of course there is always a chance that he will think the adopted discount rate is too low, and it is a good practice to check the robustness of the investment by performing two or three NPV calculations at different test discount rates. Always specify the discount rate used for the particular calculation.

Example 9.5.1 illustrates an NPV calculation with constant annual revenues and costs.

Quick calculation of NPV **Example 9.5.1**

A pre-feasibility study makes quick estimates of the finances for a hydro scheme as follows:

Revenue: $20000 each year for 15 years **Expenditure:** start-up cost $120000. Yearly expenditure $8000.

Calculate the NPV for the scheme assuming a 12% discount factor, and comment.

The annual net income is $12000. The total present value (PV) of receiving this annuity each year for n years is:

$$PV = \text{annuity}\ \frac{(1+r)^n - 1}{r(1+r)^n} = \$12000\ \frac{(1.12)^{15} - 1}{0.12\,(1.12)^{15}} = \$12000 \times 6.8 = \$81600$$

Notice that the discount factor 6.8 can be found from Table 9.3.1 (or by using a calculator).

The net present value of the hydro scheme is this present value sum of earnings ($81600) with the original investment of $120000 subtracted:

NPV (r = 12%) = PV – investment = $81600 – $120000 = – $38400

The conclusion is that the scheme is not viable from the investor's point of view, since the NPV is negative.

The investor would do better to use the finance in other projects offering a real return of 12% or more rather than to finance a hydro scheme. It is of course possible that the investment in hydro will have other benefits which will help indirectly to create financial returns not visible in this calculation.

Example 9.5.2 also calculates NPV but this time expenditure and revenue are not assumed to be constant annual sums. Instead of using the annuity equation, the single sum discount equation is used (as given on Table 9.2.1) to discount separately the net earning of each year.

Notice that in all NPV calculations we are using today's prices for future costs and revenues. There is no need to allow for inflation. This is possible because we are using the real discount rate 'r' which is equivalent to the market rate 'm' with the inflation element removed (for an explanation of this see Section 9.2 above).

Example 9.5.2 **Net present value (NPV)**

1 Calculate the Net Present Value of a scheme which has annual expenditures and revenues as presented below (units are $1000). The start-up cost of the scheme is presented as a $100000 expenditure in year 0. In this case a lifetime of only 12 years is assumed.

Year	0	1	2	3	4	5	6	7	8	9	10	11	12
Expenditure	–100	–15	–5	–5	–5	–5	–5	–7	–7	–7	–37	–7	–7
Revenue	0	28	28	28	30	31	31	31	31	31	25	31	31
Annual net earnings	–100	13	23	23	25	26	26	24	24	24	–12	24	24
Discount factor (Discount rate 12%)	1	0.89	0.80	0.71	0.64	0.57	0.51	0.45	0.40	0.36	0.32	0.29	0.26
Annual present values	–100	11.6	18.3	16.4	15.9	14.8	13.2	10.9	9.7	8.7	–3.9	6.9	6.2

NPV (r = 12%) = Sum of Annual Present Values = –100 + 129 = $29000. Since this is positive and is a substantial proportion of the investment, it indicates that the scheme is financially viable.

It is usually fairly tedious to calculate present value for each separate year of a cash flow table. The annuity equation can speed things up. Notice that the net earnings for years 6, 7, 8 and 9 keep the same value; $24000. All these years can be considered together in terms of their total value in year 6. To do this, first discount years 7, 8, and 9 back to year 6 using the equation:

$$PV = \text{annuity} \times \text{discount factor} = \text{annuity} \times \frac{(1+r)^n - 1}{r(1+r)^n}$$

where in this case n is 4 years into the future and the annuity is 24000. Use Table 9.3.2 or a programmed calculator to find the discount factor. Then add this to the year 6 net cash flow and discount the result back to year 0.

9.6

Internal rate of return (IRR)

The internal rate of return is the discount rate at which NPV = 0. The net present value of a project's earnings might be, for example, $1500 at a discount rate of 12%. If the discount rate is higher, the NPV is less. In this case a discount rate of 15% may yield a NPV of zero. The project is therefore said to have an IRR of 15%.

In example 9.5.1 the NPV was calculated to be – $38400 at a discount rate of 12%. Example 9.6.1 calculates the IRR of this scheme.

Example 9.6.2 shows how IRR can also be calculated on a cash flow table of single annual sums. Example 9.6.3 compares the IRR and NPV indicators and Note 9.6.1 lists their main advantages and disadvantages.

IRR and NPV are the most useful indicators of a project's worth and it is recommended that they are both presented (in preference to the payback period indicator).

Quick calculation of IRR — Example 9.6.1

Considering the same scheme as in Example 9.5.1, the capital cost is $120000, and the annual net cash flow is $12000. The project life is taken as 15 years. At a 12% discount rate, the NPV was found to be negative (– $38400). What is the IRR of this scheme?

The IRR will be equal to the discount rate at which NPV is zero. You can iterate quickly toward the answer by using the annuity equation, especially if it is held in the memory of your calculator. Simply try various values of discount rate, until the annuity is $12000. This discount rate is the IRR.

It is equally possible to find IRR by using Table 9.3.1, since the discount factors are Present Value divided by Annuity or 'C/A'. In this example A is 12000 and C is 120000, so C/A (the discount factor) is 10. Look at the 15 year row, and you find that a factor of 10 lies in between interest rates of 5 and 6%, a little closer to 6%. The IRR for the scheme is therefore, in round numbers, 6%.

Internal rate of return (IRR) — Example 9.6.2

Calculate the IRR of the scheme projected in Example 9.5.2.

Step 1 Choose a discount rate. Looking at Example 9.5.2, a positive NPV was achieved using a discount rate of 12%, which indicates that a value higher than 12% is needed to achieve a NPV value of zero. Let's try 15%.

Step 2 Draw up a cash flow table as shown below for the project. Calculate NPV, using the first guess discount factor of 15%. In this case we find the NPV is $13000 (see the table below).

Step 3 Now apply the principle:
"If NPV is more than zero, increase the discount rate until NPV is as near to zero as you can get, using whole numbers for discount rates. If NPV is less than zero, increase the discount rate".
Doing this (see below) we find that 17% gives us a NPV of $4000. We then try 18%, which gives us an NPV of zero.

Year	**0**	**1**	**2**	**3**	**4**	**5**	**6**	**7**	**8**	**9**	**10**	**11**	**12**
Expenditure	–100	–15	–5	–5	–5	–5	–5	–7	–7	–7	–37	–7	–7
Revenue	0	28	28	28	30	31	31	31	31	31	25	31	31
Annual net earnings	–100	13	23	23	25	26	26	24	24	24	–12	24	24
Discount factor 1st guess discount rate 15%	1	0.89	0.80	0.71	0.64	0.57	0.51	0.45	0.40	0.36	0.32	0.29	0.26
Annual present values	–100	11.3	17.4	15.1	14.3	12.9	11.2	9.0	7.8	6.8	–3.0	5.2	4.5

NPV = – 100 + 113 = $13000 (discount rate 15%)

Discount factor 2nd guess discount rate 17%	1	0.89	0.80	0.71	0.64	0.57	0.51	0.45	0.40	0.36	0.32	0.29	0.26
Annual present values	–100	11.1	16.8	14.4	13.4	11.9	10.1	8.0	6.8	5.8	–2.5	4.3	3.6

NPV = – 100 + 104 = $4000 (discount rate 17%)

Discount factor 3rd guess discount rate 18%	1	0.89	0.80	0.71	0.64	0.57	0.51	0.45	0.40	0.36	0.32	0.29	0.26
Annual present values	–100	11.0	16.5	14.0	12.9	11.4	9.6	7.5	6.4	5.4	–2.3	3.9	3.3

NPV = – 100 + 100 = $0 (discount rate 18%)

Step 4 The IRR is the discount rate at which NPV is nearest to zero.
Therefore in this case the IRR is 18%.

Example 9.6.3 — IRR and NPV: Comparing two projects

A micro-hydro scheme is proposed which has a capital cost of $8000. The annual revenue is expected to be $2400 and the annual operating costs are $1000. A fifteen year life is assumed. Prepare a simple graph showing how NPV varies with discount rate (Plot NPV on the vertical scale and discount rate up to about 28% on the horizontal). A diesel generator is proposed for the same site which is expected to achieve the same revenue. It costs $2000 to purchase and is expected to require $1900 in maintenance and fuel costs over the next fifteen years. (In practice diesel generators tend to have shorter lives but we will assume a long life here to keep the example clear.) Plot NPV on the same graph as above, and indicate IRR for both projects.

Solution: Net annual revenue for the hydro is 2400 – 1000 = 1400 and for the diesel is 2400 – 1900 = 500. Use Table 9.3.1 to calculate NPV for various discount rates:

NPV = PV – C = (1400 × factor) – 8000

For the diesel generator the same procedure yields the expression:

NPV = PV – C = (500 × factor) – 2000

The IRR for the hydro is 16% and for the diesel it is 24%. Yet for discount rates of 12% and less the hydro has the higher NPV.

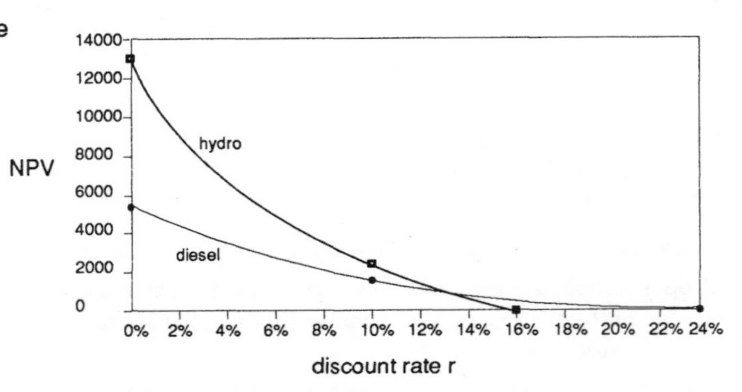

eg			
Micro-hydro	IRR	=	16%
	NPV (10%)	=	$2650
Diesel	IRR	=	24%
	NPV (10%)	=	$1800

Which project should we choose?
The IRR indicator would suggest the diesel; the NPV (at r = 10%) the micro-hydro!

To answer this question we first of all need to find out what the figure $2650 means when we state that the hydro project's NPV (r = 10%) = $2650. It can be shown that $2650 is the net benefit, less costs of running the project, *provided that monies released from the project are immediately put to work elsewhere at a return of 10%.* If the investor is certain that his own opportunities elsewhere will give a return of 10% then the NPV calculation, at the time discount rate he personally prefers to use, will usually indicate to him which is the better project to invest in. The hydro project should be chosen.

In contrast to the NPV calculation the IRR value tells the investor the annual rate of return on monies *whilst they remain tied up in the project.* The IRR indicates which project will maximize return on monies whilst they are in the project; it is not concerned with what happens to monies once they are released. The IRR in this respect is very useful for an investor with very few opportunities – he does not need to know his own time preference discount rate. If the investor is very poor he is likely to have a high discount rate, in as much as he will tend to discount the future. IRR calculations will very often anticipate his ranking of alternative projects. It will also guide him as to what maximum interest rate he could borrow money and run the project successfully. Being poor most of the money will have to be borrowed (see Section 9.8).

Note 9.6.1 — NPV and IRR compared

NPV

1 Discount rate 'r' needs to be chosen before calculation.
2 Calculation is easy to perform.
3 Cannot be used directly for ranking projects of different lengths.
4 Final value depends on the size of the project. (This may be a help of course.)

IRR

1 Discount rate 'r' is not needed for calculation.
2 Calculation is not easy except for the simplest projects.
3 Can be used for ranking projects of differing lengths.
4 The best indicator if you are poor and have a high time preference discount rate.
5 Will provide a guide as to the maximum interest rate at which monies can be borrowed to finance the project.

9.7

Simple and discounted payback periods

A very common financial indicator is the pay-back period. This is often used in pre-feasibility studies when the finances of a scheme are seen in terms of fixed annual sums which do not vary year by year. Example 9.7.1 shows how to calculate payback.

It is important to distinguish between 'simple payback' and 'discounted payback'. Simple payback does not take the time-value of money into account, while discounted payback does. This is illustrated in Example 9.7.1. To use the discounted payback equation it is necessary to estimate an appropriate discount rate.

Simple and discounted payback periods **Example 9.7.1**

A proposed hydro scheme has a capital cost of $60000. An annual revenue of $18000 is expected. Expenditure is $4000 per year.

1 Calculate the simple payback period of this scheme

$$\text{Simple payback} = \frac{\text{total capital cost}}{\text{annual revenue} - \text{annual expenditure}} = \frac{60}{18 - 4} = \frac{60}{14} = 4 \text{ years}$$

2 Calculate the discounted payback for the same scheme, assuming an appropriate discount rate is 12%.

Use Table 9.3.2. The discount factors are Present Value divided by Annuity, or C/A. In this case C is 60000 and A is 14000, so the factor is 4.3. Looking down the 12% column, the discounted payback period is between 6 and 7 years, but nearer to 7; so we therefore say it is 7 years.

9.8

Bank loans and interest

We have already seen that money expected to arise in a future year is discounted in value, that is, it is considered to be equivalent to a smaller value in hand in the present. We have also seen that this means that a sum in hand now is equivalent to a larger sum in the future. In other words a present value has an increased future value, the increase being the 'interest'. The word 'interest' therefore expresses exactly the same idea as the word 'discount'. Just like a discount rate, a market interest rate takes both inflation and the cost of capital into account. A bank will quote a market rate of interest, because it expects repayments which compensate it for lending the money, as well as to compensate it for the devaluation due to inflation. We can give the market interest rate charged by a bank the symbol 'i' as the equivalent of the market discount rate (m). In the same way we can correct market interest for inflation (f) to achieve a figure for real interest (r):

$$1 + r = \frac{1 + i}{1 + f}$$

For very approximate financial evaluations it is usually sufficient to say $r = i - f$.

When arranging a loan from a bank, it is usual to negotiate a repayment schedule. The simplest schedule consists of repaying the loan and interest as a series of constant annual payments. The value of these can be easily found from annuity equation:

Constant annual loan repayments

$$= \text{loan} \times \frac{i(1 + i)^n}{(1 + i)^n - 1}$$

Other repayment schedules often mention the 'principle' and the 'interest' separately. The principal is the original sum borrowed, which must be returned, as opposed to the interest, which we have seen is the additional compensatory payment.

Suppose the micro-hydro scheme described in Example 9.6.1 is to be implemented with the whole of the \$8000 capital cost being raised with an annuity-type bank loan (ie constant repayments of capital and interest) spread over 10 years at 12% per year ($i = 0.12$).

This will give a constant annual loan repayment

$$= \$8000 \times \frac{0.12\ (1.12)^{10}}{(1.12)^{10} - 1} = \$1416 \text{ per year}$$

The original evaluation to obtain NPV and IRR indicators was as shown in Fig 9.8.1. None of the future costs and benefits were increased for general inflation; the latter being taken care of in the choice of discount rate.

With the annuity loan from the bank the \$1416 per year needs to be reduced to take account of inflation at 8% per year. This allows future costs and benefits on the table to be presented without allowance for general inflation. The resulting cash flow can be seen in Fig 9.8.2. The net benefit at the end of the first year is almost eliminated in bank repayments. In contrast, at the end of the tenth year the net benefit is more than double the bank repayments. Notice that the sum of 1416 is what is actually paid to the bank; the extra expense of inflation would have to be met by inflation of the costs and benefits. It would be possible to present all costs and benefits in inflated form and keep the \$1416 repayment figure unadjusted. (This approach is used in the next section, on cash flow analysis.) Notice also that the bank loan could be some part of the original investment, for instance it might be \$4000, in which case a capital investment of \$4000 would show in year 0.

The annuity loan of course is not the only possibility. For instance \$1000 borrowed over 10 years might incur a repayment of principal of \$100 each year. Interest payments will be additional to this sum. After year 1, the size of the loan will have reduced to \$900; after year 2, it will be \$800, and so on. Because the loan is reducing in size, interest payments will also reduce, since they will be a certain percentage of the remaining loan. In this case, the total loan repayment at 5% interest will \$150 for the year 1, \$145 for year 2 and so on.

In many cases the borrower will prefer not to repay any principal for the first few years. This is because it may take some years before sufficient funds are generated by the project. The bank may allow this, but may insist on annual or monthly interest payments throughout.

The 'repayment schedule' details which repayment sums are expected each year. An important part of the financing of a hydro scheme is the negotiation of a suitable repayment schedule for loans; this is done with the help of a cash flow table, as described below in Section 9.9.

The cash flow table, with or without bank loan repayment details, is part of the 'cash flow analysis' and is usually not necessary in prefeasibility studies. It is often necessary in full feasibility studies.

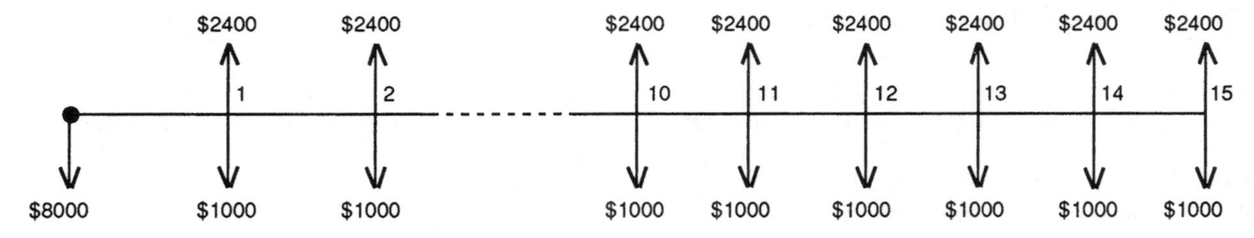

Fig 9.8.1 Profitability analysis with initial capital of \$8000

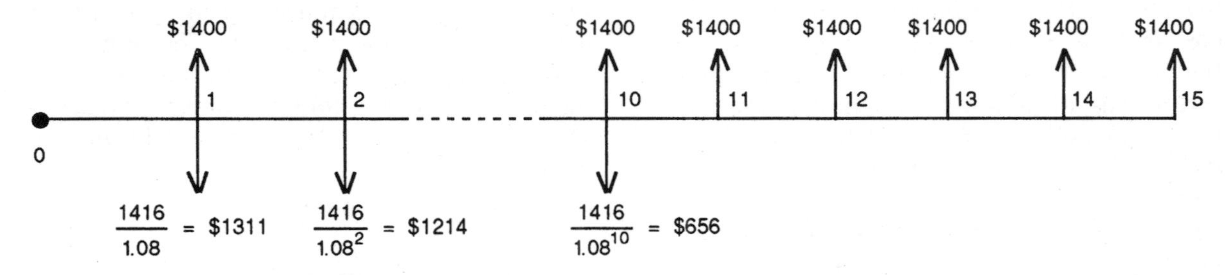

Fig 9.8.2 Profitability analysis with \$8000 bank loan

9.9

Cash flow analysis

In a pre-feasibility study expenditures and revenues are often estimated as constant each year, but in a full feasibility study it is accepted that they will vary with time, and it is useful to prepare a table predicting this variation. For instance, in Example 9.9.1, a cash flow table is drawn up for a scheme which has a start-up cost of $100000. In this type of table, bank loan repayments are not included as expenditures. Instead, the table gives quick guidance as to a suitable loan repayment schedule.

It is assumed in this example that general inflation is zero, and that there are no prospects of exceptional inflationary effects. This assumption helps keep the cash flow table simpler to understand, but in practice it will be necessary to adjust all the figures on the table to allow for expected inflation rates: for instance, wages may rise, repairs may cost more, and tariffs will be increased.

In the first year of operation, predicted expenditure on O+M is high ($15000) because of training and initial management costs. O+M expenditure then settles down to a steady $5000 per year. In the seventh year O+M expenses are predicted to rise to $7000 per year to take account of increased wear and tear on machinery, rising operator wages, extension of new services to customers, and increases in the prices of spare parts. It is considered likely that in the tenth year complete refurbishment of the turbine and alternator will be required at a cost of $30000. Meanwhile revenue is fairly steady, rising from $28000 per year to $31000 except for a drop while refurbishment takes place.

In the example the 'annual net cash flow' is calculated – this is the money remaining in hand each year once expenditure has been subtracted from revenue. Notice that in year 0 (the 'start-up' year) and in year 10 (the refurbishment year) the net cash flow is negative. Some form of external investment may be needed to cover the start-up cost; this may be a loan with interest. It may be that only a portion of the initial cost needs to be raised as a loan, and some of it can be found from the investor's or the hydro user's own resources. The questions then arise, will the money earned in other years be sufficient to recover this investment, and secondly, will revenues cover for the negative balance of $12000 in year 10?

A quick way to answer the first question is to calculate a 'simple running total' for each year – this is the sum of the annual net cash flows of previous years. This can also be referred to as the cumulative net cash flow, but the term 'simple' is a useful reminder that the time value of money is not yet taken into account. The simple running total calculated in the example shows that the investment can be recovered within five years. This should be interpreted to indicate that a recovery period of about twice as long will be needed, since the discounting effect will usually double the cost of start-up capital. (The five year period is also referred to as the 'simple payback period' – this is discussed in Section 9.7.)

The second question, how to cover for shortfall in year 10, is not so easily answered. If calculated for the whole period the simple running total will show a healthy accumulation of funds in year 9 which would easily finance the refurbishment costs in year 10, but no conclusions should be drawn from this, since the real (time-value) cost of finance is not yet considered. A second table is needed before a realistic assessment of accumulated funds can be made.

The second table in Example 9.9.1 is drawn up by guessing, or by discussing with a bank, a possible loan and repayment schedule for the same project. It is assumed here that the full start-up cost of $100000 is procured as a loan on the basis of constant annual repayments. It would be equally possible to borrow another sum, say $60000, and leave -$40000 in the net cash flow column for year 0, representing an investment of internal resources.

Table 1 indicates that a loan repayment period of about ten years is probably necessary, since this is twice the simple payback period. Using the annuity equation for a bank interest rate of 12%, the result is an annual repayment of $18000. Now consider if this is a sensible sum – can the project afford it? Since the net cash flows are around 23 to 26000, it is an affordable sum. A shorter period than 10 years could be chosen and the annuity equation used to calculate a higher repayment sum. Conversely, if the repayment is too high for revenue to cover it, choose a longer period. It is important to leave net cash flow figures of about $3000 since this provides a safety margin in case the predictions turn out to be inaccurate.

Annual net cash flow figures are calculated for Table 2 exactly as in Table 1, but this time loan repayments are included as expenditures. The result is that we again find negative cash flows occur, this time in years 1 and 10. These are unacceptable, since they imply that the project will not be able to meet its operating costs, and will therefore be forced to stop, failing to generate the planned revenue.

The shortfall of funds in year 10 shows a need for 'working capital' – this is the phrase used to refer to additional finance requirements which appear on a cash flow table. One of the main purposes of the cash flow analysis is to reveal working capital requirements.

One source of working capital is revenue accumulated in previous years.

Accumulated funds, or 'cumulative cash flow', is a running total calculated in the same way as in the previous table. The cumulative cash flow in year 9 on Table 2 is $46000 which is sufficient to cover for the refurbishment costs the following year. The positive cumulative sum of $16000 the following year demonstrates this. The problem of negative cash flow in year 10 is therefore solved, the required working capital having been found from accumulated funds. It should be remembered that the investor will also regard accumulated cash as a 'cushion' or safety margin protecting against unexpected problems or errors in prediction, so it would not be acceptable to use all of the accumulated funds as working capital. For instance, if the working capital demand for year 10 had been $45000 (rather than $30000) the remaining cumulative sum would be $1000 which would be considered an insufficient margin. The investor might decide to reject the scheme as financially unsound on this basis.

The problem showing on Table 2 of negative cash in year 1 can be solved by revising the repayment schedule. To do this, consider the year 1 cash flow shown on Table 1. Since this is $13000 it is clear that loan repayments in year 1 should not exceed $10000. This would allow a $3000 safety margin. The original loan to cover start up costs is $100000. Suppose $10000 is repaid at the end of year 1, and that this is regarded as a payment of principal only, with no allowance for interest. The remaining principal is then $90000. Interest of $12000 (12% of $100000) will also be owed for the first year. The total remaining debt is therefore $102000. This could be paid off with a constant annual repayment over the following ten years. Following the annuity equation (Section 9.3), it can be calculated that a constant annual repayment of $19000 would be needed for the next 9 years. Often an investor has more confidence in a scheme with as short as possible loan repayment period. An eight year period would demand a constant repayment of $21000 each year (calculated from the same equation). Anything shorter than this would create cash flow problems in the first few years of operation, but of course it would be possible to negotiate further 'soft-start options' with the lender. As one example of several possible solutions, Table 3 shows an eight year constant repayment following the 'soft-start' in the first year. It is noticeable that the loan is cleared one year earlier than in Table 2, so that the refurbishment in year 10 takes place without the extra burden of loan repayments; consequently the net cash flow deficit in year 10 is less severe. As in Table 2 accumulated funds are required to accommodate the year 10 deficit.

When preparing a table, make sure that all the following points are considered:

- Is the expected life of the machinery a realistic estimate? Are replacement costs included and increasing O+M costs as the age of the machinery begins to tell?
- Training costs. New operators will need further training. Are these costs included?
- End-use equipment. Initial purchase, replacement and maintenance of wires, lights, grain mills, and so on, will need finance.
- Has the cost of organizing and managing the end-use equipment and the hydro been properly included?
- Are subsidies or grants involved? These can be subtracted from the costs.

Cash flow tables — Example 9.9.1

Prepare cash flow tables for a project which has a start-up cost of $100000 and has annual expenditures and revenues as predicted by Table 1 below. The tables should be used to help plan a scheme and to assist negotiations for a loan and a suitable repayment schedule. They should also be used to help a potential investor evaluate the future viability of the schedule. Assume zero inflation.

1 Cash flow table without loan repayments (units of $1000)

Year	0	1	2	3	4	5	6	7	8	9	10	11	12
Expenditure	–100	–15	–5	–5	–5	–5	–5	–7	–7	–7	–37	–7	–7
Revenue	0	28	28	28	30	31	31	31	31	31	25	31	31
Annual net cash flow	–100	13	23	23	25	26	26	24	24	24	–12	24	24
Simple running total	**–100**	**–87**	**–64**	**–41**	**–16**	**10**							

2 Cash flow table for the same project including a proposal for a loan repayment schedule

The loan is for $100000 and the schedule consists of constant annual repayments of $18000 for 10 years, calculated from the annuity equation for an interest rate of 12%. Notice that this table shows a cash flow problem in year 1 which indicates that either the repayment schedule is unacceptable or additional external funds must be introduced. The shortfall in year 10 shows a working capital requirement of $30000 which could be covered by previous accumulation of funds.

Year	0	1	2	3	4	5	6	7	8	9	10	11	12
Expenditure	–100	–15	–50	–5	–5	–5	–5	–7	–7	–7	–37	–7	–7
Loan repayment	0	–18	–18	–18	–18	–18	–18	–18	–18	–18	–18	–0	–0
Revenue	100	28	28	28	28	30	31	31	31	31	25	31	31
Annual net cash flow	0	–5	5	5	7	8	8	6	6	6	–30	24	24
Cumulative cash flow	**0**	**–5**	**0**	**5**	**12**	**20**	**28**	**34**	**40**	**46**	**16**	**40**	**64**

3 Cash flow table with revised repayment schedule to overcome negative cash flow in year 1

This is achieved by repaying an 'affordable' sum of $10000 at the end of the first year, and then recalculating constant payments on the remaining debt. The period of repayment has been shortened as far as possible, to nine years, without allowing annual net cash flows to drop below $2000 which may be accepted by the investor as a minimum safety margin for a viable scheme. Cumulative cash flow can also be regarded as providing a contingency fund for the scheme.

Year	0	1	2	3	4	5	6	7	8	9	10	11	12
Expenditure	–100	–15	–5	–5	–5	–5	–5	–7	–7	–7	–37	–7	–7
Loan repayment	0	–10	–21	–21	–21	–21	–21	–21	–21	–21	0	0	0
Revenue	100	28	28	28	30	31	31	31	31	31	25	31	31
Annual net cash flow	0	3	2	2	4	5	5	3	3	3	–12	24	24
Cumulative cash flow	**0**	**3**	**5**	**7**	**11**	**16**	**21**	**24**	**27**	**30**	**18**	**42**	**66**

Operation and maintenance 10

10.1 Introduction

The importance of planned maintenance to micro-hydro power schemes cannot be stressed highly enough.

The implementation of a scheme will involve a large capital commitment. In order to repay this investment the scheme needs to run efficiently and continuously throughout its design life (typically more than 15 years). Efficient and continuous running will only be possible with skilled operation of the scheme and a well planned maintenance programme.

O+M procedures must be planned and put into action in the initial stages of any scheme to prevent breakdowns and reduced power outputs. The failure to use O+M procedures will result in financial problems and loss of confidence in the value of micro-hydro.

Safety in the place of work is also a major consideration. The use of O+M procedures will ensure choice of the correct spare parts and correct tools. Safety equipment should be installed in the plant and should be correctly rated, and never overridden or bypassed. The O+M procedures should include checks on the condition of safety equipment and training in the proper use of safety equipment. Operators should practise safe procedures, such as always shutting down the plant before undertaking a maintenance task.

This chapter considers the O+M requirements of each component of the scheme in turn, from weirs through to electrical equipment. It also considers spare parts, tools and O+M schedules.

A fault diagnosis table is placed at the end of the chapter.

10.2 Responsibilities

The design engineer, the equipment installers and the users of the scheme, all have important parts to play in the O+M of micro-hydro schemes.

Designers

The designer must have O+M in mind throughout the design process. He or she must have an understanding of the skill levels, motivation, availability and costs of O+M staff in order that appropriate designs are used. For example, an unlined earth channel may be appropriate where access is good and full-time maintenance staff are envisaged, but totally inappropriate for an unattended plant in difficult terrain. The design engineer must write the O+M schedules described in section 10.4 below. He or she must get to know the users of the scheme very well, in order to make sure that the skills, availability and motivation of the O+M staff will be commensurate with the schedules.

The schedules must specify clearly which kinds of tasks should be undertaken by directly resident O+M staff, which tasks should be undertaken by local repair workshops, and which tasks should be referred to the manufacturers of the equipment in question. The design engineer must make sure that the local workshops in question have sufficient written (as well as verbal) information. For instance, if turbine and alternator bearings are to be replaced periodically, the correct procedure must be written on paper in order to facilitate later verbal explanations and reminders.

It is the responsibility of the design engineer to get to know the skills and resources of mechanics, workshops and manufacturers, and to distribute maintenance tasks accordingly.

Most important is the designer's role in choosing appropriate equipment which can be repaired easily and for which spare parts are readily available. Well-trained staff may be available, but if the designer specifies equipment without good spares availability, they may be powerless to achieve reliable operation.

The design engineer must supervize installation work in order to make sure that the quality standards specified are adhered to.

The design engineer must arrange and supervize proper training courses for the O+M staff, both at initiation of the scheme, and at regular stages in the future. Often experienced O+M staff can be asked to train new O+M staff recruited for new schemes, which can be an opportunity to refresh and upgrade the skills of the older staff.

The design engineer must write the training manuals described below in the section on O+M documentation.

Installers

In larger hydro schemes, the installers will have tightly supervized specifications to work to. With smaller schemes it often happens that the responsibility for detailed design and quality control lies with the installer. This can result in schemes which are difficult to operate and maintain. For example, a switchboard may be within broad specifications but may be untidy, difficult to access, labelled in the wrong language, and use switches for which there is no local agent.

It is essential always that this situation is avoided and that quality work is requested by the designer. The installer should be present during a quality check stage and during commissioning and testing stages, and should be bound by contract to make good any inadequacies.

It is common practice for the training of operators to be left in the hands of installers. The result is that the operators are often not trained and are ignorant of essential procedures (such as lubrication and sources of spare parts) with disastrous results. To avoid this, it is essential that the installers provide full O+M documentation associated with specialist equipment. Secondly they can be requested to engage in formal training of operators which is arranged and supervized by the design engineer.

Users

The users of micro-hydro schemes are usually themselves responsible for O+M duties or for recruitment of O+M staff.

Operation and maintenance of a modern micro-hydro can be a monotonous job – in many cases involving long periods of inactivity. Motivation can be a problem. In many cases it is important for the designer to avoid the cost and complexity of labour-saving devices in order to ensure that the operator has a satisfying and responsible job. The users should brief the designers at an early stage on the skills, motivation and availability of the O+M staff envisaged. Are they to live in the powerhouse? (design for low noise, space, low pressure water supply etc). Is the plant to be unattended some of the time? In this case automated controls, such as remote alarms, and extra safety trips, have an important role to play.

The users have to implement the O+M schedule and make sure that spares are stocked and replaced in time.

They will be responsible for the recruitment and the payment of wages to the O+M staff and their well-being as employees. They will have to implement the training schedule specified by the design engineer.

Often maintenance and repair tasks will occur which cannot be undertaken by the O+M staff directly. The staff and the users should be clearly instructed as to which level of outside assistance to call in for which job. Some jobs are best referred to local workshops and some jobs must be referred to the manufacturer of the equipment in question. The user will assist in making these referrals and will make sure that the local workshop is following the written recommendations of the design engineer concerning procedure.

More complex repairs (for instance, repairs to a runner) will need to be done by the manufacturer. In some cases maintenance contracts may be set up. Maintenance contracts have the advantage of continuity with the same contractor, but create a monopoly situation which can be a disadvantage if the chosen contractor becomes complacent or unresponsive.

Because micro-hydro schemes are usually located in remote areas it can be very difficult for the user to successfully and quickly negotiate with specialist manufacturers.

The user should insist that the scheme contains a minimum of complex components and should look to local workshops for almost all repairs. The design engineer should seek to upgrade the skills of local manufacturers and should also open effective channels of communication between user and specialist manufacturer where these are needed.

10.3

The O+M documents

Even for small informal installations, a structured system for O+M is essential. This is expressed in a complete set of the following documents:

1 **Operation manual**
This will contain basic operating instructions, such as start-up, maximum loads, descriptions of meters, trips etc. This manual should be readily available to O+M staff, and in the right language.

2 **Component manuals**
Typically, manuals are available for turbines, alternators, switches, belts, couplings, governors and so on. Copies of all manuals must be readily available in the right language.

3 **Installers' manuals**
These should cover any equipment not covered by the manuals; for example, the penstock and the powerhouse. These documents should give drawings, dimensions, design information and O+M schedules.

Sample portion of maintenance schedule **Example 10.3.1**

Maintenance schedule	All year	Drought	Floods
Weir and intake			
1 Check for boulder damage	monthly		daily
2 Check for leaks, undercutting		once	
Regulating sluice			
1 Check operation	monthly		
2 Grease screw	monthly		
3 Adjust		as needed	as needed
Settling tank			
1 Grease flushing sluice screw	monthly		
2 Drain and clean	monthly		daily
Channel			
1 Inspect for leaks, overflowing	weekly		daily
2 Drain and clean	every 3 months		
3 Clean culverts	monthly		daily
4 Sealing leaks and general repairs	annually		
Forebay tank			
1 Clean screen	daily		
2 Check screen	weekly		twice daily
3 Grease valves	monthly		
4 Drain and clean	weekly	monthly	twice weekly
Penstock			
1 Visual check for flange leaks	monthly		
2 Repaint	every two years		
3 Visual corrosion check		once	
4 Inspection of supports	every 6 months		

4 **Maintenance schedule**
This describes in detail the various maintenance tasks and when they should be carried out. Typical items will be greasing bearings and emptying silt basins. Example 10.3.1 shows a typical schedule for civil works. Most problems will occur during times of heavy rainfall and more frequent inspection may be necessary during the rainy season. Repair work on the other hand is more easily carried out during the dry season. The schedule takes account of these variations.

Each scheme must have schedules drawn up which are specifically designed for the equipment used in the particular scheme.

5 **Repairs manual**
This document lists sources of technical assistance. It will give the name and address of the design engineer and of other suitable technical advisers. For each component it gives the address of the manufacturer and of the local workshop familiar with that component. It will have space for notes as to the current state of any service contract, works carried out by that particular repair workshop, costs incurred, and so on.

6 **Log book**
This may vary from a small notebook to a more formal record book which can be checked and countersigned by a supervisor. It is an essential document for continuity and monitoring of O+M activity. Without a log book, for example, oil changes might be missed or the fact that a certain bearing fails far too often might not be noticed.

7 **Training manual**
This contains the content of both the initial training courses offered to O+M staff, and of annual or bi-annual follow up courses. The information is presented clearly in the right language so that it can be used to help initiate further training for new staff members. It also contains training notes addressed to administrative staff associated with the scheme.

Example 10.3.2 **Sample page from a log book**

	Staff		Supervisor	
	Remarks	Signed	Remarks	Signed
Monday 4th August				
Check bearing temperatures:	*OK*	*M.S.*	-	*AJS*
Check belt tension:	*OK*	*M.S.*	-	*AJS*
Check ELC meters:	*OK*	*M.S.*	-	*AJS*
Check ballast tank supply:	*OK*	*M.S.*	-	*AJS*
Inspect channel:	*Leak near intake*	*M.S.*	***Cement ordered for repair***	*AJS*
Clean screens:	*OK but screen allowing leaves past*	*M.S.*	***M & G asked for quote to repair***	*AJS*
Tuesday 5th August				
Visual check of changeover switches:	*All OK*	*M.S.*	-	*AJS*
Inspect channel:	*Leakage causing erosion*	*M.S.*	***Bring channel cleaning forward to 22/9/99 and repair leak***	*AJS*
Check alternator, ELC, ballast tank connectors for tightness	*Two connectors loose*	*M.S.*	***Monthly checks from now on***	*AJS*

10.4

Weirs

Temporary weirs are designed for seasonal repair after damage by heavy rains. The traditional practices of the region, used for irrigation weirs, should be followed.

In most cases a concrete weir will need very little maintenance. Heavy rains bring large loads of vegetation, silt and boulders, which collect upstream of the weir. It is not necessary to remove these since they do not stop the weir from functioning, although they may block the channel intake next to the weir and so need partial removal. In fact it is better not remove all the silt and boulders collected at a weir as they can help to keep a good water seal, and prevent damage to the weir.

If water is leaking through the weir it will show up during the dry periods of the year. It is during this time of year that it is necessary to get as much water into the channel as is available in the stream. If there are leaks through the weir this will not be possible. When detected in the dry season these leaks can be stopped as a temporary measure with sand bags.

The dry periods of the year are also the best time to inspect the weir and wing walls for cracks and carry out repairs. Flood water can erode the banks if the wing walls are damaged, and erosion of the banks could have serious implications on the working of the hydro plant.

Some weirs are designed with sluice pipe or flushing gate which will help to flush the silt collected at the weir. This opening will also help divert the water when work is done on the weir.

Particular care should be taken when repairing high weirs as they must remain strong after repair. It is also advisable not to raise the height of the weir as the foundations may not be strong enough to withstand the higher level of water.

Note 10.4.1 — **Suggested cement mixtures for repairs**

To repair leaks resulting from cracks on channels, weirs, etc.

Use one portion of cement and two or three portions of sand.

For rendering work on surfaces with contact with water
(for example: channels, forebay, tank, silt tank, weir, intake):

Use one portion of cement and three portions of sieved sand.

For rendering work on external surfaces which are not in contact with water it is possible to use any of the following combinations depending on the strength required:

One portion of cement to four, five or six portions of sand.

For repairs and patch work on concrete, not in contact with water:

Use one portion of cement, two portions of sand and four portions of aggregate. The aggregate should be half (1/2) or three-quarter (3/4) inch.

For repairs and patch work on concrete which is in contact with water:

Use one portion of cement, one and a half portions of sand and three portions of aggregate, size as above.

10.5

Intakes

Inspect the intake at least daily, especially during the wet periods of the year. During the rainy season the stream will carry more debris and silt which could block the intake. Some intakes have built-in steel bars or trash racks which will need clearing from time to time.

The intake may also have a wall above the channel which will limit the flow of water during flood. In some cases it may be a sluice gate that does this job. It is very important to check the intake opening has not become larger than designed, or that the sluice gate is not left raised higher than the minimum, to avoid damage to the channel caused by flood flows. If excess water is allowed into the channel it will overflow along the channel causing the sides of the channel to wash away.

10.6

Overflows and spillways

Overflows and spillways should be inspected periodically like the rest of the civil works for various damages which may happen with time, eg cracks. These should be promptly repaired or the damage will extend and will be more expensive to repair later.

In some schemes there is provision to close overflows and spillways with stop logs during the dry periods of the year. This is to get the maximum flow in the channel. One of the dangers of wooden stop logs is that they can be stolen, so some form of fixing is useful.

10.7

Channels

During daily maintenance, the channel should be inspected for blockages. Any stones, silt or vegetation should be removed. It may be possible to flush silt toward the silt tank and remove it from there.

It is important to repair any damage as soon as it appears. If leaks are not attended to, they get worse and sometimes wash away the ground that holds the channel.

Drains to carry rainwater away from the channel should always be inspected and cleaned or repaired as necessary. In general, drains over the channels are better than drains under the channels, because underground culverts get blocked during heavy rains. It is always better to drain the rainwater away from the channel than to allow it to get into the channel.

Any particular section of a channel should carry water at a correct speed. An earth channel for instance will erode if the water runs too fast, causing the sides to collapse, and any channel will block with silt deposits if the water runs too slow. A concrete lined channel can take a fast flow.

When repairing a channel always refer to Chapter 3 to check that you know the speed limits of that section of channel, as the repair may change the water speed and so cause worse damage later.

Aqueducts

It can happen that people use the water in the channel for cultivating the land, and very often the channel is broken or blocked for this purpose. If this is a problem one solution is to install one or more 50 mm pipes in the side of the channel to supply water for cultivation. This is far more desirable than the wholesale damage that can result from makeshift modifications.

The footpath along the channel should be maintained in good condition so that inspections can be carried out regularly without inconvenience.

Inspect the supports and the structure of the aqueduct. Steel aqueducts should be painted annually with anti-corrosive paints. If they leak as a result of corrosion they should be repaired or replaced. Some steel aqueducts have rods across the top to strengthen them and these too should be inspected for corrosion weakness.

Sometimes pipes are used as part of the channel to bring the water to the forebay tank. They may get blocked and need to be cleaned. Also inspect the supports and joints along the pipe, and check the hillside above and below for signs of falling stones, as these can damage a plastic pipe.

10.8

Silt tanks and forebay tanks

Silt tanks form an important part of a micro-hydro scheme as it is the silt tank that determines the wear on the turbines. It is therefore wise to keep the silt tank in good condition at all times. The collected silt should be removed daily through the flushing gate; if not it will collect up to its limit and any excess will be passed into the turbine. During rainy periods the tank will have to be emptied frequently as there will be more silt in the water, washed from the soil.

Other than cleaning the silt out daily, the tank needs very little maintenance. The occasional masonry repair can be carried out during the dry period of the year. Flushing valves may need attention as there are moving parts which need to be lubricated about once a week. Care should be taken to direct the water leaving the flushing valve away from the tank, to avoid erosion of the foundations.

Sometimes a silt tank designed in previous years becomes insufficient in capacity. This may be because the land upstream is cleared of forest, which causes more silt to flow in the stream, or because the wear on the turbine over the years means that the machine now needs more water. The silt tank will therefore have to be enlarged, or cleared of silt at very frequent intervals.

The forebay tank will contain a trash rack which will need daily maintenance or cleaning, as it is here that all the water-borne vegetation and debris is prevented from entering the penstock and turbine. The trash rack should be cleaned as often as possible. During rainy periods it may be necessary to clean it twice daily.

If the trash rack is not designed for easy cleaning, such as vertical raking, it is better to redesign and replace it than to carry on with inadequate equipment. This is because a badly cleaned trash rack often results in a broken trash rack, which allows foreign objects to enter the penstock, reducing power, or damaging the turbine.

10.9

Penstocks

Cast iron pipes need very little maintenance, although sometimes the joints may leak. This is common with the spigot and socket type of joint which can be easily repaired with lead wool. Flanged joints need no maintenance other than inspection and replacement of any missing bolts.

Mild steel penstocks should be repainted annually or bi-annually, depending on conditions. PVC penstocks on the other hand should be kept away from direct sunlight, by painting the exposed parts of the penstock, by covering the pipe with vegetation, or by burying the pipe.

Penstock supports should be inspected as often as possible and any damage should be repaired immediately, as neglecting this could cause extensive damage to the penstock itself. If the supports are broken or damaged, the pipe is unable to support itself, and breaks under its own weight.

The drainage around penstock supports should be such that the water is moved away from the supports, as running water will erode the foundation of the supports. On some hydro schemes bad drainage has caused a few pipe lengths to move due to earth slips. Penstocks can easily be damaged by falling rocks and wherever possible these rocks should be removed from above the penstock.

10.10 Valves

Valve seals generally tend to leak with time which is not a major problem. Water leaks into the powerhouse can be stopped by repairing the sealing arrangement on the valve, but if the valve does not shut off the water completely, it is a skilled job to repair it.

Most gate valves have felt "stuffing box" seals. The felt needs to be kept moist by a small water leakage to prevent deterioration. The amount of leakage through the seal is usually adjusted by two bolts. The adjustment should be checked periodically.

Valves installed on the penstock should be fully opened while in operation and should not be used partly open as flow control valves. Flow control valves are designed differently and are generally installed on the turbine.

10.11 Turbines

If turbines are new the manufacturer's instructions should be followed during maintenance and operation.

The turbines themselves will need very little maintenance as long as the water supply is kept clean. However, if things do get into the turbine it is necessary to open the inlet valve in the case of the Pelton and Turgo and the inspection covers in the case of the Francis turbine. Foreign objects should be removed.

Many turbines have some type of mechanical link to the governor for speed control and these links should be greased and their nuts and bolts should be checked to see that they are tight.

Bearings should be kept dry. In most cases turbine bearings are placed a few inches away from the casings. There is sometimes a small hole positioned to drain any water which may come towards these bearings. This should be checked for blockage. A common mistake is to overgrease bearings (they should only be half full of grease). The excess grease will block the drain hole, and so result in water entry to the bearing.

It is important to keep the turbine tailrace clean, as otherwise water leaving the turbine will find it difficult to move out and will ultimately flood the casing, reducing the power output.

Francis turbines

Silt in the water is a major danger to the Francis as it wears out the internal parts. It is important to keep the water clean by designing an effective silt tank and emptying it regularly.

Francis turbines tend to get blocked with vegetation, which reduces the power output of the machine. The vegetation is removed by opening the inspection plate.

Francis turbines tend to leak along the shafts that are connected to the guide vanes. To a certain extent this leak is allowable as the water acts as a lubricant for the seals. However, if these seals leak too much the seals should be tightened.

The draught tube is an integral part of the Francis turbine. Its open end should be immersed in water during operation. Check that the tail water sill is in place. In some Francis turbines an air vent is provided to remove the air from inside the casing during start-up. Check that this is not blocked.

Crossflow turbines

Crossflow turbines can be catastrophically damaged by sticks because of the relatively fragile blades and small case-to-runner clearances. Good trash racks are essential. Some turbines are provided with inspection plates on the nozzle which helps to remove any dirt collected in this area.

Crossflows are more affected by erosion than Peltons because of the relatively thin blades, often fabricated from mild steel. Good silt control and regular inspection of the runner are therefore important.

10.12

Drives, bearings and belts

A direct drive needs very little maintenance, only the occasional inspection to check that the nuts and bolts are tight. Flexible couplings contain leather or rubber pieces which need inspection and replacement at intervals.

Plain bearings

'Plain' or 'lined' bearings are found on old machines. Usually these bearings are partially immersed in oil and provided with oil rings that lubricate them during operation. The oil levels should be checked before the turbine is started. The rings should be checked to see that they are free to rotate. The oil in these bearings should be changed periodically, at least every 4000 hours of operation. During operation these bearings should work at a low temperature, below 50°C, which means that the housing should be warm to the touch and not hot. If the bearings are in good condition they will run quietly, so listen for noisy operation.

Plain bearings should be dismantled and inspected at least once a year – the inspection will reveal the state of the lining. It can be remetalled if necessary.

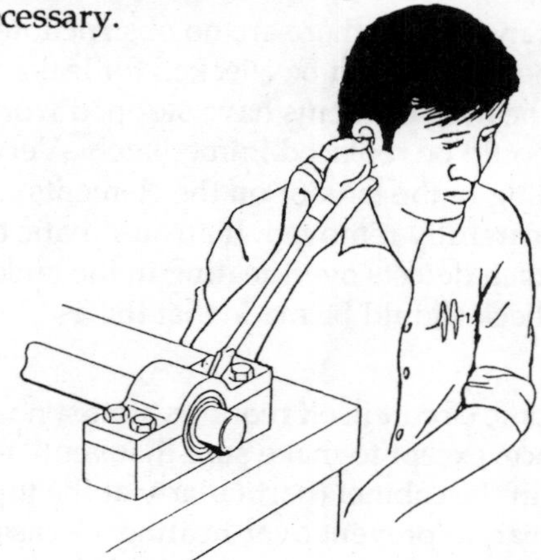

Fig 10.12.1 Listening for bearing noise

Rolling bearings

A large screwdriver makes a good listening device for detecting bearing trouble. The bearing should sound smooth – any grittiness, rumble or clicking indicates deterioration. Roller bearings should operate below 60°C. Where thermometers are not fitted, check by touching: 60°C feels hot but not too hot to allow contact for a few seconds without discomfort. Sound and touch tests should be made at least weekly. Make sure replacement bearings are at hand and fit them whenever deterioration occurs.

Most rolling bearings need grease for lubrication. In a few cases oil is needed. The bearings should be greased with the recommended grease. If no instructions are available they should be regreased every 4000 hours. Care should be taken not to use too much grease as this tends to make the bearings run hot. If grease guns are used, care should be taken, as bearing seals can be damaged if excess pressure is applied. Always remove the old grease and never reuse it. Remove all grit. Keep grease clean.

Belts

For good performance the belts should be:

- clean and free of oil or grease
- tightened to the correct tension
- correctly aligned.

Canvas belts are affected by the weather and they tend to tighten if the weather is wet and slacken if the weather is dry. Sometimes this has an effect on the performance of the drive but can be overcome by the use of belt dressing (paste) which is applied to belt.

Belt tensioning is generally done by moving the alternator on slide rails. Belt tensioning devices are used for Vee belts. Modern flat belts usually have specified extensions eg 104% of original length. This is checked by making marks on the belt and measuring extension between marks with steel rule.

If the alternator is moved on slide rails the alignment should be checked after tightening of the belts. If the machines are not correctly aligned the belts can be damaged.

Occasionally Vee belts should be inspected as they tend to wear and crack with use. When replacing these belts it is necessary to replace the complete set.

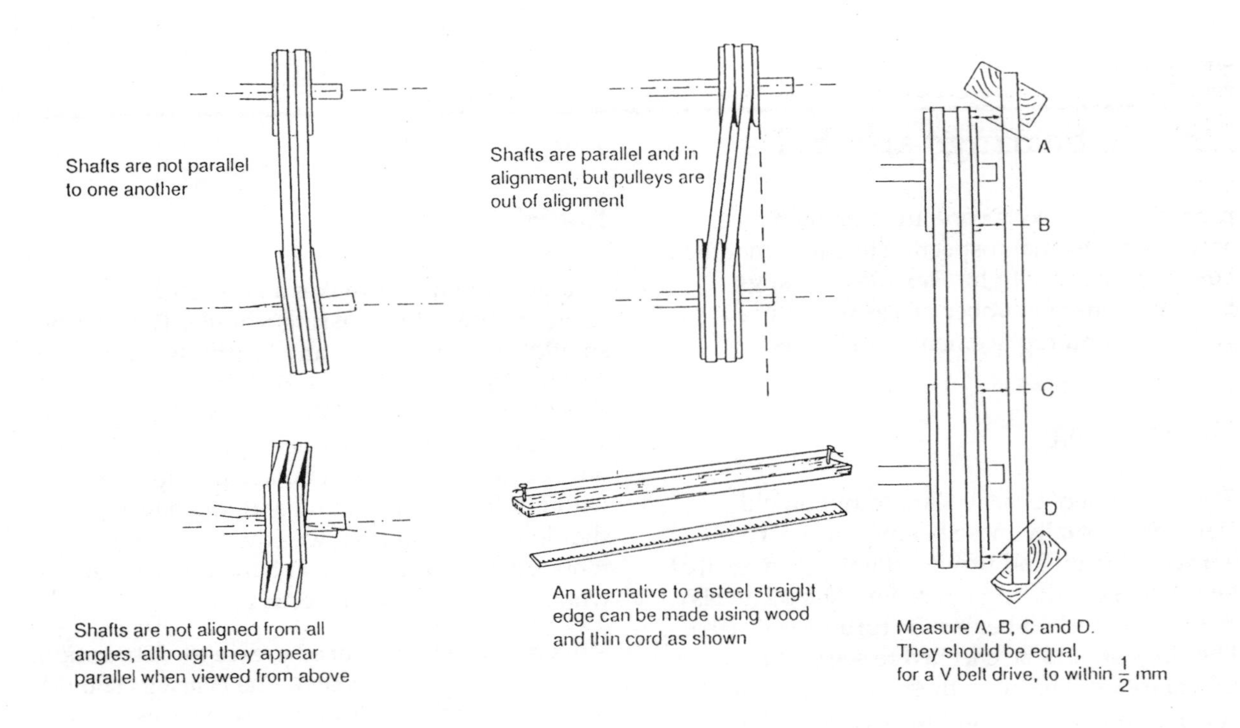

Shafts are not parallel to one another
Shafts are parallel and in alignment, but pulleys are out of alignment
A
B
C
D
Shafts are not aligned from all angles, although they appear parallel when viewed from above
An alternative to a steel straight edge can be made using wood and thin cord as shown
Measure A, B, C and D. They should be equal, for a V belt drive, to within $\frac{1}{2}$ mm

Fig 10.12.2 Alignment of belt drive pulleys

10.13 Governors

Mechanical governors should be inspected daily to check that belts are in good condition and link mechanisms are well lubricated. The governors have various adjustments and springs, which should be adjusted only after reference to the manufacturer's literature. Some mechanical governors work on oil pressure. The oil should be topped up and changed when required with the recommended oil. Dash pots are included in most of the oil pressure governors and these should be filled with oil.

Electronic load control governors consist of an electronic circuit connected to a ballast tank. A daily inspection must be made of the water supply to the ballast tank and the water flow through it; make sure there are no obstructions to flow. The tank should be checked for leaks. If any of the heating elements have stopped working they should be replaced immediately. Very severe scaling in the tank or on the elements should be carefully removed. If an automatic trip is fitted which detects over-heating in the tank, a monthly check should be made that this is working.

The electronic circuit itself requires almost no maintenance, except to make sure that ventilation holes in the cabinet (particularly at the top) are kept clear, to prevent over-heating. Occasionally check tightness of connections and clean or replace connections showing signs of corrosion.

10.14
Electrical generators

In terms of ease of maintenance, there are three grades of generator, the induction generator, the modern alternator, and early alternators, the latter containing brushes or slip rings.

The induction generator (reversed motor) is the easiest type of generator to maintain; it only requires an occasional check that the cooling air flow is not blocked, that the bearings are not overheating or running noisily, and that it is not vibrating or shaking heavily. The first two problems are rare, since motors have a simple arrangement for external air flow, and the bearings are usually heavy duty. Comment on bearing maintenance is given below.

Heavy vibration indicates phase imbalance or the need to change the connections on the phases.

A modern alternator is prone to damage from dust, because it has an internal cooling air flow which can suck dust or dirt into the windings. The dust absorbs moisture which can cause electrical problems where there are defects in the winding insulation. The dust also causes overheating of the windings with possible breakdown of insulation.

It is therefore essential to inspect the alternator windings frequently, and to clean whenever necessary, usually monthly but at least once every 6 months. A blower can be used to remove dust. Chemical solvents should not be used as they will damage the winding insulation.

The manufacturer's literature will indicate how often the bearings should be dismantled, cleaned, and regreased. Corroded components can be replaced or the complete bearing replaced if inspection reveals severe wear or pitting on the rolling surfaces. Often the bearing will be considered sealed for life, otherwise if information is lacking, assume that regreasing is required every 4000 running hours, or every 6 months. Use the recommended grease, and do not over grease (more that half-fill the bearing) as excess grease causes the bearing to run hot and may splash on to the alternator windings. These comments apply to all types of generators.

Weekly checks on electrical connections are necessary for all generators. Loose connections must be tightened. The frame of the generator should be earthed and this connection checked as well. Ventilation in the powerhouse should be adequate so that the generator does not run hot.

Early alternators with brushes, commutators and slip rings, require constant inspection and maintenance. Brushes become shorter as they wear, causing damage through sparking, and so need to have the brusher holder position adjusted at frequent intervals to take up the wear.

When the brush is completely worn it must be replaced with the correct size and grade; incorrect grades will cause spark damage and rapid wear. Commutators and slip rings, brush holders, and insulators, must be kept clean, for instance by daily wiping with a soft cloth before starting the machine. Commutators and slip rings should also be cleaned with fine sandpaper every 1000 running hours. Emery paper should never be used. If wear is very bad, the surfaces can be machined back to true so that the brushes run smoothly on them.

10.15
The switchboard

The switchboard contains various switches, fuses, electrical connections and a set of meters. Switches will burn out if they are not rated correctly. The most common O+M fault is the replacement of fuses with the incorrect fuse rating. A monthly check should be made of fuse ratings. If copper wire is used in simple fuse holders, Table 10.15.1 gives guidance on the correct sizing.

Wire connections should be tight as loose connections cause heating and damage to the switches. Switches also get hot if the contacts in them are worn, in which case the contacts must be replaced.

Meters and other instruments on the switchboard should be working as these indicate the condition of the plant. Safety trips and protection devices should be tested monthly.

Table 10.15.1 **Size of fuse elements**

Plain or tinned copper wire for semi-enclosed fuses

Current rating of fuse Amps	Nominal size of wire (diameter) mm	SWG
3	0.15	38
5	0.20	35
10	0.35	29
15	0.50	25
20	0.60	23
25	0.75	22
30	0.85	21
45	1.25	18
60	1.53	16
80	1.80	15
100	2.00	14

10.16 Transmission lines

Overhead transmission lines should be inspected monthly. Trees or branches within 2 m of the line should be cleared. Connections along the overhead lines should be checked and tightened if necessary.

It is best to include a set of lightning arrestors on the overhead line. If they are already installed they should be checked and replaced when necessary.

Some turbine installations have transformers and high voltage transmission lines. Specialist engineers should be called to maintain these. The user must make sure that high voltage transformers are always enclosed by a fence and some indication given to show that they are dangerous. The user can also monitor the state of the dessication unit on the transformer, and the oil level, since these are easily visible from the outside.

10.17 Recommended spare parts

This is a vital and often neglected part of maintenance. For example, in a typical US$30,000 installation a full set of tools and spares will cost around US$1,500, about 5 per cent of the capital cost. The penalty for not having the right bearing or the right tool in stock is typically, for remote sites, two months down-time. This might cost US$1000 in lost energy. The toolkit will therefore pay for itself after only two breakdowns.

One complete set of spares and essential tools should be specified by the design consultant and supplied by the contractor to the client. Each system must have a carefully thought out inventory and a re-ordering system which minimizes the chances of down-time. One system is to reorder as soon as a part is used. Where delivery times are long, a stock adequate cover multiple replacements during this period must be kept.

Spares should be kept in a cool, dry and clean place. Bearings should be stored in the original packing, or if this is not available they should be coated with grease to prevent rust and stored in a dry place. It is recommended by the Vee belt manufacturers that these are hung on a former and not left coiled on the floor. Electrical spare parts should be packed, labelled and stored away from moisture. Oil and grease should be protected from contamination by moisture and dirt.

Inventory of spares **Example 10.17.1**

This is just one example. An inventory must be specially drawn up for each scheme.

Spares	Quantity required	Delivery time	Supplier
1 Multi-purpose grease in dust-free container	2 litres	1 month	ADM
2 V belts, SPC, 270mm	6 off	2 months	BRH
3 C2163 roller bearings	2 off	6 months	STC
4 CX2116 ball bearings	2 off	6 months	STC
5 6BX001 diodes	6 off	6 months	SS
6 AVR No 233	1 off	12 months	SS
7 High speed fuses	3 off	6 months	BC
8 One 3m penstock length	1 off	9 months	OTT
9 Penstock paint	2 litres	2 months	FF
10 Multi-grade oil	2 litres	1 week	FF

10.18 Tools

Good quality tools are essential for good maintenance. Cheap, poor quality tools will break and will be expensive in the long run. Poor quality tools will also tend to damage the equipment that is to be maintained. If the turbine installation is new, then the tools should also be ordered along with the equipment. However, if it is an old installation the necessary tools will be: flat and crosshead screwdrivers, pliers, a set of open-ended spanners, hacksaw, files, grease gun, and any other special tools which may be required for the plant – for example, a puller to remove bearings.

Torches or portable lamps are very useful when working in the dark both in the powerhouse and outside. The outside work includes cleaning the trash rack and inspecting the channel.

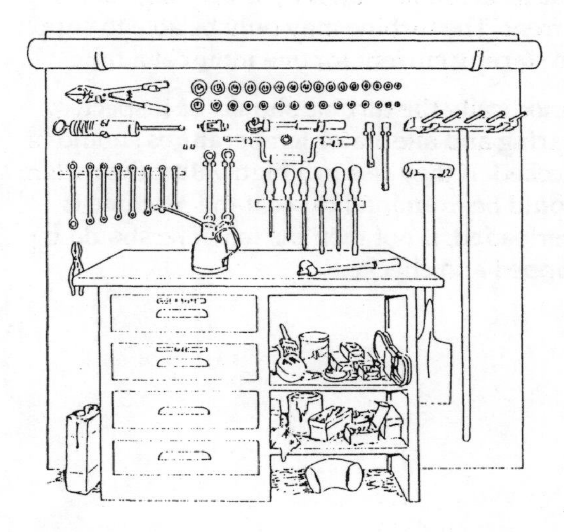

Fig 10.18.1 Tool rack

10.19

Operation

Operation of plants vary from site to site. In general the water supply to the machine should be checked before the plant is put into operation. For this, it is necessary to inspect the channel, regulating valves, silt tank, forebay tank and trash racks, to see that they are clean and the required amount of water is flowing in the channel.

Before the turbine is started it is good practice to rotate the shaft by hand to see whether it is free to rotate, and check the bearings, belts, nuts and bolts, etc. If the plant has an ELC then the water supply to the ballast tank should also be checked. The pressure gauge on the turbine should indicate the maximum static value.

The turbine should be started gradually and brought up to speed. The operator should observe the pressure gauge and the revolution meter as the turbine comes up to speed. If the pressure drops below its normal running value it is possible that there is not enough water or there is something blocking the penstock. If the speed exceeds the rated value then it may be that the governor is not working properly. At rated speed the voltage also should reach the nominal value (eg 415V) and in some cases it may be set at a higher or lower value, which should be in the range of ±3%.

Once the turbine is operating at its rated speed and voltage it is possible to load the machine. Select the required load and connect it onto the hydro. Motor loads should be started one at a time as the motor takes up a very high starting current. The turbine may only be able to supply the starting current for one motor at a time.

Periodically the turbine should be inspected. Bearing and alternator temperatures should be checked. If they rise to a high value, the meters should be examined to see if the machine is overloaded, if not then the machine should be stopped and checked.

The following could be taken as guidance for starting and stopping a micro-hydro plant.

Starting

1. Visual inspection of all equipment
2. Clean trash rack and silt tank if necessary
3. Turn on water at intake as required
4. Inspect oil levels
5. Check pressure gauge
6. Check that all switches are in the OFF position
7. Inform factory or village that turbine is to be started
8. If a bypass valve is installed, open it
9. Gradually let water into turbine and observe the pressure gauge, rpm meter, and the voltage meter
10. Listen to machine as it runs; if there is any abnormal noise or vibration, stop turbine and check
11. Continue to increase water flow gradually and observe pressure gauge; do not allow the pressure gauge to fluctuate rapidly
12. Once turbine speed and voltage come to the normal value, increase water as required
13. Switch ON loads by connecting hydro onto switchboard.
14. Observe machine and meters, check bearings; if load is too much, increase water or change part of load back to other source of power.

Stopping

1. Inform factory or village of stopping turbine
2. Switch OFF all loads connected to hydro
3. Reduce water gradually and observe pressure gauge; do not allow pressure to fluctuate rapidly
4. Allow turbine to stop and check bearing temperature
5. Stop water at intake if necessary.

10.20

Micro-hydro fault diagnosis

Turbine problems

Fault	*Cause*	*Remedy*
Turbine does not start	Insufficient water.	Check water supply.
	Rotating parts not free to rotate.	Stop water and rotate shaft by hand. Something may be stuck inside the runner.
	In Pelton and Turgo the deflector may be in the path of the water jet.	Could be a governor fault, where the deflector is not returning to the open position. It may be necessary to move deflector manually.
	Spear valve is not opening.	Check water outflow in the tailrace. If valve is not opening then less water will flow in tail-race. Repair valve if necessary.
	In the case of a Francis turbine it may be that the guide vanes are not opening.	Check movement of levers as water is turned on. Repair if necessary.
	Guide vanes open, turbine does not run. This could be due to vegetation etc. which get into the turbine and wraps itself around the guide vanes and runner.	Open inspection covers and clean if necessary. It may be necessary to remove the draught tube too.
Turbine starts but does not come up to speed.	Insufficient water.	Check and adjust water supply as necessary.
	Governor not functioning.	Adjust if necessary.
	RPM meter reading incorrect.	Check with another meter.
Turbine starts but operates at a higher speed.	Defective governor.	Repair.
	Belt slip on governor.	Check and adjust.
	RPM meter reading incorrect.	Check with another meter.
	For ELC type governors.	Defect in ballast circuit, eg fuse blown or elements defective.
Turbine starts but drops speed as it is loaded.	Insufficient water.	Check and increase if necessary.
	Trash rack blocked.	Check and clean if necessary.

Fault	*Cause*	*Remedy*
Turbine starts but drops speed as it is loaded *(continued)*	Foreign matter in penstock which blocks the flow of water.	Check pressure gauge. If pressure reading drops with increase in flow of water and needle vibrates, this could be due to something stuck in penstock.
	Defective governor.	Repair.
Turbine operates for a short time and then drops in speed or stops.	Insufficient water in stream or reservoir.	Check and reduce water and/or load on turbine, to match water available.
	In the case of the Pelton and Turgo it could be that the tailrace is flooded.	Check and clean if necessary
Turbine speed varies up and down on no load. (known as hunting)	This is often observed on mechanical governors.	Sometimes could be reduced by fine adjustment of governor. Not a major problem up to a 5% variation.
	Could be belt slip on the governor.	Tighten belts.
Turbine speed varies up and down on load.	Could be due to a pulsating load.	Check on different loads.
	Defective governor.	Check and repair.
	Oscillation set up by a long flat belt and governor.	Tighten flat belts or adjust governor.
	Same as above with Vee belts.	Due to belt slip. Tighten belts.
Noise from inside the turbine.	Could be small stones carried by the water.	Not a major problem as the water carries them out.
	Something has become loose.	Stop turbine and check. Repair if necessary.
Noise from bearings.	Bearings worn out.	Replace.
	Something loose or touching inside.	Inspect and repair.
Bearings run too hot.	Excess grease.	Reduce and check.
	Plain bearings. Worn out.	Inspect and repair.
	Low oil.	Top up oil.
	Cooling device not functioning.	Check and repair.
	Belt(s) too tight.	Check tension.

Fault	*Cause*	*Remedy*
Bearings run too hot *(continued)*	If bearings have been remetalled or housings or shafts repaired.	A high degree of skill is needed to sort this type of problem. The correct tolerance should be maintained.
	If bearings are incorrectly loaded.	Same as above.
	Incorrect alignment.	Check and correct.
Belt slip.	Belt too loose.	Check and tighten.
	Lack of belt paste.	Check and apply belt paste (only on flat belts).
	Belts too old.	Replace if necessary.
	Number of belts not sufficient.	Check and replace belts.
	Load too large.	Check load and reduce if necessary.
	Incorrect type of belts.	Check and replace.

Alternator problems

No voltage from the alternator.	AVR fuse ruptured.	Check and replace with identical type.
	Residual magnetism is too small due to heavy shocks during transport or after a prolonged period of standing idle.	With the alternator at rest apply a 12 or 24 volt dc supply to the field winding, for a short time. This should bring back the residual magnetism. It is also possible to run the alternator with the battery connected to the field. This is known as the 'Battery Test'.
	A break or bad connection on the exciter wires, field windings, or main stator windings. Battery test will not excite.	Check cable connections exciter windings and field windings.
	Speed is too low.	Measure speed and bring to rated speed.
	Heavy load already connected to alternator before starting.	Check all switches are in the off position. If ELC is used then it could be a short on the electronic switches.
	Exciter diodes defective. Will not excite with the battery test above.	Replace diodes if necessary.

Fault	*Cause*	*Remedy*
No voltage from the alternator *(continued)*	AVR defective.	Replace if necessary. If the alternator excites on the battery test, it is a clear indication that the AVR is defective.
	Internal short in windings. Battery test will not generate a voltage.	Test windings.
	Brush gear worn, or not free to touch rings.	Replace if necessary. Clean if necessary.
	Manual regulator defective	Repair.
	Defective slip rings.	Megger test slip rings with all electronic components disconnected.
	Field wires are connected to the wrong polarity of the exciter or AVR .	Field wires have to be interchanged.
	Short circuit on field surge protector. The field surge protector is across the field winding to protect the exciter diodes against a high voltage. If this is short circuited, the field current will not build up.	Test and if necessary replace surge protectors.
Alternator voltage too low with no load.	Speed not correct.	Measure and correct.
	AVR defective.	Check using battery test.
	Voltage preset too low.	Adjust preset.
	Single Transformer type AVR.	Adjust air gap or tappings.
	Three Transformer type AVR.	If three transformers are used in the AVR it may be that one transformer is defective.
	Manual regulator.	Stuck or open at some points.
	Brush gear not clean.	Clean brush gear.
	Brush gear moved.	Rotate brush holder to best position on commutator to get highest dc voltage.
	Rotating diodes defective.	Disconnect and check.
	Connections not good.	Make better connections electrically, the bad connection will be hot.

Fault	*Cause*	*Remedy*
Alternator voltage too low with no load *(continued)*	AVR sensing wires connected to wrong terminals.	If electronic AVR it may be sensing a high voltage, eg phase to phase instead of phase to neutral. If transformer type AVR it may be sensing a low voltage, eg phase to neutral instead of phase to phase. Check circuit for low voltage winding tappings.
Alternator voltage too high with no load.	Speed too high.	Correct speed.
	Voltage preset too high.	Adjust preset.
	External preset.	Defective or not in circuit.
	AVR defective.	Replace AVR.
	AVR sensing wires open.	Check wiring.
	AVR sensing wires connected to wrong terminals	If electronic AVR it may be sensing a low voltage, eg phase to neutral instead of phase to phase. If transformer type AVR it may be sensing a high voltage, eg phase to phase instead of phase to neutral. Check circuit for low voltage winding tappings.
Alternator voltage drops with load.	Belt slip.	Tighten belts.
	Alternator is over loaded.	Check current and reduce if necessary.
	Bad power factor.	Try some other load.
	If current sensing is available on the AVR it may be connected the wrong way.	In transformer type AVRs interchange the voltage sensing wires.
	AVR setting is not correct.	AVR setting may be under-compensated.
	Quadrature droop kit.	If wrongly connected , disconnect if not necessary.
	In the case of an alternator with a separate exciter the load dependent excitation is incorrectly connected, so the excitation decreases instead of increasing with load.	Compare the separate exciter unit with the wiring diagram With a meter unit, it should be observed that this voltage should increase with load.

Fault	*Cause*	*Remedy*
Alternator voltage drops with load *(continued)*	Brief voltage collapse when large loads are switched ON.	A brief voltage collapse of 15% is allowed. If it is more than this loads should be started in a different manner, eg Star/delta starting.
	A diode may be defective.	Check and replace.
	Internal winding may be short circuited.	This may not show on simple battery test. Isolate all electronics and megger test windings.
	Brush gear worn-out.	This can be seen on the brushes as sparking may occur, and if the brushes are pressed down the sparking should reduce and voltage should improve.
	Unbalanced loads	Check loads.
	Severe waveform distortion.	Check type of load and change if possible.
Alternator voltage hunts continuously.	Speed not stable.	Check governor and speed.
	Belt slip.	Check belt and tighten if necessary.
	Flat belts.	The flap of long flat belts could cause this. Adjust stability of governor or AVR.
	AVR	Electronic AVRs have an adjustment for stability. Electro-mechanical AVRs the springs need to be adjusted.
	Bad connection	A bad connection to the AVR can cause hunting due to sparking in the connection.
	Slip rings worn-out due to irregular surface.	Machine or replace
	Large loads. Due to large pulsating loads.	Test on different load.
	External interference.	Screen all alternator control wires.
	Defective bearings. (Plain bearings only)	Due to uneven air gap, check and change bearings.
Neutral conductor gets too hot.	Various star points connected with one another and large circulating currents of triple mains frequency is flowing in the neutral wire.	Connect star point choke to suppress circulating currents in neutral wire. Alternatively leave star points without connecting together.

Fault	*Cause*	*Remedy*
Neutral conductor gets too hot *(continued)*	There is a highly unbalanced load that leads to overload of the conductor.	Measure neutral conductor currents and distribute load symmetrically on all three phases.
	Type of governing	If an ELC is used which has a waveform control type switching, then a neutral current will be present between alternator and ballast. Use a large cable.
	Type of load eg a 3 phase star connected transformer.	It may be due to triple mains frequency currents in the primary of the transformer. Disconnect neutral and check.
Alternator becomes too hot (NB: modern alternators can have specified case temperatures of around 100°C).	Inlet or outlet of the alternator cooling system is obstructed or partially throttled.	The inlet and outlet of the cooling air should not be obstructed. Clean mesh if necessary.
	The warm outlet air is free to enter the inlet openings for cold air, resulting in a re-circulation of warm air.	This may occur below the alternator eg in the bed of a base plate. A re-circulation of cooling air must be prevented by means of partition sheets (baffles).
	The ambient temp. is too high, caused by cramped space, lack of ventilation or other factors.	The installation site must be ventilated, and the ambient temp. should not be above alternator rated temperature.
	Alternator winding is badly contaminated causing cooling to be ineffective.	Clean alternator with dry compressed air when at rest and clean windings.
	Alternator overloaded or power factor is too low.	Reduce output current to rated value, correct power factor with capacitors.
	Waveform distorted.	Due to a bad load, or the use of thyristors.
Alternator runs roughly, bearings produce noise or overheat.	Alternator under strain and incorrectly aligned.	Re-align alternator.
	Alternator has been transported with locked rotor causing 'Brinelling' of the bearings.	Replace bearings.
	Bearing has too little grease.	Replace grease filling immediately.

Fault	*Cause*	*Remedy*
Alternator runs roughly, bearings produce noise or overheat *(continued)*	Bearing has too much grease and is therefore overheating.	Alternators with automatic grease feeder do not suffer from excess grease, this is only possible in machines not having this device. If the bearing temp. does not drop after a period of running some grease should be removed from the bearings. Note that the permissible bearing temp.increase is rather high. It amounts to 50° C so at an ambient temp. of 25° C a bearing temp. of 25 + 50 = 75° C is permissible.
	Bearing is defective due to normal wear and tear.	Install new bearing of same grade.
	Main rotor has a short circuit between windings.	Test; alternator runs faultlessly without excitation on no load, vibrations appear and intensify with load. Measure the resistance of main rotor windings.
	If the shaft or housing has been filled or sleeved.	The shaft diameter is too large and bearing fit too tight on shaft. If housing diameter too small and bearing fit too tight to housing.
	Bearing locks or sleeves touching housing.	Check housing.
	Bearing loose in housing.	Check housing and bearing for marks. If the bearing is free in the end plate housing it will rotate in the housing causing heat.
	Plain bearings – if bearings have been remetalled	Scrape bearing to suit shaft.
	Excess belt slip. Due to belt slip the pulley gets hot and this heat flows along the shaft to the bearing.	Adjust belt.
Alternator vibrates foundation	Loose bolts	Tighten all fixing and bolts.
	If alternator has been rewound the rotor is out of balance.	Remove belts and rotate alternator shaft, if the shaft returns to the same position it is out of balance.
	The pulley may be out of balance.	Remove pulley from alt. and carry out above test. Balance alternator with pulley.

Problems that may arise with an Electronic Load Controller type of governor

Fault	*Cause*	*Remedy*
Speed too high	Ballast fuse blown.	Check and replace with same type.
	Ballast elements burnt.	Check and replace.
	Defective ELC.	Needs repair.
	No voltage from alternator.	See above on alternators.
	Defective meter.	Check with another meter.
Speed too low	Defective ELC.	Needs repair.
	Insufficient water.	Check and increase.
	Alternator over loaded.	Reduce load and check.
	Defective meter.	Check with another meter.
	Belt slip.	Tighten belts.
Speed hunts.	Defective ELC.	Needs repair.
	Belt slip or belt flap of flat belts.	Check and tighten. Apply belt paste if necessary to flat belts.
	Interacting with automatic voltage regulator. (New installations only)	Needs skilled personnel.
	Type of load.	Due to large pulsating load.
Ballast voltmeters unequal when load is not connected.	Defective ELC.	Needs repairs. Do not run plant if there is more than 20% variation.
	Ballast fuse blown.	Check and replace.
	Ballast elements burnt.	Same as above.
	Defective meter.	Check and replace.
Ballast voltage rises with alternator volts during start-up.	Defective ELC.	Needs repairs.
Unable to connect load.	Defective trips.	Needs repairs.
	Incorrect voltage, frequency, etc.	Check on meters.
	Defective switch.	Needs repairs.
ELC gets hot.	Ventilation blocked.	Check and clean.
	Fans not working.	Check and replace.

Switchboard problems

Fault	*Cause*	*Remedy*
Unable to connect load	Defective trips.	Needs repairs.
	Trips need to be reset.	Check and reset all trips.
	Incorrect voltage frequency, etc.	Check on meters.
	Defective circuit breaker.	Needs repairs.
	Load too large.	Check with small loads.
	Could be an earth fault.	Check earth fault relay or trip.
	Defect in transmission line.	Power supply not coming into switchboard from powerhouse.
	Broken insulators.	Check and repair.
	Loose connections.	Check and repair.
Switch gets hot.	Defective switch.	Replace.
	Switch overloaded.	Check currents or loads.
	Loose connections in switch.	Clean and tighten all connections. However if it still gets hot replace switch.

Commissioning and testing

11

On average a new micro-hydro scheme will take one to two years from the time of the original design to completion, and a rehabilitation scheme may take as much as one year before power can be produced. The last stage of the micro-hydro installation will attract most attention and also will be the stage when the purchaser and contractor alike will apply pressure to carry out performance tests.

Performance tests must be carried out carefully and recorded carefully. The record made of initial performance will then act as an essential bench mark against which future faults and maintenance requirements can be referred. These tests are also an opportunity to monitor figures arrived at during the design of the scheme, and so to improve design accuracy in later schemes.

It is very important to carry out the testing process with trained personnel. Ample time should be spent on doing so. Manufacturer's literature should be read and if there are any doubts the content should be clarified. Time and a budget should be allocated to this process at the planning stage.

Head works

The channel should be cleaned and flushed out so that all loose soil, stones, etc, will not enter the turbine. It may be better to let the water run along the channel with the flushing valve open for some time, for example overnight. This will flush out all the stones, sticks, etc, and also will wash away any loose areas of soil from sides of the channel.

It is also possible to take note of any leaks along the channel and observe the manner in which silt may settle along the way. Observe if the overflow is of the correct size, and if the channel as a whole is functioning as designed.

If the penstock is left empty for some time, there will invariably be some stones and other objects inside it. If possible it must be inspected or if not some water should be allowed to flow in the penstock so that these objects will move to the bottom where they can be removed.

With impulse type turbines any objects in the penstock can be removed fairly easily but with a Francis or crossflow turbine removing objects from the penstock is very difficult. These objects could do more damage to the latter type of turbines. The removal of inspection plates on these turbines will help to remove whatever is inside.

The penstock should be pressure tested by closing the main valve and letting the pipe fill with water. There will usually be a 200% overpressure test specified for larger plants with a procedure given. It is important to allow the penstock to fill at a slow rate as this will allow the air to escape. There must be no air in the penstock during tests.

If secondhand equipment has been used in the penstock, care must be taken during the pressure test, as these items could fail if not rated for the static pressure of this particular plant. Under all circumstances stay clear of this equipment until fully tested. It is also safe to keep all doors and windows open when these tests are carried out as experience has shown that if a failure occurs the powerhouse will be flooded within a very short time. It may also be wise to design the powerhouse to have all doors opening outwards so that they will tend to let the water out if flooding occurs.

Electro-mechanical equipment

Before the machines are run it is best to check all the equipment.

- Check that all the nuts and bolts are tight,
- That the rotating parts are free to rotate,
- That the bearings are lubricated,
- Couplings or drives are correctly aligned and tightened.
- All wiring is according to circuit diagrams provided by the alternator, switchboard, ELC, manufacturers.
- Ventilation is not obstructed.
- The governor is filled with oil if necessary.
- If there is an ELC, that the ballast load is full of water.
- That all switches are in the off position.
- That the tailrace is clear.

Alternator insulation levels should be checked and if low it should be dried out in accordance to the manufacturer's requirements.

It is always better to run up one item at a time and observe how it works. When doing so it is best to keep the main gate valve partially closed so that only the required amount of water flows through it. This is because it can take a long time to shut this valve if anything goes wrong.

If possible disconnect the turbine from the alternator by removing the belts and run the turbine at a very low speed, eg. as much as 5% of the rated speed. If it is a problem to disconnect the alternator then run them together with the exciter or AVR fuse disconnected, since this will prevent the voltage from building up. The governor should be on manual control.

Observe any noise, misalignment on the drive, or any uneasy or knocking movement. If detected these should be rectified. Check bearing temperature with time. There may be oil pumps circulating oil into the bearings, these should also be checked.

After about half an hour of running, if there is no sign of anything wrong, raise the speed gradually in 10% intervals up to rated speed and repeat all the observations described above and in the manufacturer's literature.

If after running at rated speed for some time, there is no indication of anything going wrong, (for example the bearings are not overheating, rotating parts not touching anything) then the governor (if it is a mechanical type) can be tested by putting it back to automatic. It may need some adjustment to bring the turbine to rated speed, but having done this, the speed should remain stable at a preset value.

Testing mechanical governors on an unloaded turbine will indicate to some extent whether the governor is working in the proper manner. This can be done by running the turbine to 90% of rated speed, and to 110% of rated speed, with the governor on manual. If, when the governor is switched back to auto, the speed drops or rises to the rated value then the governor is functioning in a proper manner. If not, check the setting up procedure.

On flow deflection type governors (as opposed to flow control) it is more difficult to do such a test as there may be no auto-to-manual switch.

If the mechanical governor is functioning satisfactorily it is time to test the alternator.

Alternator

Once the insulation level of the alternator is above the recommended level the alternator can be tested.

Stop the turbine and reconnect the exciter wires or AVR fuse. All connections and the arrangement of the electrical equipment must be carefully checked against the wiring diagrams. The connection of monitoring and control equipment must be checked. Perfect wiring is essential for a successful commissioning. If the alternator manufacturer has indicated a direction of rotation, the machine should run in this direction because the alternator fan will supply sufficient cooling air only in this direction. It is important to see that warm exhaust air from any machine should not be drawn into the intake opening of the alternator.

With limited water flow, run up the turbine to about 5% of the rated speed and observe that the complete equipment runs smoothly in this condition. At this stage it may be observed that a small voltage appears on the voltmeter. Gradually increase the speed to about 95% of rated speed on ELC type governing or 100% on mechanical type governing. This is because, in ELC type governing, as soon as the machine reaches rated speed the ballast load is switched on.

At the above alternator speed the voltage should remain steady and it should be equal on all three phases. Again observe that the machine runs smoothly. On an ELC system raise the speed so that it is just on rated speed. Check that this gives a small ballast power.

On new alternators the AVR is factory set and it is not necessary to change these settings. However, if there is a need to change the voltage it is best to follow alternator manufacturer's instructions. If any changes are carried out they should be within the nameplate values or the manufacturer should be informed of the final no load voltage at the terminals, the no load exciter voltage and current and also the full load values. This is because if these values are beyond the design value damage to the alternator or AVR could result.

Electronic AVRs will hold the voltage steady under the load condition to within ± 2% or as stated in the specifications. Transformer type AVRs will have a variation of ± 5%. It will also be noticed that a transformer AVR will drop about 10 volts on phase to neutral when the machine is hot.

On testing the machine it may be observed that there is an oscillation on the voltage at this stage. It is important to locate the cause of this oscillation and then rectify it. The oscillation in the voltage may be due to various reasons, for example it may be due to the governor and the speed variation, or to the AVR, or to the belts, so it is important to check them before changing the preset on the alternator AVR.

The alternator should be observed for some time before loading into the ballast or main load. It is also necessary to start taking various readings of voltage, current, frequency, exciter current, temperature, etc, along with time of recording, ambient temperature and even flow in the channel.

If the governor is an ELC it should be set up in accordance with the manufacturer's instructions, and various readings taken with time. If the alternator is running into ballast it is necessary to keep the current in all three phases balanced and also take a reading of the neutral current. The machine should respond to the adjustment of the frequency control of the ELC.

The water flow through the turbine can now be increased to load the machine. In the case of an ELC type governor, the machine should be load tested on the ballast load first. In the case of a mechanical type governor the machine must be tested on the main load or on a dummy load. The loading procedure must be recorded and all readings entered against time.

Loading machine on main load

If a transmission line is included on the scheme this should be 'megger tested' before energizing. All relevant wiring in the switchboard should be checked against the circuit diagrams.

On connecting up the hydro power to the switchboard the meters on this board should indicate the presence of the power. The loads can be switched on individually and the currents recorded. If there are any electric motors to be run on the hydro the direction of rotation of the motors should be checked. When heavy loads such as motors are connected on to the hydro it is necessary to wait a few seconds between starting each motor for the turbine to stabilize after the first start.

Taking readings

While readings are being taken it is important to keep everything in a steady state, as it may take as much as 15 minutes to take all the readings. It is also important to use calibrated meters.

Flow measurements can be taken in the channel or in the tailrace and used in the calculation of the gross hydro power available at the site or forebay tank. If flow measurements are made in the channel the flow through the channel should be adjusted so that all the water flows through the turbine without any overflow. An alternative is to stop flow temporarily into the forebay and then time a decrease in water volume in the forebay.

For various flow readings the pressure drop in the penstock can be recorded at the pressure gauge on the penstock.

Turbine speed (rpm) should be recorded for various loads and flow conditions.

Bearing temperatures should be recorded over a period of 8 hours against ambient temperature.

If the alternator is belt driven, the speed of the alternator should be recorded with various loads.

Alternator temperature at various points on the case and the inlet and outlet air temperature should be recorded with time and against the ambient temperature.

If the governor is a hydraulic type the temperature of the oil should be recorded over a period of 8 hours.

Alternator voltages and currents on the three phases and the exciter voltage and current should be recorded.

The power factor of the load and kWh meter readings should be recorded.

If governing is by an ELC special voltage and ammeters are preferred to read the voltage and current flowing out of the alternator into the ballast. These should read the 'True RMS' values of the current. The neutral current should also be recorded.

If a transmission line is used the voltage at both ends must be recorded along with the load currents and the power factor.

Some turbine manufacturers require that an overspeed test also be carried out on the turbine and alternator.

With the above readings it is possible to work out the power available at site and all the power losses along the system. Each efficiency can be worked out and also the total efficiency.

These readings will help at a later date as reference on the project, as any faults can be picked out from a new set of readings. For example if the alternator is running hot, the temperature can be measured and referred back to the original test report.

Setting up trips

Once the turbine and alternator have been tested and their performance is satisfactory, the trips can be set up so that the plant can be handed over for use by the client.

The trips should be set up in conjunction with the test report as they will require specific time delays depending on how long it takes for certain loads to start. (During starting the speed may drop below the required trip preset value.) The trips should be tested individually and also together. It may be better to allow the trips to trip the circuit often rather than under set them, as damage to the equipment will result if they are not functioning when needed.

Appendix 1 Glossary

A

Units

This glossary lists the symbols used in the manual and the units which should be used when applying the equations given. In most cases SI units have been used, a note is made below if a non-SI unit is used.

For instance, a useful equation is:

$P_o = e_o \times Q \times H \times 10$

The glossary below gives the correct units for this equation as follows:

P_o	Power	kilowatts	kW
e_o	Efficiency	no units	
Q	Flow	metres³/second	m³/s
H	Head	metres	m

Symbol	Description	Units of measurement	
a	Coefficient of thermal expansion and contraction (see Example 3.9.1)	metres per metre per degree centigrade	m/m C
a	Part of length of shaft (fig 7.7.1)	metres	m
a	Velocity of a pressure wave of water in a pipe of specified diameter, wall thickness and material	metres per second	m/s
A	Area	square metres	m²
ADF	Average daily flow. The nominal average flow of a river, based on daily flow measurements over many years (see Example 2.2.1)	cubic metres per sec	m³/s
Ah	Battery capacity	Ampere hours	Ah
α	Angle	degrees	°
B	The width of a channel at its bed (see Fig 3.6.2)	metres	m
β	Angle	degrees	°
C	Capacitance	farads	F
C	Constants		

Symbol	Description	Units of measurement	
CG	Centre of gravity		
CI	Cast iron		
c_d	Coefficient of discharge (see Note 5.3.2)		
c_v	Coefficient of velocity (see Note 5.3.2)		
c_w	Coefficient of discharge for a weir		
C	Temperature in centigrade	centigrade	C
d, D	Depth	metres	m
d, D	Diameter for drive components	metres	m
d	Diameter of a pipe	metres	m
T	Change in temperature	degrees centigrade	C
e	Eccentricity or the extent of off-balance of a mass. (see Note 3.12.4)	metres	m
e	Efficiency defined as $\frac{\text{output}}{\text{input}}$		
E	Modulus of elasticity or Young's modulus	newtons per metre^2	N/m^2
F	Force	newtons	N
F	Freeboard allowance. A safety factor such as 1.2 by which the design flow is increased when calculating the size of a channel cross-section.		
g	Acceleration due to gravity. This is a constant, 9.8 m/s^2	metres per second per second	m/s^2
G	Gearing ratio. The output speed of a pulley and belt arrangement, or a gearbox, divided by the input speed.		
H	Head or height	metres	m
h	Head or height	metres	m

Symbol	Description	Units of measurement	
HDPE	High density polyethylene		
I	Moment of inertia for bending	metres^4	m^4
k	The roughness of the inside of a pipe wall, used to calculate a pipe's resistance to flow (see Note 3.11.2). k is used as part of the ratio k/d, roughness over diameter, which must be calculated using the same units for diameter and roughness.	millimetres (SI unit: metres)	mm
L	Distance or length	metres	m
L	Inductance	henrys	H
L, l	Length of shafts (fig 7.7.1)	metres	m
L_{10h}	The expected life of a bearing in hours.	hours	h
M	Mass	kilograms	kg
M	Bending moment	newton metres	Nm
MDPE	Medium density polyethylene		
MS	Mild steel		
μ	Coefficient of friction (see Example 3.12.2)		
μ	Electrical conductivity (see Note 2.5.1)	1/ohms or siemens	ohm^{-1} or S
n	Roughness coefficient, as used in Manning's equation for flow in open channels (see Table 3.6.4)		
N, n	Rotational speed, for instance of a turbine shaft or generator.	revs per minute (SI unit: radians/sec)	rpm
N	Side slope of a channel wall (see Fig 3.6.2)		
n_j	The number of jets fitted to a Pelton turbine.		
N_s	Specific speed of a turbine (see Note 5.1.1)	revs per minute (SI unit: radians/sec)	rpm

Symbol	Description	Units of measurement	
P	Apparent power, product of voltage and current where these are out of phase	volt-amp	VA
P	Mechanical, thermal, or electrical power. Power is the rate of energy transfer. Watts are joules per second, kilowatts are 1000 joules per second	kilowatts (SI unit: watt)	kW
P	Pressure. The SI unit of pressure is the pascal (Pa). This is identical to newtons per square metre.	newtons per metre^2	N/m^2
P	Wetted perimeter of a channel cross-section	metres	m
Pf	Power factor, cosine of the angle between voltage and current		
Q	Flow (volume per unit time)	cubic metres/seconds	m^3/s
Q (number)	A hydro scheme is designed for a flow (Q) together with a Q number. The number is the proportion of the year during which the river has a larger flow than the design flow. For instance, a 20m^3/s Q25 scheme only works at 20m^3/s for 25%, or 3 months of the year. The rest of the year it works on part flow or is shut down.		
θ	Angle	degrees	°
R	Hydraulic mean radius of a channel (see Note 3.6.1)	metres	m
R	Reaction force at bearings	newtons	N
rpm	Revolutions per minute	1/minutes	min^{-1}
ρ	Density, or mass per unit volume	kilograms per cubic metre	kg/m^3
s	Seepage loss resistance of a particular type of soil	volume flow per square metre of soil surface	$\frac{m^3/s}{m^2}$
S	Slope, the vertical distance divided by the horizontal		

Symbol	Description	Units of measurement	
S	Ultimate tensile strength	newtons per metre^2	N/m^2
SF	Safety factor		
Σ (quantity)	Sigma, meaning the sum total of all the values of the quantity. For instance, ΣH is the sum of all the horizontal forces on a support block.		
t_p	Permissable shear stress (Ex 7.6.1: N/mm^2)	newton per metre^2	N/m^2
t	Thickness	metres	m
t_{jet}	Thickness of the nozzle aperture of a crossflow turbine	metres	m
t	Time	seconds	s
t_{wall}	Wall thickness of a pipe	metres	m
T	The width of a channel at the top of the cross-section (see Fig 3.6.2)	metres	m
T	Torque	newton metres	Nm
uPVC	Unplasticized polyvinylchloride		
V	Volume	cubic metres	m^3
v	Velocity	metres per second	m/s
w, W	Width	metres	m
W	Weight, or force downwards imposed by a mass. Weight is mass multiplied by 'g'.	newtons	N
x	Small change in length of a pipe due to temperature change	metres	m
Z	Impedance (combined effect of resistance, inductance and capacitance)	ohms	Ω

Conversion table

Unit	Abbreviation	Conversion factor	
		To metric unit	To SI unit
Length			
millimetre	mm		1 mm = 1×10^{-3} m
inch	in	25.4 mm	1 in = 25.4×10^{-3} m
foot (= 12 in)	ft		1 ft = 0.3048 m
mile (= 5280 ft)	—	1.609 km	1 km = 1000 m
(= 1760 yd)			1 yd = 0.9144 m
Area			
square millimetre	mm^2		1 mm^2 = 1×10^{-6} m^2
square inch	in^2	6.45 cm^2	1 in^2 = 6.45×10^{-4} m^2
square foot	ft^2		1 ft^2 = 0.0929 m^2
square yard	yd^2		1 yd^2 =0.836 m^2
acre (= 43560 ft^2)	ac	0.405 ha	1 hectare = 10000 m^2
square mile (= 640 ac)	—	2.59 km^2	1 sq. mile = 2.59×10^6 m^2
Volume			
litre	l		1 l = 1×10^{-3} m^3
cubic inch	in^3	16.39 cm^3	1 in^3 =16.39×10^{-6} m^3
imperial gallon (= 1.20 US gallon)	imp gal	4.55 l	1 imp gall = 4.55×10^{-3} m^3
US gallon (= 0.833 imperial gallon)	US gal	3.79 l	1 US gall = 3.79×10^{-3} m^3
cubic foot	ft^3	28.32 l	1 ft^3 = 0.0283 m^3
cubic yard	yd^3		1 yd^3 = 0.765 m^3
acre-foot (= 43560 ft^3)			1 acre-foot = 1233.5 m^3
Mass			
ounce	oz	28.35 g	1 g = 1×10^{-3} kg
pound (= 16 oz)	lb	0.454 kg	1 lb = 0.454 kg
ton (= 2240 lb)	ton	1.016 tonne	1 ton = 1016 kg
tonne	tonne		1 tonne =1000 kg
Flow			
cubic foot per second (= 2 acre-feet per day approx.)	ft^3/s	28.32 l/s	1 l/s = 1×10^{-3} m^3/s 1UK gal/hr=1.26×10^{-6} m^3 /s 1 ft^3/s = 28.32×10^{-3} m^3/s
Density			
pound per cubic inch	lb/in^3	27.68 g/cm^3	1 lb/in^3 = 27680 kg/m^3
pound per cubic foot	lb/ft^3		1 lb/ft^3 = 16.018 kg/m^3
Pressure			
pounds-force per sq. in (= 2.31 ft water)	lb/in^2 (psi)	1 bar = 10^5 N/m^2	1 lb/in^2 = 6895 N/m^2 1 kgf/cm^2 = 98067 N/m^2
Power			
horsepower (= 550 ft lb/sec)	hp	0.746 kW	1 hp = 746 W 1 ft lb/sec = 1.3558 W

Appendix 2
Pocket computers

A

The following three sample programs are written for the Casio FX-730P Pocket Computer. They are given here as examples of what you can do yourself. It is quite simple to improve these programs and to write new ones, following guidance given in the 'user manual' of the pocket computer.

Pocket computers can be easily carried anywhere and can serve as both a design tool for micro-hydro schemes and as a calculator. The programs are written in a version of Basic but may be simply modified for use with other computers or programmable calculators.

The first program, *Chanflow*, automates the tedious part of the calculation described in Note 3.6.1 in Chapter 3. It is not identical, and you may wish to improve it; for instance, by inserting a freeboard allowance under the 'design' option.

The second program, *Pipelos3*, automates the calculation of Note 3.11.2. It would be possible to amend it to include the wall thickness calculation of Note 3.11.3.

The third program, *OHL4*, speeds up cable sizing calculations as described in Note 8.8.1.

Remember that a device like the FX-730P also has a 'memo' and 'calc' facility which allows many of the more straightforward procedures in this book to be stored quickly, without taking the trouble of writing full programs like these. A full program is useful if 'iterations' are involved (an iteration is a repeated calculation involving going back with a 'better guess' a number of times). But even iterative calculations can be done manually using the 'calc' and 'memo' facilities. The user manual for the Casio FX-730P explains how to use these facilities.

A2.1
Chanflow

```
1Ø    PRINT "CHANFLOW"
2Ø    INPUT "Flow, l/s", F, "Length, m", L
3Ø    INPUT "Concrt → 1, Soil → 2, Design → 3", Z
4Ø    GOSUB 2ØØ + Z
5Ø    X = 2 * SQR (1 + N↑2) - 2 * N
6Ø    A = Q/V
7Ø    H = SQR (A/(X + N))
8Ø    W = H * X
9Ø    P = W + 2 * H * SQR (1 + N↑2)
1ØØ   R = A/P
11Ø   S = (K * Q/A/R↑.67) ↑2
12Ø   SET F2
13Ø   PRINT "Slope = 1 in" ; 1/S
14Ø   PRINT "Height = " ; H, "Bed Width = " ; W,
      "Top Width = " ; W + 2 * N * H
15Ø   IF W< .2; PRINT "TOO NARROW!"
17Ø   PRINT "Excavation =" ; A * L * 1.1; "m↑3":
      GOTO 2Ø
2Ø1   V = 1.5 : K = .Ø15 : N = .577 :
      Q = 1.2 * F/1ØØØ : RETURN
2Ø2   V = .4 : K = .Ø3 : N = 1.5
      Q = 1.2 * F/1ØØØ : RETURN
2Ø3   INPUT "Speed, m/s", V, "n", K,
      "N (opt.577)", N : Q = F/1ØØØ : RETURN
```

This program sizes water channels (as in Note 3.6.1) by using Manning"s equation. To use it, input the expected flow in litres per second and length of channel. When asked at line 30 to choose between channel lining materials, specific values for variables are specified.

For Concrete (press 1) a speed of 1.5m/s, Manning coefficient of Ø.Ø15, side slope of Ø.577 and a freeboard allowance of 2Ø% of the flow are specified.

For Soil (press 2) Ø.5m/s, Ø.Ø3, 1.5 & 12Ø% are specified for the above quantities respectively.

If Design (press 3) is selected, you can insert your own values with no freeboard allowance. To obtain a freeboard allowance, increase the original flow specification.

Dimensions and slopes are calculated in lines 5Ø to 11Ø and rounding off of values is achieved in line 12Ø.

If the channel is less than Ø.2m it will block easily. Line 15Ø prints a warning of this. Line 17Ø calculates the excavation required for the channel. At first it is best to do some manual calculations as in Example 3.6.1 and compare them with the program, to make sure everything is working correctly.

Example A2.2 **Pipelos3**

Calculate the pressure drop and efficiency in a PVC penstock, length 65Øm, internal diameter 237m, flow 92 l/s. Use a k factor (roughness factor) of Ø.Ø6mm for PVC. Gross head is 15Øm.

Screen display	Input
PIPELOS3	
Dia, mm?	237
Flow, l/s?	92
Gross Hd, m?	15Ø
Lth, m?	65Ø
Coef list?	Ø
Coef, k, mm?	Ø.Ø6
Δh = 9.7Ø5 m	
Nt Hd = 14Ø.295 m	
eff = 93.53Ø%	

So, the head loss is 9.7m and the efficiency is 93.5%.

Example A2.1 **Chanflow**

We have to join an intake to a forebay tank using an earth channel. The distance is 6ØØm, the design flow is 4ØØ l/s. What are the dimensions of the channel, and its slope?

Screen display	Input
CHANFLOW	
Flow, l/s?	4ØØ
Lgth, m?	6ØØ
Concrt ->1,Soil ->2, Dsgn ->3?	2
Slope = 1 in 1882	
Ht = Ø.75	
Bed width = Ø.46	
Top width = 2.72	
Excavation = 792.ØØ m↑3	

So, the channel is Ø.75m high, has a bed width of Ø.46m and a top width of 2.72m. The slope is 1 in 1882.

A2.2

Pipelos3 – friction losses in penstock

1Ø	PRINT "PIPELOS3"	
2Ø	INPUT "Int. Dia, mm",D, "Flow, l/s", Q, "Gross Hd, m", G, "Lth, m", L	Inputs the main parameters: flow; gross head and length
25	INPUT "Coef list", A : IF A = 1; GOSUB 3ØØ	Allows access to a very brief list of roughness coefficients – usually called "k", here called "K". Input "Ø" for "No" and go to the next line. Input "1" for "Yes" and bring up the list on line 3ØØ
27	INPUT "Coef, k, mm", K	Inputs coefficient "k"
3Ø	H = G*.1	"Guesses" the head loss as 1Ø% of gross head.
4Ø	Q = Q/1ØØØ : D = D/1ØØØ : K = K/1ØØØ	Changes the units
5Ø	V = Q * 4/π/D↑2	Calculates velocity from flow/area
55	S = H/L	Calculates hydraulic slope
6Ø	X = SQR(2 * 9.81 * D * S)	Works out the term $\sqrt{2g\,DS}$
7Ø	U = -2 * X * LOG((K/3.7/D) + 2.51 * 1.Ø1E-6/D/X)	Calculates velocity from Darcy/Colebrook-White, but with "S" still based on a guessed head loss
8Ø	SET F3	Sets outputs
9Ø	F = π *D ↑2/4 * U	Calculates flow (F) based on the velocity (V) from line 7Ø
1Ø5	IF F> 1.3 * Q ; H = H * .5	Iterate through lines 55 to 12Ø to get F = Q to within about Ø.1%
11Ø	IF F> Q + .ØØ1 ; H = H* .93 : GOTO 55	
115	IF F< Q * .7 ; H = H * 1.5	
12Ø	IF F<Q - .ØØ1 ; H = H * 1.Ø7 : GOTO 55	
125	PRINT "Δ h =" ; H ; "m"	Prints head loss
13Ø	B = G - H : PRINT "Nt Hd ="; B ; "m"	Prints net head
14Ø	PRINT "eff =" ; B/G * 1ØØ ; "%"	Prints efficiency
16Ø	SET N ; GOTO 2Ø	Returns to 2Ø to repeat with new parameters.
3ØØ	PRINT "MS = .Ø15to . 25 : CI = .6 to 1Ø : PVC = .ØØ3" : RETURN	MS refers to mild steel pipe CI refers to cast iron pipe "PVC" refers to *new* PVC

The method used is the simultaneous solution of two equations for water velocity. This is done iteratively, by varying the head loss until the two velocities are equal.

The Colebrook-White equation for velocity (U) is:

$$U = -2\sqrt{(2gDS)} \times \log\left[k/(3.7D) + 2.51n/[D\sqrt{(2gDS)}]\right]$$

Also velocity (labelled V here) can be expressed:

$$v = \frac{Q \times 4}{\pi \times D^2}$$

where

D = internal diameter (m)

S = hydraulic gradient or head loss

k = roughness coefficient (m*10^{-6})

n = kinematic viscosity of water, taken as 1.Ø x 10^{-6} m^2/s at 2Ø°C

Q = flow (m^3/s)

A2.3

OHL4

Line	Statement	Comment
5	PRINT "OHL4"	
1Ø	INPUT "Lth,m", L, "kW", P, "pf", F, "3ph", B, "No of wires",W	inputs actual line length, the power, power factor and number of wires. It asks "3ph?". Press 1 for 3 phase, Ø for single phase.
12	IF B = Ø ; L = 2 * L	doubles the effective line length for single phase.
15	INPUT "Cu = 1, Al = Ø", Z	press 1 for Copper, Ø for Aluminium.
2Ø	INPUT "MaxV, p/n", X, "MinV, p/n", N, "Spacing, m", D	inputs maximum and minimum allowable voltages, measure phase to neutral.
21	SET F1	
22	IF B=1; D = (D * D * (2 * D)) ↑.333	works out spacing for 3 phase, assuming conductors are in a straight line – the worst case.
3Ø	GOSUB 3ØØ + B	calculates currents from lines 3ØØ, 3Ø1.
4Ø	PRINT I ; "Amps"	
45	GOSUB 2ØØ + Z	guess a value for csa of wire
5Ø	A = K * L * I /(X-N)	
55	R = SQR(A/7/ π) * 3E-3	works out conductor radius, assuming 7 strands
6Ø	H = L * (5 + (46 * LOG(D/R))) * 1E-8	works out inductance
65	E = 2 * π * 5Ø * H	calculates inductance reactance
7Ø	IF Z = 1 THEN J = 16 * L / 1ØØØ / A : GOTO 75	calculate resistance
71	J = 28.64 * L/1E3/ A	
75	M = SQR (J↑2 + E↑2)	calculates total impedance
8Ø	V = M * I	calculates volt drop
85	IF X-N > V+.1; A = A * .95 : GOTO 55	iterate to find cross-sectional area
9Ø	IF X-N ≤ V ; A = 1.Ø6 * A : GOTO 55	
95	PRINT A ; "mm ↑2 "	
96	Q = A * L * W * .32	calculates cost (based on 1985 Sri Lanka Aluminium conductors)
97	IF B = Ø THEN Q = Q/2	
98	PRINT " Cbl = Rs " ; Q	
99	INPUT " Tech Detail" , C	press 1 for technical detail, which is self explanatory.
1Ø2	IF C = Ø ; GOTO 2Ø	
1Ø5	SET E2	
11Ø	PRINT " Induc = " ; H ; "H"	
115	PRINT "Ind. Reac = "; E ; "Ω"	
12Ø	PRINT " Resist = " ; J; "Ω"	
13Ø	PRINT "Imped = " ; M ; "Ω"	
14Ø	SET F1 : GOTO 2Ø	
2ØØ	K = .Ø7 : RETURN	
2Ø1	K = .Ø4 : RETURN	
3ØØ	I = P * 1E3/X/F : RETURN	
3Ø1	I = P * 333/X/F : RETURN	

The method is described in Section 8.8.

Example A2.3 **OHL4**

We have a 6Ø kw, 3 phase, 5Ø Hz set, 35Øm from the load. The power factor is Ø.5. Maximum allowable voltage is 42ØV, minimum 38ØV, measured line to line.

What is the best size of aluminium conductor to use?

Note that, to convert 'line to line' voltages to 'line to neutral' voltages as required by the program, divide by $\sqrt{3}$.

Screen display	Input
OHL4	
Lgth, m?	35Ø
kW ?	6Ø
pf ?	Ø.5
3ph ?	1
No of wires	4
Cu = 1, Al = Ø	Ø
Max V, p/n ?	42Ø/3 ↑.5
Min V, p/n ?	38Ø/3 ↑.5
Spacing, m ?	Ø.3
164.8 Amps	
[This step takes about 1Ø seconds]	
97.8 mm^2	
Cbl = Rs 43,824	
Tech detail ?	1
Induc = 3.ØE-Ø4H	
Ind.Reac = 9.5E-Ø2 Ω	
Resist = 1.ØE-Ø1 Ω	
Imped = 1.4E-Ø1 Ω	

So, choose 7/4.29 aluminium which has a csa of 1Ø6 mm^2.

Appendix 3 Bibliography

A

General references and Chapter 1

Casley and Lury (1981)
Data collection in developing countries

A standard work for field surveys.

Fraenkel et al, Micro-Hydro Power (1991)
A guide for development workers
IT Publications, 103/105 Southampton Row, London WC1B 4HH, ISBN 1 8 5339 029 1. 150pp

Suitable for development workers requiring a detailed briefing.

Fritz, J F (1984)
Small and mini hydropower systems
McGraw-Hill, London, ISBN 0 07 0224 70 6. 200pp

A collection of experts have authored chapters covering many aspects of small and mini hydro, much of which is applicable to micro-hydro work.

Holland R E (1983)
Micro-hydro electric power
IT Publications, 103-105 Southampton Row, London WC1B 4HH. 46pp

A good brief introduction to micro-hydro power.

Inversin, Allen (June 1986)
Micro-hydropower sourcebook
NRECA Foundation, Washington DC

A complete reference covering similar ground to this text, but with greater emphasis on civil works.

Meier U (1981)
Local experiences with micro-hydro technology
SKAT, St Gallen, Switzerland

Describes local micro-hydro technology in Nepal.

Meier U, Arter A, MHP (1990)
Information package: A selected, annotated and classified bibliography on micro-hydropower development
SKAT, St Gallen, Switzerland

An information package for both 'beginners' and skilled practitioners. Useful for those seeking a more comprehensive bibliography of the subject.

Quayle, J P
Kempe's engineers year book
Morgan-Grampian Book Publishing Co. Ltd, Calderwood Street, London SE18 6QH

This handbook is revised annually and has sections on penstocks, channels, transmission lines, drives etc. A very useful general technical reference.

Chapter 2

Crawford Dr H, and Thurin S M (Sep 1981)
Hydrologic estimates for small hydroelectric projects
National Rural Electric Co-operative Association (NRECA), Washington. 49pp

Very useful methods for peak flows, flow duration curves, and a simple computer program for calculating monthly run-off.

Inversin, Allen (June 1986)
Micro-hydropower sourcebook
NRECA Foundation, Washington DC

Covers methods for estimating minimum flow and comparing catchments.

Medina R, Juvenal (1991)
Fenomenos geodinamicos
ITDG Peru

A very useful exposition of geotechnical aspects of micro-hydro design.

Pearce E A, Smith C G (1990)
The world weather guide
Hutchinson and Co, 20 Vauxhall Bridge Road, London SW1V 2SA. 490pp

Covers rainfall, humidity temperature and a summary for every country. Very useful for preliminary work.

Ramsahoye S I (Dec 1982)
Flow duration curves in tropical equatorial regions
Water Power and Dam Construction. pp 66 to 871
Interesting rapid method of obtaining flow duration curves with tests of the method and conclusions.

Whyte W S (1976)
Basic metric surveying
Newnes-Butterworths, 88 Kingsway, London WC2B 6AB, ISBN 0 408 00197 6. 312pp
Covers all the surveying techniques needed for micro-hydro civil works.

Wilson, E M (1984)
Engineering hydrology
Macmillan, Third Edition, ISBN 0333 353188. 301pp
Probably the most useful text book for micro-hydro hydrology. Covers Meteorological Data, floods, and various forecasting methods.

Chapter 3

Arter A, Meier U (1990)
Hydraulics engineering manual
SKAT, St Gallen, Switzerland, ISBN 3 908001 13 7. 73pp
Useful nomograms and diagrams on weirs, penstocks etc.

Brett P (1988)
Formwork and concrete practice
Heinemann Professional Publishing, 22 Bedford Square, London WC1B 1HH, ISBN 0 434 90177 6. 204pp
Practical guide to construction of formwork and placing concrete. Useful diagrams.

Chadwick A, Morfett J (1986)
Hydraulics in civil engineering
Harper Collins, 77-85 Fulham Palace Road, Hammersmith, London W6 8JB, UK, ISBN 0 04 627004 3. 492pp
Comprehensive detailed text covering surge, weirs, intakes, cavitation and spillways.

Design Standards No. 3 (1967)
Canals and related structures
Bureau of Reclamation, PO Box 25007, Denver, Colorado, 80225-00007/USA. 365pp
Manual with plans and drawings of hydraulic structures including penstocks.

Inversin A R (1986)
Microhydropower sourcebook
NRECA, 1800 Massachusetts Ave NW, Washington DC 20036/USA. 285pp
Highly recommended text for civils and penstocks, much of the methodologies used are adopted in the above chapter.

Lauterjung H and Schmidt G (1989)
Planning of intake structures
GATE/GTZ, ISBN 3 528 02042 3. 122pp
Very comprehensive text on small intakes, includes many diagrams and calculations. Probably the best single publication on this topic.

McKinney et al (Jan 1983)
Microhydropower handbook
Technical Information Center, US Dept of Energy, Volume 1. 200pp
Section 4.4 of this bulky publication gives clear calculation methods for all major components.

Young et al (1986)
Pipe joints, Part 3: metallic pipe joints
Institute of Mechanical Engineers, London, ISBN 0 85298 6114. 100pp
Covers calculations for flanges, welding, tests and joints with full reference to BS, ASTM, ASME, and DIN standards.

Chapter 5

Arter A, Meier U (1990)
Hydraulics engineering manual
SKAT, St Gallen, Switzerland, ISBN 3 908001 13 7. 73pp
Covers theory and design of crossflow turbines.

Eisenring M (1992)
Micro Pelton turbines
SKAT, St Gallen, Switzerland
Some designs and manufacturing hints for Pelton turbines.

Various authors
Crossflow turbine T1, Crossflow turbine T3, Crossflow turbine T7
SKAT, St Gallen, Switzerland
Three separate sets of detailed drawings from which crossflow turbines can be fabricated. Developed in Nepal, these machines have been successfully built and used in many countries.

Victorov, G V (1986)
Guidelines for the application of small hydraulic turbines
UNIDO, available from SKAT, St Gallen, Switzerland
Provided detailed guidelines for the application of small hydraulic turbines, description of turbines and ways of standardization of turbines.

Chapter 6

Fisher, Arter, Meier, Chapallaz (1990)
Governor production information
SKAT, Tigerberstrasse 2, CH-9000, St Gallen, Switzerland, ISBN 3 908001 19 6. 115pp
The best available text on governing for micro-hydro, this includes coverage of many types of governing, selection and product information.

Inversin A (1986)
Microhydropower sourcebook
NRECA, 1800 Massachusetts Ave NW, Washington DC 20036/USA. 285pp
Good coverage of governing aspects, particularly applicable to micro hydro.

Chapter 7

Quayle, J P
Kempe's engineers year book
Morgan-Grampian Book Publishing Co. Ltd, Calderwood Street, London SE18 6QH
This handbook is revised annually and always includes comprehensive chapters on bearings, drives, and mechanics.

Chapter 8

Hughes, Edward (1977)
Electrical technology
Longman. 710pp
A standard text covering most aspects of single and 3-phase generating systems, including first principles. Many useful worked examples.

Meier, U
Electrical components of micro hydropower installations
SKAT, St Gallen, Switzerland. 42pp
Examples of electrical switchboards and wiring diagrams of micro-hydro installations in Nepal.

Ryss, P S
Electrical machinery
Prentice Hall
Fairly advanced text on electrical machines – theory and construction

Say, M G (1973)
Electrical engineers reference book
13th Edition, 1000pp
Comprehensive text covering protection, transmission, instrumentation and generators. Good diagrams and charts.

Chapter 9

Fritz J J (1984)
Small and mini hydropower systems
McGraw-Hill. 300pp
Includes a comprehensive chapter on feasibility with examples and case studies.

Gulliver J S, and Arndt R E (1991)
Hydropower engineering handbook
McGraw-Hill. 300pp
Includes up-to-date chapter on cost estimation and financial and economic analysis, very well referenced.

Holland R E (1983)
Micro-hydro electric power
IT Publications, 103-105 Southampton Row, London WC1B 4HH. 46pp
Includes a section on basic economics for micro-hydro.

Lumby, Stephen (1983)
Investment appraisal
Van Nostrand Reinhold (UK) Co. Ltd, Molly Millars Lane, Wokingham, Berkshire, England. 261pp
For those who need to understand more fully the concepts explained in Chapter 9.

Appendix 4
Suppliers index

This section gives a limited list of turbine manufacturers capable of supplying equipment under 100 kW in capacity.

The listing draws on product guides produced by the periodicals *Modern power systems* (Wilmington House, Wilmington Dartford, Kent, UK), and *Water power and dam construction* (Quadrant House, The Quadrant, Sutton, Surrey, UK). Both magazines produce listings from time to time.

More comprehensive lists can be found in *Micro-hydro power* by P Fraenkel et al (particularly good for developed country manufacturers) and *The power guide* by J Clancy et al (particularly good for developing country data). Both are available from IT Publications.

Australia

Tamar Designs Pty Ltd
195 Foreshore Road
DEVIOT 7275, Tasmania
Telephone (+61) 3 947 357
Fax (+61) 3 947 565

Turbine type	Francis	Kaplan	Pelton	Turgo
Power output (kW) max	300	600	620	640
Power output (kW) min	0.1	7	0.1	0.1
Head (m) max	200	15	800	300
Head (m) min	10	5	3	20

Austria

Gugler GmbH & Co KG
A-4085 Nierderranna 41

Telephone (+43) 7285 514
Fax (+43) 7285 6242
Telex 7531 0982

Turbine type	Deriaz	Francis	Kaplan	Pelton
Power output (kW) max	3000	3000	3000	5000
Power output (kW) min	20	1	1	0.5
Head (m) max	50	200	30	1000
Head (m) min	20	2	0.5	10

Kossler GmbH
A-3151 St Georgen

Telephone (+43) 2746 82 72
Fax (+43) 2746 2626
Telex 15652

Turbine type	Francis	Kaplan	Pelton
Power output (kW) max	5000	5000	5000
Power output (kW) min	10	10	10
Head (m) max	150	40	800
Head (m) min	2	1	20

Belgium

Willot JLA
Rue Pierre Jacques 72
B-52840 MOHA
Telephone (+32) 8521 1394

Turbine type	Crossflow
Power output (kW) max	150
Power output (kW) min	2
Head (m) max	80
Head (m) min	2

Canada

Dependable Turbines Ltd
7–3005 Murray Street
PORT MOODY BC V3H 1X3
Telephone (+1) 604 461 3121
Fax (+1) 604 461 3086
Telex

Turbine type	Francis	Pelton	Propeller	Turgo
Power output (kW) max	5000	5000	2000	2000
Power output (kW) min	10	1	1	10
Head (m) max	150	500	25	250
Head (m) min	5	30	2	30

China

CMEC Sichuan Co
27 Hong Zhoa Bi Street
CHENGDU, Sichuan Province
Telephone (028) 29547, 23057
Fax (028) 660 629
Telex 60137 pcmec CN

Turbine type	Francis	Kaplan	Pelton
Power output (kW) max	30000	36000	12500
Power output (kW) min	320	735	75
Head (m) max	300	405	700
Head (m) min	10	6	60

France

CIMH
42 Ave Jacques Desplats
F-81100 CASTRES
Telephone (+33) 63 35 63 15
Fax (+33) 62 59 57 21
Telex 520 200

Turbine type	Francis	Pelton
Power output (kW) max	400	100
Power output (kW) min	1	0.5
Head (m) max	120	200
Head (m) min	2	20

Leroy Somer
blvd Marcelin Leroy
F-16015 ANGOULEME
Telephone (+33) 45 62 41 11

Telex 790 0441

Turbine type	Crossflow	Francis	Kaplan	Pelton
Power output (kW) max	250	400	400	500
Power output (kW) min	20	100	20	100
Head (m) max	40	50	8.5	200
Head (m) min	10	20	1	40

Germany

Ossberger GmbH & Co.
PO Box 425, Otto Rieder Str 7
D-8832 WEISSENBURG
Telephone (+49) 9141 4091
Fax (+49) 9141 70522
Telex 624 672

Turbine type	Crossflow
Power output (kW) max	1500
Power output (kW) min	1
Head (m) max	200
Head (m) min	1

J M Voith GmbH
PO Box 1940
D-7920 HEIDENHEIM
Telephone (+49) 732 1370
Fax (+49) 732 1373 000
Telex 714 79920

Turbine type	Francis	Kaplan	Pelton
Power output (kW) max	10000	5000	5000
Power output (kW) min	50	50	50
Head (m) max	380	74	1100
Head (m) min	2	2	50

Wasserkraft Volk GmbH
Gefall 45
D-7809 SIMONSWALD
Telephone (+49) 7683 844
Fax (+49) 7683 805
Telex 772688

Turbine type	Crossflow	Francis	Pelton
Power output (kW) max	750	5000	5000
Power output (kW) min	10	100	10
Head (m) max	200	250	1000
Head (m) min	5	3	30

Hungary

Ganz Machinery Works
PO Box 62, Kobanyal ut 21
BUDAPEST
Telephone (+36) 1 140 840
Fax (+36) 1 143 864
Telex 22 3318

Turbine type	Francis	Kaplan	Pelton	S-Type
Power output (kW) max	5000	5000	5000	5000
Power output (kW) min	50	50	50	50
Head (m) max	400	40	1000	20
Head (m) min	10	4	100	3

India

Flovel Private Ltd
A–219 Okhla Industrial Area, Phase 1
NEW DELHI
Telephone (+91) 683 6036
Telex 031 75100 flov in

Turbine type	Francis	Kaplan	Pelton
Power output (kW) max	10000	10000	10000
Power output (kW) min	15	15	15
Head (m) max	500	500	500
Head (m) min	2	2	2

Joyti Ltd
Industrial Area
VADODARA 390 003, Gujarat
Telephone (+91) 320 440
Telex 0175 214 jm IN

Turbine type	Francis	Kaplan	Pelton	Turgo Impulse
Power output (kW) max	5000	4000	6000	3000
Power output (kW) min	50	500	50	10
Head (m) max	107	20	390	200
Head (m) min	24	2	50	24

Indonesia

PT Barata Indonesia
Jl Ngagel 109
SURABAYA 60246, E Java
Telephone (+62) 31 69 075
Fax (+62) 31 69 079
Telex 47316 Barata JA

Turbine type	Francis	Kaplan	Pelton	Other
Power output (kW) max	2500	625	1500	
Power output (kW) min	736	45	–	
Head (m) max	5	5	–	
Head (m) min	5	5	–	

Irish Republic

Belton Ironworks
Ardagh
LONGFORD
Telephone (+353) 43 75017
Fax (+353) 43 75017

Turbine type	Francis	Crossflow
Power output (kW) max	300	100
Power output (kW) min	5	5
Head (m) max	4	50
Head (m) min	15	5

Italy

Hydroart
PO Box 1739, via Stendhal 34
1-201144 MILANO
Telephone (+39) 2 479 104
Fax (+39) 2 479 398
Telex 332 291 hy art

Turbine type	Francis	Kaplan	Pelton
Power output (kW) max	3000	3000	3000
Power output (kW) min	100	100	100
Head (m) max	300	65	1000
Head (m) min	15	2.5	85

IREM Spa
via Vaie 42
1-10050 S. Antonio di Susa TORINO
Telephone (+39) 11 964 9133
Fax (+39) 11 964 9933
Telex 212 134 iremto I

Turbine type	Pelton	Crossflow
Power output (kW) max	100	100
Power output (kW) min	0.05	0.5
Head (m) max	140	60
Head (m) min	6	3

Orengine Srl
via H Staglieno 10-14
1-16129 GENOA
Telephone (+39) 10 592 011
Fax (+39) 10 532 719
Telex 271 035 hysu I

Turbine type	Crossflow	Francis	Kaplan	Pelton
Power output (kW) max	1500	2000	3000	5000
Power output (kW) min	10	100	100	50
Head (m) max	200	150	60	1000
Head (m) min	2	15	1	50

Japan

Fuji Electric Co Ltd
New Yurakucho Building, 12-1 Yurakucho 1-chome
Chiyoda, TOKYO 100
Telephone (+81) 3 211 7111
Fax (+81) 3 211 7988
Telex j22331 fujielec

Turbine type	Crossflow	Francis	Kaplan	Pelton
Power output (kW) max	2000	10000	10000	10000
Power output (kW) min	50	50	50	50
Head (m) max	150	450	80	1200
Head (m) min	10	25	4	100

Mexico

Turbinas y Equipos Industriales Ltd
Apartado Postal S-412,
06500 DF MEXICO
Telephone (+52) 5511 7533
Fax (+52) 5511 1654
Telex 176 3020

Turbine type	Francis	Kaplan	Pelton
Power output (kW) max	–	–	–
Power output (kW) min	100	100	100
Head (m) max	800	65	1700
Head (m) min	2.8	1.5	45

Nepal

Balaju Yantra Shala (P) Ltd
PO Box 209, Balaju Ind Est.
Kathmandu
Telephone (+977) 1 - 272147, 270894, 412844
Fax (+977) 1 - 272379
Telex 2429 bys NP

Turbine type	Crossflow
Power output (kW) max	250
Power output (kW) min	3
Head (m) max	100
Head (m) min	3

Butwal Engineering Works
PO Box 1
Butwal
Telephone (+977) 3 - 20212

Turbine type	Crossflow
Power output (kW) max	50
Power output (kW) min	5
Head (m) max	50
Head (m) min	3

Kathmandu Metal Industries
Kha 3-812, Naghal, Quadon, Block 12/514
Kathmandu 3
Telephone (+977) 1 - 224069

Turbine type	Crossflow	Pelton	Propeller	Turgo
Power output (kW) max	50	100	5	10
Power output (kW) min	1	1	1	2
Head (m) max	30	150	5	20
Head (m) min	5	30	1	4

National Structure & Engineering
Patan Industrial Estate, Lagankhel
Patan
Telephone (+977) 1 - 521405

Turbine type	Crossflow	Turgo
Power output (kW) max	50	10
Power output (kW) min	5	2
Head (m) max	50	20
Head (m) min	5	4

Nepal Hydro & Electric Co
PO Box 1
Butwal
Telephone (+977) 7 - 320212
Fax (+977) 7 - 320465

Turbine type	Pelton	Francis
Power output (kW) max	1000	1000
Power output (kW) min	100	100
Head (m) max	1000	150
Head (m) min	100	5

Nepal Machine & Steel Structures
Rajat Jayanti Park
Butwal

Turbine type	Crossflow
Power output (kW) max	50
Power output (kW) min	5
Head (m) max	50
Head (m) min	5

Nepal Yantra Shala
Patan Industrial Estate, Lagankhel
Patan
Telephone (+977) 1 - 522167
Fax
Telex

Turbine type	Crossflow	Pelton
Power output (kW) max	50	50
Power output (kW) min	5	1
Head (m) max	50	150
Head (m) min	5	25

Thapa Engineering Industry
Kalika Nagar, Rupandehi
Butwal
Telephone (+977) 7 - 320417

Turbine type	Crossflow	Pelton
Power output (kW) max	50	50
Power output (kW) min	1	1
Head (m) max	–	150
Head (m) min	12	25

Portugal

Sorefame SA
Apartado 5
2701 AMADORA codex
Telephone (+351) 1 - 976 051
Fax (+351) 1 - 976 696
Telex 16 101 sorefam P

Turbine type	Francis	Kaplan	Pelton
Power output (kW) max	15000	15000	15000
Power output (kW) min	500	100	500
Head (m) max	250	50	1000
Head (m) min	20	2	100

Singapore

Elmar Engineering Services
PO Box 132, 8 Gul Drive
JURONG 2262
Telephone (+65) 861 2777
Fax (+65) 861 1576
Telex RS 37322

Turbine type	Crossflow	Francis	Kaplan	Pelton
Power output (kW) max	1000	6000	1600	7500
Power output (kW) min	18	25	35	5
Head (m) max	100	185	15	500
Head (m) min	8	6	2	28

Sri Lanka

Brown & Co Ltd
481 TB Jayah Mawatha
COLOMBO 10
Telephone (+94) 697111, 698411
Fax (+94) 698489
Telex 21111 METAL Colombo

Turbine type	Crossflow	Pelton
Power output (kW) max	5	300
Power output (kW) min	1	5
Head (m) max	15	250
Head (m) min	4	25

Systems Engineers
131 Anderson Road
DEHIWALA
Telephone (+94) 714 777
Fax (+94) 1 502 850

Turbine type	Pelton
Power output (kW) max	30
Power output (kW) min	1
Head (m) max	100
Head (m) min	20

Sweden

Flygt AB

PO Box 1309, Svetsarvaegen
S–17125 SOLNA
Telephone (+46) 8 733 65 00
Fax (+46) 8 29 58 16
Telex 43887flygt ab S

Turbine type	Semi-Kaplan (submersible)
Power output (kW) max	530
Power output (kW) min	30
Head (m) max	20
Head (m) min	2.5

Turab Turbin

Forradsgalan 2
S–57139 NASSJO
Telephone (+46) 380 155 10/15
Fax (+46) 380 155 30

Turbine type	Francis	Kaplan	Pelton
Power output (kW) max	500	500	–
Power output (kW) min	–	–	–
Head (m) max	100	30	150
Head (m) min	–	–	–

Turbosun AB

PO Box 71
S–66900 DEJE
Telephone (+46) 552 210 85
Fax (+46) 552 210 32
Telex 66176

Turbine type	Francis	Kaplan	Pelton	Other
Power output (kW) max	500	500	500	500
Power output (kW) min	10	2	10	0.5
Head (m) max	100	20	300	200
Head (m) min	10	2	–	10

UK

F Bamford & Co Ltd

Ajax Works, Whitehill
STOCKPORT, Cheshire SK4 1NT
Telephone (+44) 61 480 6507
Fax (+44) 61 480 7693
Telex 668 518

Turbine type	Propeller
Power output (kW) max	2000
Power output (kW) min	–
Head (m) max	17
Head (m) min	3

Biwater Hydro Power Ltd

Millers Road
WARWICK CV34 5AN
Telephone (+44) 926 411740
Fax (+44) 926 410740
Telex 317473

Turbine type	Crossflow	Francis	Kaplan	Pelton
Power output (kW) max	1000	15000	5000	15000
Power output (kW) min	50	50	50	50
Head (m) max	100	300	30	1000
Head (m) min	5	5	15	30

Evans Engineering & Power

Trecarrell Mill, Trebullet
LAUNCESTON, Cornwall PL15 9QE
Telephone (+44) 566 82285
Fax (+44) 566 82285

Turbine type	Crossflow	Kaplan	Kaplan	Pelton
Power output (kW) max	30	500	250	2000
Power output (kW) min	5	5	10	5
Head (m) max	50	10	20	400
Head (m) min	5	1	3	30

Gilbert Gilkes & Gordon

KENDAL
Cumbria LA9 7BZ
Telephone (+44) 539 720028
Fax (+44) 539 732110
Telex 65125

Turbine type	Francis	Pelton	Turgo Impulse
Power output (kW) max	7500	10000	8000
Power output (kW) min	5	5	5
Head (m) max	200	1000	400
Head (m) min	3	30	10

MacKellar Engineering Ltd
Stratspey Industrial Estate
GRANTOWN-ON-SPEY, Morayshire, Scotland
Telephone (+44) 479 25 77
Fax (+44) 479 24 36
Telex 75532

Turbine type	Kaplan	Crossflow
Power output (kW) max	65	75
Power output (kW) min	3	5
Head (m) max	9	80
Head (m) min	2.5	3

Portmore Engineering Ltd
48 Dunadry Road
ANTRIM BT41 4QN, Northern Ireland
Telephone (+44) 84 94 32 444

Turbine type	Crossflow
Power output (kW) max	150
Power output (kW) min	3
Head (m) max	30
Head (m) min	3

Water Power Engineering
Coaley Mill, Coaley
DURSLEY, Gloucestershire GL1 5DS
Telephone (+44) 453 89 376

Turbine type	Francis (Modular polymer)
Power output (kW) max	60
Power output (kW) min	0.5
Head (m) max	10
Head (m) min	1

Weir Pumps Ltd
Newlands Road, Cathcart
GLASGOW G44 4EK
Telephone (+44) 41 637 7141
Fax (+44) 41 637 7358

Turbine type	Francis	Kaplan	Pelton	Reverse Pump
Power output (kW) max	20000	20000	20000	2000
Power output (kW) min	100	100	100	30
Head (m) max	400	75	10000	–
Head (m) min	5	3	80	10

USA

Canyon Industries Inc.
5346 Mosquito Lake Road
DEMING, WA 98244
Telephone (+1) 206 592 5552
Fax (+1) 206 592 2235
Telex 650 352 6659

Turbine type	Crossflow	Francis	Pelton
Power output (kW) max	500	1000	5000
Power output (kW) min	4	4	100
Head (m) max	75	100	1000
Head (m) min	6	10	30

Cornell Pump Co
2323 SE Harvester Drive
PORTLAND, Oregon 97222
Telephone (+1) 503 653 0330

Turbine type	Reverse Pump
Power output (kW) max	350
Power output (kW) min	5
Head (m) max	100
Head (m) min	10

Flygt Corporation
PO Box 857, 129 Glover Avenue
NORWALK, CT 06856-9998
Telephone (+1) 203 846 2051
Fax (+1) 203 849 0679
Telex 965 975

Turbine type	Semi-Kaplan (submersible)
Power output (kW) max	530
Power output (kW) min	30
Head (m) max	20
Head (m) min	2.5

Hayward Tyler Inc.
PO Box 492, 46 Roosevelt Hway
BURLINGTON, VT 05401
Telephone (+1) 802 655 4444
Fax (+1) 802 655 4682
Telex 954646

Turbine type	Pelton
Power output (kW) max	2500
Power output (kW) min	25
Head (m) max	1000
Head (m) min	150

McKay Water Power
61 Village Green
West LEBANON, New Hampshire
Telephone (+1) 603 298 5338

Turbine type	Francis	Kaplan	Pelton
Power output (kW) max	350	500	500
Power output (kW) min	6	8	5
Head (m) max	70	65	1200
Head (m) min	20	15	50

Blue Water Hydro
80 Rust Street
HAMILTON, MA 10982
Telephone (+1) 508 468 1598

Turbine type	Crossflow	Francis	Kaplan	Pelton
Power output (kW) max	1000	20000	20000	20000
Power output (kW) min	–	50	5	15
Head (m) max	120	150	45	450
Head (m) min	2	2	2	30

Peabody Floway
PO Box 164, 2494 S Railroad Avenue
FRESNO, CA 93707
Telephone (+1) 209 442 4000
Fax (+1) 209 442 3098

Turbine type	P.A.T. (Pump-as-turbine)
Power output (kW) max	1000
Power output (kW) min	5
Head (m) max	300
Head (m) min	10

Small Hydroelectric Systems
5141 Wickersham Street
ACME WA 98220
Telephone (+1) 206 384 8086

Turbine type	Kaplan	Pelton	Turgo
Power output (kW) max	100	5000	450
Power output (kW) min	1	1	1
Head (m) max	10	373	152
Head (m) min	2.4	12	12

Thomas Brothers Hydro In
PO Box 2040, 115 Snapping Shoals Road
COVINGTON GA 30209
Telephone (+1) 404 787 0898

Turbine type	Francis	Pelton	Propeller
Power output (kW) max	10000	500	10000
Power output (kW) min	3	3	3
Head (m) max	300	600	100
Head (m) min	25	75	25

Yugoslavia

Dem Tozd Elektrokovinar
Ciljska costa 52
63270 LASKO, Slovenia
Telephone (+38) 63 730 041
Fax
Telex 33 121 dem

Turbine type	Francis	Pelton	Tubular
Power output (kW) max	630	1000	400
Power output (kW) min	16	16	16
Head (m) max	100	300	8
Head (m) min	6	25	2

Zimbabwe

Hydro Power Zimbabwe
PO Box A31, Avondale
HARARE
Telephone (+263) 4 884196

Turbine type	Crossflow
Power output (kW) max	50
Power output (kW) min	2
Head (m) max	25
Head (m) min	5